T0310241

THE PDMA HANDBOOK OF INNOVATION AND NEW PRODUCT DEVELOPMENT

THE PDMA HANDBOOK OF INNOVATION AND NEW PRODUCT DEVELOPMENT

FOURTH EDITION

Edited by
Ludwig Bstieler
Charles H. Noble

CONTENTS

INTRODUCTION

Thank you for purchasing this book. We are delighted and honored to bring you the 4th edition of the *PDMA Handbook of Innovation and New Product Development*. Our partners at the Product Development and Management Association (PDMA) and John Wiley & Sons Publishing are as enthusiastic as we are to see this almost completely new edition of a very popular and useful book come to life.

The goal of this Handbook, as in prior editions, is to provide an introduction, a thought-provoking series of essays, and a practical guide for managers and thought leaders working in new product development (NPD) and innovation. Astute readers of our prior editions may notice that the title has slightly changed in this 4th edition. We now appropriately highlight *innovation* as a key focus of what we are to explore in addition to our continuing emphasis on new product development.

Our deepest thanks go to the wonderful collection of authors who have contributed to this book and to prior editions. In this collection, they share their thoughtful insights and broad experiences to bring you a compilation that represents the state-of-the-art in the field. One of the unique features of this book series has always been that our authors represent a balance of outstanding practitioners and consultants, and leading edge academic researchers. Thus, you will find here both applied, first-hand insights from the field, and findings from academics who have been studying innovation and new product development using the latest techniques. These diverse perspectives combine to bring you knowledge that is powerful, practical, and thought-provoking.

This book is a collection of 32 chapters representing a host of different voices. The style of the chapters presented here also varies greatly. Some chapters are extremely hands-on and step-by-step, providing practical tools to apply in your organization. Other chapters may present new concepts, ideas, and approaches that are meant to stimulate your thinking and to help you find creative ways to study or apply these thoughts in your workplace or in your research.

The *Handbook* has been the most popular publication from the PDMA and has proved indispensable for many practicing in the field. The chapters we bring you in this edition revisit familiar topics with fresh approaches and insights and will likely introduce you to entirely new concepts that take you to the leading edge of this exciting world of new product development and innovation.

The Intended Audience for This Book

As with prior editions this *Handbook* is written for two very different groups of readers. First, we hope to appeal to managers who have a strong interest in increasing their knowledge of these important topics. This is not "*NPD for Dummies,*" and our goal is not to instruct the absolute novice on every detail of the day-to-day functioning of a new product development unit. We leave that to other authors and their instructional books. Our primary target for the work here is the manager who is moderately experienced in new product development and is looking for pockets of exceptional knowledge that can ultimately lead to superior performance in the workplace. Second, we also offer this book to academics researching NPD and innovation. These scholars should find the insights from both practitioners and other researchers to be thought provoking stimuli for future research efforts that, we hope, will lay the groundwork for the 5th edition of this *Handbook*!

How to Use This Book?

Our goal in organizing this book was to present something that provides the reader with as much flexibility as possible, understanding that different readers may look to this book with very different goals in mind. Some readers may choose to read this cover to cover, and we commend them for that journey! Others may choose to skim the contents, identifying those topics which pique their interest most, while perhaps relegating others to a later time. We have done several things to help the reader determine the best reading strategy. First, we have organized these chapters into sections that provide a flow and a general theme for the diverse writing that readers will find within. Also, if readers see certain topics or concepts highlighted in the Table of Contents that pique their interest, the Glossary in the back of this book provides a basic understanding of those ideas. The Glossary can also serve as an important reference for the active product manager working in this field.

The Book's Organization

Section 1 of this handbook includes chapters that are both strategic and foundational. Beginning with Bob Cooper, who many know as the father of the stage gate concept, he brings fresh ideas that also lay the groundwork for what is to come. The chapters in the first section cover big issues and important concepts that any product manager needs to consider. Section 2 examines the new product development process and essential tools and concepts such as perspectives on portfolio management, managing the front end of innovation, gaining insights from consumers, and other important issues. Section 3 considers topics that have exploded since the last edition of this *Handbook*, focused on user participation and value creation in the new product development process. We're excited to introduce work on open innovation, co-creation, crowdfunding, and related ideas. In Section 4 we cover what we call transformative forces in new product development and innovation. This section includes topics related to digital transformations, artificial intelligence, and design thinking, all of which are already shaping new product development and innovation in different ways. Section 5 is also a new offering in this edition of the *Handbook* with a clear focus on service innovation concepts, an area that any practicing manager knows has grown increasingly critical in recent years. Finally, Section 6 on applications in new product development includes key considerations, tools and techniques to help managers excel in the practice of NPD and innovation.

The Appendix in this book provides a description and contact information for the Product Development and Management Association (PDMA). This wonderful organization is an essential knowledge resource and networking opportunity for any professional trying to excel in this field. As mentioned earlier, the Glossary offered in this book should be quite helpful in understanding the many and changing terms associated with this field.

In summary, the key sections in this Handbook are:

Section 1. Getting started with innovation and new product development.

Section 2. New product development process.

Section 3. User participation and value creation in new product development.

Section 4. Transformative forces of new product development and innovation.

Section 5. Service innovation.

Section 6. Applications in new product development.

Appendix: About the Product Development and Management Association (PDMA)

PDMA Glossary of New Product Development Terms

Acknowledgments

As with all PDMA projects, this work is done for the good of the field and not for the individuals involved. This is an all-volunteer project from the writing to the editing, with many talented people donating numerous hours of effort to create the resource that you have before you. We are particularly grateful that our authors have found the time to make these contributions just as the business world is emerging from the incredible challenges of the COVID pandemic and the many time and resource drains associated with it. Responsibility for chapter content lies with the author(s). The editors played a key role in managing the review process plus shaping the overall direction of this collection. We encourage readers to consider the many resources that PDMA has to offer to help in their professional development. These can be found at www.pdma.org and include the award-winning *Journal of Product Innovation Management* and various other publications and online resources that the organization provides. Additional information on the PDMA is presented in the Appendix. In particular, we would like to thank the leadership of the PDMA including the former Chair, Mark Adkins, and Executive Director Eric Ewald, for all that they have done to support this project. Finally, we appreciate the support that our publisher, John Wiley & Sons has provided. Their editorial team has provided excellent guidance and suggestions throughout this process. We hope you find as much enjoyment in reading this book as we did in working with our outstanding author teams and curating this collection for you!

PDMA Handbook 4th Edition Editorial Team

Ludwig Bstieler (PhD, University of Innsbruck) is Professor and Chair of Marketing at the Paul College of Business and Economics at the University of New Hampshire. Prior to joining the Peter T. Paul College, he was Assistant Professor of Marketing at the University of Innsbruck, and held visiting positions at MIT's Sloan School of Management, the Michael DeGroote School of Business at McMaster University, and the European School of Management (ESCP Europe). He has provided in-house trainings, trainings in manager programs, (international) marketing research and consulting services for various companies and institutions in Europe. He studies the design and marketing of new products with a particular emphasis on the influence of close buyer-supplier collaboration and how relational and contractual governance mechanisms facilitate or hinder successful collaborative new product development. His recent work examines whether sustainability-minded firms can achieve better innovation outcomes and the circumstances that lead to these results. His research is published in leading journals, including the *Journal of Product Innovation Management, Journal of Business Research*, and *Technovation*. Dr. Bstieler is Associate Editor for the *Journal of Product Innovation Management* and serves on PDMA's Board and the IPDMC's Scientific Committee.

Charles H. Noble (PhD, Arizona State University) is the Henry Distinguished Professor of Business in the Haslam College of Business at The University of Tennessee. In addition, he currently serves as Editor-in-Chief of the *Journal of Product Innovation Management*, the premier research journal in the areas of new product development and innovation. He is also a Research Faculty member with the Center for Services Leadership (Arizona State University) and an Advisory Board Member for the Snyder Innovation Management Center at Syracuse University. His research interests focus generally on front end design and development processes, as applied to both products and services. He has published in many leading journals including the *Journal of Marketing, Strategic Management Journal, Journal of the Academy of Marketing Science, Journal of Product Innovation Management, IEEE Transactions on Engineering Management, Sloan Management Review* and many others. He is on the editorial review boards of the *Journal of the Academy of Marketing Science, the Journal of Product Innovation Management*, and the *Journal of Business Research*. In his corporate life, Charles worked in strategic planning and corporate finance for leading retailers in the consumer electronics, home improvement, and warehouse club sectors.

GETTING STARTED WITH NEW PRODUCT DEVELOPMENT AND INNOVATION

CHAPTER ONE

NEW PRODUCTS: WHAT SEPARATES THE WINNERS FROM THE LOSERS AND WHAT DRIVES SUCCESS

Robert G. Cooper

1.1 Introduction

The central role of product innovation in business strategy coupled with poor innovation performance by many firms has resulted in an urgent quest for solutions – how to improve new product (NP) practices and performance. That begs the questions: What are the factors that underlie NP success? Why are some new products so successful? And why do some companies achieve outstanding performance in product innovation? In short, *what makes a winner* – that's the topic of this chapter.

1.1.1 Critical to Firms, but High Failure Rates

Product innovation – the development of new and improved products and services – is crucial to the survival and prosperity of the modern corporation. New products launched in the last three years account for 27 percent of company sales and 25 percent of profits (Buffoni et al. 2017).

The secrets to success in new product development (NPD) are complex, however, and elusive for many firms. Witness the *high failure rate of new products* and the *poor innovation performance in industry – 40 percent* of *launched products fail* (but not the common assertion that 80 to 90 percent of new products fail; Castellion and Markham 2013). If one starts measuring earlier in the NP

The PDMA Handbook of Innovation and New Product Development, Fourth Edition. Edited by Ludwig Bstieler and Charles H. Noble.

process, out of every 7 to 10 NP concepts, on average only one becomes a commercial success (Barczak et al. 2009). Further, only 53 percent of businesses' NP projects achieve their financial objectives; 44 percent are launched on time (Cooper and Edgett 2012; Edgett 2011); and only 13 percent of firms' new product efforts achieve their annual profit objectives (Cooper 2019). Wide variances exist around performance metrics, however, with the *best performing firms doing dramatically better than the rest* (the top 20 percent of businesses in terms of NPD performance results, as measured by a number of metrics).

This Chapter's Scope: This chapter is based on studies from the world of physical or manufactured new products. While service and software are important sectors, the fact that in the US manufacturers account for 58 percent of R&D spending means that physical products are very much a vital area in NPD (National Science Foundation 2021), hence the current focus. Some success factors also have applicability to the other two sectors; but not all do, nor are the relative impacts the same.

1.1.2 Some Definitions

New product development covers the complete process of bringing a new product to market, renewing an existing product, or introducing a product in a new market. Thus, there are different "degrees of newness" of a new product, as shown in Figure 1.1. Here, both the average breakdown by numbers of projects (Barczak et al. 2009) and by resources spent by project type (Cooper and Kleinschmidt 2021) are shown.

The concept of *innovation* is more complex and includes three major facets (Joubert and Van Belle 2012). First, an innovation is *not the same as an invention*; it must be successfully introduced and adopted in the marketplace (Garcia and Calantone 2002). Second, innovation is *an iterative process*. Cycles of innovation repeat, and so improved versions of innovations are continuously introduced to the market. Third, the definition includes *many different types of innovation* in an organization, such as product and service innovation, and also process and technological innovation. This chapter only focuses on one aspect of innovation, *product innovation*.

1.2 The Critical Success Factors in Product Innovation

Twenty-five success drivers have been singled out in this chapter. Each has been cited in several notable studies. For reading convenience, these 25 drivers are arbitrarily divided into three categories (although some drivers cut across categories):

1. Success drivers of individual new product projects: These are tactical and capture the characteristics of the NP project or the product itself (see Table 1.1).

FIGURE 1.1 TYPES OF NEW PRODUCTS ON TWO DIMENSIONS – "NEW TO THE COMPANY" AND "NEW TO THE MARKET."

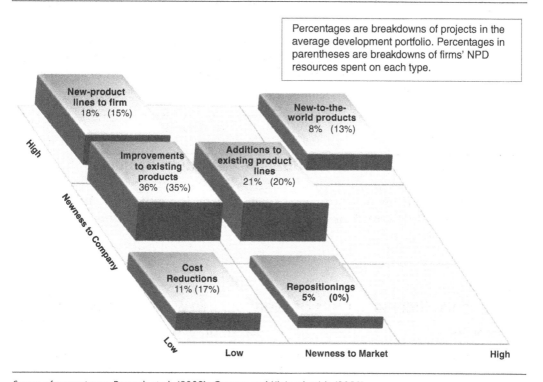

Percentages are breakdowns of projects in the average development portfolio. Percentages in parentheses are breakdowns of firms' NPD resources spent on each type.

Source of percentages: Barczak et al. (2009); Cooper and Kleinschmidt (2021).

2. Drivers of success for the business, including the business's innovation strategy, climate, and culture, and how the firm organizes for NPD (see Table 1.2).
3. The systems and methods that the firm has in place for managing NPD (see Table 1.3).

The Quest for the Critical Success Drivers

The keys to new product success outlined in this chapter are based on numerous research studies and reviews of why new products succeed, why they fail, comparisons of winners and losers, and benchmarking studies of best-performing businesses. Many of these investigations have been reported in the PDMA journal, the *Journal of Product Innovation Management (JPIM)*.

Some of the most revealing studies have been the large-sample quantitative investigations of successful versus unsuccessful new product projects. They began with Project SAPPHO in the early 1970s, followed by the NewProd series of studies, the Stanford Innovation Project, and, subsequently, studies in countries outside of North America and Europe (Mishra et al. 1996; Song and Parry 1996). Then,

(Continued)

several large benchmarking studies of best practices in firms have provided other insights into how to succeed at product innovation (American Productivity & Quality Center, APQC 2003; Barczak et al. 2009; Cooper et al. 2004; Edgett 2011; Griffin 1997).

In more recent years, a number of relevant meta-analyses and literature reviews have been undertaken, and have provided additional useful insights to success drivers: Cankurtaran et al. (2013); Dwivedi et al. (2021); Evanschitzky et al. (2012); Joubert and Van Belle (2012); Salmen (2021); and Storey et al. (2016).

1.3 Success Drivers of Individual New Product Projects

Why do so many new products fail and why do some succeed? And is there a pattern? Ten drivers of success at the *NP project level* have been identified (Table 1.1).

TABLE 1.1 SUCCESS DRIVERS OF INDIVIDUAL NEW PRODUCT PROJECTS.

Driver Name	Success Driver Description
1. Product Advantage	A unique superior product – a differentiated product that delivers unique benefits & a compelling value proposition to the customer or user; also the role of "smart products"
2. VoC	Building in the voice-of-the-customer – market-driven & customer-focused NPD
3. Up-front Homework	Predevelopment work – doing the homework & front-end loading the project; doing the due diligence before Development gets underway
4. Product Definition	Sharp, early & fact-based product definition to avoid scope creep & unstable specs; leads to higher success rates & faster to market
5. Iterative Development	Building into the project a series of "build & test" iterations; putting something in front of the customer, early & often, to get the product right; early & frequent technical validations of versions of the product
6. Speed & Order of Entry	The impact of speed and time to market; timing & the order of market entry (first to market)
7. Agility	Responsive to making rapid changes in the portfolio of projects; able to react to changing customer needs; ability to pivot from the original plan of action
8. Proficient Launch	A well-conceived, properly executed market launch, driven by a solid, properly resourced marketing plan
9. Global Orientation	A global or "glocal" product (global platform, tailored for local markets) and targeted at international markets (as opposed to the product designed to meet home-country needs)
10. Expectations of Success	Expect success, get success – success is a self-fulfilling prophecy: when project team members expect success, they are empowered & thus realize success

1.3.1 Product Advantage – A Unique Superior Product

Gaining competitive advantage by delivering *differentiated products with unique benefits* and a *compelling value proposition* for the customer and or user distinguishes new product winners from losers more often than any other single factor (Cooper 2017b, 2019; Dwivedi et al. 2021; Evanschitzky et al. 2012; Salmen 2021; Storey et al. 2016). Such superior products have five times the success rate, over four times the market share, and four times the profitability of "me too," copycat, reactive and ho-hum products with few differentiated characteristics (Cooper 2019; McNally et al. 2010.)

What do superior products have in common? Winning products…

- Are superior to competing products in terms of meeting users' needs, offer unique features not available in competitive products, or solve a problem the customer has with competitive products;
- Feature good value for money for the customer, reduce the customer's total costs (high value-in-use), and boast high price/performance;
- Provide excellent product quality relative to competitors' products (in terms of how the user measures quality); and
- Offer product benefits or attributes that are highly visible, easily perceived as useful, and are meaningful to the customer.

Other product characteristics have been investigated, with little or insignificant connection to new product success, either positive or negative (Evanschitzky et al. 2012). These less relevant product characteristics include:

- Product price;
- Product technological sophistication; and
- Product innovativeness (new to the market).

The term "superior product" used above is defined from an *external* or *market perspective*, that is, differentiated from (or better than) *competitive products*, consistent with the definitions of "innovativeness" by Garcia and Calantone (2002). This external perspective is different than an *internal perspective of innovativeness*, namely "new to the firm" products, which may have lower success rates due to loss of synergy or lack of experience in the new market or technology.

Innovativeness or differentiation alone is not necessarily the key: A product might be "novel and different" in the eyes of the customer – the first of its kind, never been seen before – yet deliver little of benefit to that customer. Product *meaningfulness* is what is vital: the *benefits that users receive* from buying and using a new product (Rijsdijk et al. 2011).

Note also that "product" is broadly defined here. It includes not only the evident or physical product, but also the "extended product" – the *entire bundle of benefits associated with the product*, including the system supporting the product, product service, and technical support, as well as the product's image or branding.

Smart products: The Digital Transformation underway in industry has introduced a relatively new category of physical products, namely "smart products" – physical products with embedded software. Smart products cover a broad range from home appliances (a smart kitchen stove) to a smart car (a Tesla) to IoT production equipment (connected to the Internet to enable monitoring of performance; Raff et al. 2020).

The hardware component still dominates physical product development, however, with more than 50 percent of product composition being hardware for almost two-thirds of manufacturing firms (Schmidt et al. 2019). On average, *embedded software is now about 30 percent* of the composition of manufacturers' new products, a significant percentage (Cooper and Fürst 2020a).

The questions become: Are smart new products really more successful? And what are their success drivers? Research into smart new product success is in its infancy, however. To the extent that connected or smart products offer new and desired functionality versus traditional physical products, these would be seen as benefits by customers, and thus one might expect a higher success rate. But no concrete evidence exists on relative success rates of smart new products versus traditional ones.

Some of *success drivers of smart new products* are consistent with those for physical new products, such as:

- A strong market orientation and understanding users' unmet needs;
- Proficiency of the new product process (how well executed both marketing and technical tasks are during the project); and
- The existence of internal competences (Shashishekar et al. 2022).

But only a few potential drivers have been investigated. Smart or connected new products are an area that merits research.

1.3.2 Market-driven Products and Voice-of-the-Customer (VoC) Built In

A thorough understanding of customers' needs and wants, the competitive situation, and the nature of the market is an essential component of new product success (Calantone and Di Benedetto 2007; Cooper 2019; Fang 2008; Florén et al. 2018; Joubert and Van Belle 2012; Kim et al. 2016; Verhaegde and Kfir 2002). This tenet is supported by *virtually every study of product success factors.* Conversely, failure to adopt a strong market orientation in product innovation, unwillingness to undertake the needed market assessments, and leaving the customer out of the development project spell disaster. These culprits are found in almost every study of why new products fail, starting back with the first study on NP failures! (National Industrial Conference Board 1964).

A strong *market orientation is missing* in the majority of firms' new product projects, and detailed market studies are frequently omitted from NP projects (Edgett 2011). In general, marketing activities are the most poorly executed activities of the entire new product process, rated far below corresponding technical (engineering, design, R&D) activities. Moreover, relatively few resources are spent on marketing actions (except for the launch), accounting for less than 20 percent of total project costs.

A market focus is relevant throughout the entire new product project – see Figure 1.2:

- Idea generation: The *best ideas come from customers*. Market-oriented idea generation activities, such as focus groups and VoC research (ethnography and site visits) to determine unmet needs or problems lead to superior ideas (Cooper and Dreher 2010). Robust ideas also come from innovative users and web-based customer inputs (open innovation). Customer participation in the ideation stage also improves new product financial performance (Woojung and Taylor 2016).

- Product design: Customer inputs often have a vital role in the design of the product, helping to define the product's requirements and specifications. But the value of customer involvement in the early stages of an NP project is somewhat controversial: Customers may not provide useful product suggestions or solutions nor rich information. The conclusion is that firms learn more from *listening to the voice of the customer* – getting customers to talk about the problems they face, their requirements and expectations, and the benefits they seek – rather than asking for specific product solutions (Florén et al. 2018).

- Before pushing ahead with development: Best performers *test the product concept with the customer* by presenting representations of the product – via models, mock-ups, "protocepts,"[1] and even virtual prototypes – and gauging the customer's

FIGURE 1.2 A STRONG CUSTOMER FOCUS INCLUDES KEY ACTIONS FROM BEGINNING TO END IN THE INNOVATION PROCESS.

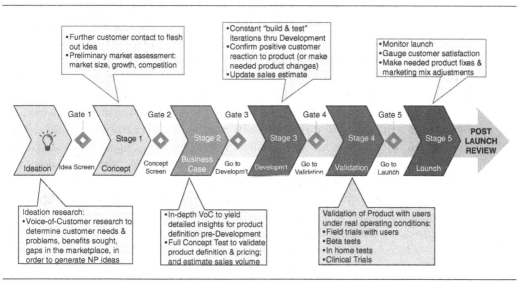

liking and purchase intent. It's much cheaper to test and learn before development begins than to develop the product and then begin customer testing (Cooper and Fürst 2020a).

- Throughout the Development stage of the project: Customer inputs shouldn't cease at the completion of the predevelopment market studies. Testing product designs with the user is very much an *iterative process*: By bringing the customer into the process to view facets of the product via a series of rapid prototypes-and-tests, or customer tests of pretotypes,[2] the developer verifies assumptions about the winning product design. Feedback shared by the customer can also help identify and resolve potential problems, which minimizes development delays and improves speed to market (Fang 2008). (See 1.3.5 below.)

- Product testing and validation stage: Customer participation in the testing or validation stage also improves NP financial performance directly, and indirectly through acceleration of time to market (Florén et al. 2018). Validation includes activities such as prototype testing (beta tests, field trials, consumer tests) and market testing (test market or trial selling; selling an MVP[3]) that provide first-hand feedback from customers on product usability, product performance, potential product problems, and the positioning and marketing mix for the new product.

Some literature emphasizes involving customers in *high-tech industries*, thereby gaining the benefits of new knowledge from customers. Evidence exists, however, that customer participation could be an even more effective strategy in *low-tech industries* because of the ease of knowledge integration and utilization in such industries (Woojung and Taylor 2016).

Open innovation: This goal of involving the customer in product development has moved beyond the relatively low-key need to "listen to the voice of the customer" to *more intense customer-involvement methods* such as crowdsourcing and open innovation (Chesbrough and Crowther 2006; Cui and Wu 2016; Docherty 2006; Enkel et al. 2009; Product Development and Management Association 2014; Poetz and Schreier 2012). Through *open innovation*, the developer obtains resources and knowledge from sources external to the company: ideas for new products; intellectual property and outsourced development work; marketing and launch resources; and even licensed products ready to launch.

Open innovation is one way that the customer is built into the process and has proven effective. But a firm must be able to acquire, transform, assimilate, and exploit the knowledge to maximize any impact that open innovation may have on competitive advantage (Morgan et al. 2018). Firms with *higher absorptive capacity* are better equipped to apply customer knowledge to newer, differentiated products; and they are more able to internalize external knowledge from customers and exploit it commercially, which enhances the relationship between customer participation and new product performance.

Codevelopment: Here, customers or users are invited *to help the product developer design the next new product*, and in so doing, provide many suggestions for significant product improvements (von Hippel et al. 2011). Indeed, consumers were found to be 2.4 times *more efficient* at developing significant innovations than producers, and much more prolific and efficient product developers when the project is in its early stages (Hienerth et al. 2012). Co-development is more feasible today, in part because of IT and Internet tools.

1.3.3 Predevelopment Work – The Up-front Homework

Homework is critical to winning. Studies reveal that the *steps preceding the development* of the product make the difference – the "game is won or lost in the first five plays" – in Stages 1 and 2 in Figure 1.2 above (Cooper 2011, 2019; Evanchitsky et al. 2012; Storey et al. 2016). Quality of execution is key to success, but the front-end of the innovation process tends to be where most of the weaknesses occur (Edgett 2011).

Successful firms spend about twice as much time and money as unsuccessful firms on these vital front-end activities (Cooper 2019):

- Preliminary market assessment – a quick market study to assess market potential and desired product attributes;
- Preliminary technical assessment – the first technical appraisal of the project, assessing technical feasibility and identifying technical risks;
- Detailed market study, market research, and VoC research;
- Detailed technical assessment – in-depth technical appraisal, establishing proof of concept, intellectual property (IP) issues resolution, and an operations and source-of-supply assessment; and
- Business and financial analysis before the investment decision to go to full-scale development.

"More homework means longer development times" is a frequently voiced complaint. However, research shows that homework pays for itself in improved success rates and actually reduces development times:

1. A much higher likelihood of product failure results if the homework is omitted.
2. Better project definition, the result of sound homework, speeds up the development process. Poorly defined projects with vague targets incur time slippage as they enter the Development stage.
3. Given the inevitable product design evolution that occurs during the life of a project, ideally, most of these design changes should be made early when they are less costly to correct. Predevelopment homework anticipates these changes and fosters their occurrence earlier in the process (Cooper 2019).

As Toyota's new products handbook (Morgan and Liker 2006) recommends, "Front-end load the project." That is, undertake a higher proportion of the project's work in the early stages.

1.3.4 Sharp, Early, and Fact-based Product Definition

Two of the worst time wasters in NPD are *project scope creep* and *unstable product specifications*. Scope creep means that the definition of the project constantly changes. The project might begin as a single-customer initiative, then be targeted at multiple users, and finally ends up being a new family of products. *Unstable product specs* means that the product definition – product requirements and specifications – keeps changing throughout the Development stage. Thus, the technical people chase elusive development targets – moving goalposts – and take forever to get to the goal (Cooper 2019).

Sharp, early, and fact-based product definition during the homework phase is one solution. How well the product is defined before the Development stage begins is a major success factor, impacting positively on both profitability and reduced time to market. This definition includes:

- The project's scope;
- The target market;
- The product concept and the benefits to be delivered to the user (including the value proposition);
- The positioning strategy (including the target price); and
- The product's features, attributes, requirements, and high-level specifications (Cooper 2019).

Unless this product definition is in place and fact-based, the odds of failure increase:

1. Building in a definition step *forces more attention to the* front-end homework, a key success driver.
2. The definition serves as a *communication tool:* All functional areas have a clear definition of the product.
3. This definition provides *clear objectives* for the technical team members, so they can move more quickly to their objective.

In today's dynamic world however, things change! In order to accommodate new information and inputs, smart firms allow *fact-based modification* of the original product definition by building in a series of "build and test" iterations…. next.

1.3.5 Iterative Development – Build, Test, Feedback, and Revise

The world moves too fast to make a stable and rigid product definition always possible. Often customers are not clear on what they want (or need) in the first place, so it's difficult to get an accurate product definition prior to development; and sometimes, customer requirements change in the time that passes between the beginning and end of development. As Steve Jobs, never a proponent of traditional market research, famously said, "People don't know what they want until you show it to them" (Isaacson 2011, p. 567).

Smart businesses have made the idea-to-launch system *much more adaptive*, allowing adjustments on the fly via *iterative development* (Cooper 2014, 2019). When there exists a high degree of customer participation at the prototype development stage (feedback, testing, opinions, and comments), the new product is much more likely to be successful in the market (Tih et al. 2016). That is, during development, efforts should be made to obtain customer feedback on early prototypes in order to increase the chances of the market success of the new product.

Smart project teams build in a series of *deliberate iterative steps* whereby successive versions of the product are shown to the customer to seek feedback and verification, as shown as the curved arrows in Figure 1.3. Each iteration consists of:

- *Build*: Build something to show the customer – a representation of the product, such as computer-generated graphics, a simulation, a virtual-reality prototype, a protocept, a rapid prototype, a crude working model, an early prototype or beta version, or a pretotype … each version closer to the final product.
- *Test*: Test that product version with the customer; and also test it internally (technically).
- *Feedback*: Gather feedback from the customer or user – what they like or don't like, what value they see, and purchase intent. And check the results of internal technical tests.

FIGURE 1.3 ITERATIVE OR SPIRAL DEVELOPMENT – A SERIES OF "BUILD-TEST-FEEDBACK-REVISE" ITERATIONS WITH CUSTOMERS – GETS THE PRODUCT RIGHT.

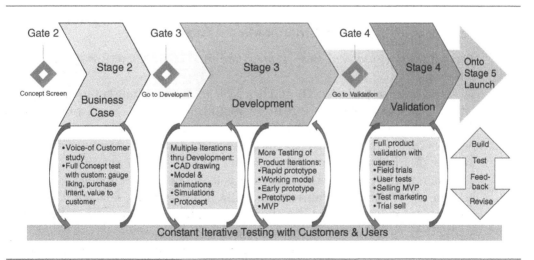

Source: Cooper (2017b).

- *Revise.* Reset your thinking about the value proposition, benefits sought, the product's design, and the technical solution; pivot if necessary; and then move to the next iteration (Cooper 2014).

This *iterative approach promotes experimentation*, encouraging project teams to fail often, fail fast, and fail cheaply.

Not only do iterations reduce market uncertainties, they also can be used to reduce technical uncertainties by seeking technical solutions in an experimental, iterative fashion. Moreover, there is strong evidence that this iterative development is *both feasible and works*: 44.8 percent of best-performing businesses practice these build-and-test iterations with customers (but only 26.3 percent of average firms do; APQC 2003). And a study of leading B2B European manufacturers revealed that, on average, between 3 and 4.5 versions of the product are presented to validate the design with customers through the Development and Testing stages, while product ideation-and-design contractors, such as IDEO, iterate on average *15 times with the customer* per project (Sandmeier et al. 2010).

Words of warning, however: Involving customers in via a series of iterative customer tests during the Development stage does take time, may increase time to market, and could diminish new product financial performance, particularly where the market is rapidly changing and interdependency among a product's parts is critical (Woojung and Taylor 2016). Thus, there is a trade-off between getting the product right versus the extra time needed to undertake iterative testing of the product with customers during development. A meta-analysis of speed of development, however, reveals that building in frequent build and test iterations has *no negative effect on speed* (Cankurtaran et al. 2012), while an earlier such meta-study revealed that frequent iterations actually improved time-to-market (Chen et al. 2010). Later in this chapter comes the topic of Agile Development, a set of principles and methods from the software world; Agile Development builds in frequent and very short build-and-test iterations.

1.3.6 The Impact of Speed and Order of Market Entry

Two concepts – *speed* and *order of entry* – are time related, and this section deals with both.

Speed or Time to Market. The theory is clear: Faster to market means higher profits (sales and profits are realized sooner; and money has a time value). Being in the market first results in higher market share, hence more sales and profits: "first mover advantage"; and being late to market means missing the window of opportunity, or incurring other penalties (such as annoying valued customers). Moreover, with rapid technology change occurring, speed has become an economic requirement of the NDP process: Delaying a product introduction can affect the sales of the product by up to 35 percent (Dwivedi et al. 2021).

The logic appears sound, *but strong research evidence on the positive impact of speed is lacking* (Cooper 2021a). Indeed, a review of empirical studies reveals inconsistent, even conflicting results on the relationship between accelerated development and project success (Chen et al. 2005; Griffin 2002; Langerak 2010). Some research on accelerated development looks at easily-measured dependent variables, such as the impact of speed on development costs; these studies often yield negative results (Cooper 2021a). While "development cost" is easily measured, it may not be a valid metric: One usually moves quickly not to save money, but in order to gain competitive advantage. Other metrics also produce mixed results (Langerak 2010). For example, the relationship between cycle time and product quality is also unclear. However, shorter cycle time is correlated with an increase in the new product's sales when speed is managed across all of the stages of the process (Eling et al. 2013).

Conflicting outcomes regarding the importance of accelerated development for new product success may also occur because the impact of cycle time reduction may pale when compared to other key success drivers (Langerak 2010). Cycle time reduction may also yield major negatives (Crawford 1992): In order to cut cycle time, *a firm may avoid bolder innovations* – too many product modifications and line extensions that can be done quickly, but result in few innovative products (Cooper 2011). Additionally, project teams, faced with an aggressive time-line, *may cut some vital tasks short* (such as VoC work or product testing). Finally, project teams, driven by time, may also become *too committed to their project and its plan*, and fail to pivot when needed.

One of the few large-sample meta-analyses on this speed topic concludes that *generally development speed is associated with increased new product success* (Cankurtaran et al. 2013). Even here, one cannot be sure about whether faster development speed leads to improved NPD success, or whether companies that are more successful at developing new products are also speedier due to their development capabilities. More on the topic of how to achieve accelerated development is in 1.5.3.

Order of market entry: Three NPD project types exist in terms of order of market entry (Marinov 2020):

1. *First mover* products are the first on the market; they have no direct competition, but compete with other forms of the product available, as well as with substitute products that satisfy the same need.
2. *Fast follower* products rival one or several products that are already on the market. This product type enters the market in the first stage of the product life cycle, before sales have grown rapidly.
3. *Late mover* products follow the first mover and the early entrants, and enter a growth or mature market, that is, after the market has grown.

The firm with the early-entry new product has the advantage *of establishing a market position, reputation, and direct relations* with customers, and also gaining better knowledge of the technology and application. But the late entrant *learns from the pioneer's mistakes*, and has more time to gather information about competitors, suppliers, and market trends, hence can choose who to compete with or how to differentiate from early entrant rivals (Marinov 2020). Note that 50 percent of best performing firms pursue a first-to-market strategy (Barczak et al. 2009).

Introducing a new product first, and getting the largest market share and hence the highest profitability, is one reason why firms seek to reduce time-to-market. But here the evidence is inconsistent: Some studies show that market pioneers perform better than late entrants, whereas others find the reverse to be true (Marinov 2020); for example, a major meta-analysis on NP success factors looked at 233 independent company samples (204 studies) and found only five that probed the impact of order of entry on NP success; the effect was insignificant (Evanschitzky et al. 2012). One explanation is that research often *focuses on the wrong variable*: on "order of market entry" rather than on the "getting the timing right." The timing, and not so much order of entry, is key: Early entrants may not generate sales because they enter the market *too early* when the window is not yet open, while followers may fail because they enter the market after the window is closed.

1.3.7 Agility in NPD

Agility refers to an organization's ability to sense and respond to changes quickly: *a responsive organization.* Agility has been widely embraced as a management approach for the modern corporation. According to a *Harvard Business Review* article, "agile methodologies – which involve new values, principles, practices, and benefits, a radical alternative to command-and-control-style management – are spreading across a broad range of industries and functions, and even into the C-suite" (Rigby et al. 2016). Some examples in NPD include:

- Responding to *changing or new customer needs quickly*, for example redefining a product already under development or defining a new product opportunity quickly.
- *Reprioritizing the portfolio of projects* if conditions change, including quickly adding new projects to the active list (Kock and Gemünden 2016).
- Being *able to pivot from the original project plan* when circumstances change.

Considerable evidence exists that shows the positive impact of agility on NP success (Evanschitzky et al. 2012; Rigby et al. 2016; Storey et al. 2016); while other studies reveal that being responsive to customer needs also enhances the odds of success (Zhan et al. 2018). But what factors improve agility in NPD?

Agility in portfolio management is enhanced by having a clear innovation strategy, formal portfolio processes, frequent portfolio monitoring, and a climate that fosters innovation and open communication of risks (Kock and Gemünden 2016). Iterative development, outlined above in 1.3.5, provides for a fast and *agile response to changing customer* needs throughout the entire NP process, leading to success (Tih et al. 2016). And big data – the effective use of data aggregation and data analysis tools – plays a role in developing agility (Hajli et al. 2020): Big data enables the firm to identify previously unrecognized customer needs and respond speedily to opportunities for innovation.

1.3.8 Planning and Resourcing the Launch

A quality launch is strongly linked to new product profitability, and effective after-sales service is central to the successful launch of the new product (Calantone et al. 2011; Evanschitzky et al. 2012; Joubert and Van Belle 2012; Kuester et al. 2012; Salmen 2021; Storey et al. 2016).

Good products don't sell themselves, and the launch should not be treated as an afterthought to be handled late in the project. A well-integrated and properly targeted launch is the result of a *finely tuned marketing plan, proficiently executed.* The launch must be *properly resourced* in terms of both people and funds (Kim et al. 2016; Kyriakopoulos et al. 2016); too often, an otherwise great new product fails to achieve its sales goals simply because of an under-resourced launch.

Those who execute the launch – the sales force, technical support people, and other front-line personnel – should be engaged in the development of the market launch plan and *therefore should be members of the project team.* This ensures valuable input and insight into the design of the launch effort, availability of resources when needed, and buy-in by those who must execute the launch – elements critical to a successful launch (Hultink and Atuahene-Gima 2000).

1.3.9 The World Product – A Global Orientation

Corporate growth and profitability depend on a *global business strategy married to product innovation.* In global markets, product development plays a primary role in achieving a sustainable competitive advantage (Kleinschmidt et al. 2007): Multinational firms that take a global approach to NPD outperform those that concentrate their R&D spending on their home market (de Brentani et al. 2010; de Brentani and Kleinschmidt 2015; Kleinschmidt et al. 2007; Le Meunier-FitzHugh et al. 2021). International products designed for and targeted at world and nearest neighbor export markets are the best-performing new products. By contrast, products designed for only the domestic or home market, and later adjusted and sold to nearest neighbor export markets, fare much worse. The magnitude of the differences between international new products and domestic products is striking: 2 or 3:1 on various performance gauges.

The management implications of these and other studies is that globalization of markets demands *a global innovation culture* and *a global innovation strategy* (de Brentani and Kleinschmidt 2015).

Sadly, this international dimension is often overlooked or, if included, is handled late in the development process or as a side issue. This global orientation translates into defining the market as international and *designing products to meet international requirements,* not just domestic ones. The result is either a global product (one version for the entire world) or a "glocal product" (one development effort, one basic product or platform, but several product variants to satisfy different international regions).

A global orientation also means undertaking VoC research, concept testing, and product testing in multiple countries rather than just in the home market, and tailored launch plans for different countries. It means employing a *global project team* with team members in multiple countries (but only one new product project team in five is reported to be a global development team, de Brentani et al. 2010; Kleinschmidt et al. 2007). Finally, the coordination and collaboration between marketing in the home country and the sales force in international markets is critical; devolving power while using cross-functional teams and formal processes to encourage communication results in successfully adapted new product launches for each country-market (Le Meunier-FitzHugh et al. 2021).

1.3.10 Expectations – "Expect Success, Get Success"

One of the strongest predictive drivers of new product success, only recently uncovered, is the existence of *expectations of success* by employees (usually members of the project team; Watson et al. 2021). A model of NP success consisting of well-known predictors, such as market orientation and product advantage, has a dramatic increase in predictive ability when "expected success" is included.

Organizational psychology and behavioral science researchers have long recognized that *expectations impact actual performance*: a self-fulfilling prophecy or the "Galatea effect." Findings from social psychology are noticeably absent from the NPD literature, however. A self-fulfilling prophecy occurs when individuals' expectations about themselves result in actions that confirm those expectations – they rise to the occasion. This Galatea effect stems from self-expectancy rather than supervisory expectations.

Recent research shows that product developers who understand customer and competitive actions and then develop products that are superior (have competitive advantage) *will expect, and thus realize, greater levels of product success* than those who simply carry out management orders. However, development teams cannot simply *think a successful product into existence* because that is what is expected of them. Self-fulfilling prophecy is powerful, but only when *leaders empower employees to act* and give them the latitude to engage in innovative and market-oriented behaviors, such as customer research (VoC), competitor analysis, and product testing (Watson et al. 2021).

This Galatea effect is positive self-expectancy, likely triggered by a supervisory source who communicates high expectations that motivate subordinates (Watson et al. 2021). But Galatea effects do not require sustained supervisor support beyond the communication of high expectations, because employees have internalized the motivation to perform. Senior management support alone will not guarantee product success, especially if the product developers themselves lack confidence in their new product. Employees need self-confidence to successfully promote a solution or idea; thus companies should foster confidence among product developers, not simply among senior management. While project teams require senior management support in order to obtain needed resources, such support may become counterproductive in instances where a heavy-handed approach stifles product developers' confidence, their creative output, and their success expectations.

The surprising and pronounced impact of this Galatea or "self fulfilling prophecy" effect comes as *new news* to many investigators of new product success drivers. But it is a well-known phenomenon in other fields. This is an area requiring more investigation in the field of NPD.

1.4 Drivers of Success for Businesses: Organizational and Strategic Factors.

Why are some businesses so much more successful at product innovation than others? Huge differences in product development productivity exist between the best and worst firms (Arthur D. Little 2005). The top 25 percent of *firms have 12 times the productivity* in NPD, realizing a huge $39 in new product sales per R&D dollar spent, while the bottom 25 percent of firms achieve only $3.3. The quest for the drivers of NPD success continues, but now with a *focus on the business* as the unit of analysis – see Table 1.2.

TABLE 1.2 DRIVERS OF SUCCESS FOR BUSINESSES – ORGANIZATIONAL & STRATEGIC FACTORS.

Driver Name	Success Driver Description
1. A Product Innovation Strategy	Having a product innovation strategy for the business: clear NPD objectives; defined strategic arenas (markets, product types) for the business to focus its NPD efforts on; and the product roadmap to define placemarks for future products
2. Sustainability Focus	Social responsibility, sustainability & a green orientation; launching green new products
3. Leveraging Core Competencies	Leverage, synergy & familiarity: getting a good fit between the firm's core competencies & the resource needs of the project to gain competitive advantage
4. Targeting Attractive Markets	Targeting large, growing, high-need markets with good long-term potential; avoiding intensively competitive markets with low margins & heavy price competition
5. Resources	Adequately resourcing projects; having dedicated project teams
6. Effective Teams	Cross-functional teams, with the right skills; a strong capable team leader; having clear goals & accountability for the project
7. Integrating the Supply Chain	Building the supply chain into NP projects; good communication with suppliers; seeking ideas & solutions from suppliers
8. Climate & Culture for Innovation	The right environment – a climate that supports innovation activities, risk taking & freedom to experiment; also: rewards teams & idea-generators; allows time for creative work; & no micromanaging of teams
9. Organizational Design	An organic structure – fluid & flexible in task execution, open channels of communication, decentralized decision making & few formal procedures
10. Top Management	Transformational leadership that is committed to NPD; develops the organization's vision, mission, NPD objectives & innovation strategy; commits the necessary resources to NPD; is engaged in the NP process, makes Go/Kill decisions & supports committed product champions

1.4.1 A Product Innovation Strategy for the Business

A product innovation strategy for the business charts the way forward for NPD; and having a new product strategy is strongly linked to positive NPD performance (APQC 2003; Cooper 2011; Dwivedi et al. 2021; Joubert and Van Belle 2012; Kock and Gemünden 2016; Song et al. 2011a). The *ingredients of such a strategy* with the strongest positive impact on performance include (Cooper 2011; Cooper and Edgett 2010):

- Clearly defined *product innovation goals and objectives*; for example, what percentage of the business's sales will come from new products?
- The *role of product innovation in achieving the overall businesses goals*, to link the product innovation goals to the business's overall goals.
- Strategic arenas defined – *areas of strategic focus on which to concentrate* new product efforts. The goal is to select strategic arenas rich with opportunities for innovation – those that will generate the business's future engines of growth (Cooper 2011, 2019; Florén et al. 2018). The great majority of businesses do designate strategic arenas – markets, product areas, industry sectors, or technologies – but evidence suggests that many are focused on *traditional and sterile areas* that fail to yield the needed opportunities (Cooper 2005).
- A *product roadmap* in place, which maps out a series of planned development, initiatives over time, often five years into the future. A roadmap is simply management's view of how to get to where they want to be, or to achieve their desired objective (Albright and Kappel 2003; McMillan 2003; Vishnevskiy et al. 2016), and provides *placemarks* for future development projects.

1.4.2 Social Responsibility, Sustainability, and a Green Orientation

Sustainability is now a key factor in firms' pursuit of long-term competitive advantage. The empirical evidence suggests a moderate and positive relationship between sustainability orientation and corporate financial performance (Chan et al. 2012; Dixon-Fowler et al. 2013).

In a similar vein, the growing environmental concern among customers has triggered many firms to adopt innovative strategies to meet the green requirement, offering "green products," thereby gaining competitive advantage. Thus, sustainability has been identified as a driver of technological innovation (Nidumolu et al. 2009).

Although a sustainability orientation impacts the corporation's financial performance, the *evidence is less clear* about the direct role of sustainability and launching green products on new product success. The development of sustainable products and services is still one of the *least understood areas* in sustainability management (de Medeiros et al. 2014).

A firm's sustainability orientation generally has a *positive influence* on new product success for the firm, and this relationship holds regardless of firm size, market, and product type (Claudy et al. 2016). A sustainability orientation includes both culture (how important environmental or social sustainability is to the company) and practices (how important sustainability practices, such as carbon footprint analysis, are during the NP process; Papadas et al. 2017). But some research shows that *sustainable new products achieve lower sales* than their conventional counterparts (van Dorn et al. 2021): Adding a sustainability claim to a new product leads to a higher likelihood of product failure. Why? If a firm's social responsibility reputation is negative, the success of its new sustainable or green products will also be low. But investing in corporate social responsibility activities compensates for this negative effect and is therefore a viable strategy to boost sales of new sustainable products.

While having a positive social responsibility reputation appears to have a positive impact on new product success, it is an *indirect impact* (Borah et al. 2021; Song et al. 2019). Corporate social responsibility impacts the firm's sustainability or green orientation (Song et al. 2019), helping firms to be environmentally conscious (Li et al. 2018). A green orientation in turn promotes investments in green innovation (technological capabilities in pollution prevention, green product design, energy-saving, waste recycling, and corporate environmental management). These green technological capabilities then influence the success of new products introduced: The technological capabilities result in new products that are accepted by green-conscious customers. That is, it is not just enough to have a sustainability or green orientation as a strategic direction, but *also to develop the technological capability to transform* those ideals into successful innovative products (Borah et al. 2021).

In summary, corporate social responsibility, a sustainable orientation, and introducing green new products have complex interlinkages with limited direct impacts on ultimate new product success, either at the firm or at the product level. More research is required here to better understand whether green products are indeed more successful, and under what conditions; and also, what drives the success – both directly and indirectly – of green new products.

1.4.3 Leveraging Core Competencies – Synergy and Familiarity

"Attack from a position of strength" may be an old adage, but it applies to new product management. A core competency is a major strength of an organization based on the organization's unique skills and experiences. The term was coined by Hamel and Prahalad (1994), who argued that a business should utilize or leverage its core competencies in order to gain competitive advantage. A variety of resources, such as skills, physical assets, intellectual property, and brand equity contribute to a company's core competencies.

In the context of new products, the ability to leverage means having a strong fit between (or have synergy with) the resource needs of the new product project and the core competencies and strengths of the firm in order to gain competitive advantage or be faster to market (Cankurtaran et al. 2013; Cooper 2019;

Evanschitzky et al. 2012). When *synergy with the base business* is lacking, new products fare more poorly on average. Areas of leverage include:

- R&D or technology resources – ideally the new product should leverage the business's existing technology competencies, IP, or patents;
- Marketing, sales force, and distribution (channel) resources;
- Branding, image, and marketing communications and promotional assets;
- Manufacturing, operations, or source-of-supply capabilities and resources;
- Technical support and customer service resources; and
- Management capabilities.

These six synergy or leverage ingredients become *important criteria* to evaluate and prioritize new product projects. If the "leverage score" is low, then there must be other compelling reasons to proceed with the project. Leverage is not essential, but it does improve the odds of winning.

"Familiarity" is a parallel concept and has its roots in the popular familiarity matrix (Roberts and Berry 1985). Some new product projects *take the company into unfamiliar territory* – a product category new to the firm; new customers with unfamiliar needs; unfamiliar technology; new sales force, channels and servicing requirements; or an unfamiliar manufacturing process – see Figure 1.4. And the business often pays the price: Step-out projects are riskier

FIGURE 1.4 NEWNESS OR FAMILIARITY CHART SHOWING PROJECT RISK LEVEL.

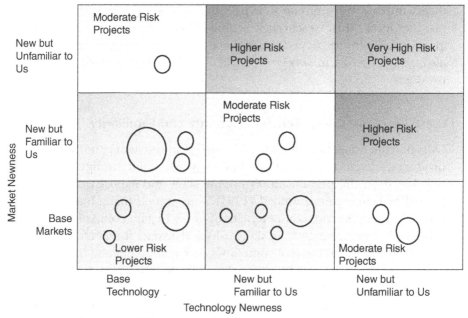

Bubbles are NPD projects; bubble size shows resource requirements per project

and have higher failure rates due to lack of strengths, experience, knowledge, and skills.

The negative impact here is not as strong as for most success drivers, however. New and unfamiliar territory certainly results in lower success rates and profitability on average, but the success rates are not dramatically lower.

Further, for such step-out projects, strategies such as *collaborative development* and *open innovation* may help the developer acquire the missing resources, skills, and knowledge (Chesbrough 2006; Docherty 2006; PDMA 2014). Indeed, resources from partner firms – from customers, other developer-firms, and even suppliers – may also have a positive impact on other success factors, such as effective cross-functional teams (team members available from the partner), voice-of-customer work, and effective launches.

Partnering does pose risks, however: Open innovation arrangements and collaborative developments are not always a "win win" situation for both firms. Conflicts and misalignments can occur due to misunderstandings, cultural differences, and even a lack of trust. Additionally, there is no strong evidence to suggest that partnering or collaborative NPD projects are more successful than those done alone (although some evidence exists that the project may not have been done at all were it not for the partnership; Campbell and Cooper et al. 1999; and Håkansson and Waluszewski 2002).

1.4.4 Targeting Attractive Markets

Market attractiveness is an important strategic variable and plays a role in notable strategy models such as Porter's "five forces" model and the BCG[4] matrix (market growth and market share) business portfolio grid. Market attractiveness is also important for new products: New products targeted at more attractive markets are more successful (Cooper 2019; Evanschitzky et al. 2012; Salmen 2021).

Market attractiveness has two facets:

- *Market potential*: Positive market environments, namely large and growing markets with large long-term potential and where the purchase is important to the customer.
- *Competitive situation*: Negative markets characterized by tough competitors with strong products, and capable sales forces and channel systems; and by intense price competition and low margins.

Both facets of market attractiveness – market potential and competitive situation – impact new product fortunes and both should be *considered as criteria* for project selection and prioritization.

1.4.5 The Resources in Place

Too many projects suffer from *a lack of time and financial commitment*. The predictable result is much higher failure rates (APQC 2003; Dalton 2016; Thomke and Reinsertsen 2012). As the quest for profits has intensified, companies often have responded by restructuring and cost-cutting – doing more with less – and so resources are limited. Also, many firms try to do too many projects for the resources available: an inability to say "no" to mediocre development projects, resulting in *pipeline gridlock* (Cooper 2021b). The resulting *resource crunch* takes its toll and is the root cause for much of what ails product development: too long to market; a lack of VoC; inadequate front-end homework; and an overemphasis on smaller, simpler projects.

Best-practice companies *commit the necessary resources* to new products, much more so than most firms. While new product resources are limited across the board – with less than 30 percent of businesses indicating that they have sufficient NPD resources in the key functional areas – the best performers are much better resourced (APQC 2003; Cankurtaran et al. 2013; Cooper and Edgett 2003; Dwivedi et al. 2021; Joubert and Van Belle 2012; Salmen 2021).

Equally important, these resources are *focused and dedicated*, with project team members heavily dedicated to one or a few development projects. Totally dedicated project teams are rare among manufacturers, however: Typically, project teams are thinly spread and multi-tasking across multiple projects and other duties (Cankurtaran et al. 2013; Joubert and Van Belle 2012). Dedicated human resources (largely R&D people) *is one of the strongest drivers of new product success* (Evanschitzky et al. 2012; Storey et al. 2016). Similarly, a dedicated project organization structure is more likely to facilitate successful products (Joubert and Van Belle 2012). Indeed, about half of the best performers have a dedicated product innovation group whose full-time job is to work on new product projects (Cooper 2019).

1.4.6 Effective Cross-functional Teams

Product innovation is very much a team effort. Do a post-mortem on any bungled new product project and invariably you'll find each functional area doing its own piece of the project with very little communication between functional areas (a fiefdom mentality), and no real commitment of team members to the project. Many studies concur that *the way the project team is organized and functions* strongly influences project outcomes (APQC 2003; Cooper 2019; Nakata and Im 2010; Sivasubramaniam et al. 2012). A great many facets of a high-performing NPD team have been uncovered, and include:

- *A cross-functional project team*: New product projects that have *clearly assigned project team members* do better. Further, NP results are better when the team is *multidisciplinary and cross-functional*, with team members from R&D, Sales, Marketing, and Operations, a practice now embraced by the majority of businesses (Cankurtaran et al. 2013; Cooper 2019; Dwivedi et al. 2021; Sivasubramaniam et al. 2012). Cross-functional project teams seem to operate as an organization on their own, simply because they possess a combination of entrepreneurial traits that complement one another to boost the NP process's performance and results (Dwivedi et al. 2021).
- *Team skills*: Innovation effectiveness is enhanced when team members have diverse skills and experience from several areas; high levels of education and

self-esteem (Joubert and Van Belle 2012); and the skills to deal with complex NPD projects, including general intelligence and previous team experience (Sivasubramaniam et al. 2012). Finally, project team members should have tenure; that is, remain on the project *from beginning to end*, not just for a short period or a single phase; and have worked together for a long time with little team turnover (Sivasubramaniam et al. 2012).

- *A capable, strong project leader.* A clearly identified *project leader* or *project manager* – a team member who is in charge and responsible for driving the project – enhances NP success (Joubert and Van Belle 2012). The project leader helps to define goals, prioritizes work, and provides leadership; and influences the product definition, promotes teamwork, facilitates strategic alignment, and creates a sense of joint team mission.

- *Characteristics of a good team leader*: A project team leader who is receptive to the ideas, needs, and wants of his/her team members is more likely to have highly motivated, conscientious people working to achieve the team's objectives (Sivasubramaniam et al. 2012). The team leader must be task-aware and emotionally intelligent in understanding the team members' work mannerisms, strengths, and weaknesses (Dwivedi et al. 2021). Most importantly, the leader must not be burdened with more than one project at a time in order to strengthen focus and enable efficiency in one direction. A good team leader requests support, lobbies for resources, and manages technical problems and design issues (Florén et al. 2018).

- *Clear goals*: Goal clarity and internal communication have a very strong relationship with NPD outcomes (including speed to market), suggesting that improved internal communication and a shared understanding of project objectives are key to success (Cankurtaran et al. 2013; Sivasubramaniam et al. 2012). Goal clarity is measured by the level of goal consensus within the project team. Clarity of goals contributes to employee motivation, and when teams engage in mission analysis, they develop a shared vision of the team's purpose and objectives, thus achieving greater clarity of team goals.

- *Accountability*: Project teams should be *accountable* for their project's end result – for example, for ensuring that their project meets profit, revenue targets, and time targets. A post-implementation review (or post-launch review) is recommended (Joubert and Van Belle 2012). This review, which involves the entire project team, and is presented to senior management, is where the accountability issues are finally addressed: Did the team deliver what they promised when the project was approved?

Project Leader versus Project Manager

Differences do exist between a *project leader* and *project manager* in different firms, and indeed a single project may have *both roles*. A consistent definition between what a project manager and a project leader do is elusive because often their jobs overlap; and sometimes the terms are used interchangeably. In many situa-

(Continued)

tions, a project leader *is* a project manager (just with extra responsibilities). But generally, a project manager is responsible for the completion of a project, while the team leader is responsible for the productivity and morale of their team (TechDee 2022).

1.4.7 Integrating the Supply Chain into NPD

Given an increasingly turbulent global environment, *supply chain management* has become a major issue for many corporations. Further, there exists mounting evidence that supplier relationships are important to NPD success (Florén et al. 2018; Johnsen 2009). Firms coordinate with their supply chain participants in NPD both to improve input quality and to reduce production costs. Numerous empirical studies have documented that supplier involvement in a firm's NPD positively influences the firm's innovation performance (Kou and Lee 2015; Moon et al. 2018; Song et al. 2011b).

In the front end of NPD, suppliers can offer assistance and useful input for concept development (Harvey et al. 2015). Suppliers may even share new product ideas with their customers (Wagner 2012). Further, suppliers can offer suggestions regarding product's design and component simplification to improve new product performance (He et al. 2014); partnering with competent suppliers in the front end may also reduce technological uncertainty.

Beyond the front end, collaboration with suppliers has been shown to improve innovation and indeed can be crucial (He et al. 2014; Joubert and Van Belle 2012). Effective supplier cooperation decreases time-to-market, reduces development costs, and improves product quality (Kim and Wilemon 2002).

A number of studies show no significant results or even negative effects of supplier involvement, however (Moon et al. 2018). One explanation is that the roles of suppliers, and hence their impact, *varys across the stages of NPD*. Supplier's activities are more or less relevant in certain stages than in others. For example, too much supplier involvement during the Concept Development stage slows the development of the product concept and design, since the product offering is still experimental and standards are not yet established. Similarly, too much involvement of suppliers in the Development stage may cause managerial problems and delay decisions about moving to the next stage, ultimately hurting the number of new products introduced in the market (Moon et al. 2018).

1.4.8 The Right Environment – Climate and Culture

A positive climate for innovation is one of the *top three success factors that distinguishes top-performing businesses in new product development,* with a huge impact on performance results (APQC 2003; Cooper 2011; Edgett 2011). Such a climate has been found to have many attributes, including (Round et al. 2020):

- Transformational leadership: senior management strongly and passionately supporting innovation in the business (Round et al. 2020) – see 1.4.10 below.

- "Intrepreneurs" (internal entrepreneurs) and risk-taking behavior encouraged;
- A propensity to take risks – not reckless risk taking, but informed decision making (defined as the willingness to confront opportunities, tolerate failure, and learn from mistakes; Joubert and Van Belle 2012);
- Allowing people the freedom to experiment, important for both group and individual autonomy (Joubert and Van Belle 2012);
- Team characteristics, including a highly motivated and ambitious team (Round et al. 2020);
- New product success rewarded or recognized (and failures not punished); team efforts recognized rather than individuals; and high morale and motivation fostered through rewards and the use of criteria to measure job satisfaction (Joubert and Van Belle 2012);
- Idea generators recognized or rewarded;
- Senior managers refraining from micromanaging projects and second-guessing project teams;
- Open project review meetings with senior people (the entire project team participating);
- Time allowed for creative people to work on projects of their own choice (projects on-the-side); and
- Employing skunk works and allowing unofficial projects (projects done "outside the system").

Most businesses are quite weak on almost all of these elements of a positive climate, with typically less than one-third of businesses employing each of these practices (APQC 2003; Cooper 2011, 2019). Best performers embrace these practices much more so.

1.4.9 The Right Organizational Design

The successful commercialization of new products is fostered when the firm's organizational structures are designed to unleash personnel to overcome both internal and external resistance to innovation (Walheiser et al. 2021). Organizational structure is the "enduring allocation of work roles and administrative mechanisms that allow firms to conduct, coordinate, and control their work activities and resource flows" (Droge et al. 2008).

Organic organizational structures (that is, with less centralization and formalization) have a performance-enhancing effect on new product outcomes (Walheiser et al. 2021). An *organic structure* is characterized by fluidity and flexibility in task execution, open channels of communication, decentralized decision making, and few formal procedures. Similarly, *organizational flexibility* (agile and responsive to change) and flexibility of personnel (the willingness to experiment and try new procedures to improve the product) enhance innovation outcomes (Joubert and Van Belle 2012).

One caution: Organizational structure affects performance in a unique way *during the commercialization phase* where a more mechanistic and formal structure may work better for some departments, such as Operations. Thus, firms are urged to consider a *selective* approach in choosing an organizational structure. Given the unique requirements of certain departments or tasks, firms might choose a structure that fits those requirements instead of following a "one-size-fits-all" approach (Walheiser et al. 2021).

1.4.10 Top Management

One of the highest-ranked critical success factors in NPD is transformational leadership with a strong commitment to product innovation (APQC 2003; Cooper 2011, 2019; Dwivedi et al. 2021; Edgett 2011; Reid et al. 2015). Senior management's main role is to define the organization's vision, mission, NPD objectives, and innovation strategy, which communicate a futuristic perception of the organization in the minds of its people, who in return drive initiatives to pursue these goals.

In best-performing businesses, senior management makes a long-term commitment to product innovation as a source of growth (Cooper 2011, 2019). It makes *available the necessary resources* for product development and ensures that they aren't diverted to more immediate needs in times of shortage. Most important, senior management is engaged in the new product process, reviewing projects, making timely and firm Go/Kill decisions, and if Go, making resource commitments to project teams. Senior management support thus improves time to market (Cankurtaran et al. 2013), and also enhances NP success rates: If senior management accepts personal accountability for an NP project, the chances of a successful product increase, and the likelihood of the project being terminated decreases (Joubert and Van Belle 2012).

Senior management *empowers project teams* and supports committed champions by acting as mentors, facilitators, or sponsors of project leaders and teams. Indeed, one role of senior management is "power promoters" or product champions (Joubert and Van Belle 2012). Finally, senior management's technology vision has a significant positive impact on NPs' competitive advantage and early success with customers (Reid et al. 2015).

1.5 The Right Systems, Processes, and Methodologies in Place

The *systems, methods, and procedures* that businesses put in place often hold the key to NP success. Newer methods, such as Agile Development from the software world but employed by developers of physical products, and Design Thinking to generate new product ideas, are positive methods that firms are adopting. Table 1.3 lists some of these methodological success drivers.

1.5.1 A Multistage, Disciplined Idea-to-Launch Gated Process

A *systematic idea-to-launch methodology*, such as a stage-and-gate system, is the solution many companies have adopted in order to overcome the deficiencies

TABLE 1.3 THE RIGHT SYSTEMS, PROCESSES, AND METHODOLOGIES.

Driver Name	Success Driver Description
1. An NP Idea-to-Launch Gated Process	An effective gating process: a multistage, disciplined idea-to-launch system, such as Stage-Gate, with defined Go/Kill decision points or gates & stages with success drivers and best-practices built in
2. Portfolio Management	Focus (doing fewer development projects) & making sharp project selection decisions (picking better projects & getting the right mix & balance of projects in the portfolio); using effective portfolio & project selection methods
3. Accelerated Development	Including: Lean Development; "build-and-test" iterations; & Agile Development. Also success drivers from this chapter, such as an innovative climate; effective cross-functional teams; & resourcing projects adequately (prioritizing projects, focusing resources)
4. Integrating Agile into the NP Process	Borrowing Agile Development from the software world & applying Agile to physical products (includes sprints, scrums, demo's, retrospects, new roles & the Agile values & mindset)
5. Effective Ideation	VoC methods; Open Innovation & co-development; lead user analysis; strategic approaches; & Design Thinking

that plague new product efforts (Barczak et al. 2009; Cankurtaran et al. 2013; Cooper 2011, 2017b, 2019; Dwivedi et al. 2021; Edgett 2011; Griffin 1997; Joubert and Van Belle 2012). A gating system, such as Stage-Gate,[5] is simply a roadmap or "playbook" for a project team to successfully and efficiently drive their NP project from idea through to launch and beyond. Other terms for such a gating process are phase-gate, phase review, and gating systems – see Figure 1.2.

Such gating systems are quite popular in business. Introduced in the 1980s, Stage-Gate remains the most widely adopted NP process (Cocchi et al. 2021). An APQC benchmarking study revealed that 88 percent of US businesses employ such a process, and identified the stage-and-gate process as *one of the strongest best practices*, employed by almost every best-performing business (Cooper and Edgett 2012).

The goal of a robust idea-to-launch system is to integrate the best practices outlined above into a single model (Cooper 2017b). A typical gating system for major projects (Figure 1.2) breaks the innovation process into five or six stages (Cooper 2014). The stages have *success drivers and best practices built in*: VoC, front-end homework, and build-and-test iterations. Stages also have *defined deliverables*, so that project teams have clear objectives.

Preceding each stage is a gate. These gates are the *quality control* and *Go/Kill decision points* in the system:

- Has the project been executed well; are the data generated reliable?
- Is the project still a good investment – technically feasible, customer demand, good financial prospects?

The gates thus open the door for the project to proceed and also commit the necessary resources – people and funds – to the project team. Weaker projects get killed at gates, and their resources are re-allocated to better projects.

Gating systems have evolved over the years and now include newer practices such as (Cooper 2014):

- A *scalable process* – for example, Lite and XPress versions of Stage-Gate for lower-risk and smaller projects; and unique versions of Stage-Gate for different types of projects, such as technology platform developments (Ajamian and Koen 2002; Cooper 2003).
- Adapting the system to accommodate open innovation (Grölund et al. 2010).
- Integration with the *total Product Life Cycle management* – from idea all the way through to product exit and disposal.
- An *adaptive and iterative process* – by using iterative development (see 1.3.5 above).
- An *automated* idea-to-launch system, via software solutions that handle idea management, the development process, portfolio management, and resource management.

1.5.2 Portfolio Management to Make the Right R&D Investment Decisions

Many businesses have moved to more *formal portfolio management systems* to help allocate resources effectively and to prioritize new product projects (Barczak et al. 2009; Cooper et al. 2004). Most companies suffer from *too many projects*, often the *wrong projects*, and *not enough resources* to mount an effective or timely effort on each (Cooper 2011, 2021b; Dalton 2016; Joubert and Van Belle 2012). This deficiency stems from a lack of adequate prioritization, with very negative results:

- First, scarce and valuable resources are wasted on poor projects. Thus, the *productivity of the business's R&D resources suffers.*
- Second, the truly deserving projects don't receive the resources they need, and so the good projects, starved for resources, move at a crawl, or just don't get done. *Time to market suffers.*

Portfolio management: The accepted definition is that "new-product portfolio management is a dynamic decision process, whereby a business's list of active new-product (R&D) projects is constantly updated and revised. In this process, new projects are evaluated, selected, and prioritized; existing projects may be accelerated, killed or deprioritized; and resources are allocated and reallocated to the active projects" (Cooper et al. 1999; Meifort 2016).

One goal in portfolio management is *to achieve focus* – focusing development efforts on those projects that deliver on the firm's long-term goals, such as growth or profitability. Indeed, focus is strongly connected to achieving a high value portfolio (Cooper 2011, 2013; Evanschitzky et al. 2012). This need to focus limited resources on the best projects coupled with the desire to cull out bad projects means that *tough Go or Kill and prioritization decisions* must be made; this results in fewer but better projects, and shorter times to market.

Project evaluations, however, are consistently cited as being poorly handled: Decisions involve the wrong people from the wrong functional areas; no consistent criteria exist to screen projects; and no will to kill projects exists. Smart firms have built in *tough gates* (Cooper 2009), and they have redesigned their idea-to-launch systems and created a funneling process that successively weeds out poor projects. Visible Go/Kill decision criteria employed at these gates, such as the use of a *scoring model*, improves decision effectiveness (Cooper 2011, 2021b; Cooper and Edgett 2006; Edgett 2013). And at periodic portfolio reviews, where *all the projects* are considered, projects should be rank-ordered – from best to worst – by using financial metrics such as the *Productivity Index* (Cooper 2021b).

Selecting high-value new product projects is only part of the task. Other portfolio goals are selecting the right *mix and balance of projects* in the development portfolio – across project types (improvements versus new products, for example) and the right risk/reward profile (Evanschitzky et al. 2012). In order to ensure the right mix and balance, some firms have adopted Strategic Buckets, earmarking *buckets of resources* targeted at different project types or strategic arenas (Cooper 2011, 2019; Edgett 2013). Ensuring *strategic alignment* in the portfolio is another goal: that the business's spending on product innovation *mirrors its strategic priorities* (Evanschitzky et al. 2012).

1.5.3 Accelerated Development

Speed offers the competitive advantage of being first on the market, namely "first mover advantage." But speed and success are not necessarily closely linked, as noted in 1.3.6. Nonetheless, speed or time to market remains an important performance metric for many firms.

Sound principles exist that firms embrace in order to reduce time-to-market, most already highlighted in this chapter, and include:

- Fostering an innovative climate in the firm (Cankurtaran et al. 2013); see 1.4.8.
- Employing *effective cross-functional teams*: a strong, influential leader; dedicated team members; clear project goals; team stability; proficiency of problem-solving skills on the team; team learning; and teams accountable for the end result (Cankurtaran et al. 2013); see 1.4.6.
- Using *parallel processing*: undertaking tasks concurrently, such as in *concurrent engineering*; overlapping stages in Figure 1.2 by moving long lead-time activities forward (such as purchasing production equipment); and moving ahead with only partial information. The relay race, sequential, or series approach to product development is inappropriate for today's fast-paced projects (Cooper 2021a).
- Using iterative development: These *build-and-test iterations* get the product right earlier and make needed adjustments long before formal product trials begin; see 1.3.5.
- Resourcing projects adequately by prioritizing and focusing – undertaking *fewer projects with higher value* (Cankurtaran et al. 2013); see 1.5.2.

- Employing Lean approaches to remove unnecessary work in the NP process (Fiore 2005; Salgado and Dekkers 2018). By mapping the NP process from end-to-end, and using *value stream analysis* to remove all waste, the result is a *leaner process.*
- Utilizing Agile Development from the software world, but with modifications to suit manufactured products (Cooper and Sommer 2016a, 2018).... next.

1.5.4 Integrating Agile Development into the NP Process

Agile development is a group of *software development methodologies* based on *iterative and incremental development*, where requirements and software solutions evolve through collaboration between self-organizing, cross-functional teams. Agile Development promotes adaptive planning and evolutionary development and delivery; it utilizes a time-boxed iterative approach; and it encourages rapid and flexible response to change. The Agile values are outlined in the *Agile Manifesto* (Beck et al. 2001).

Some features of the *Scrum version*[6] of Agile:

- A software development project is broken down into a number of iterations called *sprints*, which are *time-boxed* and very short, typically lasting *two weeks.*
- Sprints begin with a *sprint planning meeting.* Daily *scrums* or stand-up meetings (development team meetings) are held during the sprint, facilitated by a *Scrum Master.*
- Each sprint produces a *working product* (executable software code) and ends with a *product demo* to management and customers, followed by a *sprint retrospect.*
- Agile is a very visual process with many charts, such as the burndown, backlog, and Kanban charts.
- Agile has *new roles*: the Scrum Master and the Product Owner (but eliminates the traditional project manager role).
- The development team is *100 percent dedicated* to the one project, and is *physically co-located* in the same room.

Agile Development is more than just new methods and roles, however. It is *a new way of working* and features *new values* and an *Agile mindset*.

Larger *software developers* with existing development systems began integrating Agile into their traditional gated development processes with considerable success (Karlström and Runeson, 2005, 2006). More recently, *manufacturers of physical products* have successfully built elements of Agile Development into their gating models (Ovesen and Sommer 2015; Sommer et al. 2015). These firms simply *integrated Agile methods into the stages* of their gating process, replacing traditional project management methods, such as Gantt charts, timelines, and milestones, with Agile project management. Agile is typically initially employed in the two technical stages, namely, Development and Testing; with experience, manufacturing firms also apply Agile to the *entire idea-to-launch process* to create a true *Agile-Stage-Gate hybrid model* (Cooper 2016; Cooper and Sommer 2016a, 2016b, 2018). Agile-Stage-Gate encourages experimentation, so is best suited to NPD projects where there are many ambiguities and unknowns, as shown in the hybrid model matrix (Cocchi et al. 2021).

Adjustments must be made when applying Agile to physical products (Cooper and Fürst 2020b). When contrasted to software development, hardware development is usually *not as divisible;* thus it's not possible to have anything that actually functions within a few weeks, as in software development. So a sprint *does not* build a working product, but a *product version* somewhere between concept through to a ready-to-trial prototype – something to show the customer, as described in *iterative development* above (1.3.5). And most manufacturing firms lengthen the two-week sprint to *four weeks,* some as long as *eight weeks* (Cooper and Sommer 2016b, 2018).

The advantages of Agile-Stage-Gate are speed (sprints are time-boxed with no relaxation of the time); dedicated or focused teams (core team members are close to 100 percent dedicated to the one project); much better communication within the team (via frequent scrums); and constant customer feedback with strategic pivots (product revisions) if needed (Sommer et al. 2015). Early adopters of this new hybrid system report positive results (Figure 1.5; Invention Center 2018; Schmidt et al. 2019). But implementation challenges do exist (Cooper and Sommer 2016b, 2018).

FIGURE 1.5 PERFORMANCE IMPROVEMENTS REALIZED USING AGILE FOR NPD FOR PHYSICAL PRODUCTS.

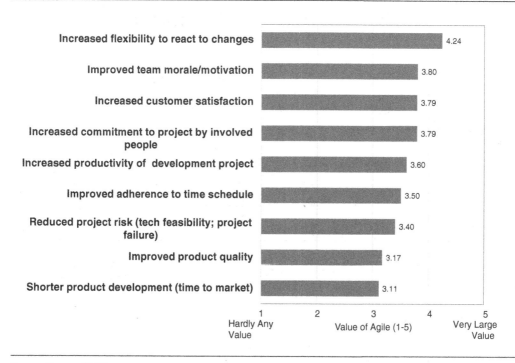

Source: Cooper 2019.

1.5.5 Effective Ideation

Great ideas are the foundation for great new products. Thus, increasingly more attention is being devoted to the "fuzzy front-end" of the innovation process. Idea generation and idea evaluation are key components. Successful ideation requires the simultaneous application of both:

- Open action strategies that increase the number and variability of ideas – for example, granting employees autonomy to develop new ideas and supporting them in their creative efforts.
- Closed action strategies that focus, integrate, and select ideas – formal processes to evaluate and select innovation ideas to increase the likelihood of success (Kock et al., *JPIM* 2014).

The rated effectiveness by users of a number of ideation methods and sources is shown in Figure 1.6, along with their popularity (Cooper and Dreher 2010). Note how effective VoC methods are rated – most are near the top of Figure 1.6.

FIGURE 1.6 POPULARITY VERSUS EFFECTIVENESS OF IDEATION METHODS.

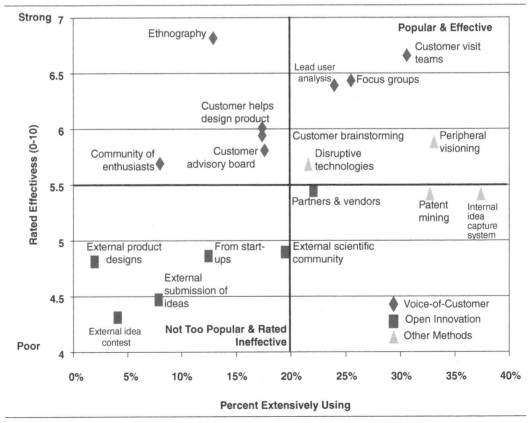

Source: Adapted from Cooper & Dreher, 2010.

And although internal idea capture methods (soliciting ideas from one's own employees) are the most popular, they are not the most effective (Cooper 2011; Cooper and Edgett 2008).

Voice-of-customer. VoC methods generally *are rated the most effective for generating breakthrough ideas*:

- Customer visit teams – typically a cross-functional team of 2–3 people undertaking a systematic visitation program with key purchase influencers in a limited number of representative customers.
- Lead user analysis – identifying leading or innovative users (ahead of the wave) and working with them (typically in workshop format) to develop new product concepts (von Hippel et al. 2011).
- Ethnography – camping out with customers to observe behaviors (cultural anthropology), and in so doing, understanding their unspoken, unmet, and often unknown needs.
- Focus groups with groups of customers (often consumers) to identify problems, desires, needs, and wants.
- Design thinking, whereby users' needs are understood through VoC (for example, ethnography) and a series of product versions are immediately tested with users (Brown 2008; Mahmoud-Jouini et al. 2019).

Lead user analysis: Much literature recognizes the importance of customer involvement in NPD (Fang 2008; Woojung and Taylor 2016; Cui and Wu 2016), indicating that customers not only are passive buyers but also *act as co-creators of new products and NP ideas*. Many commercially important products are initially thought of and even prototyped by users rather than by suppliers (Lilien et al. 2002; von Hippel et al. 1999). Such products tend to be developed by "lead users" – innovative companies, organizations, or individuals that are well ahead of market trends and even have needs that go far beyond the average user. The challenge is to track down lead users, who are by definition rare. Social networks are a source of lead users, and identifying them is facilitated by using web-mining tools and netnography methods (Kratzer et al. 2016).

Open innovation: Looking outside one's company has a positive impact on the firm's product innovativeness, customer satisfaction, and NP profitability (Cheng and Huizing 2014). Open innovation is *valuable source of NP ideas* (Chesbrough 2006; Docherty 2006; PDMA 2014). But most firms are not well positioned to solicit or handle outside ideas and IP. Thus, it is important to adapt the firm's processes and systems for open innovation in order to encourage the inclusion of ideas, IP, R&D work, and even fully developed products from outside the firm (Docherty 2006; Grölund et al. 2010).

Strategic methods: These also are positively rated and include exploiting disruptive technologies (Christensen 2000) and peripheral visioning (Day and Schoemaker 2005).

1.6 Looking Ahead

Philosopher and statesman Edmund Burke once said, "Those who don't know history are destined to repeat it." Thirty years of research has led to many more insights about NP success drivers than we had when such research began in earnest circa 1970 (Cooper 2017a).

Today however, product developers face many new challenges: The world is faster, more global, less predictable, and more ambiguous than it was when those early research studies began. And there have been many new practices introduced to NPD since then in order to deal with these challenges: practices such as Agile Development, Open Innovation, Lean Product Development, and developing smart products and green products, whose impacts have not yet been thoroughly investigated. And so, research into NP success drivers and new and promising NPD practices must continue, simply because product innovation is *so important to business prosperity*, and yet many of the keys to success still remain a mystery.

Notes

1 Protocept: Something between a concept (words and drawings) and a working product prototype.
2 Pretotype: A mock up or stripped-down version of a product, used to validate the product with customers or users.
3 MVP or minimum viable product: a feature-limited product that can actually be sold and thus generate revenue (Ries 2011). An MVP contains more detail than a pretotype and is also usable; it provides enough value that customers will actually pay to buy it. The term MVP is often misused in industry – calling a not-for-sale early prototype an MVP.
4 BCG: Boston Consulting Group.
5 Stage-Gate® is a registered trademark of R. Cooper, R.G. Cooper & Associates Inc., and Stage-Gate International Inc., in the USA, Canada, and Europe.
6 A "scrum" is a meeting on the field of the players in the game of British rugby. Scrum is the most popular version of Agile Development, and is also the version used by most manufacturing firms.

References

Ajamian, G. and Koen, P.A. (2002). Technology Stage-Gate: a structured process for managing high risk, new technology projects. In: *PDMA Tool Book for New Product Development* (ed. P. Belliveau, A. Griffen, and S. Somermeyer), 267–295. New York, NY: John Wiley and Sons.

Albright, R.E. and Kappel, T.A. (2003). Roadmapping in the corporation. *Research-Technology Management* 46 (2): 31–40.

American Productivity and Quality Center, APQC (2003). *Improving New Product Development Performance and Practice.* Houston, TX: APQC.

Arthur D. Little (2005). *How Companies Use Innovation to Improve Profitability and Growth, Innovation Excellence Study.* Boston, MA: A.D. Little Inc.

Barczak, G., Griffin, A., and Kahn, K.B. (2009). Trends and drivers of success in NPD practices: results of the 2003 PDMA best practices study. *Journal of Product Innovation Management* 26: 3–23.

Beck, K., Beedle, M., van Bennekum, A., Cockburn, A., Cunningham, W., Fowler, M., Grenning, J., Highsmith, J., Hunt, A., Jeffries, R., Kern, J., Marick, B., Martin, R.C., Mellor, S., Schwaber, K., Sutherland, J., and Thomas, D. (2001). *Principles Behind the Agile Manifesto, Manifesto for Agile Software Development.* http://agilemanifesto.org/principles.html

Borah, P.S., Dogbe, C.S.K., Pomegbe, W.W.K., Bamfo, B.A. and Hornuvo, L.K. (2021 September 7). Green market orientation, green innovation capability, green knowledge acquisition and green brand positioning as determinants of new product success. *European Journal of Innovation Management.* https://www.emerald.com/insight/content/doi/10.1108/EJIM-09-2020-0345/full/html.

Brown, T. (2008 June). Design Thinking. *Harvard Business Review* 86(6): 85–92.

Buffoni, A., de Angelis, A., Grüntges, V., and Krieg, A. (2017). *How to Make Sure Your Next New Product or Service Launch Drives Growth.* McKinsey & Company. https://www.mckinsey.com/capabilities/growth-marketing-and-sales/our-insights/how-to-make-sure-your-next-product-or-service-launch-drives-growth

Calantone, R. J. and Di Benedetto. C.A. (2007). Clustering product launches by price and launch strategy. *Journal of Business & Industrial Marketing* 22(1): 4–19.

Calantone, R. J., Di Benedetto, C.A., and Song, M. (2011). Expecting marketing activities and new product launch execution to be different in the U.S. and China: an empirical study. *International Journal of China Marketing* 2(1): 14–44.

Campbell, A.J. and Cooper, R.G. (1999). Do customer partnerships improve success rates? *Industrial Marketing Management* 28 (5): 507–519.

Cankurtaran, P., Langerak, F., and Griffin, A. (2013). Consequences of new product development speed: a meta-analysis. *Journal of Product Innovation Management* 30 (3): 465–486.

Castellion, G. and Markham, S.K. (2013). New product failure rates: influence of argumentum ad populum and self-interest. *Journal of Product Innovation Management* 30 (5): 976–979.

Chan, R.Y.K., He, H., Chan, H.K., and Wang, W.Y.C. (2012). Environmental orientation and corporate performance: the mediation mechanism of green supply chain management and moderating effect of competitive intensity. *Industrial Marketing Management* 41(4): 621–30.

Chen, J., Reilly, R.R., and Lynn, G.S. (2005). The impacts of speed-to-market on new product success: the moderating effects of uncertainty. *IEEE Transactions on Engineering Management* 52(2): 199–212.

Chen, J., Damanpour, F., and Reilly, R.R. (2010). Understanding antecedents of new product development speed: a meta-analysis. *Journal of Operations Management* 28 (1): 17–33.

Cheng, C.C.J. and Huizing, E.K.R.E. (2014). When is Open Innovation beneficial? The role of strategic orientation. *Journal of Product Innovation Management* 31(6): 1235–1253.

Chesbrough, H. (2006). "Open Innovation" myths, realities, and opportunities. *Visions* XXX (2): 18–19.

Chesbrough, H. and Crowther, A. (2006). Beyond high tech: early adopters of Open Innovation in other industries. *R&D Management* 36 (3): 229–236.

Christensen, C.M. (2000). *The Innovator's Dilemma.* New York, NY: Harper Collins.

Claudy, M.C., Peterson, M., and Pagell, M. (2016). The roles of sustainability orientation and market knowledge competence in new product development success. *Journal of Product Innovation Management* 33(1): 72–85.

Cocchi, N., Dosi, C., and Vignoli, M. (2021). The hybrid model matrix: Enhancing Stage-Gate with Design Thinking, lean startup, and Agile. *Research-Technology Management* 64 (5): 18–30.

Cooper, R.G. (2003). Managing technology development projects – different than traditional development projects. *Research-Technology Management* 49(6): 23–31.

Cooper, R.G. (2005). Your NPD portfolio may be harmful to your business's health. *Visions*, XXIX (2): 22–6.

Cooper, R.G. (2009). Effective gating: make product innovation more productive by using gates with teeth. *Marketing Management Magazine* (March–April): 12–17.

Cooper, R.G. (2011). The innovation dilemma – how to innovate when the market is mature. *Journal of Product Innovation Management* 28 (7): 2–27.

Cooper, R.G. (2013). Where are all the breakthrough new products? Using portfolio management to boost innovation. *Research-Technology Management* 156(5): 25–32.

Cooper, R.G. (2014). What's next? After Stage-Gate. *Research-Technology Management* 157 (1): 20–31.

Cooper, R.G. (2016). Agile-Stage-Gate hybrids: The next stage for product development. *Research-Technology Management* 59(1): 1–9.

Cooper, R.G. (2017a). We've come a long way baby. *Journal of Product Innovation Management* Special Virtual Issue 34(3): 387–391.

Cooper, R.G. (2017b). *Winning at New Products: Creating Value through Innovation*, 5e. New York, NY: Basic Books, Perseus Books Group.

Cooper, R.G. (2019 January). The drivers of success in new-product development. *Industrial Marketing Management* 76: 36–47.

Cooper, R.G. (2021a). Accelerating innovation: lessons from the pandemic. *Journal of Product Innovation Management* 38 (2): 1–11.

Cooper, R.G. (2021b November 8). Unlocking "pipeline gridlock": effective portfolio management is the key. *Innovation Management*. https://innovationmanagement.se/2021/11/08/unlocking-pipeline-gridlock-effective-portfolio-management-is-the-key/.

Cooper, R.G. and Dreher, A. (2010 Winter). Voice of customer methods: what is the best source of new product ideas? *Marketing Management Magazine* 38–43.

Cooper, R.G. and Edgett, S.J. (2003). Overcoming the crunch in resources for new product development. *Research-Technology Management* 46 (3): 48–58.

Cooper R.G. and Edgett, S.J. (2006). Ten ways to make better portfolio and project selection decisions. *Visions*, XXX (3): 11–15.

Cooper, R.G. and Edgett, S.J. (2008). Ideation for product innovation: what are the best sources? *Visions*, XXXII (1): 12–17.

Cooper, R.G. and Edgett, S.J. (2010). Developing a product innovation and technology strategy for your business. *Research-Technology Management*, 53 (3): 33–40.

Cooper, R.G. and Edgett, S.J. (2012). Best practices in the idea-to-launch process and its governance. *Research-Technology Management* 55 (2): 43–54.

Cooper, R.G., Edgett, S.J., and Kleinschmidt, E.J. (1999). New product portfolio management: practices and performance. *Journal of Product Innovation Management* 4 (16): 333–350.

Cooper, R.G., Edgett, S.J., and Kleinschmidt, E.J. (2004). Benchmarking best NPD practices–2: strategy, resources and portfolio management practices. *Research-Technology Management* 47 (3): 50–60.

Cooper, R.G. and Fürst, P. (2020a March 11). Digital transformation and its impact on new-product development for manufacturers. *InnovationManagement*. https://innovationmanagement.se/2020/03/11/digital-transformation-and-its-impact-on-new-product-management-for-manufacturers/.

Cooper, R.G. and Fürst, P. (2020b November 10). Agile development for manufacturers: The emergent gating model. *InnovationManagement* https://www.innovationmanagement.se/2020/11/10/agile-development-for-manufacturers-the-emergent-gating-model.

Cooper, R.G. and Kleinschmidt, E.J. (2021 January 21). *Portfolio Management: An Update to the 2004 Benchmarking Best NPD Practices Study*: Washington DC: IRI, Innovation Research Interchange, National Association of Manufacturers.

Cooper, R.G. and Sommer, A.F. (2016a). The Agile-Stage-Gate hybrid model: a promising new approach and a new research opportunity. *Journal of Product Innovation Management* 33 (5): 513–526.

Cooper, R.G. and Sommer, A.F. (2016b November). Agile-Stage-Gate: new idea-to-launch method for manufactured new products is faster, more responsive. *Industrial Marketing Management* 59: 167–180.

Cooper, R.G. and Sommer, A.F. (2018). Agile-Stage-Gate for manufacturers – changing the way new products are developed. *Research-Technology Management* 61 (2): 17–26.

Crawford, C.M. (1992). The hidden costs of accelerated product development. *Journal of Product Innovation Management* 9 (3): 188–199.

Cui, A.S. and Wu, F. (2016). Utilizing customer knowledge in innovation: antecedents and impact of customer involvement on new product performance. *Journal of the Academy of Marketing Science* 44: 516–538.

Dalton, M. (2016 August 23). Manage pipeline bandwidth to avoid derailing new prod ucts. *Industry Week*. https://www.industryweek.com/innovation/process-improvement/article/22007299/manage-pipeline-bandwidth-to-avoid-derailing-new-products.

Day, G.S. and Schoemaker, P.J.H. (2005). Scanning the periphery. *Harvard Business Review* 83 (11): 135–148.

de Brentani, U. and Kleinschmidt, E.J. (2015 February–March). The impact of company resources and capabilities on global new product program performance. *Project Management Journal* 46 (1): 12–2. 9.

de Brentani, U., Kleinschmidt, E.J., and Salomo, S. (2010). Success in global new product development: impact of strategy and the behavioral environment of the firm. *Journal of Product Innovation Management* 27 (2): 143–160.

de Medeiros, J.F., Ribeiro, J.L.D., and Cortimiglia, M.N. (2014). Success factors for environmentally sustainable product innovation: a systematic literature review. *Journal of Cleaner Production* 65: 76–86.

Dixon-Fowler, H. R., Slater, D. J., Johnson, J. L., Ellstrand, A. E., and Romi, A.M. (2013). Beyond "Does it pay to be green?" A Meta-analysis of moderators of the CEP-CFP relationship. *Journal of Business Ethics* 112 (2): 353–366.

Docherty, M. (2006). Primer on "open innovation": principles and practice. *Visions* XXX (2): 13–17.

Droge, C., Calantone, R.J., and Harmanciogu, N. (2008). New product success: is it really controllable by managers in highly turbulent environments? *Journal of Product Innovation Management* 25: 272–286.

Dwivedi, R., Karim, F.J., and Starešinić, B. (2021). Critical success factors of new product development: evidence from select cases. *Business Systems Research* 12 (1): 34–44.

Edgett, S.J. (2011). *New Product Development: Process Benchmarks and Performance Metrics.* Houston, TX: American Productivity and Quality Center (APQC).

Edgett, S.J. (2013). Portfolio management for product innovation. In: *PDMA Handbook of New Product Development*, 3e (ed. K.B. Kahn), 154–166. Hoboken, NJ: Wiley. Ch 9.

Eling, K., Langerak, F., and Griffin, A. (2013 July). A stage-wise approach to exploring performance effects of cycle time reduction. *Journal of Product Innovation Management* 30 (4): 626–641.

Enkel, E., Gassmann, O., and Chesbrough, H. (2009). Open R&D and Open Innovation: exploring the phenomenon. *R&D Management* 39(4): 311–316.

Evanschitzky, H., Eisend, M., Calantone, R.J., and Jiang, Y. (2012). Success factors of product innovation: an updated meta-analysis. *Journal of Product Innovation Management* 29 (S1): 21–37.

Fang, E. (2008 July). Customer participation and the trade-off between new product innovativeness and speed to market. *Journal of Marketing* 72: 90–104.

Fiore, C. (2005). *Accelerated Product Development: Combining Lean and Six Sigma for Peak Performance.* New York: Productivity Press.

Florén, H., Frishammar, J., Parida, V., and Wincent, J. (2018). Critical success factors in early new product development: a review and a conceptual model. *International Entrepreneurial Management Journal* 14: 411–427.

Garcia, R. and Calantone, R.J. (2002). A critical look at technological innovation typology and innovativeness terminology: a literature review. *Journal of Product Innovation Management* 19 (2): 110–132.

Griffin, A. (1997). *Drivers of NPD Success: The 1997 PDMA Report.* Chicago, IL: Product Development and Management Association.

Griffin, A. (2002). Product development cycle time for business-to-business products. *Industrial Marketing Management* 31 (4): 291–304.

Grölund, J., Rönneberg, D., and Frishammar, J. (2010). Open Innovation and the Stage-Gate process: a revised model for new product development. *California Management Review* 5 (3): 106–131.

Hajli, N., Tajvidi, M., Gbadamosi, A., and Nadeem, W. (2020). Understanding market agility for new product success with big data analytics. *Industrial Marketing Management* 86: 135–143.

Håkansson, H. and Waluszewski, A. (2002). Path dependence: restricting or facilitating technical development? *Journal of Business Research* 55 (7): 561–570.

Hamel, G. and Prahalad, C.K. (1994). *Competing For the Future.* Boston, MA: Harvard Business Review Press.

Harvey, J. F., Cohendet, P., Simon, L., and Borzillo, S. (2015). Knowing communities in the front end of innovation. *Research-Technology Management* 58 (1): 46–54.

He, Y., Lai, K.K., Sun, H., and Chen, Y. (2014). The impact of supplier integration on customer integration and new product performance: the mediating role of manufacturing flexibility under trust theory. *International Journal of Production Economics* 147: 260–270.

Hienerth, C., von Hippel, E.A., and Jensen, M.B. (2012). Efficiency of consumer (household sector) vs. producer innovation. SSRN eLibrary. https://www.researchgate.net/ publication/228197597_Efficiency_of_Consumer_Household_Sector_vs_Producer_ Innovation

Hultink, E.J. and Atuahene-Gima, K. (2000). The effect of sales force adoption on new product selling performance. *Journal of Product Innovation Management* 17 (6): 435–450.

Invention Center (2018). *Consortium Benchmarking 2018, "Agile Invention": Evaluation of the Study Results.* Rheinisch-Westfälische Technische Hochschule Aachen (RWTH).

Isaacson, W. (2011). *Steve Jobs: The Exclusive Biography.* New York, NY: Simon & Schuster.

Johnsen, T.E. (2009). Supplier involvement in new product development and innovation: taking stock and looking to the future. *Journal of Purchasing and Supply Management* 15(3): 187–197.

Joubert, J. and Van Belle, J.-P. (2012). Success factors for product and service innovation: a critical literature review and proposed integrative framework. *Management Dynamics* 12 (2): 01–26.

Karlström, D. and Runeson, P. (2005 May-June). Combining Agile methods with Stage-Gate project management. *IEEE Software.* 43–49.

Karlström, D. and Runeson, P. (2006). Integrating Agile software development into Stage-Gate managed product development. *Empirical Software Engineering* 11: 203–225.

Kim, J. and Wilemon, D. (2002). Focusing the fuzzy front-end in new product development. *R&D Management* 32 (4): 269–279.

Kim, N., Shin, S., and Min, S. (2016). Strategic marketing capability: mobilizing technological resources for new product advantage. *Journal of Business Research* 69 (12): 5644–5652.

Kleinschmidt, E.J., de Brentani, U., and Salome, S. (2007). Performance of global new product development programs: a resource-based view. *Journal of Product Innovation Management* 24 (5): 419–441.

Kock, A. and Gemünden, H.G. (2016). Antecedents to decision-making quality and agility in innovation portfolio management. *Journal of Product Innovation Management* 33 (6): 670–686.

Kou, T.C. and Lee, B.C.Y. (2015). The influence of supply chain architecture on new product launch and performance in the high-tech industry. *Journal of Business and Industrial Marketing* 30 (5): 677–687.

Kratzer, J., Lettl, C., Franke, N., and Gloor, P.A. (2016). The social network position of lead users. *Journal of Product Innovation Management* 33 (2): 201–216.

Kuester, S., Homburg, C., and Hess, S.C. (2012). Externally directed and internally directed market launch management: the role of organizational factors in influencing new product success. *Journal of Product Innovation Management* 29 (S1): 38–52.

Kyriakopoulos, K., Hughes, M., and Hughes, P. (2016). The role of marketing resources in radical innovation activity: antecedents and payoffs. *Journal of Product Innovation Management* 33 (4): 398–417.

Langerak, F. (2010). *Accelerated Product Development. Product Innovation and Management, Part 5*. Hoboken, NJ: Wiley.

Le Meunier-FitzHugh, K., Cometto, T., and Johnson, J. (2021). Launching new global products into subsidiary markets: the vital role of sales and marketing collaboration. *Thunderbird International Business Review* 63: 543–558.

Li, Y., Ye, F., Sheu, C., and Yang, Q. (2018). Linking green market orientation and performance: antecedents and processes. *Journal of Cleaner Production* 192: 924–931.

Lilien, G. L., Morrison, P.D., Searls, K., Sonnack, M., and von Hippel, E.A. (2002). Performance assessment of the lead user idea-generation process for new product development. *Management Science* 48 (8): 1042–1059.

Mahmoud-Jouini, S.B., Fixson, K.S., and Boulet, D. (2019). Making design thinking work. *Research-Technology Management* 62 (5): 50–58.

Marinov, K. (2020). Order of market entry as a success factor in product innovations. *Trakia Journal of Sciences* 3: 230–237.

McMillan, A. (2003). Road mapping – agent of change. *Research-Technology Management* 46 (2): 40–47.

McNally, R.C., Cavusgil, E., and Calantone, R.J. (2010). Product innovativeness dimensions and their relationships with product advantage, product financial performance, and project protocol. *Journal of Product Innovation Management* 27 (1): 991–1006.

Meifort, A. (2016). Innovation portfolio management: A synthesis and research agenda. *Creativity and Innovation Management* 25 (2): 251–269.

Mishra S., Kim, D., and Lee, D.H. (1996). Factors affecting new product success: cross-country comparisons. *Journal of Product Innovation Management* 13 (6): 530–550.

Moon, H., Johnson, J.L., and Mariadoss, J. (2018). Supplier and customer involvement in new product development stages: implications for new product innovation outcomes. *International Journal of Innovation and Technology Management* 15 (1): 1–21.

Morgan, J.M. and Liker, J.K. (2006). *The Toyota Product Development System*. New York, NY: Productivity Press.

Morgan, T., Obal, M., and Anokhin, S. (2018). Customer participation and new product performance: Towards the understanding of the mechanisms and key contingencies. *Research Policy* 47: 498–510.

Nakata, C. and Im, S. (2010). Spurring cross-functional integration for higher new product performance: a group effectiveness perspective. *Journal of Product Innovation Management* 27 (4): 554–571.

National Industrial Conference Board, NICB (1964). Why new products fail. *The Conference Board Record*. New York, NY: NICB.

Nidumolu, R., Prahalad, C.K., and Rangaswami, M.R. (2009). Why sustainability is now the key driver of innovation. *Harvard Business Review* 87 (9): 56–64.

National Science Foundation, NSF (2021 November). Businesses reported an 11.8% increase to nearly a half trillion dollars for U.S. R&D performance during 2019. *InfoBrief, National Center for Science and Engineering Statistics*, Table 2. https://www.ncses.nsf.gov/pubs/nsf22303

Ovesen, N. and Sommer, A.F. (2015). Scrum in the traditional development organization: adapting to the legacy. *Modeling and Management of Engineering Processes, Proceedings of the 3rd International Conference 2013*, Berlin and Heidelberg, Germany: Springer-Verlag, 87–99.

Papadas, K.K., Avlonitis, G.J., and Carrigan, M. (2017). Green marketing orientation: conceptualization, scale development and validation. *Journal of Business Research* 80: 236–246.

Poetz, M.K. and Schreier, M. (2012). The value of crowdsourcing: can users really compete with professionals in generating new product ideas? *Journal of Product Innovation Management* 29(2): 245–256.

Product Development & Management Association, PDMA (2014). *Open Innovation: New Product Development Essentials from the PDMA* (ed. C. Noble, S.S. Durmusoglu, and A. Griffin). Hoboken NJ: Wiley.

Raff, S., Wentzel, D., and Obwegeser, N. (2020). Smart products: conceptual review, synthesis, and research directions. *Journal of Product Innovation Management* 37 (5): 379–404.

Reid, S.E., Roberts, D., and Moore, K. (2015). Technology vision for radical innovation and its impact on early success. *Journal of Product Innovation Management* 32 (4): 593–609.

Ries, E. (2011). *The Lean Start-up: How Today's Entrepreneurs Use Continuous Innovation to Create Radically Successful Businesses.* New York, NY: Crown Publ.

Rigby, D.K., Sutherland, J., and Takeuchi, H. (2016 May). Embracing Agile. *Harvard Business Review* 48–50.

Rijsdijk, S.A., Langerak, F., and Jan, E. (2011). Understanding a two-sided coin: antecedents and consequences of a decomposed product advantage. *Journal of Product Innovation Management* 28 (1): 33–47.

Roberts, E.B. and Berry, C.A. (1985 Spring). Entering new businesses: selecting strategies for success. *Sloan Management Review* 26: 3–17.

Round, H., Wang, S., and Mount, M. (2020). Innovation climate: a systematic review of the literature and agenda for future research. *Journal of Occupational and Organizational Psychology* 93: 73–109.

Salgado, E.G. and Dekkers, R. (2018 Oct). Lean product development: nothing new under the sun? *International Journal of Management Reviews* 20: 903–933.

Salmen, A. (2021). New product launch success: a literature review. *Acta Universitatis Agriculturae Et Silviculturae Mendelianae Brunensis* 69 (1): 151–176.

Sandmeier, P., Morrison, P.D., and Gassmann, O. (2010). Integrating customers in product innovation: lessons from industrial development contractors and in-house contractors in rapidly changing customer markets. *Creativity and Innovation Management* 19 (2): 89–106.

Shashishekar, M.S., Anand, S., and Paul, A.K. (2022). Proactive market orientation and business model innovation to attain superior new smart connected products performance. *Journal of Business & Industrial Marketing* 37 (3): 497–508.

Schmidt, T.S., Atzberger, A., Gerling, C. et al. (2019). *Agile development of physical products: an empirical study about potential, transition and applicability.* Report, University of the German Federal Armed Forces, Munich, Germany. https://www.researchgate.net/publication/331952292_Agile_Development_of_Physical_Products_-_An_Empirical_Study_about_Potentials_Transition_and_Applicability.

Sivasubramaniam, N., Liebowitz, S.J., and Lackman, C.L. (2012). Determinants of new product development team performance: a meta-analytic review. *Journal of Product Innovation Management* 29 (5): 803–820.

Sommer, A.F., Hedegaard, C., Dukovska-Popovska, I., and Steger-Jensen, K. (2015). Improved product development performance through Agile/Stage-Gate hybrids – the next-generation Stage-Gate process? *Research-Technology Management* 58 (1): 1–10.

Song, X.M. and Parry, M.E. (1996). What separates Japanese new product winners from losers. *Journal of Product Innovation Management* 13 (5): 422–439.

Song, W., Ren, S., and Yu, J. (2019). Bridging the gap between corporate social responsibility and new green product success: the role of green organizational identity. *Business Strategy and the Environment* 28: 88–97.

Song, X.M., Im, S., van der Bij, H., and Song, L.Z. (2011a). Does strategic planning enhance or impede innovation and firm performance? *Journal of Product Innovation Management* 28 (4): 503–20.

Song, L. Z., Song, M., and Di Benedetto, C.A. (2011b). Resources, supplier investment, product launch advantages, and first product performance. *Journal of Operations Management* 29 (1): 86–104.

Storey, C., Cankurtaran, P., Papastathopoulou, P., and Hultink, E.J. (2016). Success factors for service innovation. *Journal of Product Innovation Management* 33 (5): 527–548.

TechDee (2022 March). Team leader vs. project manager – responsibility and difference. *TechDee.* https://www.techdee.com/team-leader-vs-project-manager

Thomke, S. and Reinertsen, D. (2012). Six myths of product development. *Harvard Business Review* 90 (5): 84–94.

Tih, S., Wong, K.-K., Lynn, G.S., and Reilly, R. (2016). Prototyping, customer involvement, and speed of information dissemination in new product success. *Journal of Business and Industrial Marketing* 31 (4): 437–448.

van Dorn, J., Risselada, H., and Verhoef, P.C. (2021 December). Does sustainability sell? The impact of sustainability claims on the success of national brands' new product introductions. *Journal of Business Research* 137: 182–193.

Verhaegde, A. and Kfir, R. (2002). Managing innovation in a knowledge intensive technology organisation (KITO). *R&D Management* 32: 409–417.

Vishnevskiy, K., Karasev, O., and Meissner, D. (2016). Integrated roadmaps for strategic management and planning. *Technological Forecasting & Social Change* 110: 153–166.

von Hippel, E.A., Thomke, S., and Sonnack, M. (1999). Creating breakthroughs at 3M. *Harvard Business Review* 77 (5): 47–57.

von Hippel, E.A., Ogawa, S., and de Jong, P.J. (2011). The age of the consumer-innovator. *MIT Sloan Management Review* 53 (1): 27–35.

Wagner, S.M. (2012). Tapping supplier innovation. *Journal of Supply Chain Management* 48 (2): 37–52.

Walheiser, D., Schwens, C., Steinberg, P.J., and Cadogan, J.W. (2021 March). Greasing the wheels or blocking the path? Organizational structure, product innovativeness, and new product success. *Journal of Business Research* 126: 489–503.

Watson, A., Obal, M., and Kannan, R. (2021). Expect success, get success: how self-fulfilling prophecy can impact new product development. *Research-Technology Management* 64 (4): 29–36.

Woojung, C. and Taylor, S.A. (2016 January). The effectiveness of customer participation in new product development: a meta-analysis. *Journal of Marketing* 80: 47–64.

Zhan, Y., Tan, K.H., Li, Y., and Tse, Y.K. (2018). Unlocking the power of big data in new product development. *Annals of Operations Research* 270: 577–596.

Dr. Robert Cooper is ISBM Distinguished Research Fellow at Penn State University's Smeal College of Business Administration; and Professor Emeritus, DeGroote School of Business, McMaster University, Canada. He was named "The World's Top Innovation Management Scholar" in the *Journal of Product Innovation Management*; and #21 Scientist in the World in the entire field of Business and Management in 2022 by https//Research.com. He is a Crawford Fellow of the Product Development & Management Association. Cooper is the creator of the *Stage-Gate*® *system*, and has published over 150 articles and 11 books on new product management, including *Winning at New Products*.

CHAPTER TWO

AN INNOVATION MANAGEMENT FRAME-WORK: A MODEL FOR MANAGERS WHO WANT TO GROW THEIR BUSINESSES

Paul Mugge and Stephen K. Markham

2.1 Introduction

When asked in IBM's yearly survey, the majority of CEOs cite "growth through new products and services" as their number one strategic objective. The globalization of markets has created a highly competitive arena where survival depends on a continuous stream of successful new products. Barriers to competition have fallen precipitously as regulations have eased and markets have become more global. After three decades of cost cutting and restructuring in response to a formidable set of global competitors, firms are turning their attention to growth. Their CEOs realize that the winners will be those companies that distinguish their products and services, that is, that create a competitive advantage.

According to the Product Development and Management Association (PDMA), successful high-technology companies have found that more than 50 percent of their current sales are coming from new products (Barczak et al. 2009). In the case of the most successful, this figure is over 60 percent. The next round of competitive positioning will be based on innovation, and a company's innovation capabilities will determine its future growth potential. This is creating a special challenge for senior management. Only innovation increases the size of the pie, which means that its mastery is vital to a company's long-term well-being. Unfortunately, many managers may be better at, and even more comfortable with, controlling costs than creating products that fuel top-line growth.

The PDMA Handbook of Innovation and New Product Development, Fourth Edition. Edited by Ludwig Bstieler and Charles H. Noble.

Companies have invested considerable resources and energy in becoming leaner and more nimble. The quest for productivity, quality, and speed has spawned a remarkable number of management tools and techniques: total quality management (TQM), reengineering, outsourcing, Six Sigma, and so on. However, many of the companies that have applied these techniques are frustrated by their inability to translate gains into sustainable, profitable growth. The products and services of these firms are indistinguishable. Bit by bit, these management tools have actually taken them away from viable competitive positions.

To get ahead of the pack, managers at leading companies are asking fundamental questions of themselves: How can they move beyond producing only incremental innovations and create more radical innovations? Which emerging technologies have the greatest potential to be disruptive and generate breakthrough results? What adjacent market segments could they enter to leverage existing platforms? What is the resulting risk of these actions? Managers are also asking which internal capabilities they need to be successful innovators and which business operations are critical to conceiving of, producing, delivering, and supporting their products and services.

2.2 The Innovation Management Framework

The Innovation Management (IM) Framework is offered to help new product development managers identify those activities required to be a successful innovator. The IM Framework (Figure 2.1) describes a systematic way to think about managing innovation. It demystifies innovation management by breaking it down into elements that can be learned, practiced, measured, and ultimately improved – that is, managed.

The IM Framework was developed from an exhaustive meta-analytical study of 25 years of technology management research sponsored by the Center for Innovation Management Studies (CIMS). Over those years, CIMS sponsored research at 65 different universities and with 110 companies. Each study was categorized, and the categories were then sorted into three aspects that make up the model in Figure 2.1. These aspects include competencies, dimensions, and levels.

2.2.1 Competencies

Based on the meta-analysis of the innovation management research, the IM Framework identifies five organizational competences that successful innovation companies possess: Idea Management, Market Management, Portfolio Management, Platform Management, and Project Management. A company must be proficient in all five competencies to reliably and repeatedly produce differentiated products and services.

FIGURE 2.1 INNOVATION MANAGEMENT FRAMEWORK.

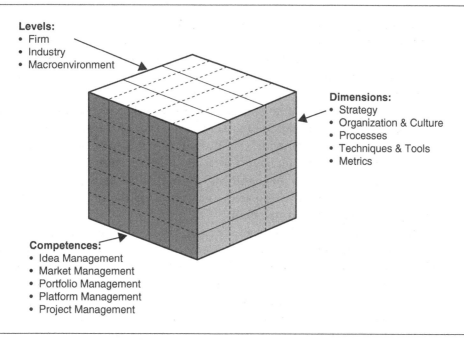

Competencies are not processes. Innovation processes can vary greatly based on the size and maturity of the firm, the industry that the firm operates in, and so on. IM competencies, by contrast, are the basic capabilities of innovation. They represent the inclination, aptitude, and practice of the organization in achieving specific IM objectives, such as managing ideas, markets, portfolios, and so on. Proficiency in IM competencies is essential to any innovation process (see Section 2.3). Moreover, organizations can learn and master IM competencies, and it is management's job to ensure that they do.

2.2.2 Dimensions

To assist managers in this task, the IM Framework breaks down each competence into five dimensions: Strategy, Organization and Culture, Processes, Techniques and Tools, and Metrics. A critical insight from the meta-analysis is that the dimensions "cross-cut" each IM competence and represent the management activities needed to build strong, durable innovation management capabilities. Taken together, IM competences and dimensions provide a complete and integrated system for helping organizations realize results from innovation.

2.2.3 Levels

The primary purpose of the IM Framework is to help managers build strong and competitive innovation capabilities in their firms. For many of these managers, just keeping up with the day-to-day demands of customers, suppliers, channel partners, and creditors can be a full-time job. Yet, events take place outside the boundaries of the firm that also determine the course of innovation, such as the acts of standards bodies, the policies of governments, and the actions of new industry entrants. Successful innovation organizations look beyond their boundaries and pay attention to their environment. The last part of the IM Framework deals with the unique innovation challenges and activities found at three levels of the environment: the firm, the industry, and the macroenvironment.

Taken together, competencies, dimensions, and levels form a cube. Section 2.3 presents the categorization of topics found in the CIMS studies. Categorizing the research not only provides an overview of the innovation management research, it also supplies managers with a logical structure with straightforward tools and techniques to manage innovation. It provides managers with a straightforward means to identify gaps or weaknesses in their organization and to develop roadmaps for improvement.

Each dimension and competence has been operationalized to form the Innovation Management Maturity Assessment (IMMA). The IMMA asks respondents to assess their level of maturity for each competence and dimension on a 1 to 5 scale where 1 = ad hoc, 2 = defined, 3 = managed, 4 = leveraged, and 5 = optimized. The results are displayed on a 5×5 grid with areas of maturity identified in a heat map. See http://cims.ncsu.edu/index.php/assessments/imma for a working model and further description of the tool.

2.3 The IM Framework Structure

In this section we present the essential competences, or capabilities, that leading organizations possess enabling them to recognize new opportunities, select appropriate technologies, and design and efficiently develop new and attractive solutions. The five IM competencies represent the inclination, aptitude, and practice of organizations to produce truly differentiated offerings, that is, to successfully innovate. Mastering these competences is vital to the growth and long-term health of the organization. After discussing the competences, we present the dimensions and then the levels of innovation management.

2.3.1 Competencies

2.3.1.1 Idea Management *Activities include: Boundary Spanning, Technology Scouting and Evaluation, Applied Research, Collaboration with Early Adopters/Scientific Centers, and so on.*

Idea Management begins to answer the question of competitive advantage by simultaneously examining technology and market factors. Ideas about possible market opportunities made feasible by new technologies are often the

> *Idea Management is the ability of a firm to effectively identify, assimilate, and qualify information regarding new technologies or ideas that can lead to highly differentiated, breakthrough products and services.*

point where innovation begins. In an all-out race to gain an advantage over their competitors, managers of leading companies personally go to great lengths to make sure that the climate of the organization for ideation is optimum. They personally sit on review boards, offer monetary incentives, and free employees from their day-to-day routine – all in an effort to glean ideas that ultimately have commercial value.

To systematize Idea Management, these firms often divide their R&D portfolio and management attention into multiple *horizons* to make sure that they have the investments and talents aimed at the early identification and maturation of new, promising technologies.

Some studies address how to manage innovation through the *fuzzy front end*, that is, when information regarding the market application of new technologies is scant, causing forecasting headlights to dim (Koen et al. 2001). A relatively few models for managing nascent technologies through this period are available (Markham et al. 2000). For some technologies – and budding entrepreneurs – this can be the "valley of death" (Markham et al. 2010).

For example, the race to leverage nanotechnologies will place even more demand on a firm's Idea Management capabilities. These technologies are barely emerging and are largely in the domain of science.

In their seminal report, the Chemical Industry Technology Partnership (2003) predicted that many of these technologies may be 20 years away from broad commercial use. Nevertheless, the allure of what potentially can be created with nanotechnologies is so great that companies around the world are building Idea Management capabilities.

2.3.1.2 Market Management *Activities include: Determining Customer Buying Preferences, Market Segmentation, Creating Market Attack Plans, Pricing, Advertising and Promotional Activities, Account Management, and so on.*

The long-term competitiveness of any company depends ultimately on the acceptance and attractiveness of its product and service offerings in the marketplace. A differentiated offering is pivotal for profits, improving the market position, creating new standards, and creating new niche markets.

> *Market Management selects market opportunities and plans offerings in response to them that represent the greatest value to both the organization and its customers.*

Market Management provides a "market-understanding" framework that allows the organization to focus on profitable markets, customers, and business opportunities to pursue. It relies on developing insight – through research and fact-based analysis of market data – that identifies and anticipates potential market opportunities related to the organization's strategic direction. Market information, however, exists in many forms.

A comprehensive Market Management capability defines the types of information, the owners of this information, and how this information will be used in defining and analyzing market characteristics. It requires segmenting macro markets in a manner that provides insight into how to define product characteristics or features that promise to provide a competitive advantage in existing or new markets. Lastly, the capability needs to spell out exactly how the attractiveness of potential market segments will be described and prioritized to support investment decisions in these segments.

2.3.1.3 Portfolio Management *Activities include: Risk/Reward Assessment, Real Option Analysis, Periodic Portfolio Review, Project Evaluation and Selection, and Pipeline Loading.*

Portfolio Management is fundamental to creating true business value. The investment choices made today determine the business value realized in the future. Often a firm's portfolio of projects is crammed with too many "me-too" projects, which steal valuable resources and divert management attention from those few good projects that will really differentiate the firm in the marketplace. Having a portfolio of high-value projects that is properly balanced and is directly tied to the business strategy is essential to optimizing the value realized.

While Project Management is dynamic and milestone driven, Portfolio Management tends to be more stable, with much longer, more strategic objectives. It is not, however, a static process. What many firms call Portfolio Management is in fact a misnomer. In these companies,

> *Portfolio Management allows a firm to manage a set of investment projects that are aligned with the business strategy, balanced, and generate the greatest economic return.*

Portfolio Management is often relegated to a once-a-year "project prioritization" event, usually to feed the annual budgeting cycle.

Leading firms realize that effective Portfolio Management is much more than this; it is a continuous process of allocating resources to best achieve the firm's business objectives. These firms constantly strive to balance the portfolio, determining the optimal investment mix between risk and return, maintenance versus growth, and short-term versus long-term gains. Portfolio Management keeps a firm's portfolio fresh and responsive to market and strategy shifts.

2.3.1.4 Platform Management *Activities include: Reference Platforms, Architecture Review Process, Modular Design, Parts/Subsystem Rationalization, and Integrated Design/Sourcing Decisions.*

Platform Management is the ability to simultaneously design and plan a line, or family of products or services, from a set of common building blocks. Single-product development approaches lack efficiency in that they fail to exploit

> *Platform Management establishes the platform strategy, reference architecture, set of modular subsystems, and development plan for an entire product or service line.*

the benefits of commonality among different products and product lines. A platform design approach provides multiple benefits by:

- Lowering total costs (R&D, production, inventory, maintenance, etc.) due to the need for fewer part numbers and the ability to achieve more frequent reuse of parts
- Reducing product development risks and expense by using proven building blocks
- Increasing market share through reduced cycle times and faster time to market.

Platform Management is inextricably linked to Market Management. It involves understanding the market attractiveness for niche-specific platforms where product differentiation is a key and leveraging horizontal platforms where adjacent market segments could be exploited.

2.3.1.5 Project Management Activities include: A Structured Project Development Process, Stage-gate Reviews, Fact-based Go/Kill Decisions, Integrated Financial Management, and Life-cycle Planning.

It is important to do Project Management well; otherwise, the flow of development projects to their successful completion, launch, and realization of objectives may be impeded. Thus, a chief concern of senior management is to make sure

> *Project Management reliably and predictably guides projects through their phases of development so that they deliver the economic and strategic values originally intended for them.*

that nothing constricts the flow of the project pipeline. Having too many projects imposes high demands on critical resources, which extends lead times and requires frequent and unscheduled management interventions.

It is important that managers not meddle with projects but limit their intervention to well-defined checkpoints. Their focus should be on provisioning projects with adequate resources – both human and capital. They should empower project teams to *run their project like a business*. Managers should make decisions only on the basis of facts and kill marginal projects as early as possible to keep the number of remaining projects matched to the firm's development capacity. Clearly, effective Portfolio Management is necessary to drive effective Project Management.

Finally, the product evolves as it progresses through its life cycle and eventually is replaced by a newer product. This must also factor into the planning because it requires resources and careful decision-making to determine when the product should be retired and/or replaced and how. Life-cycle decisions are much more than the last phase of development – or the problems and tasks an organization encounters in the field. Leading companies plan for the full life cycle of a product from its conception to its withdrawal from service. They also consider not only the financial impact of full life-cycle management but also how it impacts the firm's promise of brand value and ultimately its reputation in the marketplace.

2.3.2 Dimensions

The meta-analysis of the literature reveals much more than the essential competences of Idea, Market, Portfolio, Platform, and Project management. The framework demonstrates that each competence must be managed in multiple dimensions. Taken together, these dimensions provide managers with a prescription for improving their organization's innovation management proficiency.

2.3.2.1 Strategy Activities include: *Targeted Business Arenas/Markets, Barriers to Entry, Value Proposition, Strategic Control Points, Strengths, Opportunities, Weaknesses, and Threats (SWOT) Analyses, Benchmarking and Competitive Evaluation, and so on.*

The core of any business strategy – connecting a company's internal processes to improved outcomes with customers – is the *value proposition* delivered to the customer (Kaplan and Norton 2001). The value proposition describes the unique mix of product, price, service, relationship, and image that the provider offers its customers. A clearly stated value proposition provides the ultimate target for focusing a business's strategy.

> *Strategy defines the specific goals of the organization and exactly how the organization will achieve them. Strategy is only valuable when it creates change in the marketplace.*

Business strategies tell the story of the organization by answering the following questions:

- What is our market position?
- How will we sustain/grow this position?
- What makes our products and services different from those of our competitors?
- How do we measure success?
- What organizational capabilities do we need to acquire/develop to be successful?

The value proposition must also lay out the innovation strategy in support of the business strategy, for indeed the two are intertwined. For example:

- What percentage of revenue is to come from new products and services?
- Are we to be first to market or a fast follower?
- Will we even do R&D internally or will we form alliances?

Good strategies are long on detail and short on vision. Good strategies start with massive amounts of quantitative analysis: hard analysis that is blended with wisdom, insight, and risk taking.[1]

2.3.2.2 *Organization and Culture* *Activities include: Authority Relationships, Human Resources, Skills Acquisition and Development (Organization); Organization's Basic Beliefs, Values, and Behaviors; Leadership, Motivation, and Rewards (Culture).*

Culture isn't one aspect of the game; it is the game. In the end, an organization is nothing more than the collective ability of its people to create value (Gerstner 2002).

Organizations must organize for and promote a culture of innovation to survive.

A company's orientation, business focus, type of people, and core competences can influence the way innovation is embraced and the degree to which it is leveraged.

> *Culture is the common language and background of how things get done. It is developed over time as people in the organization learn to deal with problems of adaptation and integration. Organization, on the other hand, is the formal alignment and direction of a firm's resources and skills toward achieving common goals.*

Most companies recognize the need for innovation in order to be successful in their respective markets. However, few companies make this recognition central to their corporate culture. And the fact that the organization's culture, that is, its capacity to innovate, can be measured and managed is completely lost on these companies.

Management's job is to create a culture that supports risk taking and invokes a common sense of urgency. Managers must ensure that all employees have meaningful work and establish a climate where employees speak out and are empowered to make decisions. Most of all, managers must realize that all employees can innovate and create value (Goodrich and Aiman-Smith 2007).

2.3.2.3 *Processes* *Activities include: Workflow Optimization, Time-blocking Activities, Task Definition, and Roles and Decision Delineation.*

Process assets enable consistent performance across the organization and provide a basis for cumulative long-term benefits to the organization. The organization's process asset library supports organizational learning and process improvement by allowing the sharing of best practices and lessons learned across the organization. It contains descriptions

> *Processes define the patterns of interaction, coordination, communication, and decision-making that people use to get work done. Processes are agreements or political alliances between management and staff in which resources are promised to do work in a certain way.*

of processes and process elements, descriptions of life-cycle models, process tailoring guidelines, process-related documentation, and data. In a very real sense, it is a blueprint of the business.

Companies that lack a process discipline may achieve a fleeting but not sustainable advantage. Managers must continuously focus on their core business processes, making sure that they are streamlined, documented, and followed. And no process is more "core" to a firm than its innovation process.

*2.3.2.4 **Tools and Techniques*** *Activities include: Virtual Workspaces, Team Rooms that Facilitate Collaboration as Well as Forecasting Models, Project-Scoring Hierarchies, Competitive Evaluation Templates, Unstructured Text Analytics, and so on.*

A vast array and range of tools are available to help managers manage innovation more effectively and efficiently. Sophisticated information technology (IT)-based collaboration tools can synchronize communications across a firm's extended enterprise – literally 24 × 7. For example, international members of a product development team can simultaneously evaluate the design of a product or service from their respective regional points of view. As a consequence, the resulting offering is stronger, costs less, and is easier and faster to produce.

An important new tool for early-stage product innovation is unstructured text analytics. This tool provides rapid insight into the market viability of new ideas before expensive development work begins. This tool is also particularly useful as an open innovation tool to find solutions for existing problems.

> *Tools and techniques provide a mechanical or mental advantage in accomplishing a task. They facilitate communications and help process, analyze, and present data to aid management in decision-making.*

Similarly, a host of decision support tools is available to the same product development team to determine the product's sourcing, order demand, life-cycle costs, and overall competitiveness. The chief challenge for management is to select the minimum set of tools that provides the organization with the information it needs in a timely manner – and then relentlessly to institutionalize them. Employees need access to the tools and must understand when and how to use them. Only after a tool set has been integrated with the firm's innovation process and has proven useful to decision-making should management attempt to automate it. In their haste to find the "silver bullet," companies often waste precious resources and time trying to digitize tools before they are properly understood and tested.

*2.3.2.5 **Metrics*** *Activities include: Key Performance Indicators, Balanced Scorecards, Compensation Plans, and so on.*

Firms use a variety of metric types to gauge their proficiency as innovators. They use traditional customer outcomes (e.g., market share growth, customer loyalty) and augment these with in-line operational metrics (e.g.,

> *Metrics are a powerful management tool and are used to both motivate and measure the organization's IM proficiency.*

time to profit and percent of preferred/common parts). Other firms add indicators of knowledge or learning (e.g., number of patents).

Regardless of the scorecard of metrics selected, it is important that:

1. The set of metrics chosen articulates the firm's innovation strategy. For example, if a company's strategic intent is to be first to market, it does little

good to measure traditional cycle times. What counts is the number of times the company is first. If a company wants to be the low-cost producer, then perhaps measuring the percent of preferred/common parts used across its platform is a better gauge of success.

2. Employees' compensation is tied to their results. Many firms spend considerable resources benchmarking best practices, designing and documenting a new innovation process, then leave reward systems unconnected to the innovation process. It's no wonder that business performance doesn't improve; employees know that innovation really isn't the priority.

Properly used, metrics can propel the organization toward improving its IM proficiency and ultimately winning in the market.

2.3.3 Levels

A central problem of management is orchestrating organizational activities to meet the challenge of the environment (Narayanan 2001). Organizations can be viewed as systems that are intricately linked and in constant interaction with their environment. Depending on how attuned the organization is to the environment, its economic performance can be greatly altered. The last part of the IM Framework deals with the unique innovation challenge and activities caused at three levels of the environment: the Firm, the Industry, and the Macroenvironment.

2.3.3.1 Firm Activities include: Opportunity Recognition, Technology/Market Evaluation, Solution Development, and Commercialization.

These activities represent the major stages in the problem-solving process firms use to manage innovation.[2] Hopefully, at this point, the role and importance that IM competences play in this process is clear. Opportunity recognition depends on being proficient in Idea Management; Technology/Market evaluation depends on Idea Management as well as possessing effective Market Management capabilities; they both depend on rigorous Portfolio Management, and so forth.

Management can have a great effect on innovation at this level. With the organization structures they put in place, managers can include employees in decision-making and gain more proprietary ownership for results. Organizational learning is a

> *The Firm Level captures the set of innovation management activities stemming from customers, suppliers, partners, and competitors directly related to the firm.*

major factor in successful innovation. How open the organization is to external information can have a significant impact: the higher the level of communication with customers and outside technical experts, the greater the probability of innovation. Managers should also encourage and support the informal flow of communications across the firm. This will result in a freer flow of information and exchange of ideas, thus helping the process of innovation.

Again, at this point, the role IM dimensions play in establishing the firm's environment should be evident. IM dimensions represent the necessary management activities needed to motivate, build, and improve innovation throughout the firm.

2.3.3.2 Industry *Activities include: Competitive Analysis, Participation in Industry Organizations Such as Trade Associations, and Spatial Clustering.*

Beyond the firm is the environment of new entrants and incumbent competitors functioning in the same industry. At this level, environmental factors directly impact all competitors in the industry. Consequently,

> *The Industry Level captures the innovation management activities caused by a firm and its competitors functioning in the same industry.*

any patterns that emerge, that is, common problems with common solutions may be useful to companies operating in that industry. For example, knowing who and what represents best practice in a particular IM competence would be valuable to the competitors of a company. For this reason, the company's innovation models are often classified as "for internal use only."

Due to a number of factors, including the difficulty and costs associated with knowledge transfer, just the opposite effect can be observed at the Industry Level. In spatial clusters, like the Research Triangle Park of North Carolina, firms in the biotech and pharmaceutical industries have located operations in this area to draw on an "information infrastructure" of large research universities, entrepreneurs, and venture capitalists. The value of knowledge to these firms is so great that it forces even the most ardent competitors to come together.

Successful innovation firms need to sort out their industry competitors, partners, suppliers, and customers. Ironically, in an economy where no one company can possibly do it all, they may be the same.

2.3.3.3 Macroenvironment *Activities include: Understanding Demographics and Lifestyle Trends; Responding to Policy Decisions, Laws, and Regulations; Scientific Discovery Surveillance.*

Successful innovation organizations continuously monitor the social, political, economic, and technological environments for issues impacting their firms. For example, innovative firms in the consumer

> *The Macroenvironment Level captures the social, political, economic, and technological innovation management affecting all industries.*

product industry are keenly aware of ethnic mixes, education levels, household formation, consumption patterns, and social habits. Successful innovation firms are also cognizant of the social acceptance of new technologies, like green technologies or genetically altered grains and vegetables. In the United States, these firms need to stay active in the national technological environment and monitor the discoveries coming out of the many national scientific centers.

This can be a time-consuming and expensive proposition for companies. And with the increasing economic interdependence of nations, the economic environment, and as a result the technological environment, are becoming more global. Nevertheless, it is well established that technological innovations can be traced to scientific research and the interplay with leading industry practitioners.

2.4 Summary

The IM Framework presented in Figure 2.1 provides a structured way of thinking about managing innovation. The framework doesn't become operational until it is populated. Firms must "fill in" or gain experience, establish practices, choose tools, and be able to teach the competences in each cell of the framework. At that point, the IM Framework becomes a useful repository for managers charged with improving their own organization's IM capabilities.

The IM Framework serves as the reference platform for training development. Multilevel education "tracks" should be developed around each essential IM competence (e.g., Idea Management). Each track might offer basic, experienced, and advanced instruction through discrete "modules," with time provided between levels for participants to apply their knowledge to their particular business operations. This type of progressive and modular design provides companies with a number of entry points based on their business strategy and current IM proficiency.

In order to create exciting new breakthrough products and services, companies must work from the premise that innovation is a multidisciplinary process and is no longer the province of R&D. Training must be aimed at managers from all business functions (sales, operations, finance, human resources, field service, etc.) in formats that allow participants to experience the diversity of viewpoints so critical to innovation.

All of these actions should be aimed at one thing: enabling managers to generate profitable growth through innovation.

Notes

1   Bruce Harreld, Chief Strategy Officer, IBM Corporation.
2   A simplified version of the model developed by Marquis (1969).

References

Barczak, G., Griffin, A., and Kahn, K.B. (2009). Perspective: trends and drivers of success in NPD practices: results of the 2003 PDMA best practices study. *Journal of Product Innovation Management* 26 (1): 3–23.

Chemical Industry Technology Partnership (2003 December). Chemical industry R&D roadmap by design: from fundamentals to function. *DOE and NSF Proceedings.*

Gerstner, L.V., Jr. (2007 February). *Making an Elephant Dance.* Collins, 2002.

Goodrich, N. and Aiman-Smith, L. (2007). Discovering the jobs your most important customer wants done: the value innovation process at Alcan Pharmaceutical Packaging. *Research Technology Management* 50 (2): 26–35.

Kaplan, R.S. and Norton, D.P. (2001). *The Strategy Focused Organization.* Cambridge, MA: Harvard Business Press.

Koen, P.A., Ajamian, G., Burkart, R. et al. (2001). New concept development model: providing clarity and a common language to the "fuzzy front-end" of innovation. *Research Technology Management* 44 (2): 46–55.

Markham, S.K., Aiman-Smith, L., Ward, S.J., and Kingon, A.I. (2010). The Valley of Death as context for role theory in innovation. *Journal of Product Innovation Management* 27 (3): 402–417.

Markham, S.K., Baumer, D.L., Aiman-Smith, L. et al. (2000 April). An algorithm for high technology engineering and management education. *Journal of Engineering Education* 89 (2) 209–218.

Marquis, D.G. (1969). The anatomy of successful innovation. *Innovation* 1 (7): 28–37.

Narayanan, V.K. (2001). *Managing Technology and Innovation for Competitive Advantage.* Hoboken, NJ: Prentice Hall.

Paul C. Mugge is Executive Director of the Center for Innovation Management Studies (CIMS) and a Professor of Innovation in the Poole College of Management at North Carolina State University. For the first 25 years of his career, Mr. Mugge worked in IBM's product development community and held a number of global manufacturing and development positions including System Manager for IBM's Mid-range S/370 Systems, VP of Manufacturing and Development for the ROLM/Siemens Corporation, VP of Development for IBM's Personal Systems Group, and Lab Director of the IBM Boca Raton, Florida, facility.

Stephen K Markham is the Goodnight Distinguished Professor of Innovation and Entrepreneurship and Executive Director of Innovation and Entrepreneurship at North Carolina State University. Dr Markham served as Toshiba's Senior Vice President of Global Strategy and Portfolio in the Global Commerce Solutions division. He co-founded BP's Innovation Board and co-founded numerous high-tech companies and served as Chair, CEO, CFO, and VP Marketing and Product Development. His research focuses on Champions of Innovation. He was director of the Center for Innovation Management Studies and the Director of the Technology Entrepreneurship and Commercialization program at NCSU.

CHAPTER THREE

SUSTAINABLE INNOVATIONS AND SUSTAINABLE PRODUCT INNOVATIONS: DEFINITIONS, POTENTIAL AVENUES, AND OUTLOOK[1]

Rajan Varadarajan

3.1 Introduction

Environmental Oath for Consumers: Do no harm to the natural environment, whenever and wherever possible. Choose ecologically less harmful substitute products to meet specific needs and wants, whenever and wherever possible. Do minimal harm to the natural environment when searching for, buying, using, and disposing products to meet specific needs and wants

Environmental Oath for Corporate Decision-Makers: Do no harm to the natural environment, whenever and wherever possible. Choose courses of action that would cause minimal harm to the natural environment, whenever and wherever possible. Continuously innovate to lower the harm caused by the firm's activities to the natural environment.

Environmental Oath for Country Leaders: Do no harm to the natural environment, whenever and wherever possible. Refrain from making politically expedient policy decisions that would harm the natural environment over the long-term. Resolve to make environmental policy decisions that are in the best long-term interests of humans and other species inhabiting the planet, even in

The PDMA Handbook of Innovation and New Product Development, Fourth Edition. Edited by Ludwig Bstieler and Charles H. Noble.

the face of opposition from interest groups, as a leader-statesman would (Varadarajan 2020).[2]

In recent years, issues relating to environmental sustainability have steadily risen in importance as one of the principal concerns of individual consumers and consumer groups, for-profit and not-for-profit organizations, governmental and non-governmental organizations (NGOs), various stakeholder groups (e.g., consumers, customers, employees, investors, and suppliers), and researchers in several academic disciplines. Relatedly, innovating for environmental sustainability has grown in importance as a major focus of firms. Against this backdrop, this chapter presents (1) definitions of the terms, "sustainable innovation," and "sustainable product innovation," (2) a framework delineating some potential avenues for sustainable innovations, (3) an exposition of some of the avenues delineated in framework in reference to sustainable product innovations, and (4) a brief discussion on the outlook and implications for practice. *Eco-innovations, environmental innovations, green innovations, sustainable innovations, sustainability-driven innovations, sustainability driving innovations, sustainability enhancing innovations,* and *sustainability-oriented innovations* are among the terms that are commonly and interchangeably used in books, journal articles, business magazine articles, and articles in the business and popular press to refer to environmental sustainability focused innovations. In the remainder of the chapter, the term "sustainable innovations" is used, with the caveat that, rather than being sustainable innovations in the truest sense, most innovations are likely to be innovations that only *lessen* the impact of a firm's activities on the natural environment.

3.2 Sustainable Development and the Innovation Imperative

Sustainable development, according to a widely cited definition, is "development that meets the needs of the present without compromising the ability of future generations to meet their own needs" (World Commission on Environment and Development 1987, p. 8). Varey (2010) points out that sustainable development entails being responsive to ecological and moral imperatives and requiring equity among the present and future inhabitants of the earth. Williams and Millington (2004) note that the continuum of thought on sustainable development spans from perspectives on altering the supply side (e.g., achieving greater efficiency in resource use) to altering the demand side (e.g., achieving significant reduction in consumption). Meeting the various needs of humanity entails use of both renewable and nonrenewable resources. *Achieving sustainability* is the goal in respect of renewable resources, and *slowing the rate of unsustainability* is the goal in respect of nonrenewable resources (Ehrenfeld 2005). The findings of several studies provide insights into the centrality of various types of innovations (e.g., from innovations in the vein of low-hanging fruits to moonshot innovations; small "i" and big "I" innovations; incremental and radical innovations; product and process innovations) for altering the supply side and demand side of sustainable development. Table 3.1 provides an overview of sustainable development, sustainability principles, sustainability-related responsibilities of stakeholders, and some innovation imperatives for environmental sustainability.

3.3 Environmental Sustainability-related Areas of Emphasis of Firms: Role of Innovation

This section provides an overview of the findings of five studies that provide insights into the role of innovation in the context of various environmental sustainability-related areas of emphasis (priorities) of firms. On the one hand, the crucial role of innovation for firms to achieve their sustainability-related goals may be viewed as self-evident. However, as highlighted in *italics* in the sections that follow, the findings of the studies summarized provide insights into various sustainability-related areas of emphasis of firms and the centrality of innovations for achieving them.

Sustainability-related priorities of multinational corporations: Based on a global survey of senior executives, in a 2008 study, the Economist Intelligence Unit (Economist Intelligence Unit 2008) reported the following to be among the top three strategic sustainability-related priorities of multinational corporations (MNCs): (1) *improving* energy efficiency; (2) *reducing* greenhouse gas emissions, waste, water, and polluting effluents; (3) *reducing* the environmental impact of products; (4) *developing* new products/services to reduce societal or environmental risk; (5) *modifying* existing products/services to reduce societal or environmental risk; (6) *implementing* stronger controls over suppliers on environmental standards; (7) *improving* the local environment around operating facilities; (8) *working* with governments to promote sustainable development in the countries where they operate; and (9) implementing stronger controls over suppliers on workers' rights standards. As evident from the words highlighted in *italics*, innovating to lower the environmental impact of the firm's activities, suppliers' activities, and other entities in the ecosystem is the common thread underlying eight of the nine strategic sustainability-related priorities of MNCs.

Major sustainability-related areas of emphasis of firms: Based on a global survey of executives, in a 2011 study, the management consulting firm, McKinsey & Company (McKinsey Quarterly 2011), reported the following as the major sustainability-related areas of emphasis of firms: (1) *reducing* energy use in operations, (2) *reducing* waste from operations, (3) *reducing* emissions from operations, (4) *reducing* water use in operations, (5) *managing* impact of products throughout the value chain, (6) *committing* R&D resources to sustainable products, (7) *responding* to regulatory constraints or opportunities, (8) *managing* portfolio to capture trends in sustainability, (9) *managing* corporate reputation for sustainability, (10) *mitigating* operational risk related to climate change, (11) *leveraging* sustainability of existing products to reach new customers or markets, (12) *achieving* higher prices or greater market share from sustainable products, and (13) improving employee retention and/or motivation related to

TABLE 3.1 SUSTAINABLE DEVELOPMENT, SUSTAINABILITY PRINCIPLES, SUSTAINABILITY RESPONSIBILITIES OF STAKEHOLDERS, AND SUSTAINABLE INNOVATION IMPERATIVES: AN OVERVIEW.

A. SUSTAINABLE DEVELOPMENT

"Development that meets the needs of the present without compromising the ability of future generations to meet their own needs" (World Commission on Environment and Development 1987, p. 8).

B. SUSTAINABILITY PRINCIPLES

1. Regeneration Capacity Principle / Replenishment Capacity Principle

Rates of use of various renewable resources should not exceed the capacity of the Earth to replenish them.

2. Assimilation Capacity Principle

Rates of emission of various wastes should not exceed the natural assimilative capacities of the ecosystems into which they are emitted (e.g., emissions into the atmosphere and effluents into oceans, rivers, lakes, and soil). (See Daly 1990.)

3. Precautionary Principle[1]

Countries should extensively employ a precautionary approach to protect the environment. Lack of full scientific certainty should not be a reason for postponing cost-effective measures to prevent environmental degradation, where there are threats of serious or irreversible damage.

C. SUSTAINABILITY RESPONSIBILITIES OF STAKEHOLDERS

1. Sustainability Responsibility of Countries (Governments of Countries)

Concurrent pursuit of a larger economic footprint and a smaller environmental footprint, with an emphasis on a steep rate of reduction of the environmental footprint. A principal sustainability-related responsibility of governments is the formulation and implementation of policies, programs, laws, and regulations that are conducive to the country achieving a *larger economic footprint* (GDP growth) and a *smaller environmental footprint* (by fostering environmentally sustainable behaviors in organizations and individuals).

2. Sustainability Responsibility of Companies[2]

Concurrent pursuit of a larger market footprint and a smaller environmental footprint, with an emphasis on a steep rate of reduction of the environmental footprint.

3. Sustainability Responsibility of Citizens (Consumers)

Concurrent pursuit of a desired quality life enabled by consumption or use of various goods and services and a concerted effort to significantly lower one's environmental footprint.

D. SOME INNOVATION IMPERATIVES FOR ENVIRONMENTAL SUSTAINABILITY: INNOVATIONS TO ACHIEVE[3]

- Significant reduction in carbon emissions, methane emissions, sulfur dioxide emissions, etc.
- Significant reduction in energy usage.
- Significant reduction in water usage.
- Significant reduction in waste during the resource extraction, production, distribution, use, and post-use disposal stages of the life cycle of a product.
- Significant reduction in the amounts of renewable resources used for producing, packaging, and distribution of goods.
- Significant reduction in amounts of various materials (nonrenewable resources) used for producing, packaging, and distribution of goods.
- Significant reduction in the amounts of renewable and nonrenewable resources used for providing services.
- Significant increase in substitution of energy generated using nonrenewable resources to energy generated with renewable resources (hydro, solar, and wind).
- Significant increase in substitution of nonrenewable materials with renewable materials.
- Significant increase in substitution of more abundant nonrenewable materials with less abundant nonrenewable materials.
- Significant increase in waste recovery during the post-use stage of the life cycle of a product.
- ...

---

[1] Priniciple # 7 in the *Ten Principles of the UN Global Compact* states: "Businesses should support a precautionary approach to environmental challenges." (UN Global Compact 2011).
[2] Priniciple # 8 in the *Ten Principles of the UN Global Compact* states: "Businesses should undertake initiatives to promote greater environmental responsibility." (UN Global Compact 2011).
[3] Priniciple # 9 in the *Ten Principles of the UN Global Compact* states: "Businesses should encourage the development and diffusion of environmentally friendly technologies." (UN Global Compact 2011).

sustainability activities. As evident from the words highlighted in *italics*, innovations are the key to achieving a firm's goals relating to 12 of the above 13 major sustainability-related areas of emphasis.

Benefits to firms from addressing sustainability-related issues: Based on a global survey of executives, a 2011 study by the MIT Sloan Management Review and the Boston Consulting Group (Haanaes et al. 2011) identified the following to be among the potential benefits that firms envisioned to realize from addressing sustainability-related issues: (1) reduced costs due to *energy efficiency*, (2) reduced costs due to *material or waste efficiencies*, (3) *better innovation* of product/service offerings, (4) *better innovation* of business models and processes, (5) *access* to new markets, (6) *improved* brand reputation, (7) *increased* competitive advantage, (8) *increased* margins or market share due to sustainability positioning, (9) *improved* regulatory compliance, (10) *reduced* risk, (11) *improved* perception of how well the company is managed, (12) *enhanced* investor/stakeholder relations, (13) *increased* employee productivity, and (14) *improved ability* to attract and retain top talent. Here again, the *italicized* words serve to highlight the role of innovations for firms to realize specific benefits from addressing sustainability-related issues.

Generating value from sustainability efforts: Practices of value creator firms: In a 2021 study focusing on the sustainability-related efforts of a global sample of firms, McKinsey & Company (McKinsey & Company 2021) distinguishes a sub-set of firms as *value creator firms* (i.e., firms that create value from their sustainability programs). The study reports that compared to other firms, value creator firms are more likely to follow certain distinctive management practices. They are more likely to (1) have a sustainability strategy with clear, focused priorities, (2) set targets or goals for sustainability initiatives, and (3) have key performance indicators for sustainability. The study further reports that value creator firms are more likely than other firms to (1) *seek* customers' inputs on sustainability attributes of goods and services, (2) *market* the sustainability attributes of goods and services, (3) *provide* information about the organization's and its product's sustainability attributes on packaging, (4) *change* product designs to manage sustainability-related impacts, (5) *shift* from product-sales model to product-as-a-service model, and (6) *offer* one or more dedicated "sustainable" brands. As is evident, a focus on innovation transcends the various distinctive management practices of value creator firms.

Climate change and environmental sustainability: Transforming businesses to meet the moment. Based on a survey of over 2,000 C-suite executives across 21 countries, a 2022 study by Deloitte (Deloitte 2022), a consulting firm, reports the following as the top five actions undertaken by firms as part of their sustainability efforts: (1) *using* more sustainable materials (e.g., recycled materials, lower carbon emitting products), (2) *increasing* the efficiency of energy use (e.g., energy efficiency in buildings), (3) *using* energy efficient or climate-friendly machinery, technologies, and equipment, (4) training employees on climate change actions and impacts, and (5) reducing the amount of post-pandemic air-travel. The study further notes that over a third of the responding organizations had not implemented more than one of the following five "*needle-moving*" sustainability actions: (1) *Developing* new climate-friendly goods and services. (2) *Requiring* suppliers and business partners to meet specific sustainability criteria. (3) Updating / relocating facilities to make them more resistant to climate impacts. (4) Incorporating climate considerations into lobbying/political donations. (5) Tying the compensation of senior leaders to environmental sustainability performance. Along the lines of findings of the other studies, the centrality of innovation for progress toward environmental sustainability is also evident in the findings of the above study.

3.4 Sustainable Innovation, and Sustainable Product Innovation: Definitions

Sustainable innovation is defined as "the creation of *value* through implementation of an *idea* for reducing harm to the natural environment due to a firm's activities by developing a new product, process, or practice, or modifying an existing product, process, or practice." *Sustainable product innovation* is defined as "the creation of *value* through implementation of an *idea* for reducing harm to the natural environment caused by a firm's product during one or more stages of the lifecycle of the product, by developing a new product or modifying an exisiting product." Sustainable innovations encompass innovations that are (1) *sustainability enhancing in the domain of renewable resources* – innovations that contribute toward achieving sustainability in the domain of renewable resources; (2) *unsustainability alleviating in the domain of nonrenewable resources* – innovations that contribute toward slowing the rate of unsustainability in the domain of nonrenewable resources; and (3) *sustainability enhancing* in the domain of renewable resources as well as *unsustainability alleviating* in the domain of nonrenewable resources.

Representative of the literature underpinnings of the proposed definition of *sustainable innovation* are the (1) conceptualization of *innovation* as "the implementation of a new product, process, or practice, or a significant modification or improvement of an existing product, process, or practice," (2) definition of *corporate environmental management* as "an effort by firms to reduce the size of their ecological footprint" (Bansal 2005, p. 199), and (3) definition of *corporate environmental strategy* as "a set of initiatives that mitigate a firm's impact on the natural environment" (Walls et al. 2011, p. 73). Representative of the literature underpinnings of the proposed definition of *sustainable product innovation* are the conceptualization of (1) *product* as "an offering by an organization that serves specific needs and wants of customers;" (2) *sustainable new product development* as "the organization-wide process of explicitly integrating sustainability concerns in new product development to minimize impacts on the natural environment, and on animal and human health" (Genc and Benedetto 2015, p. 150); and (3) *life cycle analysis* as "the environmental impacts of resource use, land use and emissions associated with all the processes required by a product system to fulfill a function – from resource extraction, through materials production and processing and use of the product during fulfillment of its function, to waste processing of the discarded product" (see Guinée and Heijungs 2005).

A defining characteristic of sustainable innovations is the creation of *environmental value* for the collective good of the society by reducing the harm caused to the natural environment (e.g., reducing the amount of greenhouse gas emissions, energy used, water used, and waste disposed) by a firm's activities relating to its product offerings over their life cycles (i.e., resources extraction, production, distribution, use/consumption, and post use/consumption disposal). A major focus of sustainable innovations by firms is the creation of economic value by generating a return on investment greater than the cost of capital. A sustainable product innovation, besides creating environmental value for society, offers *utilitarian value* to users of the product by meeting specific needs and wants (offering specific benefits, or solving specific problems). Other types of value created by sustainable innovations include *reputational value* for the firm and *non-utilitarian value* for users (e.g., emotional value, image value, symbolic value, virtue value – conspicuous consumption of environmentally sustainable products).

3.5 Potential Avenues for Sustainable Innovations: A Framework

Conceptual frameworks delineating potential avenues for sustainable innovations can provide firms with insights into specific types of innovations. For instance, Ambec and Lanoie (2008) broadly classify sustainable innovations as cost reduction-oriented innovations (cost of materials, energy, labor, and capital) and revenue growth-oriented innovations (sustainable innovations as enablers of differentiation and access to certain markets; and sale or licensing of proprietary sustainability-related technology). Mariadoss et al. (2011) broadly classify sustainable innovations as technical (incremental and radical new products) and non-technical (new marketing programs and new managerial programs) innovations. Hansen et al. (2009) propose the sustainability innovation cube as a framework for structuring the *sustainability effects of innovations*. Here, the authors distinguish between three types of innovations (business model, product-service system, and technological) across three life cycle stages of a product (manufacture, use, and end-of-life), and their sustainability effects (ecological effects, economic effects, and social effects).

In their synthesis of research on sustainable innovations, Adams et al. (2012, pp. 33–34) group *sustainable design strategies* under eight broad categories: (1) new concept development, (2) selection of low-impact materials, (3) reduction of material usage, (4) optimization of production techniques, (5) optimization of the distribution system, (6) reduction of impact during use, (7) optimization of initial lifetime, and (8) optimization of end-of-life system. In respect of each of the above, they enumerate potential substrategies. For instance, in reference to selection of lower impact materials as a sustainable design strategy, they list cleaner materials, renewable materials, recycled materials, and recyclable materials as potential sub-strategies. In reference to reduction of material usage as a sustainable design strategy, they list reduction of material weight and reduction of volume in transportation as potential sub-strategies.

FIGURE 3.1 POTENTIAL AVENUES FOR SUSTAINABLE INNOVATIONS: AN OPPORTUNITIES SPACE FRAMEWORK.

A. Sustainable Innovation Type[1]

B. Sustainable Innovation Opportunity Stage[1]

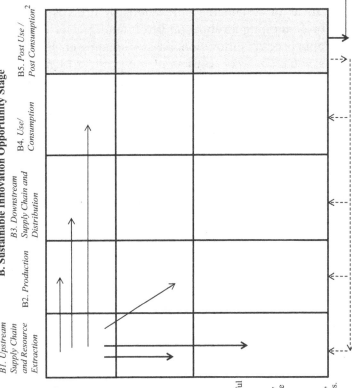

| B1. Upstream Supply Chain and Resource Extraction | B2. Production | B3. Downstream Supply Chain and Distribution | B4. Use/ Consumption | B5. Post Use / Post Consumption[2] |

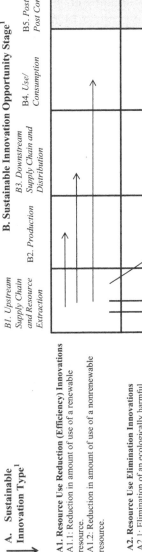

A1. Resource Use Reduction (Efficiency) Innovations
A1.1: Reduction in amount of use of a renewable resource.
A1.2: Reduction in amount of use of a nonrenewable resource.

A2. Resource Use Elimination Innovations
A2.1: Elimination of an ecologically harmful ingredient from a product.
A2.2: Elimination of a filler ingredient from a product.
A2.3: Elimination of the need to use a complementary product

A3. Resource Use Substitution Innovations
A3.1: Substitution of a nonrenewable resource with a renewable resource.
A3.2: Substitution of an ecologically more harmful nonrenewable resource with an ecologically less harmful nonrenewable resource.
A3.3: Substitution of a less abundant nonrenewable resource with a more abundant nonrenewable resource, and the substitution not having a negative impact on the overall sustainability profile of the product.
A3.4: Substitution of *below ground mined* with *above ground mined* raw material -- reuse of resources extracted from post use stage in upstream supply chain, production, downstream supply chain, and/or use stages.

[1]The scope of certain sustainable innovation opportunities may be limited to a specific innovation type in a specific innovation opportunity stage (e.g., cell "A1.2, B2" -- innovation for reduction is use of non-renewable resource during the production stage). The scope of certain other sustainable innovation opportunities may span multiple innovation stages and/or innovation types. These are illustrated with arrows from cell "A1, B1" to other cells. For example, the top-most arrow denotes an innovation spanning cells "A1.1, B1" and "A1.1, B2.

[2]Innovation types A1.1 through A3.4 are primarily in reference to innovation opportunity stages B1 to B4. Although they may have some potential during stage B5, the primary focus of sustainable innovations during the post use/post consumption stage is *value recovery*. That is, recovery of resources through *above ground mining* for use as inputs during stages B1 to B4 of the focal product (shown by dotted arrows from B5 leading into B1 to B4), and/or other products (shown by a solid arrow from B5).

Complementing the above perspectives, Figure 3.1 presents a conceptual framework for sustainable innovations. The framework specifically focuses on *resource centric sustainable innovations* and delineates a relatively large number of fine-grained sustainable innovation opportunities. The framework broadly distinguishes between three *sustainable innovation types* (resource use reduction or resource use efficiency innovations, resource use elimination innovations, and resource use substitution innovations) across five major *sustainable innovation opportunity stages* (resources extraction and upstream supply chain, production, downstream supply chain and distribution, use or consumption, and post-use or post-consumption disposal).[3] In respect of the three broad innovation types, nine finer gradations of sustainable innovation types are shown in the framework (A1.1 through A3.4). In the interests of simplicity of exposition, similar finer gradations in respect of specific innovation stages are not shown in the framework. Some of the innovation opportunities delineated in the framework are *sustainability enhancing* in the domain of renewable resources, others are *unsustainability alleviating* in the domain of nonrenewable resources, and still others are both sustainability enhancing and unsustainability alleviating, in the domains of renewable and nonrenewable resources, respectively. A brief discussion of the proposed framework follows.

3.5.1 Resource Use Reduction, Elimination, and Substitution: Sustainable Innovations for Altering the Supply Side

Resource use reduction or resource use efficiency innovations refer to innovations that lower the environmental impact of a product by achieving greater productivity in the use of a resource used as an input. Resource use efficiency innovations can be broadly distinguished as innovations focused on achieving reductions in the amounts of various renewable and nonrenewable resources that are used in a firm's activities.

Over a 60-year period, numerous innovations are credited to have resulted in a decrease in the weight of the aluminum can used for carbonated soft drinks and other beverages from 60 grams to 13 grams, and an increase in the percent of recycled aluminum used to 73 percent. The above is illustrative of environmental value creation through reduction in the amount of a resource used (i.e., aluminum), as well as reuse of the resource (i.e., substitution of below ground mined aluminum with above ground mined aluminum; i.e., recycled aluminum). The number of cans produced annually (about 500 billion) is suggestive of both the substantial economic value created for firms that are users of aluminum cans (e.g., lower material costs and transportation costs) and environmental value created for society by the innovations (i.e., reduction in harm to the natural environment) (see Moran et al. 2020).

Resource use elimination innovations refer to innovations that lower the environmental impact of a firm's activities by eliminating the use of specific resources as inputs. In specific reference to sustainable product innovations, the upper limit for resource use reduction is resource use elimination with no discernible effect on the functionality and performance of the product. Sustainable innovations in this category can be broadly distinguished as innovations focused on (1) elimination of ecologically harmful ingredients from a product, (2) elimination of filler ingredients from a product, and (3) elimination of the need to use a complementary product.

Garnier brand shampoo bar was launched in the UK in 2020 and the US in 2021, as an alternative to shampoo in liquid form. The above product form innovation (from liquid to solid form) illustrates elimination of an ecologically harmful ingredient from a product (plastic bottles as a container for shampoo in liquid form), and a substantial reduction in the amount of a renewable resource used in the product (water content in shampoo in liquid versus solid form). The bars are packaged in cardboard certified by the Forest Stewardship Council (FSC) as 100% recyclable. According to Garnier, the shampoo bars are formulated with 94% plant-based ingredients, are blended free of preservatives, silicones, soap, and dyes, and are 97% biodegradable. Compared to a liquid shampoo in a plastic bottle, the shampoo bar uses 80% less packaging and 70% less fossil energy for transportation. Unlike in shampoo in liquid form, in shampoo in solid form as a bar, the active ingredients are in concentrated form. As a result, one solid Shampoo Bar can last up to two months and save up to one bottle of water per wash due to the fast-rinse technology incorporated in the product (see *Circular* 2020).

Resource use substitution innovations refer to innovations that lower the environmental impact of a firm's activities by substituting a resource that is used as an input. In specific reference to sustainable product innovations, resource use substitution innovations can be broadly distinguished as innovations focused on (1) substitution of a nonrenewable resource with a renewable resource, (2) substitution of an ecologically more harmful nonrenewable resource with an ecologically less harmful nonrenewable resource, (3) substitution of a less abundant nonrenewable resource with a more abundant nonrenewable resource, subject to the substitution not having a negative impact on the overall sustainability profile of the product, and (4) substitution of a below ground mined raw material with an above ground mined raw material (i.e., reuse of a resource extracted during the post-use or post-consumption stage during an earlier stage).

The portfolio of sustainable innovation opportunities that a firm pursues is likely to be comprised of some whose scope is limited to a specific innovation type during a particular innovation opportunity stage (e.g., cell "A1.2, B2": a nonrenewable resource use reduction innovation during the production stage), and others whose scope spans multiple innovation stages and/or innovation types. The latter types of sustainable innovation opportunities are illustrated in Figure 3.1 with representative arrows from cell "A1, B1" into other cells. Illustrative of sustainable innovations within the domain of a single cell in Figure 3.1 are innovations in the automobile industry focused on greater resource use efficiency during the use stage (cell "A1, B4" in Figure 3.1) such as greater fuel efficiency (more miles per gallon), engine oil change at less frequent intervals (e.g., once every 10,000 miles instead of once every 5,000 miles), and replacement of spark plugs at less frequent intervals (e.g., once every 100,000 miles instead of once every 60,000 miles).

Illustrative of a sustainable product innovation spanning multiple stages is the digital camera as a substitute for a celluloid film-based camera. It's conceivable that sustainability considerations may not have been the impetus underlying the development of digital cameras. That is, the positive sustainability effects of the emergence of digital cameras as a substitute for celluloid film-based cameras may be serendipitous rather than deliberately planned. However, as pointed out by Halila and Rundquist (2011), sustainable innovations include both innovations developed with the explicit aim of reducing environmental harm, as well as those developed without the explicit aim of reducing environmental harm. What matters is, whether ex post, the introduction of the innovation results in reduced environmental harm. Compared to the amounts of various resources consumed during the manufacturing, distribution, use and post use stages of the life cycle of celluloid film-based cameras (e.g., resources used in the manufacture of cameras, photo film, photo printing paper, chemicals used for processing of film and printing of photographs on paper, resources wasted on film and paper on poor-quality pictures that are subsequently discarded, and fossil fuel used to drive to a retail outlet to drop-off the film for processing and later to pick-up the processed pictures), substantially lesser amounts of various resources are likely to be used during the manufacturing, distribution, use, and disposal stages of the life cycle of digital cameras (e.g., online storage and viewing of pictures taken; selective printing of photographs as opposed to printing the complete roll used in a film camera).[4]

While certain resource use efficiency innovations during a particular stage of the life cycle can have a positive effect during another stage of the life cycle of the product, certain other innovations can have a negative impact. Illustrative of the former is the introduction of concentrated formulations of liquid detergents (an innovation focused on resource use efficiency during the production stage of the life cycle – reduced amount of use of water in the liquid detergent, reduced amount of use of plastic resins for manufacturing containers for the liquid detergent) also having a positive sustainability effect during the distribution stage of the life cycle (e.g., lower storage and transportation costs). Illustrative of the latter is a comparative environmental life cycle assessment study of conventional and electric vehicles (Hawkins, Singh, Majeau-Bettez and Stromann 2013). The study reports that for a vehicle lifetime of 150,000 kilometers, electric vehicles (EVs) powered by the electricity mix (at the time of the study) offer a 10% to 24% decrease in global warming potential (GWP) relative to conventional diesel or gasoline vehicles. The study further notes that while for conventional internal combustion engine vehicles (ICEVs), the use phase accounts for the majority of the GWP impact, with the production phase accounting for about 10% of the GWP impact. However, the study reports that the environmental impact is greater during the production phase for EVs than for ICEVs in respect of nine of the ten environmental impact categories considered in the study (e.g., global warming potential, human toxicity potential, freshwater sustainable toxicity potential, mineral resource depletion potential, and fossil resource depletion potential). Drawing attention to the sensitivity of the results to assumptions regarding electricity source, use phase energy consumption, vehicle lifetime, and battery replacement schedule for EVs, the authors caution that it may be counterproductive to promote EVs in areas where electricity is primarily generated from coal, lignite, or oil combustion. In such a scenario, they note that at best, EVs would aggregate emissions to a few point sources (power plants, mines, etc.) from millions of mobile sources (tail pipe emissions from ICEVs).

Although not shown in Figure 3.1, the inter-dependencies between sustainable innovation types and innovation opportunity stages merit mention. For instance, the positive effect of designing products for ease of disassembly during the production stage (stage B2) on the extent of raw materials recovered during the post use or post consumption stage (stage B5) having a positive impact on resource use substitution innovation (innovation type A3.4). That is, substitution of *below ground mined raw materials* with *above ground mined raw materials* during the production, distribution, and/or use stages. While the proposed framework offers a fine-grained roadmap of resource centric sustainable innovation opportunities that may be available to firms, it is important for firms to employ multiple frameworks to identify a larger number of potential sustainable innovation opportunities. For instance, the proposed framework is of limited value for identifying sustainable business model innovation opportunities, such as *servicization or servitization of goods* (decoupling ownership of a tangible good from the benefits that consumers derive from using the good) and collaborative consumption.

3.5.2 Demand Elimination, Reduction, and Redirection: Sustainable Innovations for Altering the Demand Side

The framework presented in Figure 3.1 provides insights into sustainable innovations for altering the *supply-side* (i.e., innovations for resource use reduction, elimination, and substitution). This section provides a discussion on sustainable innovations for altering the demand-side (i.e., sustainable innovation opportunities for demand elimination, reduction, and redirection through consumption or use elimination, reduction, and redirection, respectively) (Varadarajan 2014). A *demand elimination innovation* is an innovation that eliminates the need to use a product to perform specific functions.

In recent years, various brands of coffee makers have come up with a reusable coffee filter made with a fine wire mesh that eliminates the need to use a disposable paper coffee filter. Similarly, vacuum cleaners with built-in canisters whose contents can be directly emptied into a waste basket have eliminated the need to use disposable bags for capture and disposal of dirt.

A *demand reduction innovation* is an innovation that reduces the amount or quantity of a specific product that customers consume or use to meet specific needs and wants. The demand reduction results from the innovation being a substitute for the product in certain places, times, occasions, etc. For instance, at the societal level, considerable progress has been achieved by steering the public to use refillable water bottles rather than buying drinking water in plastic bottles. However, in the face of factors such as product form utility, place utility, and time utility, the demand for bottled water for consumption at certain places, occasions, and times will persist.

Illustrative of a demand reduction focused sustainability initiative is Starbucks' plans to steer customers toward the use of reusable cups and thereby reduce the number of single-use cups used in its stores, which account for about 20% of its global waste. In selected markets, Starbucks has conducted experiments to gain insights into the relative efficacy of alternative programs for steering customers to use reusable cups instead of single-use cups (e.g., offering a discount for using reusable cups as an incentive versus charging for single-use cups as a disincentive). Under the program, customers can borrow a reusable cup from a store location when purchasing a beverage, and later return for cleaning and reuse (Calfas 2022).

Demand redirection sustainable innovations for altering the demand side refer to innovations that reduce the amount of specific renewable and nonrenewable resources that are used to meet specific needs and wants of customers by redirecting consumption or use from ecologically more harmful to ecologically less harmful substitute products.

Illustrative of an innovation for redirecting demand from an ecologically more harmful to a less harmful substitute product is the redirection of demand from incandescent bulbs to light emitting diode (LED) bulbs in numerous countries. With lighting accounting for approximately one-fifth of the world's electricity consumption, from the standpoint of environmental sustainability, the benefit of demand redirection to a significantly more energy efficient substitute is estimated to be quite substantial. Compared to incandescent bulbs that generate about 16 lumens per watt of energy input and compact fluorescent lights (CFLs) that generate about 70 lumens per watt of energy input, LEDs generate about 300 lumens per watt of energy input. Furthermore, compared to CFLs that last for about 10,000 hours and incandescent bulbs that last for about 1,000 hours, LEDs can last up to 100,000 hours (Tinjum 2014). The annual savings in energy conservation from the vastly superior energy efficiency of LEDs has been an impetus behind the ban on the production and sale of incandescent light bulbs in numerous countries.

3.6 Sustainable Product Innovations: An Exposition of the Sustainable Innovations Framework

Table 3.2 presents an exposition of the framework for sustainable innovations (Figure 3.1) in specific reference to sustainable product innovations (i.e., potential avenues for sustainable product innovation through resource use reduction, resource use elimination, and resource use substitution). In reference to sustainable product innovations through resource use reduction, the following are delineated as potential opportunities: (1) product technology innovation, (2) product miniaturization innovation, (3) product convergence innovation, (4) product versatility innovation, (5) complementary product use amount reduction innovation, (6) variable use amount facilitation innovation, (7) product substitution innovation, (8) product upgrade in lieu of replacement innovation, (9) reverse innovation, and (10) packaging innovation. In specific reference to sustainable innovations that entail product modification, potential avenues for innovations through resource use efficiency include reducing the amount of (1) an ingredient product that is used during the production stage of a product, and (2) a complementary product that is used during the use stage of a product. Potential avenues for innovations through resource use elimination include eliminating (1) an ingredient product used during the production stage of a product, and (2) a complementary product used during the use stage of a product.

TABLE 3.2 SUSTAINABLE PRODUCT INNOVATIONS: AN EXPOSITION OF THE SUSTAINABLE INNOVATIONS FRAMEWORK.

Innovation Type	Description and Illustrative Examples
A1. Resource Use Reduction (Efficiency) Innovation	Reduction in the quantities of various renewable and nonrenewable resources used during different stages of the life cycle of a product.
1. Product Technology Innovation	More resource efficient new technology
	Example: Digital cameras versus celluloid film-based cameras
2. Product Miniaturization Innovation	More resource efficient new technology (e.g., from analog to digital technology), and/or major improvements in current technology (e.g., within analog or within digital technology).
	Example: Evolution of music storage devices and players from vinyl records and record player, to cassette tapes and cassette player, to compact discs (CDs) and CD player, to an even more compact device with a hard drive for storage of a considerably larger amount music (e.g., iPod)
3. Product Convergence Innovation	A new product that subsumes in a single product several erstwhile distinct standalone products.
	Example: A smartphone which incorporates in a single device the functions of multiple standalone devices (e.g., phone, digital camera, music storage and playback, information storage and retrieval, sending and receiving e-mails, etc.).
4. Product Versatility Innovation	A new product with greater versatility that eliminates the need to use different standalone products for different use situations.
	Example: A vacuum cleaner that can be used on carpet, hardwood, and tiled floor in place of vacuum cleaners specifically designed for use on carpets versus hard surfaces.
5. Complementary Product Use Amount Reduction Innovation	A new product that lowers the amount of a complementary product needed to use the product.
	Examples: (1) Single rinse formulation of laundry detergent. Cold water formulation of laundry detergent (It is estimated that about three-quarters of the energy use and greenhouse-gas emissions from washing a load of laundry result from heating the water).(Martin and Rosenthal 2011).
	(2) Automobiles designed for greater fuel efficiency, less frequent oil change (e.g., at intervals of 15,000 miles versus 5,000 miles), and using longer lasting parts, components, and subassemblies (e.g., tires and batteries).
	(3) Designing printers/copiers to print/copy on both sides of paper.
6. Variable Use Amount Innovation	Product redesign for variable amount of use as opposed to a predetermined, fixed amount.
	Example: A paper towel roll with perforations that facilitates use of smaller amounts of paper (e.g., one-half or one-quarter of regular size sheet in a roll).

(Continued)

TABLE 3.2 (CONTINUED)

Innovation Type	Description and Illustrative Examples
7. Product Substitution Innovation	Video conferencing as a substitute for travel by plane, train, or car for in-person business meetings.
8. Product Upgrade through Module Replacement versus Full Replacement	Reduction in quantities of various resources used during the production stage of the life cycle by replacing only certain components and/or subassemblies to upgrade a product as opposed to the whole product.
	Example: As an alternative to full product replacement, designing a product so that it can be upgraded, and its useful life extended by replacing subsystems or modules (see Guiltinan 2009)
9. Reverse Innovation	Compared to products developed in and for customers in developed markets, products developed in and for customers in emerging markets at a price that is affordable by a large majority of customers in these markets tend to use substantially lesser amounts of various materials during the production and use stages. When these innovations are subsequently introduced in developed markets, they result in positive sustainability effects [see Govindarajan and Trimble (2012) for illustrative examples].
10. Packaging Innovation	Reduction in the amount of various packaging materials used.
A2. Resource Use Elimination Innovation	Elimination of specific ingredient products (e.g., ecologically harmful ingredients) during the production stage of the life cycle of the product.
1. Ingredient Product Elimination Innovation	Example: Phosphate free formulation of laundry detergent
2. Complementary Product Elimination Innovation	Elimination of the need to use a complementary product. Examples:
	(1) Vacuum cleaners fitted with canisters whose contents can be directly emptied into a waste basket, in place of vacuum cleaners that require use of disposable bags for capture and disposal of dirt.
	(2) In the recent past, several restaurant chains had announced plans to discontinue providing plastic straws along with beverages. In the aftermath of the pandemic, their implementation been deferred. In contrast, redesigning the lid of beverage cups that would eliminate the need for a straw is illustrative of an innovation for eliminating the need for the straw as a complementary product.
3. Complementary Product Carryover Innovation	Elimination of resources expended on complementary products when upgrading to a newer core product.
	Example: A universal cell phone charger that eliminates waste associated with the charger being discarded each time a customer upgrades to a newer model of a cell phone with new and/or more features, or switches from one brand of cell phone to another brand.

(Continued)

TABLE 3.2 (CONTINUED)

Innovation Type	Description and Illustrative Examples
4. Packaging Innovation	Elimination of packaging. *Example: Elimination of paper/cardboard carton packing for deodorants by incorporating all product-related information in the container of the product.*
A3. Resource Use Substitution Innovation	
1. Substitution of: (a) Nonrenewable resource ingredients with renewable resource ingredients. (b) Scarce nonrenewable resource ingredients with relatively more abundant nonrenewable resource ingredients. (c) Ecologically more harmful resource ingredients with ecologically less harmful resource ingredients.	*Examples:* *(1) Substitution of disposable polystyrene foam plates in fast food restaurants with paper plates made with natural fibers, a byproduct of wheat harvest and a renewable resource, that disintegrates and biodegrades swiftly and safely in professionally managed composting facilities.* *(2) Substitution of paper napkins made using wood pulp with paper napkins made using recycled paper or a mix of wood pulp and recycled paper.*
2. Packaging Innovation	(1) Change in packaging materials used from non-biodegradable to partially biodegradable or fully biodegradable packaging materials. (2) Increase in percent or recycled paper products used as packaging material content.

Potential avenues for sustainable product innovations presented in the first column of Table 3.2 were inductively derived. That is, based on analysis of real-world examples of sustainable product innovations such as presented in the second column of Table 3.2, the underlying innovation principle that generalizes as a potential sustainable innovation opportunity for a broad cross-section of products is inferred and an appropriate descriptive label such as "product miniaturization innovation" or "product convergence innovation" is proposed. A limitation of the inductive approach is that the potential opportunities for sustainable product innovation identified using such an approach, even if relatively extensive, cannot be viewed as comprehensive. It is conceivable that sustainability considerations may not have been the impetus behind some of the illustrative examples presented in Table 3.2. However, as noted earlier, sustainable innovations include both innovations developed with and without the explicit aim of reducing environmental harm (Halila and Rundquist 2011). Even in instances where the sustainability-related benefits of innovations are serendipitous, to the extent they provide insights into potential pathways for sustainable product innovations in the context of other products, they are valuable.

Sustainable product innovations focused on resource use efficiency, resource use elimination, or resource use substitution during the production stage of a product (see Figure 3.1) manifest as *physical differentiation attributes* (e.g., concentrated liquid detergent; compact fluorescent light bulb). Sustainable product innovations focused on resource use efficiency, resource use elimination, or resource use substitution during the use stage of a product manifest as *experienced differentiation attributes* (e.g., lower power consumption of a compact fluorescent light bulb compared to a tungsten filament light bulb; greater fuel efficiency of a hybrid car relative to a comparable internal combustion engine-based car). In general, in a sustainable product innovation, the product's attributes undergirding the reduction in its environmental impact are potential bases for differentiating the innovation from competing offerings in the marketplace. With respect to resource elimination-based sustainable product innovations, as opposed to "what's in the product," "what's not in the product" (e.g., phosphate free detergent) is integral to product differentiation (as well as positioning, market segmentation, and target marketing) and the customer value proposition.

The sustainable innovations opportunities framework (Figure 3.1) and the exposition of the framework in specific reference to sustainable product innovations (Table 3.2) are of relevance to both established firms and new ventures. In fact, the genesis of several new ventures can be traced to their conducting an in-depth analysis of an industry and identifying opportunities to serve specific needs of customers that are either not met by incumbent firms or not satisfactorily met by incumbent firms.

3.7 Discussion: Implications and Outlook

Innovations of various types (e.g., product, process, and business model innovations) are central to a firm's strategy for achieving and sustaining a competitive advantage in the marketplace. Product innovations are central to a firm's marketing strategy in myriad contexts such as (1) meeting consumers' needs and wants, (2) responding to changes in consumers' preferences, (3) shaping consumers' preferences, (4) enhancing a firm's market position in presently served markets, (5) entering new markets, (6) differentiating the firm's product offerings from competitors' offerings, (7) neutralizing the effects of competitors' actions, (8) preemptively entering a market, (9) deterring entry of new competitors into markets that are currently served by the firm, (10) altering industry structure, and (11) altering the rules of competition (Varadarajan 2009). Under the broader umbrella of product innovations, the strategic role of sustainable product innovations in the above contexts can be expected to grow in importance in the future.

Kanter (2006) characterizes the innovation strategy of successful innovators as an *innovation pyramid* – a few big bets at the top, a larger number of promising mid-range ideas in the test stage, and a broad base of early-stage ideas or incremental innovations. She further notes that not every innovation idea has to be a blockbuster, and sizeable profits can also result from numerous small or incremental innovations. In a similar vein, the cumulative impact of a number of incremental sustainable innovations can enable firms to achieve their environmental sustainability-related goals such as (1) reduction in CO_2 emissions, energy consumption, water consumption, and waste disposed, (2) reduction in amount of various materials used for making a product and for packaging of the product, (3) amount of substitution of nonrenewable energy with renewable energy, and (4) amount of substitution of nonrenewable materials with renewable materials. Ex ante, viewed in isolation, the potential contribution of certain sustainable innovations to the environmental sustainability goals of the firm may be low or modest. However, ex post, the collective impact of numerous incremental sustainable innovations to the environmental sustainability goals of the firm are likely to be sizeable. Furthermore, an organizational climate that is supportive of the pursuit of a broad array of sustainable innovations (incremental and radical, sustainable innovations in the vein of low-hanging fruits and moonshot innovations, etc.) can be conducive to fostering a sustainable innovation culture in organizations. For instance, based on interviews with executives at firms that had participated in the Carbon Disclosure Project (CDP), Blanco et al. (2017) report that once initiated, the carbon reduction efforts of firms gain momentum. Firms tend to uncover more opportunities for cutting emissions, and in turn, realize more benefits than they had initially expected.

3.7.1 Marketing of Sustainable Products and Communication of Sustainable Business Practices: Guidance Principles for Environmental Claims

In recent years, the deceptive, misleading, or questionable conduct of firms pertaining to their environmental sustainability-related performance and practices have come under public scrutiny. They are also the focus of a growing body of research under the label, "greenwashing." Lyon and Maxwell (2006, p. 9) define greenwashing as "selective disclosure of positive information about a company's environmental or social performance, without full disclosure of negative information on these dimensions, so as to create an overly positive corporate image." Literature review articles (de Freitas Netto et al. 2020; Gatti et al. 2019) provide insights into extant research on issues relating to greenwashing.

Under the broad theme of *Global Guidance on Environmental Claims*, the World Federation of Advertisers (2022) recommends six key principles for firms to follow in their environmental sustainability-related claims to ensure that they are viewed by the public as trustworthy and not engaging in greenwashing. The World Federation of Advertisers (WFA) conceptualizes *environmental claims* as any claims about the environmental attributes of a product or business (e.g., uses less energy; causes less pollution) or impact (e.g., positive impact, no negative impact, or less negative impact on the environment than comparable offerings), that are based on a product's composition, how it is produced, how it can be disposed post-use, etc. The following is an adaptation of WFA's six principles as guidance for environmental claims by firms:

Principle 1. The basis for the environmental claim(s) must be clear, and not be misleading.

Principle 2. The supporting evidence for the environmental claim(s) must be robust, objective, and capable of substantiation.

Principle 3. Material information related to the environmental claim(s) should not be omitted from marketing communications. Under conditions of time and/or space limitations, alternative means should be used to make relevant information readily accessible by the audience by providing details on how and where the information can be accessed.

Principle 4. Environmental claim(s) must be based on the full life cycle of a product, unless stated otherwise in the marketing communications, and the limits of the life cycle-related claims made clear.

Principle 5. Products that are compared in marketing communications must be those which meet the same needs or are intended for the same purpose. The basis for comparisons must be clear to the audience and allow them to make an informed decision about the products compared.

Principle 6. Marketers must include all information relating to the environmental impact of the advertised product as required by laws, regulations, and/or codes to which they are signatories.

3.8 Conclusion

Firms generally state their environmental sustainability-related goals as target percent (1) reduction in greenhouse gas (GHG) emissions, (2) reduction in the amounts of various renewable and nonrenewable resources used for producing goods, performing services, and packaging of goods, (3) reduction in waste disposed, and (4) increase in the substitution of nonrenewable energy sources with renewable energy sources. Crucial to achieving these and other sustainability-related goals are innovations in products, processes, and practices. In the evolving environment characterized by a growing awareness of sustainability-related issues among the various stakeholders of firms, sustainable innovations of various types can be expected to become increasingly important from the standpoint of organizational legitimacy, reputation, and performance. An oft-repeated and prescient characterization of successful new-to-the-world products is, "products that did not exist yesterday, but most people worldwide cannot live without today." In a similar vein, in the future, successful sustainable innovations may be viewed as, "innovations that did not exist yesterday, but the world cannot live without tomorrow" in view of their effect on conserving the natural environment and ensuring the ability of future generations to meet their needs.

According to the World Bank, the combined GDP of the 15 largest economies in the world was $85.8 trillion in 2018 (about 75% of total global GDP) (World Economic Forum 2019). The environmental impact of the amount of renewable and nonrenewable resources used during various stages of the life cycles of hundreds of thousands of products (mining/raw materials extraction, manufacturing, distribution, consumption/use, and post consumption/use disposal) undergirding a global GDP of over 100 trillion dollars may be beyond the realm of comprehension of most of us. Notwithstanding various competing forces, the need for humanity to formulate policies and engage in behaviors conducive to minimizing harm to the natural environment is an imperative. Guided by these reasons, three oaths were proposed as a lead in for the chapter – *citizen consumer environmental oath, company manager environmental oath,* and *country leader environmental oath.* The oaths are a call to Homo sapiens as consumers and decision-makers in organizations to engage in behaviors that would cause minimal harm to the natural environment, and avoid causing harm to the natural environment, whenever and wherever possible, and as country leaders to formulate appropriate policies. Given the predicted severity of the impending threats to the natural environment (IPCC 2022), to the extent the proposed oaths take roots and become enshrined in the collective conscience, attitudes, and behaviors of Homo sapiens, they can result in a significant reduction in the impact of human activities on the natural environment (Varadarajan 2020).

Notes

1 Figure 3.1 and Table 3.2, and the sections of this chapter relating to the figure and table, are adapted from: Varadarajan, Rajan (2017), "Innovating for Sustainability: A Framework for Sustainable Innovations and a Model of Sustainable Innovations Orientation," *Journal of the Academy of Marketing Science*, 45 (January), 14–36.

2 The inspiration for the proposed oaths (Varadarajan 2020) is the *Hippocratic Oath* ("First, do no harm"). See Tyson (2001) for an overview of classical and modern versions of the Hippocratic Oath (a promise by physicians to act morally in their roles), and a discussion on the controversies relating to the oath. Tyson notes: "While Hippocrates, the so-called father of medicine, lived in the early 5th century B.C., the famous oath that bears his name emerged a century later. No one knows who first penned it." (http://www.pbs.org/wgbh/nova/body/hippocratic-oath-today.html.)

3 The terms used in the proposed framework in reference to some of the sustainable innovation opportunity stages differ from the corresponding terms used in the Life Cycle Analysis (LCA) literature. They were based on the following considerations. First, the scope of the terms "upstream supply chain and resource extraction" and "downstream supply chain and distribution" (the first and third innovation opportunity stages delineated in Figure 3.1) is more encompassing compared to the general construal of the scope of the corresponding terms used in the LCA literature (resource extraction and distribution, respectively). For instance, Hansen and Große-Dunker (2013) conceptualize supply chain (upstream supply chain in Figure 3.1) as encompassing all raw materials, premanufactured components, parts, and modules sourced from third-party suppliers – both directly from first-tier suppliers as well indirectly from n-tier suppliers, further upstream. Second, given the focus on potential opportunities for innovation following the use or consumption of a product (i.e., value recovery focused innovations such as reusing, recycling, and repurposing), use of the term "post use or post consumption" to refer to the last of the five sustainable innovation opportunity stages delineated in Figure 3.1 seems more appropriate compared to the terms used in the LCA literature (disposal and end-of-life).

4 The discussion pertaining to the resource implications of film-based versus digital cameras concerning the lower environmental impact of the former should be viewed as tentative. Only a quantitative data-based objective assessment of their relative environmental impact would constitute conclusive evidence.

References

Adams, R., Jeanrenaud, S., Bessant, J. et al. (2012). *Innovating for sustainability: a systematic review of body of knowledge*. Network for Sustainability. nbs.net/knowledge.

Ambec, S. and Lanoie, P. (2008). Does it pay to be green? A systematic overview. *Academy of Management Perspectives* 22 (4): 45–62.

Bansal, P. (2005). Evolving sustainably: a longitudinal study of corporate sustainable development. *Strategic Management Journal* 26 (3): 197–218.

Blanco, C., Caro, F., and Corbett, C.J. (2017). An inside perspective on carbon disclosure. *Business Horizons* 60 (5): 635–646.

Calfas, J. (2022). Starbucks wants to ditch those disposable cups for good. *The Wall Street Journal*,March 15. https://www.wsj.com/articles/starbucks-wants-to-ditch-those-disposable-cups-for-good-11647384972?mod=Searchresults_pos10&page=1.

Circular (2020). L'Oréal's Garnier launches solid shampoo bars with 'zero plastic waste'. Circular,November. https://www.circularonline.co.uk/news/loreals-garnier-launches-solid-shampoo-bars-with-zero-plastic-waste.

Daly, H.E. (1990). Toward some operational principles of sustainable development. *Ecological Economics* 2 : 1–6.

de Freitas Netto, S.V., Sobral, M.F.F., Ribeiro, A.R.B. et al. (2020). Concepts and forms of greenwashing: a systematic review. *Environmental Science Europe* 32 : 19. doi: 10.1186/s12302-020-0300-3.

Deloitte (2022). *Deloitte 2022 CxO Sustainability Report*. Deloitte. https://www2.deloitte.com/content/dam/Deloitte/global/Documents/2022-deloitte-global-cxo-sustainability-report.pdf.

Economist Intelligence Unit (2008). Sustainability across borders. *Economist Intelligence Unit Briefing Paper*. http://www.eiu.com/report_dl.asp?mode=fi&fi=123934397.PDF.

Ehrenfeld, J.R. (2005 Winter). The roots of sustainability. *MIT Sloan Management Review* 46 : 23–25.

Gatti, L., Seele, P., and Rademacher, L. (2019). Grey zone in – greenwash out. A review of greenwashing research and implications for the voluntary-mandatory transition of CSR. *International Journal of Corporate Social Responsibility* 4 : 6. doi: 10.1186/s40991-019-0044-9.

Genc, E. and Di Benedetto, C.A. (2015 October). Cross-functional integration in the sustainable new product development process: the role of the environmental specialist. *Industrial Marketing Management* 50 : 150–161.

Govindarajan, V. and Trimble, C. (2012). *Reverse Innovation: Create Far from Home, Win Everywhere*. Harvard Business Review Press.

Guiltinan, J. (2009). Creative destruction and destructive creations: environmental ethics and planned obsolescence. *Journal of Business Ethics* 89 (Supplement 1): 19–28.

Guinée, J.B. and Heijungs, R. (2005). Life cycle assessment. In: *Kirk-Othmer Encyclopedia of Chemical Technology*. http://onlinelibrary.wiley.com/doi/10.1002/0471238961.lifeguin.a01/full.

Haanaes, K., Balagopal, B., Arthur, D. et al. (2011). First look: the second annual sustainability and innovation survey. *MIT Sloan Management Review* 52 (2): 77–83.

Halila, F. and Rundquist, J. (2011). The development and market success of environmental innovations: a comparative study of environmental innovations and "other" innovations in Sweden. *European Journal of Innovation Management* 14 (3): 278–302.

Hansen, E.G. and Große-Dunker, F. (2013). Sustainability-oriented innovation. In: *Encyclopedia of Corporate Social Responsibility, Vol. I* (ed. S.O. Idowu, N. Capaldi, L. Zu, and A. Das Gupta), 2407–2417. Heidelberg, Germany and New York: Springer.

Hansen, E.G., Große-Dunker, F., and Reichwald, R. (2009 December). Sustainability innovation cube: a framework to evaluate sustainability-oriented innovations. *International Journal of Innovation Management* 13 : 683–713.

Hawkins, T.R., Singh, B., Majeau-Bettez, G., and Strømman, A.H. (2013 February). Comparative environmental life cycle assessment of conventional and electric vehicles. *Journal of Industrial Ecology* 17 : 53–64.

IPCC (2022). *Climate change 2022: mitigation of climate change*. IPCC Sixth Assessment Report. https://www.ipcc.ch/report/ar6/wg3.

Kanter, R.M. (2006). Innovation: the classic traps. *Harvard Business Review* 84 (11): 73–83.

Lyon, T.P. and Maxwell, J.W. (2006 January). Greenwash: corporate environmental disclosure under threat of audit. *Journal of Economics and Management Strategy* 20 (1): 3–41.

Mariadoss, B.J., Tansuhaj, P.S., and Mouri, N. (2011). Marketing capabilities and innovation-based strategies for environmental sustainability: an exploratory investigation of B2B firms. *Industrial Marketing Management* 40 (8): 1305–1318.

Martin, A. and Rosenthal, E. (2011). Cold-water detergents get a cold shoulder. *The New York Times*, September 16. https://www.nytimes.com/2011/09/17/business/cold-water-detergents-get-a-chilly-reception.html.

McKinsey & Company (2021). How companies capture the value of sustainability: survey findings. McKinsey & Company, April. https://www.mckinsey.com/capabilities/sustainability/our-insights/how-companies-capture-the-value-of-sustainability-survey-findings.

McKinsey Quarterly (2011). The business of sustainability: McKinsey global survey results. *McKinsey Quarterly*, October.

Moran, L., Garza, A., and Gallegos, D. (2020). Dispelling the myth of sustainability vs. profit. Viewpoints, November. https://kalypso.com/viewpoints/entry/dispelling-the-myth-of-sustainability-vs-profit/?utm_source=SmartBrief&utm_medium=Featured Content&utm_campaign=sustainability.

Tinjum, A. (2014). Commentary: here's three reasons why LED lights matter. *The New York Times*, October 26. https://www.washingtonpost.com/business/capitalbusiness/commentary-heres-three-reasons-why-led-lights-matter/2014/10/24/3c26c072-589f-11e4-8264-deed989ae9a2_story.html?itid=lk_readmore_manual_10.

Tyson, P. (2001). The Hippocratic Oath Today. March 27. http://www.pbs.org/wgbh/nova/body/hippocratic-oath-today.html.

UN Global Compact (2011). *The ten principles of the United Nations global compact.* https://www.unglobalcompact.org/what-is-gc/mission/principles.

Varadarajan, R. (2009 January-February). Fortune at the bottom of the innovation pyramid: the strategic logic of incremental innovations. *Business Horizons* 52: 21–29.

Varadarajan, R. (2014). Toward sustainability: public policy, global social innovations for base-of-the-pyramid markets, and demarketing for a better world. *Journal of International Marketing* 22 (2): 1–20.

Varadarajan, R. (2017 January). Innovating for sustainability: a framework for sustainable innovations and a model of sustainable innovations orientation. *Journal of the Academy of Marketing Science* 45 : 14–36.

Varadarajan, R. (2020). Market exchanges, negative externalities and sustainability. *Journal of Macromarketing* 40 (3): 309–318.

Varey, R.J. (2010). Marketing means and ends for a sustainable society: a welfare agenda for transformative change. *Journal of Macromarketing* 30 (2): 112–126.

Walls, J.L., Phan, P.H., and Berrone, P. (2011). Measuring environmental strategy: construct development, reliability, and validity. *Business & Society* 50 (1): 71–115.

Williams, C.C. and Millington, A.C. (2004 June). The diverse and contested meanings of sustainable development. *The Geographical Journal* 170 : 99–104.

World Commission on Environment and Development (1987). *Our Common Future.* Oxford University Press.

World Economic Forum (2019). The $86 trillion world economy – in one chart. https://www.weforum.org/agenda/2019/09/fifteen-countries-represent-three-quarters-total-gdp.

World Federation of Advertisers (2022). *Global guidance on environmental claims.* https://wfanet.org/knowledge/item/2022/04/04/Global-Guidance-on-Environmental-Claims-2022.

Dr. Rajan Varadarajan is University Distinguished Professor and Distinguished Professor of Marketing, Regents Professor, and Brandon C. Coleman, Jr. '78 Endowed Chair in Marketing in the Mays Business School at Texas A&M University. His primary teaching and research interests are marketing strategy, innovation, international marketing, and environmental sustainability. He has published over 125 journal articles and book chapters. Dr. Varadarajan served as editor of the *Journal of Marketing* from 1993 to 1996, and the *Journal of the Academy of Marketing Science* from 2000 to 2003. He is a Fellow of the American Marketing Association and Distinguished Fellow of the Academy of Marketing Science.

CHAPTER FOUR

ORGANIZATIONAL DESIGN FOR INNOVATION: LEVERAGING THE CREATIVE PROBLEM-SOLVING PROCESS TO BUILD INTERNAL INNOVATION EFFECTIVENESS

Wayne Fisher

4.1 Introduction

This chapter outlines a robust approach to improving innovation effectiveness by combining best practices from Organizational Design and Creative Problem Solving.

Jay Galbraith (2002) is considered by many to be the father of organizational design. The Galbraith Star Model™ organizational design framework has been used by thousands of companies worldwide to improve organizational effectiveness and business results. Organizational Design for Innovation is the application of the Star Model™ to an organization's innovation management system.

The three phases and key steps of the Organizational Design for Innovation (ODI) work process are summarized below in Figure 4.1. The three phases of ODI – Assess, Design, Mobilize – correspond to the three phases of Creative Problem Solving – Problem Definition, Ideation, Action Planning (see Basadur 2021 for an excellent primer on Creative Problem Solving).

Organizational Design for Innovation begins with an objective assessment of the organization's current innovation capability, identifying gaps and barriers

The PDMA Handbook of Innovation and New Product Development, Fourth Edition. Edited by Ludwig Bstieler and Charles H. Noble.

FIGURE 4.1 SUMMARY OF THE ODI PROCESS.

ASSESSMENT PHASE
Define the ODI Level of Ambition
Innovation Readiness Assessment
Stakeholder Interviews
Gap Analysis

DESIGN PHASE
Industry Best Practice Review
Refine Existing Work Processes
Define New Roles and Responsibilities
Design the Organizational Structure

MOBILIZATION PHASE
Assess Current Capabilities
Match Skills to Roles
Retrain and Recruit
Deploy New Design through Pilots

to achieving long-term business goals. Robust ODI must answer the following kinds of questions:

- What new innovation work processes will be needed to achieve our innovation goals?
- What new organization structures, roles, and responsibilities will be needed?
- How might we get the right people with the right capabilities in the right roles to do the work?

The ODI process must acknowledge the realities of the current state while preparing a transition plan to the desired end state. The transition plan must instill a sense of urgency without overwhelming the organization – embracing a philosophy of "We can't do everything at once, but we must do something at once."
This chapter describes:

- A framework for Organizational Design for Innovation based on the Creative Problem-Solving process
- Tools for assessing and closing the gap between the current state and the desired end state of your organization's innovation capability

During Organizational Design for Innovation, our objective is to innovate how we innovate. One helpful definition of innovation is a continuous cycle of proactively seeking important new problems to solve, identifying new approaches to solve existing problems, and finding new ways to implement business-building ideas. This definition reinforces the critical role of the Creative Problem-Solving (CPS) process in innovation. In this chapter, you'll see how to apply Creative

Problem-Solving principles and tools to understand the current state of the organization, identify what changes are needed to overcome barriers to innovation, and develop a transition plan to achieve the desired future state of the organizational design.

4.2 Assessment Phase

The first phase of Organizational Design for Innovation – the Assessment Phase – corresponds to the Problem Definition phase of Creative Problem Solving. Problem Definition is the most important phase in CPS, but it is often the most challenging phase for teams to lead themselves through. During innovation workshops, roughly half of the available time is dedicated to Problem Formulation. Likewise in ODI, the Assessment Phase is the most important phase and represents roughly half of the total effort associated with an ODI project.

4.2.1 Step 1. Defining the Organizational Design for Innovation Program Level of Ambition

The objective of any innovation program is to deliver business results faster and more reliably. The Level of Ambition template is a great tool to help an organization identify potential areas for improvement and help frame the organization's appetite for change. It is best to start with the typical types of innovation projects delivered over the past 3–5 years. Then consider what additional types of innovation projects would be desirable in the next 1–2, 3–5, and 5–10 year time frame. It is always best to be divergent in the types of innovation under consideration, and then converge on how much change the organization can realistically adopt in the next 3–5 years.

In the example below, a company mapped their current business goals on a Level of Ambition chart. Their current innovation program, focused on short-term incremental cost-saving and improvements in customer service, is illustrated below in Figure 4.2. Their objective was to design an organization able to identify bigger, longer-term challenges with the potential for delivering breakthrough business results. They identified Design Thinking and Open Innovation as two potential new enabling tools. This early framing was invaluable in the design of the Assessment Phase of ODI.

Other examples of ODI program objectives include:

- Improve the collaboration between technical and commercial functions to accelerate new product commercialization cycle time by 6 months
- Transition from a technology-focused R&D Center to a customer-focused Product Innovation Center
- Reduce the time and resource requirements for new technology commercialization by 50%

FIGURE 4.2 EXAMPLE LEVEL OF AMBITION TEMPLATE.

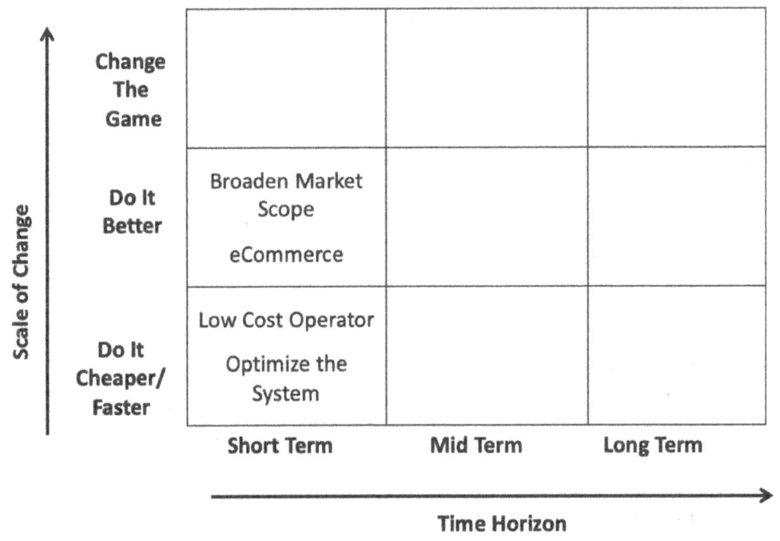

4.2.2 Assembling an ODI Design Team

Once the scope of the Organization Design team is defined using the Level of Ambition (or similar) tool, it is time to assemble a design team. The key players include:

- The **ODI Team Sponsor,** who has a span of responsibility within the organization commensurate with the scope of the ODI program. The Sponsor is also responsible for securing the time/money/resources needed to execute the design work and proposed pilot projects to demonstrate the effectiveness of the new design.
- The **ODI Team Leader,** who is responsible for leading the design team and provides regular updates to the Sponsor.
- The **ODI Design Team**, who are experienced managers representing the key functions that will likely be impacted by the new design.
- Key **Stakeholders**, who are functionally adjacent to the innovation organization (e.g., Manufacturing or Purchasing). While not active members of the design team, they provide valuable perspective on how the new innovation organization may impact overall business performance.

4.2.3 Step 2. Innovation Readiness Assessment

A key fact-finding activity in Organizational Design for Innovation is assessing the health of the current innovation processes and organizational climate. The section below provides an example of a quick assessment tool for both process and climate that ODI design teams can use to jump-start this step and identify key barriers to innovation effectiveness.

4.2.3.1 Assessing the Organization's Innovation Processes. The Product Development and Management Association (Kay et al. 2012) has found that outstanding corporate innovators consistently have Front End of Innovation (FEI) processes that incorporate the key elements shown in Figure 4.3:

- An **Innovation Strategy** that summarizes "Where to Play" and "How to Win" choices for each business unit.
- A robust customer-driven **Idea Generation** process to generate a sustained pipeline of short-term to long-term innovation projects.
- A **Portfolio Management** process that prioritizes innovation projects (against each other and competing business priorities) and maximizes the combined Net Present Value of innovation projects to achieve 3–5 year revenue and profit growth goals.
- A **Resource Management** process that ensures top innovation projects are adequately funded and staffed with (primarily) dedicated resources.
- A **Phase-Gate Review** process to evaluate the progress of innovation projects against their stated objectives. Early phase project work focuses on "killer issues" that quickly prove or disprove the viability of the project.

Conducting a Key Element Assessment (KEA) with multifunctional leadership teams is one powerful tool to identify potential barriers to innovation success

FIGURE 4.3 COMMON ELEMENTS OF AN INNOVATION PROGRAM.

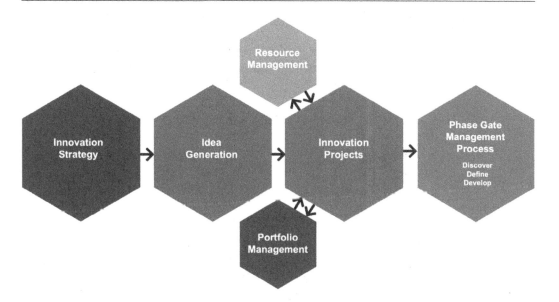

and defining important "problems to be solved" during the Design phase of ODI. The KEA lists desirable states of a successful innovation program, and members of the leadership team rate each statement using the scale:

1. – Strongly Disagree
2. – Somewhat Disagree
3. – Neutral
4. – Somewhat Agree
5. – Strongly Agree

The leadership team then meets to review their individual ratings, identify areas of agreement and disagreement, and converge on the top opportunities for improvement in the new organizational design.

Listed below are some potential KEA statements for each innovation program element listed above. Again, these are only examples and the statements should be customized based on the Level of Ambition identified for the ODI program.

4.2.3.1.1 Innovation Strategy Assessment Innovative companies typically have three primary innovation strategies shown in Figure 4.4. The most visible of these, of course, is delivering a steady stream of new product and service offerings, with meaningful new product introductions every 2–3 years.

The most innovative companies also have a very disciplined approach to cost innovation across the supply chain, bringing many millions of dollars of cost savings to their bottom line, and for their suppliers and customers as well.

A third innovation strategy – their secret sauce you might say – is innovation productivity. That is, they continually innovate how they innovate. They continually look for ways to bring bigger ideas to the market faster, with fewer resources required, and with less R&D and Engineering investment per new product introduction.

Innovative companies also make it a habit to translate business goals and strategies into a concrete innovation strategy with supporting innovation

FIGURE 4.4 THREE ELEMENTS OF AN INNOVATION STRATEGY.

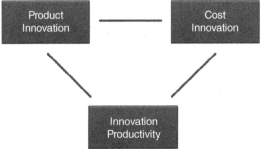

projects. As indicated in Figure 4.5, these innovation projects fall into two important categories:

- New Product initiatives that drive top line growth
- Cost-savings projects that improve profit margins

Together, better products and lower costs mean greater value for our customers and sustained competitive advantage.

Putting this all together, three potential Key Element Assessment statements are shown in Figure 4.6:

4.2.3.1.2 Idea Generation Assessment Chapters 3 and 4 of PDMA's Body of Knowledge (Anderson et al. 2020) list a wide variety of product innovation processes and tools. Innovative companies have, through trial and error, settled on proven methods that work for their industry. They also continually experiment with new approaches ("innovate how we innovate").

FIGURE 4.5 LINKING INNOVATION PROJECTS TO BUSINESS GOALS.

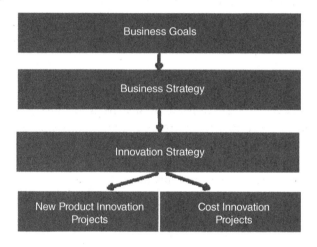

FIGURE 4.6 POTENTIAL INNOVATION STRATEGY ASSESSMENT STATEMENTS.

INNOVATION STRATEGY *(1 = Strongly Disagree; 5 = Strongly Agree)*					
Our product and cost innovation strategies and programs are clearly linked to our business strategy and sales/profit growth goals.	1	2	3	4	5
We have broadly communicated our "Where to Play" and "How to Win" choices.	1	2	3	4	5
Our innovation strategy leverages our organization's unique competitive advantage.	1	2	3	4	5

The Design Thinking work process detailed in a separate chapter is one proven approach to generate new product and service ideas to address unmet customer needs. Church and Dwight, 2020 winner of the Outstanding Corporate Innovator award, identifies targeted "drill sites" for identifying new business-building opportunities. They then "dig deeper" using a wide range of inspiration sources such as trendspotting, conferences, syndicated research, and innovative ingredient suppliers. This approach enables a much broader range of new product ideas to be fed into their idea screening process.

Open Innovation is another common process for inviting employees, suppliers, start-ups, and external researchers to offer new product or technologies ideas for consideration. There are two important best practices for a successful open innovation campaign. First, there should be a well-crafted "technology brief" that clearly describes the technical requirements and criteria for any proposed solutions to be considered. Second, there should be clear guidance on the characteristics of a "big idea" in your industry. For example, in the health care industry characteristics of a "big idea" might include:

- Shifting the site of care from hospital to clinic to doctor's office
- Enabling a patient to be discharged sooner
- Detecting a disease earlier
- Lowering the cost of care to expand the reach of treatment

Figure 4.7 offers up three potential assessment statements for Idea Generation.

4.2.3.1.3 Portfolio Management Assessment Robust Idea Generation will always result in more potential projects than a company has time, money, and resources to execute. Therefore, they need a mechanism for screening ideas to ensure that the innovation projects are adequately staffed and funded to ensure commercial success in a timely fashion. They also need to ensure that the portfolio of selected projects are aligned with the business and innovation

FIGURE 4.7 POTENTIAL IDEA GENERATION ASSESSMENT STATEMENTS.

INNOVATION PROCESS – *IDEA GENERATION* *(1 = Strongly Disagree; 5 = Strongly Agree)*					
We have a proven Idea Generation process that identifies new product and service opportunities to address compelling unmet customer needs.	1	2	3	4	5
We have clear success criteria that describe the characteristics of a "big idea" in our industry.	1	2	3	4	5
We conduct ongoing exploratory work to continually expand our vision of "what's possible" in our industry.	1	2	3	4	5

strategy and are sufficient to deliver the required business results. Portfolio management also strives for the right balance of the right projects (short term vs. long term; low risk vs. high risk; etc.)

One robust criteria for screening ideas and estimating the total value of an innovation portfolio is expected Net Present Value. There should be clear criteria for the minimum eNPV for any individual project passing the "Idea Gate" for further development. In addition to project screening, one important benefit of calculating eNPV during early phases of new product development is the ability to articulate key assumptions and conduct sensitivity analyses to focus early development activities on potential "killer issues." Having a robust backlog of high potential product ideas makes it easier to "let go and move on" when killer issues arise on an active project.

Figure 4.8 offers up three potential assessment statements for Portfolio Management.

4.2.3.1.4 Resource Management Assessment Resource management goes well beyond traditional measures of staff availability to support innovation projects. We are also concerned with the availability of experienced Subject Matter Experts with the right mix of skills and knowledge to do the new product development work, with sufficient training and experience in innovation best practices.

Likewise, innovation projects should be provided sufficient calendar time and development expense budget to execute all phases of new product development. Assuming an individual project demonstrates a technical and business right to succeed, sufficient capital and marketing budget should be readily available to fully commercialize the new offering.

One often overlooked aspect of resource management is the capacity and receptivity of downstream organizations to commercialize new product ideas. Product supply and commercial partners should have a clear voice in important design and delivery decisions that impact their performance measures.

FIGURE 4.8 POTENTIAL PORTFOLIO MANAGEMENT ASSESSMENT STATEMENTS.

INNOVATION PROCESS – *PORTFOLIO MANAGEMENT* *(1 = Strongly Disagree; 5 = Strongly Agree)*					
Expected revenues from our 3-5 year product pipeline are consistent with our long-term revenue growth goals.	1	2	3	4	5
We have clear minimum revenue thresholds for individual projects to keep the total number of projects in our portfolio to a manageable number.	1	2	3	4	5
We maintain an active backlog of new product ideas to replace projects that are launched or killed during the stage gate process.	1	2	3	4	5

Figure 4.9 offers up three potential assessment statements for Resource Management.

4.2.3.1.5 Phase-Gate Project Management Assessment While Lean and Agile are becoming very commonplace Front End of Innovation tools, most innovative companies continue to leverage some form of the Stage-Gate process originally introduced by Robert Cooper (2017) to balance risk and rigor as projects are commercialized. Figure 4.10 is a simplified Discover-Define-Demonstrate-Deliver model suitable for projects with modest technical and market risk.

PDMA's Outstanding Corporate Innovator Award program has honored companies from diverse industries that have all found a way to make phase-gate

FIGURE 4.9 POTENTIAL RESOURCE MANAGEMENT ASSESSMENT STATEMENTS.

INNOVATION PROCESS – *RESOURCE MANAGEMENT*					
(1 = Strongly Disagree; 5 = Strongly Agree)					
We have adequate funding and dedicated full-time employees to execute the key projects in our innovation portfolio.	1	2	3	4	5
We set realistic capital spending limits for individual projects and total annual spend.	1	2	3	4	5
Downstream receiving organizations (Engineering, Manufacturing, Sales) have a clear voice in choices made that impact their success measures.	1	2	3	4	5

FIGURE 4.10 DISCOVER-DEFINE-DEMONSTRATE-DELIVER PHASE-GATE MODEL.

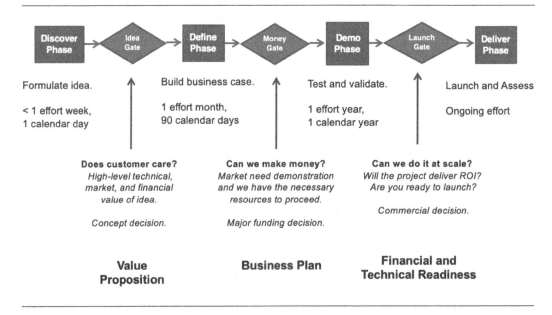

project management a key enabler of their innovation program. [Visit pdma. org/store to view OCI winner presentations at PDMA's annual conference.] Some common characteristics of their approaches to phase-gate project management are highlighted in Figure 4.11:

4.2.3.2 Assessing the Organization's Innovation Culture.

So far we have discussed the importance of Strategy and Process for innovation success. Galbraith (2002) describes additional organization design elements shown in Figure 4.12 that contribute to a culture of innovation within an organization.

FIGURE 4.11 POTENTIAL PHASE-GATE PROJECT MANAGEMENT ASSESSMENT STATEMENTS.

INNOVATION PROCESS – *PHASE GATE PROJECT MANAGEMENT* *(1 = Strongly Disagree; 5 = Strongly Agree)*					
Required inputs, outputs, and interdependencies across all multi-functional partners are defined for each stage of the development process.	1	2	3	4	5
Clear go/no go gate review decisions are made against objective project success criteria.	1	2	3	4	5
Individual projects are assigned a skilled project manager who can balance rigor with agility in developing project timelines.	1	2	3	4	5
Downstream receiving organizations are well-prepared to commercialize new products when development is complete.	1	2	3	4	5

FIGURE 4.12 ORGANIZATIONAL DESIGN ELEMENTS THAT DETERMINE AN INNOVATION CULTURE.

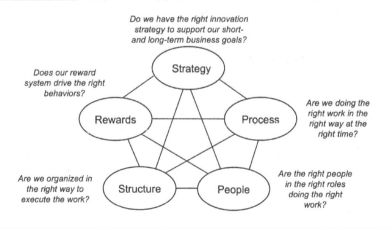

4.2.3.2.1 Innovation Structure Assessment PDMA's Outstanding Corporate Innovator award program has uncovered a wide variety of ways that companies can organize to develop and commercialize new products and services. What they all have in common are the following design elements shown in Figure 4.13:

4.2.3.2.2 Innovation Resources Assessment We touched on Resource Management as part of the innovation process assessment above, but Figure 4.14 provides some additional considerations for developing innovation resources.

4.2.3.2.3 Innovation Rewards Assessment Charlie Munger, Berkshire Hathaway's vice chairman, reminds us that "the iron rule of nature is: you get what you reward." Companies that wish to sustain a culture of innovation must deliberately include innovation measures in their executive compensation systems. Some best practices related to rewards are shown in Figure 4.15:

FIGURE 4.13 POTENTIAL INNOVATION STRUCTURE ASSESSMENT STATEMENTS.

INNOVATION STRUCTURE *(1 = Strongly Disagree; 5 = Strongly Agree)*					
We have sufficient multifunctional resources dedicated to execute innovation projects in a timely fashion.	1	2	3	4	5
We have internal processes for integrating innovation project work into the base business.	1	2	3	4	5
We have strong partnerships with key customers and suppliers for commercializing innovation project work.	1	2	3	4	5

FIGURE 4.14 ADDITIONAL INNOVATION RESOURCES ASSESSMENT STATEMENTS.

INNOVATION RESOURCES *(1 = Strongly Disagree; 5 = Strongly Agree)*					
Innovation resources have sufficient training and experience in industry best practices.	1	2	3	4	5
Innovation resources understand how their role is essential to delivering business results.	1	2	3	4	5
Innovation resources work in a sustained culture of innovation across all levels and areas of the organization.	1	2	3	4	5

FIGURE 4.15 INNOVATION REWARDS ASSESSMENT STATEMENTS.

INNOVATION REWARDS	(1 = Strongly Disagree; 5 = Strongly Agree)				
Senior leaders regularly communicate the importance of innovation to delivering long-term business results.	1	2	3	4	5
Compensation systems include rewards for driving long-term success.	1	2	3	4	5
Innovation successes and failures are celebrated.	1	2	3	4	5

4.2.3.2.4 Actionable Climate Factors that Drive a Culture of Innovation The working atmosphere or climate within an organization is as important as processes to driving innovation productivity. Climate is defined as the observed and recurring patterns of behavior, attitudes, and feelings that characterize life day-to-day in the organization. Over time, these behaviors, attitudes, and feelings translate into a sustained innovation culture.

Researchers have identified nine critical climate factors for innovation – Engagement, Idea Time, Idea Support, Autonomy and Empowerment, Multi-lensing, Risk-taking, Playfulness and Humor, Openness and Trust, and Conflict Management (Isaksen and Akkermans 2011). Improvements in the organization's innovation climate will accelerate the adoption of the new processes, roles, and structures by encouraging and supporting new behaviors.

The following climate assessment tool helps to identify specific behaviors that hinder or promote innovation in an organization (Figure 4.16).

4.2.4 Step 3. Stakeholder Interviews

Stakeholder interviews are often used to complement the leadership team self-assessments used in Step 2. Stakeholders could include internal organization members, interdepartmental stakeholders, customers, suppliers, and other external stakeholders. The Design Thinking chapter offers some guidelines for designing an interview guide that are useful here. Importantly, the interview guide should be customized for each type of stakeholder. For example, an interview with a sales manager might focus on how well our recent product launches were received in the marketplace, while an interview with a manufacturing manager might focus on the impact of the same product launches on manufacturing system reliability and total delivered costs versus target. Interviews with upstream and downstream supply chain partners might focus on what we are like to work with compared to other companies in our industry.

Stakeholder interviews often identify gaps between an organization's *prescribed* processes and *actual* practices. One common interview technique with internal stakeholders is to conduct a project postmortem on major innovation

FIGURE 4.16 POTENTIAL INNOVATION CLIMATE ASSESSMENT STATEMENTS.

	(1 = Strongly Disagree; 5 = Strongly Agree)				
ENGAGEMENT - People in my organization are highly motivated and committed to making contributions in accomplishing the purposes and goals of the team.	1	2	3	4	5
IDEA TIME - People in my organization take the time to consider and test new ideas and ways of doing things.	1	2	3	4	5
IDEA SUPPORT- New ideas are always welcome and are thoughtfully considered by bosses, peers, and others. People listen to, encourage, and try new ideas here.	1	2	3	4	5
AUTONOMY & EMPOWERMENT - People in my organization define much of their own work and frequently take independent initiatives to acquire information, make decisions, and plan.	1	2	3	4	5
RISK-TAKING - People in my organization feel as though they can go out on a limb and be first to put an idea forward. They tolerate uncertainty and ambiguity.	1	2	3	4	5
OPENNESS & TRUST - People in my organization trust each other, are open and honest, and count on each other for personal support.	1	2	3	4	5
PLAYFULNESS & HUMOR - People in my organization have fun doing work in a relaxed environment. There is a great deal of good-natured joking and laughter.	1	2	3	4	5
MULTI-LENSING - People in my organization discuss and consider opposing opinions and a diversity of viewpoints. Different perspectives on a problem or idea are deliberately sought out. Individuals have a network of peers to provide multi-functional input.	1	2	3	4	5
CONFLICT MANAGEMENT - People in my organization understand that personal and emotional tensions will arise, and they address those conflicts in a timely, professional manner.	1	2	3	4	5

initiatives to identify what's working and not working in project execution. Interviewing multifunctional project team members individually yields deep insights into what's working and not working with the current innovation work processes. Figure 4.17 is an example interview guide for multifunctional project team members on recent initiatives.

4.2.5 Step 4. Gap Analysis

The gap analysis tool provides a framework to organize all the information collected thus far and to begin projecting the ideal future state of the organization.

FIGURE 4.17 EXAMPLE PROJECT POSTMORTEM INTERVIEW QUESTIONS.

Key Project Debrief Interview Guide

Project Name:

Project Goals and Business Objectives:

When did you join the project team (at which phase gate milestone)?

What were your major deliverables?

How was your work coordinated with other functions (e.g., success criteria definition, technical readiness reviews)?

What worked well?

What didn't work well?

Largest time/money/resource drains. Extraordinary efforts.

Comparing the current reality to the desired state helps identify the most important gaps to address during the Design Phase.

Some examples of barriers to innovation in organizations include:

- A mismatch in functional time horizons – R&D is focused on the long-term while Marketing is focused on the short-term
- No clear integration of upstream and downstream innovation processes to move ideas forward quickly
- A single function driving innovation (R&D) rather than multifunctional teams
- Insufficient dedicated resources on innovation projects

The example below (Figure 4.18) is from an organization attempting to introduce an innovation program for the first time.

4.3 Design Phase

The second phase of Organizational Design for Innovation – the Design Phase – corresponds to the Solution Finding phase of Creative Problem Solving. Fortunately, the challenges associated with ODI are fairly consistent across industries. It is easy to find published case studies that provide good alternative starting points for potential new work processes, organization structures, and innovation best practices.

FIGURE 4.18 EXAMPLE GAP ANALYSIS EXITING THE ASSESSMENT PHASE.

	Current State	Future State	Key Gaps
Structure	Organization duplicates functional roles across departments.	Streamline and consolidate functional roles wherever possible.	Departments believe they need dedicated resources under their control.
Decision Making	Decisions are personality-driven and progressed through advocacy.	Decisions are data-driven leveraging multiple perspectives including customer feedback.	Lack of clear decision-making processes and owners.
Information &Technology	Information is not effectively shared across the organization.	Open, interactive channels promote communication across all levels.	No venue for communication beyond email and committee meetings.
Rewards	Reward system favors improving daily operations over delivering long-term strategy.	Reward system aligned with both long and short term performance measures.	Leadership focus and communication reinforce emphasis on short-term results.
Culture	Fear of failure perpetuates a risk-averse culture.	Operate with excellence while encouraging the risk-taking required for innovation.	Failed initiatives reflect on individual performance rather than organizational innovation processes.

4.3.1 Step 1. Industry Best Practice Review

The use of analogies – embracing that nagging feeling that "somebody, somewhere has already solved this problem" – is a powerful Idea Finding tool. A little research into the best practices of other innovative companies will likely provide insights into new ways to close the gaps identified in the Assessment Phase.

Fortunately, the Product Development and Management Association (PDMA) has done decades of research into innovation best practices. The most recent PDMA Body of Knowledge (Anderson et al. 2020) provides a wide survey of tools and techniques that can be considered as potential solutions to the key challenges identified during the Assessment Phase. PDMA has also studied well over 100 successful companies as part of its annual Outstanding Corporate Innovator Award program, with winners sharing their best practices at PDMA's annual conference.

Between the PDMA Body of Knowledge and the findings from the Outstanding Corporate Innovator Award program, ODI design teams can become overwhelmed by the scale of change needed to become a world-class innovation company. A useful tool from the Theory of Inventive Problem Solving (TRIZ) is the Ideal Final Result technique (Mann 2004). This technique invites the design team to first imagine the company fully implementing the innovation best practices in a way that makes sense for their industry. Then, the design team outlines a transition plan consisting of continuous improvement cycles over 3–5 years, based on their company's current situation and their capacity for change.

4.3.2 Step 2. Refine Existing Work Processes

More often than not, companies do not explicitly define the details of their innovation work processes. In this case, a detailed review of recent major projects can provide an accurate map of existing work processes and highlight any inconsistencies between projects or product lines. Refined work flows and processes to support innovation should clearly identify key activities, inputs, outputs, and linkages across all functions:

- *How are ideas generated?* How are they collected? Who contributes to them? In what depth are they explored?
- *How are ideas evaluated?* How are ideas assessed? Who assesses them? How do we know they're big enough opportunities? How do we mitigate the risk while maximizing the reward?
- *How are ideas implemented?* Who will transform the ideas into action, and with what resources? How do we separate the work of maintaining our core businesses while developing new, potentially game changing solutions?

4.3.3 Step 3. Define New Roles and Responsibilities

Often, entirely new roles will be needed to execute the new work processes. Existing roles will likely be clarified and include a mix of old and new job responsibilities. The old and new roles will then be mapped to the new work processes (who will do what, and what will be the key hand-offs and linkages).

A common Project Management tool that can be reapplied here is the RACI (Responsible-Accountable-Consult-Inform) Matrix. This tool can become quite complex for large programs, so it is best used to clarify roles and responsibilities

FIGURE 4.19 THE IDEAL FINAL RESULT TECHNIQUE.

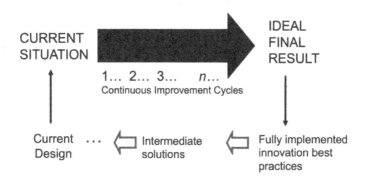

only when introducing new innovation design elements. In the example below, a company was implementing formal gate reviews for large projects for the first time. The RACI helped codify the steps needed for conducting a productive gate review, along with the agreed roles for each step in the process. In this case, the Phase-Gate Process Owner was introduced as a completely new role to ensure success of the new program. The Project Manager, by contrast, was an existing role with important new responsibilities in the new design shown in Figure 4.20.

FIGURE 4.20 RACI MATRIX FOR A NEW INNOVATION PROGRAM ELEMENT.

Example Innovation Program RACI Matrix R = Responsible A = Accountable C = Consulted I = Inform	Innovation Program Manager	Phase-Gate Process Owner	Gate Review Board	Project Manager	Project Sponsor	Project Team
Activity: Conduct Go to Development Gate Review						
Before the review:						
Assess review readiness and timing	C	A		R	R	
Schedule Go to Development Gate review and send invitations to Review Board	A	R	I	I		I
Invite additional Subject Matter Experts as needed	A	R		C	C	
Assemble and distribute pre-reading materials		A		R		R
Prepare Go to Development Gate review presentation		C		A		R
Prepare Go to Development Gate meeting agenda and timings	A	R		C		
Prepare Go to Development Gate meeting handouts		A		R		R
During the review:						
Provide and update on Killer Issues / Key Unknowns				R	A	
Review project background and prework		A		R		R
Capture input on Listening Templates		A	R			
Share input with Project Team		A	R	I	I	I
Make Stop / Start / Continue recommendation and seek concurrence		C	A	R	I	I
Recap conclusions and next steps	A	R	I	I	I	I
After the review:						
Transcribe and summarize input	A	R	I	I	I	I
Update a 90-Day Action Plan			I	R	A	R
Assign resources to action items				A	R	I
Send final report to review team			I	R	I	I
Schedule 90-day update			I	R	A	I

4.3.4 Step 4. Design the Organizational Structure

Innovation work needs to be deliberately structured within an organization, but kept separate from the organization's core business operations. Innovation organizations vary widely in their structure. Chapter 6 of the PDMA Body of Knowledge (Anderson et al. 2020) provides an especially rich overview of various types of team structures that can be used for product innovation – functional, lightweight, heavyweight, and autonomous. Chapter 6 also outlines the advantages and disadvantages of these team structures and where they are best applied.

4.4 Mobilization Phase

The third phase of ODI – the Mobilization Phase – corresponds to the Planning & Execution phase of Creative Problem Solving. The transition plan must provide a bridge from the current state to the future state.

4.4.1 Step 1. Assess Current Capabilities

This assessment answers the question, "Does the capability to do this work exist anywhere in the organization?" It also answers the question, "Do we have sufficient capacity to do the new work in addition to our current workload?"

Referring back to the Ideal Final Result approach (Figure 4.19), a realistic assessment of current organizational capabilities and resource availability should guide how much (or which elements) of the new organizational design can be implemented in the short term. It is best to focus on one new innovation capability (e.g., Design Thinking) at a time, along with its supporting work processes. Again, the RACI tool is helpful in identifying what specific skills will be needed in the new roles and which existing roles will be affected by the design change.

4.4.2 Step 2. Match Skills to Roles

One benefit of explicitly defining an innovation work process is providing a professional development plan for individuals in (new or existing) innovation roles. The Step-up Card template introduced in the Design Thinking chapter outlines the skills needed to design and facilitate a Design Thinking workshop. While Design Thinking may be a new innovation capability for the organization, there will likely be individuals with many of the required skills for the new Design Thinking roles. Assigning existing employees with "good enough" skills to these roles will help accelerate implementation of the new organizational design. The Step-up Card will also help identify key skill gaps that need to be addressed in the short term to set up the employee for success in the new role.

If no existing employee with the required skill set is available, the Step-up Card can be used to help craft a job description for an external hire or consultant.

4.4.3 Step 3. Retrain and Recruit

Unfortunately, the knowledge required to perform most innovation tasks is 20% explicit and 80% tacit. That is, the knowledge required to competently execute innovation work is largely gained from experience. Bringing in experienced staff who were part of a successful innovation program at another company is one common approach to growing innovation capability. Another approach is to engage external innovation guides (consultants) for hands-on coaching and training to accelerate innovation work on key projects while growing internal capability over time.

4.4.4 Step 4. Deploy New Design through Pilots

The new design will need to be piloted to "run water through the pipes" prior to formalizing the new best practices. This gives individuals within the organization an opportunity to "try on" their new roles, practice new behaviors, and fill in the details of the new work processes. This is often done through "hero projects" that are important to the company's strategy and are good candidates to demonstrate the business benefits of deploying the new innovation capabilities. Resources from other projects or departments contribute to these "hero projects" in exchange for first-hand experience with the new capabilities and "lessons learned" before adopting in their own organization. Annual health checks and a commitment to continual improvement will help ensure the new design continues to drive innovation effectiveness toward the Ideal Final Result.

4.5 In Summary

The three phases and key steps of the Organizational Design for Innovation work process – Assessment, Design, Mobilization – work hand in hand when attempting to build internal innovation effectiveness. Typical calendar time for completing all three phases is about six months, the primary bottleneck being the availability of stakeholders for assessment interviews, design feedback, and concurrence to deploy to the organization.

References

Anderson, A. and Jurgens-Kowal, T. (2020). *Product Development and Management Body of Knowledge, self published by PDMA.*

Basadur, M. (2021). *The Power of Innovation.* www.basadur.com.

Cooper, R. (2017). *Winning at New Products: Creating Value Through Innovation.* Basic Books.

Galbraith, J. (2002). *Designing Dynamic Organizations.* www.jaygalbraith.com.

Isaksen, S. and Akkermans, H. (2011). Creative climate: a leadership lever for innovation. *Journal of Creative Behavior* 45 (3): 161–187.

Kay, S., Boike, D., Fisher, W. et al. (2012). Lessons learned from outstanding corporate innovators. *Visions* 10–15.

Mann, D. (2004). *Hands on Systematic Innovation: For Business and Management. Edward Gaskell Publishers.*

Wayne Fisher, PhD, founded Rockdale Innovation after spending 27 years as an R&D manager at Procter & Gamble. While at P&G, Wayne created a series of popular innovation workshops for all phases of new product development. He has trained thousands of managers, providing a common language and framework for innovation, fostering collaboration across P&G's diverse business units. He served as a full-time innovation consultant and Creative Problem-Solving facilitator at The GYM, an IDEO-inspired design studio that remains a key enabler of P&G's innovation capability. Wayne serves on PDMA's Board of Directors and on the Outstanding Corporate Innovator award committee.

CHAPTER FIVE

REPURPOSING: A COLLABORATIVE INNOVATION STRATEGY FOR THE DIGITAL AGE

Bastian Rake and Marvin Hanisch

5.1 Introduction

Most academics and practitioners associate the term innovation with the creation of something new, and our general mindset is shaped by innovations that are new to the world and have a large impact across economies and societies. Less emphasis has been placed on managing innovation and new product development that is based on new applications of existing solutions, although the innovative potential of recombining existing knowledge and technologies was acknowledged early on in the innovation literature (Mastrogiorgio and Gilsing 2016; Nelson and Winter 1982). Academics and practitioners are starting to show renewed interest in the potential of reusing existing solutions to address unsolved problems and needs (Langedijk et al. 2015). The practice of transferring a known solution to a new context is called "repurposing" and represents an often untapped opportunity for companies to innovate at low cost (Allarakhia 2013).

When thinking about repurposing as an innovation strategy, the example of Sildenafil, better known as Viagra, quickly comes to mind. Pfizer's well-known drug was initially developed as a treatment for angina pectoris (chest pain). An unexpected side effect led to its very successful commercial repurposing into a drug for the treatment of erectile dysfunction (Pushpakom et al. 2019). This is just one example of successful repurposing. In fact, repurposing is one of the key innovation strategies in the biopharmaceutical industry, with up to 30% of drugs approved by

The PDMA Handbook of Innovation and New Product Development, Fourth Edition. Edited by Ludwig Bstieler and Charles H. Noble.

the United States Food and Drug Administration being the result of repurposing efforts (Kesselheim et al. 2015). The widespread recourse to repurposing was particularly evident during the COVID-19 pandemic, where 86% of drugs tested against the disease through July 2020 were repurposed drugs (Hanisch and Rake 2021).

While many examples of successful repurposing come from health care, the phenomenon is prevalent in many contexts and industries. A classic example is Richard T. James' 1946 invention of springs that supported and stabilized delicate instruments aboard ships in rough seas, which were repurposed as the wildly popular Slinky toy. A similar example is the creative arts and crafts toy Play-Doh, which was originally used as a wallpaper cleaner. More recently, during the COVID-19 pandemic, Epic Games repurposed the opportunities for interaction among the players of its game Fortnite to host a digital concert series. Other examples include the repurposing of electric vehicle batteries to power data centers, offices, and railroad crossing devices. The use of LIDAR, a laser-based technology for precise distance measurement in autonomous driving vehicles, is another example of repurposing, as this technology originated in the aerospace industry and was used on the Apollo 15 mission in 1971 to map the moon. In the IT and software industries, it is common to repurpose existing software codes for new applications. For instance, the image-sharing platform Flickr was largely built through repurposing code used for an instant messaging service within an online game.

As these examples demonstrate, repurposing occurs frequently across many industries and contexts, but there is a need for a more systematic discussion of what repurposing is and how companies can use repurposing as a product development strategy. To address this challenge, we first introduce repurposing as a viable innovation strategy (5.2) and discuss its building blocks (5.2.1), sources and applications (5.2.2), and links to the concept of open innovation (5.2.3). In the third section (5.3), we identify the strengths (5.3.1) and weaknesses (5.3.2) of repurposing as an innovation strategy. In our final section (5.4), we show how repurposing can be executed successfully through targeted search and rapid execution (5.4.1), and the use of digital technologies (5.4.2).

5.2 Repurposing as a Product Development Strategy

The innovation management literature indicates that repurposing is a frequent, yet often overlooked source of innovation. Most studies on repurposing associate the concept with drug development. Accordingly, Pushpakom et al. (2019) refer to repurposing as a strategy for "identifying new uses for approved or investigational drugs that are outside the scope of the original medical indication." Beyond this contextualized definition, the innovation literature understands repurposing in a broader sense as the reuse of a product or its parts for a functionality that differs from the originally intended purpose (Garud et al. 2021). Based on this broader definition, the literature normally focuses on the repurposing of technological innovations or scientific discoveries that may serve a new market but go beyond mere market expansion, as serving a new functionality is a defining feature of repurposing. As such, repurposing offers an alternative route to innovation in contrast to innovation processes characterized by "science push," "demand pull," or "institutional steering," which are typically commenced and driven by the desire to realize a functionality that has been identified ex ante as desirable (Garud et al. 2018).

5.2.1 The Building Blocks of Repurposing: Bricolage and Exaptation

Repurposing consists of two core elements: bricolage and exaptation (Garud et al. 2021). Bricolage is the use of existing and available resources to address new problems or opportunities (Baker and Nelson 2005; Lévi-Strauss 1966). An example of bricolage is the development of a replacement CO_2 filter with resources available in the Apollo 13 capsule – including plastic bags and duct tape – by NASA engineers and the crew on board. While this example shows that bricolage can create an ad hoc solution that in this particular case helped prevent fatalities, it is clearly not a viable or reliable innovation strategy for spaceflight in general (Rönkkö et al. 2014). Instead, bricolage may offer a "quick fix" for a pressing need but the performance of such solutions is often limited. An alternative mode of discovery is exaptation, which refers to the institutionalized process by which a product, or a part of a product, developed for a specific purpose is coopted to serve a different purpose in a different area (Andriani and Carignani 2014). Exaptation describes the institutionalized and more deliberate process by which a solution to a problem emerges serendipitously from a known solution to another problem (Andriani and Kaminska 2021). The repurposing of the CD from a system for playing high-quality sound recordings to a tool for data storage during the 1980s and 1990s is a classic example of repurposing, which in this case required additional systematic and institutionalized processes, including the creation of industry standards. Figure 5.1 illustrates the logic of bricolage and exaptation as the main mechanisms behind repurposing.

FIGURE 5.1 THE PILLARS OF REPURPOSING.

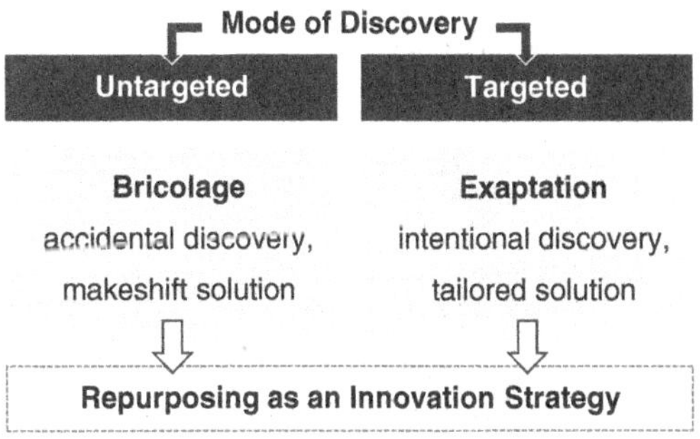

5.2.2 Sources and Applications of Repurposing

For organizations, an understanding of the sources and applications of technology repurposing is critical to its effective deployment as an innovation strategy. Figure 5.2 distinguishes repurposed innovations in terms of their source – internal or external to the focal organization – and their application, i.e., whether it is in a field familiar or unfamiliar to the focal organization. The repurposing of Dyson's vacuum cleaner technology for other types of home appliances, such as bladeless fans and heaters (Vincent 2017), provides a good example of repurposing within the focal organization for applications with which it is familiar. In this case, Dyson had mastered the technological and market sides, and was thus well equipped to repurpose its product independently without the involvement of collaboration partners. The advantage of a "do-it-alone" strategy, in which one organization has mastered all the necessary skills and controls the process, is that it enables rapid execution in terms of technology development and marketing.

The origin of an innovation that is repurposed in a familiar area may also be outside the focal organization. In this scenario, the organization identifies a technology with which it has little or no experience but that can be used in a new area in which it already operates. Recently, the Japanese optics and imaging company Nikon entered a technology partnership with the American technology specialist Aeva "to repurpose FMCW [Frequency Modulated Continuous Wave] technology for industrial automation and metrology"

FIGURE 5.2 ORIGIN, APPLICATION, AND EXECUTION OF REPURPOSING.

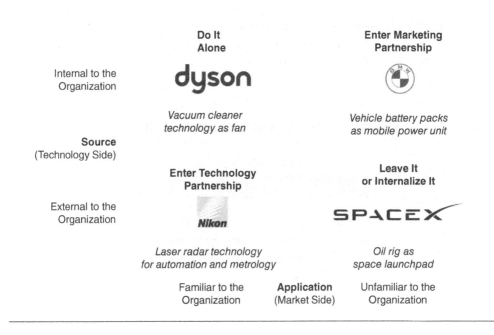

(Nikon 2021). Nikon has long-standing experience in industrial automation and metrology and is working with Aeva to apply laser radar technology, used primarily in automotive navigation, to a new field. For Nikon, it is reasonable to partner with Aeva to gain access to technological know-how and to leverage its existing market access to effectively repurpose the technology. In general, such technology partnerships are often a viable option when the technology comes from an external source, but the market know-how resides in-house. Compared to "do-it-alone" strategies, technology partnerships create additional hurdles due to learning costs on both sides and the potential fear of knowledge misappropriation on the part of the technology provider, which can slow down development.

Similarly, companies can also seek new applications for their existing technologies outside their focal markets. The repurposing of electric vehicle batteries at the end of their life cycle into mobile power sources provides a good example of technology repurposing that originates within the focal organization and is applied to an unfamiliar domain. BMW engages in this type of repurposing through a marketing partnership with Off Grid Energy Ltd., a United Kingdom-based provider of electricity supply for construction sites, events, and for other users not connected to the grid (BMW Group 2020). A (marketing) partnership based on a long-term contract or even a joint venture is often an appropriate governance structure in cases where the focal organization has access to and expertise with a technology, while the partner has knowledge of the target market for the repurposed technology. In principle, the situation is similar to the first example of a technology partnership, with the difference that the agency for repurposing (i.e., who initiates the collaboration) in the first case lies with the marketer rather than the technology provider.

Finally, the impetus for technology repurposing may come from a technology external to the focal organization with applications that the focal organization is unfamiliar with. Sometimes an "outsider" view is necessary to recognize the repurposing potential of a technology in a new market. An example of this fourth type of repurposing is SpaceX's investment in oil rigs originally designed for deep water drilling. SpaceX purchased decommissioned oil rigs in 2021 to repurpose them as floating spaceports for its Starship, a "fully reusable transportation system designed to carry crew and cargo for interplanetary flights" (SpaceX 2022). In this case, the technology was unfamiliar to SpaceX, recommissioning was carried out by a specialized service provider, and the application (floating spaceports for a novel rocket type) did not even exist when the project was launched. Given the dual risks that this type of repurposing entails, most organizations do not pursue this approach without a compelling strategic reason to develop the required competencies. If the repurposing opportunity is deemed worth pursuing, it often makes sense for a company to acquire the technology that they aim to repurpose to internalize the know-how and to collaborate with specialized service providers to realize the full potential.

5.2.3 Open Innovation and Repurposing: Collaborating for Success

Repurposing often requires cross-company collaboration to access existing technologies or markets outside the companies' current scope of operations. The literature on open innovation has long emphasized the benefits of exploiting the potential for innovation outside the firm's boundaries. This idea is equally significant for repurposing strategies. Open innovation describes the "use of purposive inflows and outflows of knowledge to accelerate internal innovation and expand the markets for external use of innovation" (Chesbrough 2006). According to this definition, open innovation includes an inside-out and an outside-in perspective (see Figure 5.3). In the context of repurposing, the inside-out approach describes cases in which an internal technology is made available to another organization for a new application, e.g., through licensing or selling intellectual property. In this case, repurposed technology is applied to an expanded market. In contrast, the outside-in approach broadens and deepens a company's internal knowledge base by integrating external knowledge for application in a new context. Hence, the repurposed technology comes from a source outside the focal organization.

The inside-out perspective of open innovation offers opportunities for companies to generate additional revenues by finding new applications for their existing product portfolio. Companies can act upon these opportunities by using different collaborative mechanisms of the open innovation toolbox, such as selling their intellectual property or licensing it, or by creating and supporting spinoff companies (Chesbrough 2012). Acting upon the opportunities provided through the application of an inside-out perspective benefits companies as they can bring their (repurposed) innovations to market more quickly by relying on a partner that has expertise in the target market. For example, when NASA developed special hydration solutions for its astronauts working in extreme conditions, it recognized that the beverage may appeal to a broader audience and licensed the technology to The Right Stuff, which now markets the product to high-performance athletes. Companies applying the inside-out perspective open up opportunities for generating additional revenues and profits that are not restricted to the markets that the company itself serves.

FIGURE 5.3 APPLYING PRINCIPLES OF OPEN INNOVATION TO REPURPOSING.

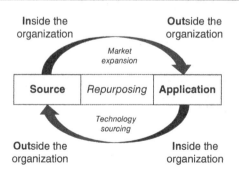

The outside-in perspective of open innovation allows companies to broaden and deepen their own knowledge spaces by integrating knowledge, competencies, or technologies that emerge outside the firm's boundaries. In the context of repurposing, the outside-in perspective of open innovation suggests that companies can expand their product portfolios by repurposing solutions that exist outside their domains, despite these solutions having been originally developed for a different context. In the 1950s, for example, French engineer Marc Grégoire and his wife Colette Grégoire recognized the potential of Teflon, a recently discovered nonsticky chemical material marketed by DuPont that was originally used as a coating for uranium-storing equipment in the production of nuclear bombs, as a potential coating for frying pans, which has since become a standard household product. With large portions of development already completed, sourcing technologies from outside a firm's boundaries and repurposing their application can provide significant benefits in terms of development costs and a shorter time to market. Furthermore, companies can build on the knowledge and developments made in industries other than their own. Because access to a solution that can be repurposed provides the originator with the benefits associated with the inside-out perspective of open innovation without increasing competition in the originator's core markets, repurposing offers great potential for mutually beneficial collaborative innovation.

5.3 Benefits and Challenges of Repurposing

5.3.1 Benefits

Cost efficiency. One of the main advantages of repurposing is its cost efficiency. As the underlying science and technology of a product that should be repurposed is already well understood and less additional testing is required compared to de novo developments, repurposing can lead to substantial cost savings and shortened development times (Breckenridge and Jacob 2019). This advantage is particularly evident with repurposed drugs that have usually proven their safety in the treatment of another disease. Despite the need to conduct additional clinical trials to demonstrate the efficacy of the drug in a new disease context, it is possible to significantly reduce development times and costs compared to new developments (Naylor et al. 2015). Estimations from the pharmaceutical industry suggest that repurposing may save up to 85% of the development costs and shorten the development cycle by up to 5 years (Paul et al. 2010). Given these financial benefits, repurposing offers an attractive opportunity to develop solutions for otherwise unserved niche markets, such as treatments for rare diseases, which are often unattractive for de novo developments due to their limited market size (Kucukkeles et al. 2019).[1] At the same time, the commercial success of repurposed innovations can be substantial and a highly attractive investment opportunity (Naylor et al. 2015). The long-term success of the antiseptic mouthwash Listerine illustrates the commercial potential of repurposed innovations. Originally developed as a surgical antiseptic in the late

19th century, it became a very profitable mouthwash during the 1920s. Today, it is still a leading brand across many markets, generating significant sales. Similarly, the annual sales of the drug Revlimid peaked at approximately USD 4.3 billion after being repurposed from an antinausea remedy to a multiple myeloma treatment (Naylor et al. 2015). Finally, repurposing is a valuable innovation strategy when external shocks or the emergence of a crisis require accelerated responsiveness with relatively low barriers to entry (Hanisch and Rake 2021; Liu et al. 2021).

Rapid execution. Another key advantage of repurposing is the rapid introduction of the repurposed innovation. In most cases, the product is already on the market, and the production facilities are in place, making it easier to launch and scale the product. Automaker Dodge, for example, repurposed the preinstalled ultrasonic parking sensors in its Dodge Charger Pursuit police cars to enhance officer safety. When the feature is turned on, the parking sensors detect movements behind the car and activate the rearview camera to alert officers of a possible ambush. Because the base hardware comes preinstalled with existing police vehicles, the feature can be deployed by simply adding a small module under the dashboard (Stellantis 2018). Due to the ease of implementation, the innovation was quickly rolled out to the existing fleet of police vehicles.

5.3.2 Challenges

Quantity rather than quality. Repurposing requires a broad search strategy that assesses a variety of potential existing solutions for a new problem based on a trial-and-error approach (Cheng et al. 2018). The generalist approach associated with repurposing may come at the expense of foregoing a deep scientific or technological understanding of the underlying causal mechanisms of the innovation that ought to be repurposed (Hanisch and Rake 2021). While short-term innovation successes are possible through haphazard search strategies, there may be negative consequences for long-term purposeful innovation that would require an in-depth understanding of the underlying scientific and technological causal mechanisms of an innovation. Finally, low barriers to entry – particularly when repurposing well-established products – may lead to a multitude of uncoordinated repurposing actions. This causes redundancies and inefficiencies as organizations focus their repurposing-based innovation efforts on low-hanging fruit and quick wins that focus on quantity at the expense of outcome quality (Hanisch and Rake 2021).

Value appropriation. One of the biggest challenges hindering repurposing is the difficulty of appropriating the value of a repurposed product. Candidates for repurposing have been on the market or at least under development for some time. Examples from the pharmaceutical industry indicate that it may take several years or even decades to discover a new use for an existing product (Pushpakom et al. 2019). Similarly, when Dr. Spencer Silver, a scientist at 3M, discovered a new adhesive that could be peeled back, he struggled for years to find a useful application for this technology (earning him the nickname "Mr. Persistent"), a process that eventually culminated in the development of Post-it® notes (3M 2022). Such long timeframes for repurposing are fraught with challenges if companies want to protect the intellectual property of the repurposed innovation. Effective patent protection is usually difficult because the corresponding patents may have already expired or will expire in the near future (Pushpakom et al. 2019). Further questions regarding the patentability of repurposed innovations may arise if parts of the relevant technological knowledge already belong to the prior art and are common knowledge. Other forms of protection – such as market exclusivity for pharmaceuticals – may not provide sufficient incentives to invest in repurposing (Pushpakom et al. 2019).

5.4 Managing Repurposing

Repurposing can complement new product development strategies and open up new avenues for innovation. Repurposing presents appealing prospects to rejuvenate products by actively pursuing new applications beyond the original area, especially for products that are close to the end of their life cycle. The innovative elements are not in a technological breakthrough, but rather in reinventing a context in which the technology can make a difference. In this way, repurposing can provide companies with an extended revenue stream. If successful, repurposing also reduces the pressure on R&D departments to develop new technologies at a rapid pace and creates the time and financial space for ambitious projects. From an implementation perspective, repurposing is less dependent on technological know-how and instead relies on broad knowledge and inspiration from other contexts. Thus, successful repurposing benefits from diverse teams that draw inspiration from a variety of sources to find new uses for existing products. In this way, repurposing does not compete with the scarce resources of the R&D team and in many cases could be supported by other departments such as marketing and sales.

5.4.1 The Repurposing Process

Repurposing requires an institutionalized process to unleash its full potential as an innovation strategy in organizations. It is important to recognize that planned repurposing goes beyond examples of serendipitous discoveries and discoveries by coincidence. The repurposing process consists of four consecutive steps in which the first two form a subprocess of targeted search, and the last two form a subprocess of rapid execution (see Figure 5.4).

From a process perspective, mapping is the central activity that initiates the subprocess of targeted search that defines the search space. Mapping describes the process of conducting a systematic assessment of the available sources which lead to repurposed innovations. Methods for creating patent or technology maps might be useful at this stage of the repurposing process (Lee et al. 2009).

FIGURE 5.4 THE REPURPOSING PROCESS.

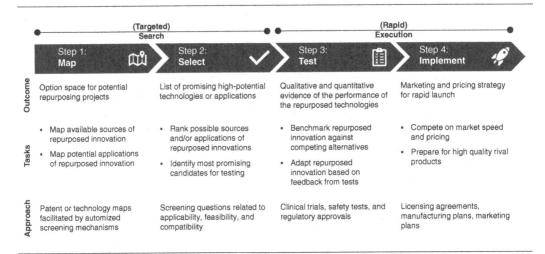

For example, the biopharmaceutical company Origenis relies on "Artificial Intelligence to navigate through the immense chemical universe, to identify islands of interesting and novel chemical matter" as their strategy for comprehensive technology mapping (Origenis 2022). Hence, the first step of mapping requires identifying known technologies that may be repurposed and then delimiting a target space of problems to which the technology could be applied.

In the second step, companies willing to repurpose must identify and select the most promising solution(s) to be developed for a new application. The innovation literature provides a rich set of useful tools for selecting the most promising innovation projects. For example, screening questions may serve as a starting point for a structured discussion to assess the potential costs and benefits of the repurposing project. These screening questions enable managers to assess the role of the customer (e.g., market, pricing, and compatibility questions), the competitive position, including an assessment of the firms' and its competitors' capabilities, and the timing and costs of the repurposing project (Schilling 2020). After assessing the potential costs and benefits, managers can use scoring methods to rank the most promising project candidates according to their evaluation of a set of criteria, including strategic fit as well as market and technology risk. Where applicable, quantitative methods can complement these qualitative methods to assess the potential future financial performance of the various repurposing projects. In addition to these more traditional methods, (internal) crowd feedback mechanisms are useful for prioritizing repurposing projects (Hoornaert et al. 2017).

Rapid execution of the repurposing project begins in step three with testing the repurposed solution against the new targeted functionality. The performance of competing products from the user's point of view may serve as a benchmark for assessing the performance of the repurposed solution. However, there may be cases in which there is no competing solution available (e.g., because the problem recently emerged or because the repurposed product should meet the needs of an unserved market). In this case, the generally accepted industry standards for testing the safety and efficacy of the repurposed solution would apply. Based on the feedback obtained in the testing step, the repurposed solution may require adaptation to enhance its performance and/or its appeal to potential customers. However, it may also turn out that the repurposed solution does not meet expectations in terms of functionality – e.g., a drug is not an effective treatment against a specific disease – or in terms of market potential. For example, Gilead Sciences and Hofmann-La Roche aimed to repurpose Remdesivir, a drug that did not show clinical efficacy against Ebola or the Marburg virus, as a treatment against COVID-19. The most recent clinical results call for further studies but do not unambiguously support moderately positive earlier assessments (Ansems et al. 2021). In cases where the tests of repurposed products lead to negative or ambiguous results, managers should stop further development of the repurposed solution, select the best suitable alternative, and start the testing phase again.

The fourth and final step is to implement and actively market the repurposed solution. To be successful, companies need to act quickly and introduce the repurposed solution as fast as possible. The reason is that the potential for protecting the repurposed solution through intellectual property rights might be limited (see Section 5.2.3), which is likely to lead to rapidly increasing competition once the commercial potential of the repurposed innovation becomes apparent. Therefore, a rapid and widespread market introduction is critical to establish an industry standard in unserved markets, to tie up strategic resources, and to achieve cost advantages through economies of scale and learning curve effects that enable an aggressive pricing strategy to preempt competitor market entry (Lieberman and Montgomery 1988). In either case, managers must be aware that competitors are closely monitoring their success and may enter the market with products that are either repurposed as well and offer the same or higher performance or even with newly developed products that are tailored to the target market and that outperform the repurposed alternatives.

5.4.2 Digitalization: Enabling Search and Validation

Emerging digital technologies provide new opportunities to leverage the full potential of repurposing. Often, repurposing is hindered by the amount of data that must be processed to map, select, and test technologies for their repurposing potential. For example, sourcing external knowledge through idea scouting or crowdsourcing quickly becomes impractical when the search space is very large and potentially detached from the context of the original solutions (Piezunka and Dahlander 2015). To illustrate the magnitude of the problem, consider the recent trend among pharmaceutical companies to open their libraries of already identified but not yet successfully developed (chemical) compounds to the broader academic research community to find new use cases. AstraZeneca alone made more than 250,000 compounds available to researchers (AstraZeneca 2021). This high number of potential solutions from just one company vividly demonstrates the impossibility of manually scanning these potential solutions and matching them with new applications. For this reason, digital analysis and processing technologies such as artificial intelligence have a very special role to play in the context of repurposing.

Artificial intelligence (AI) – particularly information extraction methods – as well as methods for analyzing and extracting hidden patterns in large amounts of (big) data, such as gene expressions, chemical structures, genotypes, proteomic data, and electronic health records, are already used to identify potential targets for drug or compound repurposing (Pushpakom et al. 2019; Zhou et al. 2020). For instance, AI can help predict the therapeutic potential of compounds. During the COVID-19 pandemic, AI helped identify promising repurposing candidates to treat the disease, demonstrating its usefulness in the repurposing context (Zhou et al. 2020). The ability to identify drug candidates that are suitable for repurposing and can move rapidly into clinical testing makes AI a core method to advance the repurposing of existing drugs while improving the effectiveness and efficacy of drug development (Zhou et al. 2020).

The use of AI for repurposing is, of course, not restricted to repurposing drugs for new disease areas. The IT and software industries provide an example where AI is used to identify software components that can be repurposed in the development of new software (Wangoo 2018). An AI algorithm is able to search forums such as GitHub to identify code that could be reused in a different context and even in other programming languages (Nguyen et al. 2020). More generally, AI opens up opportunities for conducting insightful analyses of large datasets and, therefore, supports the repurposing of existing products in areas where information processing is the main barrier to innovation, including the search for new innovation opportunities (Haefner et al. 2021). For example, Mounet et al. (2018) used AI to scan more than 100,000 materials to predict which compounds have the potential to create new and interesting materials suitable for battery development. Such advanced analytical methods also enable companies to deal with big data generated by user or customer participation in open innovation processes to identify or test products for repurposing (Del Vecchio et al. 2018).

To take advantage of the opportunities offered by AI in the context of repurposing, companies need to meet certain requirements. In addition to the availability of large amounts of adequate data with minimal biases (Lee et al. 2018), a company must have employees with data analytics skills to support the creation of new combinations of technologies as well as the reuse of existing solutions in different contexts (Wu et al. 2019). Employees with data analytic skills are required in the context of repurposing to facilitate product development that requires sifting through broad knowledge domains or vast amounts of information (Wu et al. 2020). At the same time, recognizing patterns alone is not enough. Complementing this, technical experts such as chemists or engineers must evaluate feasibility in terms of implementation costs and speed, while managers must assess the marketability of the technologies in terms of demand and pricing.

Successful repurposing requires not only data analytical skills but also a secure IT infrastructure for data exchange. Innovation repurposing is an inherently collaborative approach and often transcends organizational and technical boundaries.

Especially when the technology originates outside the focal organization's boundaries or a company wishes to open its technology portfolio to others, it is often of vital interest to the parties that the intellectual property flows are traceable and verifiable. This becomes relevant for assessing potential intellectual property infringements, determining licensing and royalty fees, or simply learning about new applications. Blockchains offer a promising solution to increase transparency in data sharing by enabling immutable and secure data records that can be used to track intellectual property flows (Goldsby and Hanisch 2022; Hanisch et al. 2022). IPwe (IPwe.com), for example, is a specialized provider of blockchain-powered intellectual property transactions that enables companies to track and manage their innovation portfolio. In addition, blockchains can provide the bedrock for secure and accountable data exchange, as recently demonstrated by a pilot project between IBM and the FDA to share oncology-related data.

5.5 Conclusion

Repurposing is a promising innovation approach that exploits untapped potential, reduces the cost of developing new products, and shortens innovation cycles. For managers, repurposing is therefore a crucial enabler to build and expand their competitive advantage and respond to new trends in a more dynamic and targeted way. With the advent of digital technologies such as AI, which automizes the effective screening of technologies on an unprecedented scale, and blockchains, which securely track data flows, repurposing is likely to become an increasingly dominant paradigm to innovation in the coming years, especially in fields that rely heavily on recombinational logic, such as materials science, chemistry, biology, and information technology. For example, a recent study has demonstrated how chemical waste products can be converted into useful drugs using complex computer algorithms that recombine chemical components (Wołos et al. 2022). It is critical for managers to recognize that the next breakthrough does not necessarily have to come from de novo developments made in their own R&D labs but that they can leverage existing technologies whose potential for another field has simply been overlooked.

Despite its merits, innovation repurposing also poses some risks that managers should keep in mind. Although repurposing can shorten development times, there is no guarantee that a particular technology will work in a new context just because it has similar properties. Thus, testing and evaluating new use cases together with technological fine-tuning can be tedious and costly. In addition, repurposed technologies may incur follow-on costs such as licensing and royalty fees, as well as lower profit margins resulting from the high risk of imitation, since repurposed technologies are often more difficult to patent or keep exclusive. For this reason, managers should view repurposing as a complementary innovation strategy alongside ambitious but inherently risky de novo developments. Overall, repurposing is an important pillar in a balanced innovation portfolio.

Note

1   The fact that the underlying product is already known and may be used and marketed for a specific purpose does not mean that repurposing is an innovation strategy that is available at almost no cost. Although the costs of repurposed innovations are likely to be lower than for de novo developments, they can still be substantial (Breckenridge and Jacob 2019). Companies may need to invest in studies that demonstrate feasibility, efficacy, or safety and conduct regulatory approval procedures. While these procedures are common in the health care industry, they also occur in manufacturing industries such as aviation. Software companies may not need regulatory approval but would need to test whether the repurposed code interacts as intended with other code in a software package.

References

3M (2022). History timeline: post-it® notes. Retrieved June 12, 2022, from: https://www.post-it.com/3M/en_US/post-it/contact-us/about-us/.

Allarakhia, M. (2013). Open-source approaches for the repurposing of existing or failed candidate drugs: learning from and applying the lessons across diseases. *Drug Design, Development and Therapy* 7: 753–766. doi: 10.2147/DDDT.S46289.

Andriani, P. and Carignani, G. (2014). Modular exaptation: a missing link in the synthesis of artificial form. *Research Policy* 43 (9): 1608–1620. doi: 10.1016/j.respol.2014.04.009.

Andriani, P. and Kaminska, R. (2021). Exploring the dynamics of novelty production through exaptation: a historical analysis of coal tar-based innovations. *Research Policy* 50 (2): 104171. doi: 10.1016/j.respol.2020.104171.

Ansems, K., Grundeis, F., Dahms, K. et al. (2021). Remdesivir for the treatment of COVID-19. *The Cochrane Database of Systematic Reviews* 8: CD014962. doi: 10.1002/14651858.CD014962.

AstraZeneca (2021). Target identification: discover advanced screening technologies to accelerate novel target identification. Retrieved March 30, 2022, from: https://openinnovation.astrazeneca.com/preclinical-research/target-identification.html.

Baker, T. and Nelson, R.E. (2005). Creating something from nothing: resource construction through entrepreneurial bricolage. *Administrative Science Quarterly* 50 (3): 329–366. doi: 10.2189/asqu.2005.50.3.329.

BMW Group (2020). BMW Group UK second-life battery solution in partnership with Off Grid Energy. Retrieved March 08, 2022, from: https://www.press.bmwgroup.com/united-kingdom/article/detail/T0318650EN_GB/bmw-group-uk-second-life-battery-solution-in-partnership-with-off-grid-energy?language=en_GB.

Breckenridge, A. and Jacob, R. (2019). Overcoming the legal and regulatory barriers to drug repurposing. *Nature Reviews Drug Discovery* 18 (1): 1–2. doi: 10.1038/nrd.2018.92.

Cheng, F., Desai, R.J., Handy, D.E. et al. (2018). Network-based approach to prediction and population-based validation of in silico drug repurposing. *Nature Communications* 9 (1): 2691. doi: 10.1038/s41467-018-05116-5.

Chesbrough, H. (2012). Open innovation: where we've been and where we're going. *Research-Technology Management* 55 (4): 20–27. doi: 10.5437/08956308X5504085.

Chesbrough, H.W. (2006). Open innovation: a new paradigm for understanding industrial innovation. In: *Open Innovation. Researching a New Paradigm* (ed. H.W. Chesbrough, W. Vanhaverbeke, and J. West), 1–12. Oxford: Oxford University Press.

Del Vecchio, P., Di Minin, A., Petruzzelli, A.M. et al. (2018). Big data for open innovation in SMEs and large corporations: trends, opportunities, and challenges. *Creativity and Innovation Management* 27 (1): 6–22. doi: 10.1111/caim.12224.

Garud, R., Gehman, J., and Giuliani, A.P. (2018). Serendipity arrangements for exapting science-based innovations. *Academy of Management Perspectives* 32 (1): 125–140. doi: 10.5465/amp.2016.0138.

Garud, R., Gehman, J., and Karnøe, P. (2021). Winds of change: a neo-design approach to the regeneration of regions. *Organization & Environment* 34 (4): 634–643. doi: 10.1177/1086026619880342.

Goldsby, C. and Hanisch, M. (2022). The boon and bane of blockchain: getting the governance right. *California Management Review* 64 (3): 141–68. doi: 10.1177/00081256221080747.

Haefner, N., Wincent, J., Parida, V., and Gassmann, O. (2021). Artificial intelligence and innovation management: a review, framework, and research agenda. *Technological Forecasting and Social Change* 162 : 120392. doi: 10.1016/j.techfore.2020.120392.

Hanisch, M. and Rake, B. (2021). Repurposing without purpose? Early innovation responses to the COVID-19 crisis: evidence from clinical trials. *R&D Management* 51 (4): 393–409. doi: 10.1111/radm.12461.

Hanisch, M., Theodosiadis, V., and Teixeira, F. (2022). Digital governance: how blockchain technologies revolutionize the governance of interorganizational relationships. In: *Digital Transformation: A Guide for Managers*, 1e (ed. B.S. Baalmans, T.L.J. Broekhuizen, and N.E. Fabian), 122–147. Groningen: Groningen Digital Business Center (GDBC), University of Groningen.

Hoornaert, S., Ballings, M., Malthouse, E.C., and van den Poel, D. (2017). Identifying new product ideas: waiting for the wisdom of the crowd or screening ideas in real time. *Journal of Product Innovation Management* 34 (5): 580–597. doi: 10.1111/jpim.12396.

Kesselheim, A.S., Tan, Y.T., and Avorn, J. (2015). The roles of academia, rare diseases, and repurposing in the development of the most transformative drugs. *Health Affairs* 34 (2): 286–293. doi: 10.1377/hlthaff.2014.1038.

Kucukkeles, B., Ben-Menahem, S.M., and von Krogh, G. (2019). Small numbers, big concerns: practices and organizational arrangements in rare disease drug repurposing. *Academy of Management Discoveries* 5 (4): 415–437. doi: 10.5465/amd.2018.0183.

Langedijk, J., Mantel-Teeuwisse, A.K., Slijkerman, D.S., and Schutjens, M.-H.D. (2015). Drug repositioning and repurposing: terminology and definitions in literature. *Drug Discovery Today* 20 (8): 1027–1034. doi: 10.1016/j.drudis.2015.05.001.

Lee, J., Davari, H., Singh, J., and Pandhare, V. (2018). Industrial Artificial Intelligence for industry 4.0-based manufacturing systems. *Manufacturing Letters* 18: 20–23. doi: 10.1016/j.mfglet.2018.09.002.

Lee, S., Yoon, B., and Park, Y. (2009). An approach to discovering new technology opportunities: keyword-based patent map approach. *Technovation* 29 (6): 481–497. doi: 10.1016/j.technovation.2008.10.006.

Lévi-Strauss, C. (1966). *The Savage Mind.* University of Chicago Press.

Lieberman, M.B. and Montgomery, D.B. (1988). First-mover advantages. *Strategic Management Journal* 9 (S1): 41–58. doi: 10.1002/smj.4250090706.

Liu, W., Beltagui, A., and Ye, S. (2021). Accelerated innovation through repurposing: exaptation of design and manufacturing in response to COVID-19. *R&D Management* 51 (4): 410–426. doi: 10.1111/radm.12460.

Mastrogiorgio, M. and Gilsing, V. (2016). Innovation through exaptation and its determinants: the role of technological complexity, analogy making & patent scope. *Research Policy* 45 (7): 1419–1435. doi: 10.1016/j.respol.2016.04.003.

Mounet, N., Gibertini, M., Schwaller, P. et al. (2018). Two-dimensional materials from high-throughput computational exfoliation of experimentally known compounds. *Nature Nanotechnology* 13 (3): 246–252. doi: 10.1038/s41565-017-0035-5.

Naylor, S., Kauppi, D.M., and Schonfeld, J.M. (2015). Therapeutic drug repurposing, repositioning and rescue: part II: business review. *Drug Discovery World* 16 (2): 57–72.

Nelson, R.R. and Winter, S.G. (1982). *An Evolutionary Theory of Economic Change.* Cambridge, MA: Harvard University Press.

Nguyen, P.T., Di Rocco, J., Rubei, R., and Di Ruscio, D. (2020). An automated approach to assess the similarity of GitHub repositories. *Software Quality Journal* 28 (2): 595–631. doi: 10.1007/s11219-019-09483-0.

Nikon (2021). Nikon and Aeva to start strategic collaboration in industrial automation and metrology. Retrieved June 14, 2022, from: https://www.nikonmetrology.com/en-us/about-us/latest-news/nikon-and-aeva-to-start-strategic-collaboration-in-industrial-automation-and-metrology.

Origenis (2022). AI innovation platform. Retrieved April 11, 2022, from: https://www.origenis.de/ai-innovation-platform.

Paul, S.M., Mytelka, D.S., Dunwiddie, C.T. et al. (2010). How to improve R&D productivity: the pharmaceutical industry's grand challenge. *Nature Reviews Drug Discovery* 9 (3): 203–214. doi: 10.1038/nrd3078.

Piezunka, H. and Dahlander, L. (2015). Distant search, narrow attention: how crowding alters organizations' filtering of suggestions in crowdsourcing. *Academy of Management Journal* 58 (3): 856–880. doi: 10.5465/amj.2012.0458.

Pushpakom, S., Iorio, F., Eyers, P.A. et al. (2019). Drug repurposing: progress, challenges and recommendations. *Nature Reviews. Drug Discovery* 18 (1): 41–58. doi: 10.1038/nrd.2018.168.

Rönkkö, M., Peltonen, J., and Arenius, P. (2014). Selective or parallel? Toward measuring the domains of entrepreneurial bricolage. In: Jerome Katz Andrew C. Corbett *Entrepreneurial Resourcefulness: Competing with Constraints*, 43–61. Emerald Group Publishing Limited.

Schilling, M.A. (2020). *Strategic Management of Technological Innovation*. New York, NY: McGraw-Hill Education.

SpaceX (2022). Starship SN15. Retrieved March 08, 2022, from: https://www.spacex.com/vehicles/starship.

Stellantis (2018). FCA US ships 10,000th officer protection package for law enforcement. Retrieved June 09, 2022, from: https://media.stellantisnorthamerica.com/newsrelease.do?id=20133&mid=.

Vincent, J. (2017). Why Dyson is investing in AI and robotics to make better vacuum cleaners. Retrieved March 08, 2022, from: https://www.theverge.com/2017/3/14/14920842/dyson-ai-robotics-future-interview-mike-aldred.

Wangoo, D.P. (2018) Artificial intelligence techniques in software engineering for automated software reuse and design. *2018 4th International Conference on Computing Communication and Automation (ICCCA)*, 1–4.

Wołos, A., Koszelewski, D., Roszak, R. et al. (2022). Computer-designed repurposing of chemical wastes into drugs. *Nature* 604 (7907): 668–676. doi: 10.1038/s41586-022-04503-9.

Wu, L., Hitt, L., and Lou, B. (2020). Data analytics, innovation, and firm productivity. *Management Science* 66 (5): 2017–2039. doi: 10.1287/mnsc.2018.3281.

Wu, L., Lou, B., and Hitt, L. (2019). Data analytics supports decentralized innovation. *Management Science* 65 (10): 4863–4877. doi: 10.1287/mnsc.2019.3344.

Zhou, Y., Wang, F., Tang, J. et al. (2020). Artificial intelligence in COVID-19 drug repurposing. *Lancet Digital Health* 2 (12): e667–e676. doi: 10.1016/S2589-7500(20)30192-8.

Marvin Hanisch is Assistant Professor in the Innovation Management & Strategy department at the University of Groningen. His research focuses on the governance mechanisms in alliances, open-source software communities, digital platforms, and blockchain networks. His work has been published in leading management journals such as the *Journal of Management Studies* and in practice-oriented outlets such as the *California Management Review*, and has received numerous international awards. Marvin Hanisch is co-affiliated with the University of Passau.

Bastian Rake is Assistant Professor at Maynooth University School of Business. His research studies innovation in knowledge-intense industries with a particular focus on the biotechnology and pharmaceutical industries, collaboration and innovation networks, as well as in the internationalization of science and R&D. His research appeared in renowned journals such as *Research Policy, Industrial and Corporate Change, Industry and Innovation,* as well as *Innovation: Organization & Management.* Bastian Rake has published practitioner-oriented contributions in outlets like *RTÉ Brainstorm* and the *Business & Society Blog.* Bastian Rake is affiliated researcher at the University of Gothenburg.

CHAPTER SIX

INNOVATION GOVERNANCE

Rod B. McNaughton

6.1 Introduction

The importance of governance in generating the conditions where innovation can flourish, and new product development thrive is often overlooked. Most attention is paid to specific new product development (NPD) processes rather than the broader corporate governance context in which they are set. However, supportive and robust governance is critical to fostering innovation. Innovation governance is "…a system of mechanisms to align goals, allocate resources and assign decision-making authority for innovation, across the company and with external parties" (Deschamps 2013). Innovation governance emphasizes the importance of the board and senior executives in setting expectations and policies, organizing the firm, delegating decision-making, and hiring the right people to support innovation. Thus, this chapter addresses company directors, C-suite executives, and the innovation and product managers who report to them.

The concept of innovation governance was popularized a decade ago by Deschamps and Nelson's (2012) book *Innovation Governance: How Top Management Organizes and Mobilizes for Innovation*. However, the influence of organization-level governance on innovation remains underappreciated. Successful innovation begins before opportunities are identified, or new products and services are developed. Many firms fail to innovate, or their innovations are unsuccessful because it is not part of their strategy. They do not allocate sufficient resources or organize themselves to foster and support innovation.

The PDMA Handbook of Innovation and New Product Development, Fourth Edition. Edited by Ludwig Bstieler and Charles H. Noble.
© 2023 John Wiley & Sons, Inc. Published 2023 by John Wiley & Sons, Inc.

Lists of new product success drivers, such as that provided by Cooper (2019), include organization-level factors like product and innovation strategy, project selection, ability to lever core competencies, resources, and top management support to achieve new product success. However, it is rarely emphasized that the fundamental parameters for these factors are determined mainly by board directors and the executives and managers they hire. Together, the board and senior leadership team govern a firm by prioritizing its activities, allocating resources, setting policies, and fostering its culture.

Much of the research literature and articles in the popular business press implicitly assume firms actively pursue innovation and that lack of innovation or NPD failure results from shortcomings in resources, capability, or creativity. However, in many firms, especially private and smaller and medium-sized ones, the primary impediment to becoming innovative is the conservative attitudes of owners and managers and the governance systems and policies that they put in place that discourage innovation (McNaughton and Sembhi 2021).

Large firms can also suffer innovation malaise. While top tech firms like Apple, Google, and Facebook are renowned for investing heavily in product development and other forms of innovation, on average, only about 10 percent of earnings and borrowing by US-listed firms go into investment, down from 40 percent in the 1960s (DePillis 2015). One reason for this decline may be increased shareholder demands for dividends and the use of financial transactions like borrowing to fund share buy-backs to maintain shareholder value. This has an opportunity cost by limiting the funds available to invest in the factors that enhance innovation.

Still, innovation is increasingly front of mind for many boards and CEOs, especially as they face the longer-term consequences of the COVID-19 pandemic, including changes in consumer behavior, supply chain disruption, inflation, and labor shortages (McClimon 2022). Some traditional industries, like oil and gas, are under intense pressure to transform in the wake of climate change legislation, or like banking and insurance must respond to technology-fueled start-ups.

This chapter makes the case that firms seeking to increase their innovation and launch successful new products and services should begin by considering the overall organizational context in which that innovation occurs. Governance offers a well-established framework for innovation, with boards and C-suite executives well-positioned to create favorable conditions where innovation processes, including NPD, can flourish.

6.2 Why Is Innovation a Governance Issue?

It may seem unconventional to view innovation as a governance issue, as historically, innovation is the purview of management and rarely appears on lists of directors' duties. Innovation might be mentioned as one of a board's fiduciary responsibilities but would be well down the list compared to risk management and compliance. Indeed, some might see innovation and governance as opposing practices, with innovation being about doing new things and governance about stability and minimizing risk (Zhu 2020). Culturally, innovation is associated with creativity, change, and agility, while governance is associated with monitoring, compliance, and risk reduction.

But innovation isn't just about new ideas, and governance isn't just about minimizing risks. Kay and Goldspink (2016) point out that risk and innovation are intertwined: "Risk management deals with uncertainty on the downside, innovation deals with uncertainty on the upside." How these two processes work together is essential, with innovation providing the fuel for growth and governance the guidance. By prioritizing innovation consistent with business objectives, governance helps maximize the business value of innovation and lessen its risks. The tension between governance and innovation needs to be balanced so that the bureaucracy of one does not stifle the other.

Many boards are spending more time on innovation as they realize innovation is a core business activity (Innovation Enterprise 2013). Failure to innovate is a risk to the longer-term sustainability of a business. Core business activities cut across all functional activities and involve financial issues like growth strategy, technology and other investments, the creation of new companies, and cultural topics like strategic orientation, learning, and knowledge management. Thus, innovation is difficult to delegate to one functional portfolio. Consequently, boards and CEOs are increasingly active in leading their company's innovation strategy while also experimenting with senior innovation leadership positions or teams of executives that cut across portfolios and with hybrid organizational forms.

The chief innovation officer is the most prominent senior leadership position devoted to innovation (often called a CINO to distinguish it from the chief information officer or CIO role). However, the roles of the chief technology officer (CTO) and the chief research officer (CRO) are related. The number of CINOs has increased over the past decade and expanded from technology companies to other sectors. However, the role, mandate, and background of those appointed remain diverse and often poorly defined (Lovric and Schneider 2019). CINOs can have a broad remit, from improving communication channels to promoting an entrepreneurial culture, and can even include oversight of marketing or IT (Innovation Enterprise 2013).

In many cases, the person holding the role is the first to be appointed and organically defines the position as they work out what needs to be done. Consequently, the function is challenging and may lack the legitimacy and authority to deliver the intended results. Commentators such as Swoboda (2020) argue that you "can't delegate innovation," and the person ultimately responsible for innovation must be the president, CEO, or Chairman.

A governance perspective takes a broad and high-level approach, seeing innovation as a holistic system that includes the innovation content (the overall strategy and selected projects) and the processes for moving those projects ahead within the organization or with partners (Deschamps 2013). The system consists of the organization's values, goals, and policies, its method for allocating resources, and assigning roles, responsibilities, and decision rights. This innovation system strongly influences the extent and nature of the innovation that happens and is thus critical to the success of NPD.

A systems approach manages all types of innovation as a portfolio, identifying synergies between them to allocate resources efficiently. Much of the discussion of innovation types focuses on whether the market and innovation are new or not. Addressing existing markets with existing technology results in incremental innovations, while new markets and technologies may result in radical innovations. But this overlooks business model innovations and productivity enhancements through process innovation. It also potentially results in fragmentation, with functions including R&D, engineering, information technology, product management, marketing, and others innovating differently.

Keeley et al., in their 2013 book *Ten Types of Innovation: The Discipline of Building Breakthroughs*, argue that creating new products is only one way to innovate and provide evidence that, on its own, NPD provides the lowest return on investment. Their study used Jay Doblin's ten types of innovation, four of which relate to the organization's configuration, two to the offering itself, and four to the customer experience. It would be difficult for one unit or function within an organization to simultaneously consider and balance such a complex portfolio of innovation types. But the organization-wide oversight of the board and C-suite executives makes coordinating a more fine-grained portfolio tractable.

Involving boards and senior executives in governing innovation is the ultimate signal of the importance of innovation to the organization. It provides visibility and transparency to shareholders and employees for innovation strategy and investments. Involving senior executives and the board shifts the personal risks of making risky decisions away from individual managers. Employees may make overly conservative decisions when their careers are on the line. By having the CEO or other senior executives sponsor riskier innovation projects and delegating decision-making for less risky or inexpensive innovations, companies can better align risk with authority in the organization (Lovallo et al. 2020).

Finally, it is worth clarifying that "innovation governance" is sometimes used to mean governance by other than boards of directors, at scales both larger than an organization or within it. Until the term is widely used to refer to the involvement of boards of directors in the innovation process, it is essential to know the context in which it is used. At the national level, the term can refer to the institutions and policies governments put in place to stimulate, support, and guide innovation across the economy. Within companies, the term can simply mean ensuring consistency and coordination between various innovation and NPD processes. For example, Haines (2012), writing in the third edition of *The PDMA Handbook of New Product Development*, identifies the challenge presented by the coupling in many people's minds between product development and a phased NPD process without a shared understanding of that process. He argues that better "governance" of inter-related innovation and NPD processes is the solution, calling for greater executive involvement and cross-functional collaboration but stops short of placing the ultimate responsibility with the board.

6.3 Top-Down and Bottom-Up

An objection to the innovation governance perspective comes from those who advocate for the benefits of bottom-up innovation. They see governance as a top-down approach to innovation. They argue that employees with the most customer contact and those with technical knowledge and experience with day-to-day processes are best positioned to recognize emerging customer needs.

There is plenty of evidence for the efficacy of bottom-up innovation. But it is a rare organization where an employee, having identified an opportunity, would be able to marshal the firm's resources on their own to pursue it. Seth Godin's (2006) widely quoted claim that "No organization ever created an innovation. People innovate, not companies." is valid in the literal sense that innovation needs people to come up with ideas. But successful innovation usually requires people to work with others within frameworks established by organizations and an organization's resources to bring those ideas to fruition.

Organizations put processes in place to vet ideas and allocate resources. Many employee-generated ideas are not aligned with the overall business strategy or are nonviable or infeasible. Moreover, for bottom-up innovation to occur, employees need to be supported by a culture that welcomes and incentivizes this behavior. Good innovation governance processes inspire bottom-up innovation focused on strategically important areas and support it with resources and a framework for managing risk.

Bottom-up and top-down innovation both play a role (Aitken 2020). Successful innovators achieve balance and carefully manage the interface between the two. Unfortunately, many firms wanting to increase their innovation focus first on the bottom-up approach, which is deceptively easier. They may think all they need is to send staff on the latest design thinking course or establish an employee "idea box" online. However, without strategic direction, the quality of employee ideas is likely to be low. Participation will be lacking without the right culture, or the ideas box will be filled with complaints rather than good ideas. Employees who share ideas but don't see them acted on, or worse, see them used by managers without giving them credit or reward, will become disenchanted.

Top-down innovation processes determine and communicate innovation focus areas, allocate resources, and foster a culture that supports innovation. They require significant attention, effort, and participation from senior leaders. And, if they fail to interface with bottom-up processes, top-down processes risk being disconnected from essential sources of knowledge about the market and internal firm capabilities. Good innovation governance balances top-down and bottom-up approaches so that innovative people are hired and encouraged, have a clear path to move ideas forward, and have their hard work pay off as frequently as possible for the company and themselves.

Company directors, C-suite executives, and employees are all "people" who can innovate. Conversely, directors and senior executives can be conservative or resistant to change, as can employees. Indeed, many CEOs are hired because of their reputation for a "steady hand" and having produced consistent financial returns. Hiring innovation-oriented leaders and changing collective values are essential for a company wanting to increase its innovation outcomes. But it may take a long time to affect individual employees' values or make changes through hiring, and change will be slow if values and incentives are misaligned, no matter how much attention is paid to governance.

6.4 How Boards Influence Innovation

Boards of directors are responsible for ensuring a firm's strategy is consistent with owners' investment expectations, appointing the CEO, advising on top management team appointments, and holding management to account by reviewing performance, managing risks, ensuring compliance, and auditing. Thus, a company's board has many opportunities to influence innovation if it chooses to act on them, beginning with the balance they strike between ambitions for growth and the owners' risk tolerance and earnings expectations. This sets the broad parameters for how much and what types of innovation a firm will undertake. Table 6.1 summarizes how boards and CEOs influence the amount and type of innovation in their organization.

TABLE 6.1 HOW DO THE BOARD AND CEO INFLUENCE INNOVATION?

The Board	The CEO and Other C-Suite Executives
• Receives instruction from owners about investment objectives.	• Sets a vision for innovation.
• Sets parameters for risk tolerance and monitors risk.	• Defines how the company will capture value from innovation.
• Participates in strategy setting including the portfolio of innovation.	• Chooses how innovation will be organized.
• With management chooses how innovation will be organized.	• Chooses innovation processes/methods.
• Monitors and audits performance.	• Defines roles and assigns responsibility.
• Oversees compliance.	• Fosters an innovation culture.
• Appoints CEO and monitors performance.	• Champions and models innovative behavior.
• Delegates decision rights.	• Sets policies and incentives and establishes management practices that support innovation.
• Nominates board members.	• Allocates resources and sets budgets.
• Sets up subcommittees or advisory boards (e.g., on innovation, or digital transformation).	• Identifies and overcomes obstacles and resistance to innovation.
	• Defines how to measure innovation, sets targets, and monitors performance.

Despite their potential to influence innovation, many boards have traditionally spent little time on it, with innovation rarely appearing on their meeting agendas (Cheng and Groysberg 2018). Instead, they relied on management to determine how much innovation and what type should be undertaken. The exception is innovations requiring significant capital expenditures that require board approval. There are numerous reasons for this, including the emphasis of boards on their regulatory and compliance responsibilities over softer aspects of strategy and culture and the sheer number of issues competing for attention during a limited number of board meetings.

But the nature of directors' responsibilities is changing, with increasing expectations, faster decision-making, and more significant interaction with the companies they govern (Hill and Davis 2017). Rather than primarily being a check on management, companies are beginning to see their board as a source of competitive advantage. Greater attention is paid to diversity in the appointment of directors, including the knowledge they bring, which may be significant to R&D efforts, or organizational innovations like digital transformation. Directors are increasingly expected to account for the interests of multiple stakeholders, not just shareholders. Many of these interests are not financial. Thus, directors are becoming more involved in strategy formulation and working with management rather than simply receiving their reports.

The composition of boards influences innovation through directors' intellectual and social capital. There is an extensive research literature on the board characteristics most strongly associated with innovation, though many findings are inconclusive or even contradictory. Sierra-Morán et al. (2021) recently summarized 96 studies, concluding that, on average, firm-level innovation is associated with board structural diversity and with board demographic diversity, but that the effects can differ whether innovation is measured by inputs (e.g., R&D) or outputs (e.g., patents). For example, larger boards are a source of more ideas, but more decision-makers can make it challenging to agree on investments. Thus, larger boards can be positive for innovation outputs but harmful for inputs.

Sierra-Morán and her colleagues found that, on average, studies show independent directors positively influence innovation performance because of their knowledge of the environment. Duality, where one person is both CEO and Chairperson, also aids innovation, possibly because of their understanding of the firm's internal processes. Directors' equity is negatively associated with innovation, suggesting owners are more risk-averse and limit their support for new ideas. The strongest association is with meeting frequency, underscoring the importance of having time to exchange ideas and make decisions without long waits between meetings. Concerning demographic diversity, heterogeneity increases the information available, so innovation is positively associated with the proportion of women on boards. When directors sit on more than one board, social ties can increase innovation. Interlocking directorates may increase firms' opportunities to access financing or collaborate on projects with other organizations. Finally, boards with a longer average tenure experience more innovation. This may result from increased knowledge of the firm's processes and more trust and better communication between board members.

6.5 Organizing for Innovation

The company's internal organization is a critical decision influencing innovation. Successful innovation requires engagement from the board and senior executives, but both are too busy with financial and operational issues to steer innovation daily. Thus, the primary way they direct innovation is through the organization's structure, which may be characterized by dimensions such as its:

1. Formalization. The degree to which rules and procedures govern the behavior of employees.
2. Standardization. The degree to which activities are undertaken similarly.
3. Centralization. The degree to which decision-making authority or activities are concentrated at the top levels of the organization.
4. Coupling. The degree to which activities are integrated to achieve coordination.

There is no best structure. Firms famed for their innovation can have very different forms. For example, Apple is highly centralized, while Facebook is decentralized. Microsoft is loosely coupled, and Google is tightly coupled. Organizational structure needs to be designed to fulfill the organization's innovation and overall corporate strategy, especially what types of innovation it means to facilitate.

Firms' organizational structure evolves. A study by Deloitte (2019) found that 90 percent of firms are thinking of redesigning their organizations. Further, there is more to organizing than just structure. To drive innovation, alignment is essential between structure and other aspects of organizational design. In their book *Designing Your Organization: Using the STAR Model to Solve 5 Critical Design Challenges*, Galbraith and Kates (2007) demonstrate using Galbraith's STAR Model to think about organizational design. The model consists of five dimensions: strategy, which determines direction; structure, which determines how decision-making rights are delegated; processes, which have to do with the flow of information; rewards, which provide incentives; and finally, people, including policies that influence mindsets and skills. The "right" combination of these dimensions can transform an organizational design into a source of competitive advantage.

Another vital governance decision is the delegation of decision-making within the organizational structure. Promoting and overseeing innovation can be entrusted to an individual, either full-time or as part of their duties, or it can be the collective responsibility of a group of managers. Deschamps (2013) surveyed 110 companies, primarily global multinationals, and identified nine different models of innovation leadership. He classified these based on their level in the organization and the size of the decision-making group. Having the top management team or a subset responsible for innovation was the most frequent. However, only 58 percent of respondents were happy with the results of this arrangement. The next most frequent was having a dedicated innovation manager or a CINO, with the lowest satisfaction at 35 percent. This was followed by the CEO solely in charge, with 58 percent satisfaction. In contrast, placing the CTO or CRO in charge was less common but had 70 percent satisfaction. This shows there is no best way to oversee innovation, as this must be aligned with other aspects of an organization's structure to achieve the desired results.

Existing organizations wanting to become more innovative have the challenge of shifting from their current organizational structure to a new one while continuing with business-as-usual to protect revenue streams. Deloitte (2019) characterizes a continuum of organizational structures that support innovation. Five archetypes range from creating cross-functional teams as the least disruptive to autonomous teams being the most disruptive. In the first, functional groups form interdisciplinary teams to manage short-term project-based outcomes. In contrast, decentralized self-managing teams set their own goals and control their budgets. Most firms have experience with cross-functional teams, but fewer have taken the step of reorganizing into autonomous ones. Somewhere in the middle of the continuum are experimentation hubs. These are centralized independent units with a mandate for a company's innovation. They can try new things and develop their innovation-friendly culture without impacting the organization.

How to design an organizational structure that supports innovation and managing the transition to it is beyond the scope of this chapter. The essential point is that while the board is ultimately responsible for innovation, it cannot oversee day-to-day activities. Thus, boards govern through the organization's structure, and the executives and managers they appoint into roles in that structure, making decisions about these critical.

6.6 Conclusion

Innovation governance is a form of organizational leadership that provides an overall framework for innovation. It determines the extent to which innovation is part of a company's strategy and thus where any innovation and NPD activities will be focused and resourced. This chapter argues that choosing the settings to stimulate and steer innovation (or not) is a matter for the board and senior management. The reasons are that innovation is a core business activity; it cuts across functional areas and benefits from being managed as an overall portfolio.

While the governance perspective appears top-down, it focuses on guiding bottom-up ideas so they align with business strategy and can be coordinated and resourced. Board directors, senior managers, and employees are all "people" who may be involved in innovation. Their orientation toward innovation, skills, competencies, and knowledge of customers and firm capabilities is essential to sparking ideas and moving them forward. However, without the guidance of an organization's strategy, and access to its resources and processes for collaborating with others, ideas are unlikely to translate into successful new products and services that create value for the organization and its various stakeholders.

A more diverse board, structurally and demographically, is generally associated with more significant innovation. Increasingly, boards play an active role, interacting with executives and managers to develop strategies and guide major innovation projects. Some directors may be chosen for their knowledge of the technology or other factors that underlie a firm's products and services. However, boards and senior executives cannot manage day-to-day innovation processes. Thus, organizational design and the managers hired into the roles in the organization's structure are critical, as are its policies, procedures, and incentives. All need to be aligned to ensure success.

This chapter offers a way forward for companies wanting to become more innovative. Rather than jumping directly into improving bottom-up innovation processes, it recommends first stepping back and examining high-level factors that may be holding the company back. These might include the owners' earnings expectations and risk tolerance, the board's composition, organizational structure and processes, and the extent of innovation orientation of those involved at all levels. Governance offers a holistic and well-understood method for evaluating such factors, and directors have the authority to do so.

The path to greater board involvement in innovation is beyond the scope of this chapter. However, Table 6.2 is a starting point, offering a list of innovation-friendly board practices that will stimulate thinking. Exercises 6.1 and 6.2 can also help by getting directors, executives, and employees to reflect on their innovation orientation and the support for innovation in the organization and its national context and start a conversation.

The popular business and academic literature on innovation and NPD are voluminous, but few beyond Deschamps and Nelson's (2012) book have taken a governance perspective explicitly. Thus, it is essential to extrapolate for yourself the implications for the board and its role when reading about innovation. Institutes of Directors are beginning to address their members' need for professional development around innovation (e.g., Kay and Goldspink 2016). Finally, many firms, from smaller privately held ones to global multinationals, have already been on this journey. Some, such as WL Gore, 3M, and Huawei, very publicly, and much can be learned by reading about these prominent cases. However, other lesser-known firms closer to home will have directors and executives willing to share their innovation journeys. Finally, the most profound learning comes from working with innovation-experienced directors and appointing them to your board.

TABLE 6.2 HOW MANY OF THESE INNOVATION-FRIENDLY PRACTICES HAS YOUR BOARD ADOPTED?

- Include innovation-orientation, experience, and/or relevant scientific/technology knowledge in the board appointments skills matrix.
- Make board appointments with independence and diversity in mind.
- Include innovation topics in board training and development opportunities.
- Make innovation a standing item on the board agenda.
- Regularly make time to review and discuss innovation strategy.
- Regularly review if organizational structure is fit for purpose and consistent with the innovation strategy.
- Report on the innovation portfolio and highlight successes and failures in communications to shareholders and at annual general meetings.
- Establish a board committee or an advisory board for innovation or related activities like digital transformation.
- Ensure the board is aware of the overall innovation portfolio and familiar with key innovation projects and their progress.
- Include innovation-related risks and mitigations on the board's risk register.
- Have clear guidance on what innovation investments the board must approve.
- Set innovation-related KPIs for management and receive regular reports on these KPIs and major projects that underpin them.
- Take innovation orientation and experience into account when appointing a CEO and other senior executives.
- Include innovation performance in the annual review of the CEO's performance.
- Link part of the compensation of the CEO and other executives to innovation performance.
- Conduct innovation audits.
- Meet with leaders of key innovation projects, star researchers, and/or visit innovation facilities or labs.
- Attend innovation-related events and participate in celebrations of innovation success (e.g., attending new product launches).

Some of these practices are adapted from Deschamps (2015).

Exercise 6.1   Attitudes to innovation

For a company wanting to increase its innovative outcomes, hiring innovation-oriented leaders, and a change in its collective values are essential.

The ability of individuals, whether on the board, in the C-suite, or employees, to behave in ways that promote innovation is an essential foundation of an innovative company. Employee innovative behaviors are ones through which employees generate or adopt new ideas and make subsequent efforts to implement them. Such behavior has multiple facets, including the generation of ideas, how they are communicated, and the ability to involve others in bringing them to life.

Lukes and Stephan (2017) developed a multi-item scale to measure employees' innovative behavior. Complete the Innovative Employee Inventory by rating each item from 1 (strongly disagree) to 5 (strongly agree). Sum your ratings, divide by 115, and multiply by 100 to derive your overall score as a percentage of the total possible points.

(Continued)

The inventory should not be interpreted as evidence that you are or are not innovative. However, it will help you think about how you may be innovative in your thinking and approach. Reflect on your results:

- Are most of your items above or below the mid-point?
- Is there a pattern (e.g., you are stronger at recognizing opportunities but weaker at communicating them or getting others alongside?)
- Are you surprised by your level of innovativeness? Do the numbers reinforce what you knew about yourself, or do they offer new insight?
- Can you think of examples from your work or other contexts that illustrate each scale component?

Innovative Employee Inventory	Rate Each Item from 1 (Strongly Disagree) to 5 (Strongly Agree)

Idea generation
1. I try new ways of doing things at work
2. I prefer work that requires original thinking
3. When something does not function well at work, I try to find a new solution

Idea search
4. I try to get new ideas from colleagues or business partners
5. I am interested in how things are done elsewhere in order to use acquired ideas in my own work
6. I search for new ideas from other people in order to try to implement the best ones

Idea communication
7. When I have a new idea, I try to persuade my colleagues of it
8. When I have a new idea, I try to get support for it from management
9. I try to show my colleagues the positive sides of new ideas
10. When I have a new idea, I try to involve people who are able to collaborate on it

Implementation starting activities
11. I develop suitable plans and schedules for the implementation of new ideas
12. I look for and secure funds needed for the implementation of new ideas
13. For the implementation of new ideas, I search for new technologies, processes, or procedures

Involving others
14. When problems occur during implementation, I get them into the hands of those who can solve them

Innovative Employee Inventory	Rate Each Item from 1 (Strongly Disagree) to 5 (Strongly Agree)
15. I try to involve key decision makers in the implementation of an idea	
16. When I have a new idea, I look for people who are able to push it through	
Overcoming obstacles	
17. I am able to persistently overcome obstacles when implementing an idea	
18. I do not give up even when others say it cannot be done	
19. I usually do not finish until I accomplish the goal	
20. During idea implementation, I am able to persist even when work is not going well at the moment	
Innovation outputs	
21. I was often successful at work in implementing my ideas and putting them in practice	
22. Many things I came up with are used in our organization	
23. Whenever I worked somewhere, I improved something there	
Sum your ratings, divide by 115, and multiply by 100 to derive your overall score as a percentage of the total possible points	

Exercise 6.2   Support for innovation

Organizational structure needs to be designed to fulfill the organization's innovation and overall corporate strategy, especially what types of innovation it means to facilitate.

Exercise 6.1 considered the effect of individuals, and their ability to behave in innovative ways, on NPD success. But individuals operate in the context of the organization that employs them, or that they own or govern. Thus, the resources and culture of that organization constrain individual behaviors and may incentivise innovative behaviours or discourage them.

Alongside their Innovative Employee Inventory, Lukes and Stephan (2017) also developed a multi-item scale to measure how supported people feel to undertake innovative behaviours – the Innovation Support Inventory. The inventory measures support by management, the organization, and more broadly, national culture.

Complete the inventory by rating each item from 1 (strongly disagree) to 5 (strongly agree). Sum your ratings, divide by 60, and multiply by 100 to derive your overall score as a percentage of the total possible points.

The inventory should not be interpreted as evidence that an organization is not innovative. However, it will stimulate reflection about your organization's and

(Continued)

national culture's influence on how supported you feel in innovating. After completing the inventory, think about these questions:

- Are most of your items above or below the mid-point?
- Is there a pattern (e.g., your manager supports innovation, but the overall organization does not?)
- Are you surprised by the level of support – either positively or negatively? Do the numbers reinforce what you already knew or offer new insight?
- Can you think of examples from your organization illustrating each scale component?

Innovation Support Inventory	Rate Each Item from 1 (Strongly Disagree) to 5 (Strongly Agree)

Managerial support

1. My manager [board or owner] motivates me to come to them with new ideas
2. My manager [board or owner] always financially rewards good ideas
3. My manager [board or owner] supports me in implementing good ideas as soon as possible
4. My manager [board or owner] is tolerant of mistakes and errors during the implementation of something new
5. My manager [board or owner] is usually able to obtain support for my proposals outside our department

Organizational support

6. The method of remuneration in our organization motivates employees to suggest new things and procedures
7. Our organization has set aside sufficient resources to support the implementation of new ideas
8. Our organization provides employees time for putting ideas and innovations into practice

Cultural support

9. Most people in [my country] come up with new, original ideas at work
10. Most people in [my country] can implement new ideas at work
11. Most people in [my country] look for new challenges at work
12. Most people in [my country] are able to improvise easily when unexpected changes happen at work

Sum your ratings, divide by 60, and multiply by 100 to derive your overall score as a percentage of the total possible points

References

Aitken, Z. (2020). Why top-down and bottom-up innovation are critical to thriving post COVID – inside HR. https://www.insidehr.com.au/thriving-post-covid.

Cheng, J.Y.-J. and Groysberg, B. (2018). Innovation should be a top priority for boards. So why isn't it? *Harvard Business Review*, September. https://hbr.org/2018/09/innovation-should-be-a-top-priority-for-boards-so-why-isnt-it.

Cooper, R.G. (2019). The drivers of success in new-product development. *Industrial Marketing Management* 76: 36–47. doi: 10.1016/j.indmarman.2018.07.005.

Deloitte (2019). The evolution of business innovation. https://www2.deloitte.com/content/dam/Deloitte/us/Documents/human-capital/us-human-capital-organizing-for-innovation-081920.pdf.

DePillis, L. (2015). Why companies are rewarding shareholders instead of investing in the real economy. *Washington Post*, February 25. https://www.washingtonpost.com/news/wonk/wp/2015/02/25/why-companies-are-rewarding-shareholders-instead-of-investing-in-the-real-economy.

Deschamps, J.P. (2013). What is innovation governance? – definition and scope. Innovation Management.https://innovationmanagement.se/2013/05/03/what-is-innovation-governance-definition-and-scope.

Deschamps, J.P. (2015). 10 best board practices on innovation governance – how proactive is your board? Innovation Management.https://innovationmanagement.se/2015/10/19/10-best-board-practices-on-innovation-governance-how-proactive-is-your-board.

Deschamps, J.P. and Nelson, B. (2012). *Innovation Governance: How Top Management Organizes and Mobilizes for Innovation.* Jossey-Bass.

Galbraith, J. and Kates, A. (2007). *Designing Your Organization: Using the STAR Model to Solve 5 Critical Design Challenges.* Jossey-Bass.

Godin, S. (2006). *Free Prize Inside: The Next Big Marketing Idea.* Penguin Books.

Haines, S. (2012). Strategies to improve NPD governance. In: *The PDMA Handbook of New Product Development* (ed. K.B. Kahn, S.E. Kay, R.J. Slotegraaf, and S. Uban). Wiley and Sons. doi: 10.1002/9781118466421.ch19.

Hill, A. and Davis, G. (2017). The board's new innovation imperative. *Harvard Business Review*, November. https://hbr.org/2017/11/the-boards-new-innovation-imperative.

Innovation Enterprise (2013). Rise of the chief innovation officer: an executive whose time has come. https://innovationdevelopment.org/sites/default/files/Rise%20innovation%20officer%20WP%20IE.pdf.

Kay, R. and Goldspink, C. (2016). *The role of the board in innovation.* Australian Institute of Directors. https://www.aicd.com.au/good-governance/organisational-strategy/long-term-strategic-plan/the-role-of-the-board-in-innovation.html.

Keeley, L. et al. (2013). *Ten Types of Innovation: The Discipline of Building Breakthroughs.* John Wiley & Sons.

Lovallo, D., Koller, T., Uhlaner, R. and Kahneman, D. (2020). Your company is too risk-averse. Harvard Business Review, March. https://hbr.org/2020/03/your-company-is-too-risk-averse.

Lovric, D. and Schneider, G. (2019). What kind of chief innovation officer does your company need? *Harvard Business Review*, November. https://hbr.org/2019/11/what-kind-of-chief-innovation-officer-does-your-company-need.

Lukes, M. and Stephan, U. (2017). Measuring employee innovation: a review of existing scales and the development of the innovative behavior and innovation support inventories across cultures. *International Journal of Entrepreneurial Behavior & Research* 23 (1): 136–158. doi: 10.1108/IJEBR-11-2015-0262.

McClimon, T.J. (2022). CEOs' most important business challenges in 2022. *Forbes.* https://www.forbes.com/sites/timothyjmcclimon/2022/01/27/ceos-most-important-business-challenges-in-2022.

McNaughton, R.B. and Sembhi, R.S. (2021). Developing an entrepreneurial orientation: capabilities and impediments. *Advances in Entrepreneurship, Firm Emergence and Growth* 22: 121–143.

Sierra-Morán, J., Cabeza-Garcíal, L., González-Álvarez, N. and Botella, J. (2021 September). The board of directors and firm innovation: a meta-analytical review. *BRQ Business Research Quarterly*. doi: 10.1177/23409444211039856.

Swoboda, C. (2020). Hiring a chief innovation officer is a bad idea. *Forbes*. https://www. forbes.com/sites/chuckswoboda/2020/02/24/hiring-a-chief-innovation-officer-is-a-bad-idea.

Zhu, P. (2020). Five aspects of innovation governance. Innovation Management. https:// innovationmanagement.se/2020/06/05/five-aspects-of-innovation-governance.

Rod McNaughton is a professor of entrepreneurship and director of innovation and professional development, a portfolio encompassing executive education and the university's entrepreneurship center within the University of Auckland Business School. As a consultant, board member, and speaker, he draws on extensive experience developing entrepreneurial ecosystems and helping start-ups launch and grow. Before joining the University of Auckland, Rod was the Eyton Chair in Entrepreneurship and Director of the Conrad School of Entrepreneurship and Business at the University of Waterloo, widely recognized as Canada's most innovative university. He received his PhD in Marketing from Lancaster University and his PhD in Economic Geography from Western University.

SECTION TWO

NEW PRODUCT DEVELOPMENT PROCESS

CHAPTER SEVEN

TOWARD EFFECTIVE PORTFOLIO MANAGEMENT

Hans van der Bij and Eelko K.R.E. Huizingh

7.1 Introduction

Already in 1997 Brown and Eisenhardt (1997) argued that due to severe competition firms are forced to continuously change. Developing new products and services must be part of the core of a firm's culture. This also requires that firms continuously adapt their strategy and, even more importantly, their strategy implementation, which is much more difficult than simply changing the formulation of a strategy (Hrebiniak 2006; Meskendahl 2010). A crucial building block to connect strategy formulation and strategy implementation is new product portfolio management (Meskendahl 2010). The firm's new product portfolio has to follow changes in strategy formulation as it represents the way a firm implements a (new) strategy.

The innovation literature has dealt extensively with the design and organization of individual New Product Development (NPD) projects (Eggers 2012). In particular, the classical work on the stage gate model of Cooper and colleagues has been very influential (e.g., Cooper 1990, 2008; Cooper et al. 2002a). However, to create a continuous flow of new products and services that match revised strategic priorities, adequate management of individual NPD projects is necessary, but not sufficient. At least as crucial is how a firm deals with the entire set of current and future NPD projects, and this is the decision-making area covered by New Product Portfolio Management (NPPM). Ideas and ongoing projects for product renewals, product line extensions, expanding current product lines to new markets, and really new products, all have to be evaluated for

The PDMA Handbook of Innovation and New Product Development, Fourth Edition. Edited by Ludwig Bstieler and Charles H. Noble.

funding at multiple instances in time (Kester et al. 2011). If not managed proficiently and in accordance with the firm's strategy, the negative impact of poor portfolio decisions on performance can be significant (Cooper et al. 2001a).

NPPM is a complex issue as it implies dealing with limited, uncertain, and changing information, dynamic opportunities, interdependences among projects, multiple goals and strategic considerations, and various decision-makers with diverging perspectives and reputations (Behrens 2016; Cooper et al. 1999; McNally et al. 2013). These factors not only add to the complexity of decision making, they also work in different directions. Environmental complexity tends to favor more radical NPD projects, while environmental instability shifts the balance toward incremental projects (Chao and Kavadias 2008). NPPM lacks a "magic" solution as there is no single right answer to the question how to arrive at an optimal set of NPD projects, any solution is temporary due to the dynamics in the internal and external environment (Paulson et al. 2007).

The current literature highlights two major NPPM problems many firms struggle with (Barczak et al. 2009; Cooper 2013; Kester et al. 2011; Repenning 2001). First, firms must shift their NPD portfolio from having predominantly incremental innovations to more radical innovations, and second, NPD portfolios tend to be overloaded, leading to a focus on urgent, short-term problems, and thereby losing overview and long-term direction. To address these problems, academic research has concentrated on methods to select and terminate NPD projects; however, research about the daily practices in dealing with the overall NPPM process is still limited (Kester et al. 2011).

In this chapter, we briefly discuss the contents and objectives of NPPM based on the existing literature. NPPM is presented as managing two related portfolios, the innovation portfolio and the NPD portfolio, while also its relation is clarified with a third portfolio, the product portfolio. Next, we present the results of our multiyear and ongoing NPPM research project based on 28 case studies of how NPPM is organized in practice and which criteria dominate decision-making.[1] The chapter ends with five guidelines for more effective NPPM and a conclusion.

7.2 NPPM State of the Art

This section starts with a clarification of NPPM by discussing its definition and objectives (Section 7.2.1). Next, we explain the confusion around NPPM by putting NPPM in the perspective of the various related portfolios that firms need to manage. These entail ideas for new products, new product development projects, and the set of current offerings (Section 7.2.2).

7.2.1 What and Why of NPPM

The purpose of NPPM is to ensure that a firm gets well-prepared for the future, which requires developing the *right* innovation projects (Chao and Kavadias 2008). Given the inherent uncertainty involving innovation, only in hindsight it becomes clear which projects would have been the right projects, NPPM is a continuous process to manage risks, and to explore and exploit opportunities. As NPPM requires constant balancing, the core of NPPM is an ongoing decision process (Cooper et al. 1999; Kester et al. 2011). Following Cooper et al. (1999, p. 335), we define NPPM as "a dynamic decision process, whereby a business's list of active new product (and R&D) projects is constantly updated and revised." Important activities include evaluating, selecting, and (re)prioritizing projects, as well as (re)allocating resources among them. Management needs to decide which projects to fund, to what levels, and at what point in time (Kester et al. 2011). Active NPD projects may be terminated, accelerated, continued, or de-prioritized. When making such choices, managers must weigh projects against the entire new product portfolio, and the available resources for innovation (Behrens 2016). As a consequence, NPPM is much more than just deciding about project selection and termination. Kester et al. (2011) therefore refer to NPPM as an integrated system of decision-making processes that asks for a portfolio mindset, focused firm efforts on the right NPD projects, and agile decision making across the portfolio's set of projects.

The "why" of NPPM concerns its objectives. Firms may prioritize different goals, but in general NPPM has four main objectives (Cooper et al. 1997a, 1997b, 2002b), see Figure 7.1.

Maximizing the value of the portfolio is often operationalized in financial terms, such as profitability, return-on-investment, or net present value (NPV). Financial success can be measured as average project success, but also needs to take into account possible synergy effects as cross-project coordination can generate value greater than the sum of the individual projects (Jonas et al. 2012). Especially for radically new products, predicting the financial value is "at best fiendishly difficult and at worst impossible" (Paulson et al. 2007: p. 18). Cooper et al. (1997a) already warn against a too strong focus on financial success measures, as this tends to lead to invest mainly in short-term, low-risk projects and/ or on a single market. It may also turn NPPM into an inward-looking issue by neglecting external stakeholders, such as customers, suppliers, subcontractors,

FIGURE 7.1 THE FOUR OBJECTIVES OF NEW PRODUCT PORTFOLIO MANAGEMENT.

and partnerships (Martinsuo and Geraldi 2020; Wheelwright and Clark 1992). External stakeholders may benefit from the portfolio, but the success of a portfolio may also depend on their contributions.

A *balanced* portfolio ensures that the firm does not put all eggs in the same basket. Balance can be achieved across many dimensions (Cooper et al. 1997a), including product categories, technologies, markets, project types (e.g., exploitation versus exploration, short-term versus long-term, high-risk versus low-risk), and amount of allocated resources. Finding the right balance is a problem for many firms, as many firms overemphasize incremental innovation efforts (Adams and Boike 2004; Cooper 2013; Paulson et al. 2007), which is detrimental to long-term firm performance.

Strategic alignment acknowledges that firms realize their strategic ambitions by investing in NPD projects which make these ambitions come to life. Strategic fit ensures that the resource allocation across projects, brands, geographies, markets, etc. is in line with the previously delineated areas of strategic focus (Cooper et al. 1997a). This prevents situations in which projects get chosen mainly because engineers found the technical problems challenging or because customers requested them (Wheelwright and Clark 1992). On the other hand, a too strict alignment can be counterproductive as it may constrain innovative choices (McNally et al. 2013), new growth avenues may well lie outside the current strategic focus areas.

Including the *right number of projects* in a portfolio was only later added as NPPM objective by Cooper and colleagues (2002b). This was done in response to the observation that most firms have too many projects in their portfolio for the limited resources available. Firms find themselves in a "pipeline gridlock," with many projects in a queue waiting for sufficient resources to take the next step. The consequence is extended time to market, or skipping or rushing important development activities (e.g., testing). The goal is to ensure a balance between resources required for the projects in the portfolio and resources available.

Obviously, all four objectives often conflict with each other, making NPPM a highly complex issue. The problem becomes even larger when taking into account that firms may undertake many different types of R&D projects (e.g., really new products, product improvements, cost reductions, fundamental research) that all compete for the same resources, and therefore need to be part of the same portfolio decision-making process.

7.2.2 NPPM in Perspective

Part of the confusion around NPPM is caused by the fact that firms need to manage multiple portfolios. These are the innovation portfolio, the new product development portfolio, and the product portfolio; see Figure 7.2. The main processes regarding in these three portfolios are idea generation (for the innovation portfolio), development (for the NPD portfolio), and sales and servicing (for the product portfolio).

The life cycle of new products starts in the innovation portfolio as early-stage concepts (Mathews 2010). The *Innovation portfolio* consists of all ideas for new offerings a firm has at a given point in time. O'Connor and Ayers (2005) refer to

FIGURE 7.2 NEW PRODUCT PORTFOLIO MANAGEMENT AS MANAGING TWO RELATED PORTFOLIOS: THE INNOVATION PORTFOLIO AND THE NPD PORTFOLIO, AND ITS RELATIONSHIP WITH THE PRODUCT PORTFOLIO.

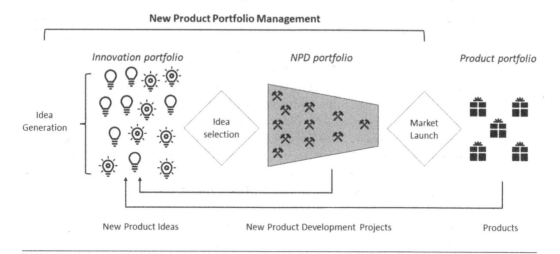

this as "the bench," the inventory of ideas and potential projects. The innovation portfolio receives new ideas and suggestions as a result of daily operational practices (e.g., customer suggestions, questions or complaints), new technologies, or dedicate activities such as hackathons (Lifshitz-Assaf et al. 2021), crowd sourcing (Allen et al. 2018), and innovation contests (Hu et al. 2020).

Some of the ideas in the innovation portfolio are regarded promising enough to warrant development, they then enter the *new product development portfolio.* This portfolio consists of all NPD projects a firm has allocated resources for at a given point in time. To monitor project progress, many firms use a method with stages and gates (Cooper and Sommer 2020), to ensure that both project progress and resource allocation get reviewed from time to time. Gate criteria include strategic fit, technical feasibility, customer acceptance, market potential, financial performance, and staying within the development budget (Carbonell-Foulquié et al. 2004; Hart et al. 2003). Not all NPD projects result in new offerings. Terminated projects drop out of the NPD portfolio, but the idea and knowledge generated during development remain available for future use, as indicated by the feedback arrow from the NPD portfolio to the innovation portfolio.

Successful NPD projects result in a market launch, and then the new offering becomes part of the *product portfolio.* This portfolio consists of all products a firm offers to customers at a given point in time. An often used model to evaluate the product portfolio is the growth-share matrix (e.g., Boyd and Headen 1978; Day 1977) developed by Boston Consulting Group (BCG) that categorizes products as stars, cash cows, question marks, and dogs. Tactical and operational product portfolio management is mainly performed by Marketing and Sales, to ensure both short- and midterm profitability of the firm's offerings. Applying marketing instruments (the classical 4 p's of price, promotion, place, and product) ensures short-term profitability, while midterm profitability often requires product adjustments or new product variants, which is represented in Figure 7.2 by the feedback arrow from the product portfolio to the innovation portfolio.

As NPPM concerns with realizing new offerings, it covers both the innovation portfolio and the NPD portfolio. Since it does not deal with existing products, the product portfolio is outside its scope. Nevertheless, ultimate business success is determined by having the three processes working in sync. Figure 7.2 shows the relations between the various portfolios: attractive ideas enter the NPD portfolio, successfully completed NPD projects enter the product portfolio, while (temporarily) halted NPD projects and suggestions from Marketing & Sales add new ideas to the innovation portfolio (both feedback loops).

One of the reasons why NPPM has the potential for conflict is in the interrelationships with the various portfolios. For example, firms make gate decisions regarding individual NPD projects and portfolio decisions. Both decisions involve partly the same criteria, and both are aimed at selecting projects and allocating resources (Cooper et al. 1997b). However, due to short-term pressures and lack of time (Wheelwright and Clark 1992), projects are often defined and managed to meet functional goals without evaluating their strategic alignment and organizational impact (Lockett et al. 2008). This is a missed opportunity as each step in evaluating an idea and developing an NPD project can be regarded as an information-gathering step (Mathews 2010, 2011) that offers insights valuable for NPPM decision-making too.

7.3 NPPM Challenges in Practice

In order to better understand how NPPM is organized and which decision criteria dominate in practice, we studied how firms carry out their NPPM decision-making. In a multiyear research project, in total 28 companies have been studied. Companies from different industries participated; our requirement was that they had a stream of innovation ideas and various projects in their portfolio. For details of this qualitative empirical research, see Table 7.1.

One general observation during the interviews was that many interviewees had difficulties in talking about NPPM; they rather discussed NPD as that was, for them, a more tangible process. Frequently the interviewer had to interrupt and refocus the interview and interviewee on NPPM.

Below, we first consider the innovation portfolio management and then NPD portfolio management.

7.3.1 Innovation Portfolio Management

Innovation Portfolio Management deals with the creation and development of innovative ideas in the firm, and the selection of these ideas for the NPD Portfolio. The interviews confirm that ideas come from many different sources, including customers, R&D employees, portfolio managers, competitors, suppliers, and (top) management after a careful comparison of the companies' product portfolio and the companies' strategy.

TABLE 7.1 DESIGN OF THE QUALITATIVE EMPIRICAL RESEARCH PROJECT.

Research method	Semi-structured interviews in 28 companies making use of an interview protocol (see below). Each interview was conducted by 2 interviewers.
	Interviewees were project leaders with insights in NPPM and decision-makers in NPPM. Interviews lasted 60–90 minutes and each interview was recorded, transcribed, and summarized. The summaries were coded with first-order and more theory-based second-order codes.
Interview protocol	Main topics:
	– General information of the company and the interviewee
	– Importance of innovation for the company
	– Description of current new product project portfolio
	– Example of a decision process concerning an idea that got funded
	– Information about portfolio meetings
	– Monitoring of projects in the NPD project portfolio
	– Evaluation of the NPPM process
	– Performance of and changes in the innovation and new product project portfolio
Participating companies	Companies from the following industries participated: food industry (6), manufacturing (4), construction industry (3), public sector (3), energy (2), pharmaceutical industry (2), publishing (2), other (6).

For the majority of the companies Innovation Portfolio Management is the most elaborated part of NPPM. Some companies distinguish multiple steps in the Innovation Portfolio Management, starting with a first evaluation of the idea generated, then evaluation of the business case, consideration of resource consequences of the idea, and ultimately the decision to fund it or not. Almost all firms without a clear procedure consider the business case and resource consequences, while most companies first allocate a small amount of resources (money or time) to develop an idea into a business case.

Many criteria are used to evaluate ideas in the Innovation Portfolio. The most frequent criteria include financial factors (revenue, profit, turnover) (22 times), followed by strategic fit (17 times), technical or operational feasibility and scalability (16 times), customer or market potential (16 times), resource fit (11 times), and balance (5 times). A product manager of a construction firm explained the use of strategic fit:

> Well, we have in our mission, vision, and strategy the starting point of innovative ideas for a changing world. That is also our slogan. That's the way it is put at the highest level of the organization.

A portfolio manager of a manufacturing firm clarified the use of technical feasibility:

> ...technical feasibility is also very important because we don't want to lose a lot of resources without creating a product.

Remarkably, as not so often mentioned in the literature, six companies use sustainability or societal relevance as a criterion. A director of a printing company explained it as follows:

> <our company> is very concerned with sustainable products and always checks whether the projects in the portfolio are future proof. Future proof is when resources are handled very consciously and the environment is burdened as little as possible.

Finally, companies also use intuition or "gut feeling" as a kind of criterion, a product manager of a printing company formulated it as follows:

> Gut feeling also plays a role because there are also cases where you don't have any data. The best example is when you have a project, and you can't calculate a business case. And that's when gut feeling is really, really important. My former division manager used to say 'I pay you to think with your gut'. His view was that if you have good managers and they can use their gut feeling well, they are worth their weight in gold.

Most companies indicated to have dominant criteria when making the final funding decision, most often this is a single criterion, sometimes two criteria. Only 2 out of the 28 firms have no dominant criterion in this stage; these firms consider all criteria when selecting ideas. A financial factor is (one of) the dominant criteria in 17 firms, in 8 it is strategic fit, in 5 technical/operational feasibility and scalability, and in two firms customer or market potential. Resource fit and balance are in none of the firms among the dominant criteria.

In Innovation Portfolio Management many functions are involved and also several hierarchical levels. So, it is multilevel and cross-functional decision-making. In 21 firms innovation or R&D staff is involved in Innovation Portfolio Management, in 13 companies the CEO, director, BU leader, or management team, in 12 companies a marketing, sales or product manager, in 7 companies the project owner, in 6 companies operations, and in 5 companies a finance manager. An R&D director of a food company appreciated this cross-functional decision making:

> I have the idea that from all angles people who can say something about a project and who need to have an opinion about it, are involved in the eventual realization of our portfolio.

Moreover, in 20 of the 28 companies two hierarchical decision levels are involved (14 included an employee and middle management, 6 middle management and top management). In 5 companies top management, middle management, and employees are all involved, and in 3 companies only the CEO, or a middle manager. A product manager of a construction firm explained the multilevel decision making as follows:

> We initially set up a policy team to discuss all these phases. It consists of the commercial director, head of product development, the product manager, project leader, and me. That is actually the preliminary discussion of the business case. If

the policy team … decides that it looks good, we bring it to the Management Team. That's where production, logistics and finance come in and it gets a bit bigger. The policy team can also decide not to bring it to the MT at all, so that is a kind of preparation phase

In conclusion, Innovation Portfolio Management is the most mature part of NPPM. Financial factors are mostly dominant in making the final funding decision. In most firms two hierarchical levels are involved in the process (often middle management and employees), and various functions, such as innovation/ R&D, general management, and marketing/sales.

7.3.2 NPD Portfolio Management

During the empirical research we learned that NPD Portfolio Management consists of and is executed on two different levels: (1) monitoring of each separate project in the portfolio, and (2) monitoring of the entire portfolio. Most companies (22 out of 28) are involved in the individual project level. Surprisingly, the majority of firms (17 out of 28) does not monitor the entire portfolio. The reason that 6 out of 28 companies did not monitor at the individual project level has mostly to do with contracts with external partners that would make it expensive to terminate projects.

On the level of individual projects, most firms (14 out of 22) monitor project progress, while using similar criteria as in Innovation Portfolio Management, but they noted that during the course of the project the information about the criteria becomes more reliable. Firms also consider possible environmental changes that could affect the criteria. Monitoring is mainly carried out in cross-functional settings. Innovation/R&D is involved in 18 companies, the CEO (director, BU leader, or management team) in 15 companies, and a marketing, sales, or product manager in 11 companies. In 11 companies again two hierarchical levels were included in the decision making, in 8 companies only one level.

For the firms that monitor the entire portfolio, the dominant criterion is balance. For instance, a food company innovation director explained the balance in risk:

What percentage is the win <extensions of already existing products in the firm> and what is the percentage of the moves <innovations initiated by competitors>. This actually indicates the risk of our portfolio.

Some companies consider the balance in the NPD funnel, as they strive for an equal distribution of projects among the various stages. Other firms are concerned about the balance between incremental and radical innovations, or between short-term and long-term projects. The latter balance appeared to be not only important for risk assessment, but also to motivate employees. Short-term projects are considered as more motivating for involved employees since they deliver results quickly. Strategic fit and resource fit (having not too many projects in the portfolio) were also mentioned as relevant criteria, as well as synergy among projects. Synergy concerns projects that can build upon each other or benefit from each lessons learned. Too much overlap between projects in the same area was mentioned in a more negative way, as it hurts the required breadth of the portfolio. One company uses a matrix of risk (incremental versus radical projects) and strategic fit, so they balance risk and strategic fit. The frequency of monitoring the entire portfolio is often between two and four times a year.

7.3.3 Evaluation of NPPM

In general, interviewees found it very difficult to evaluate NPPM. Many of the interviewees were satisfied with their NPPM, but when we asked "why" they referred to the NPD process and responded to be proud of that process. When probing further, many interviewees said that their NPPM was still partly unstructured. Almost all preferred having a more structured process in which they could be more critical with respect to the number of projects that got started, and that would allow a better overview of the progress of the projects and, thus, make it easier to stop projects that do not deliver. An innovation director of a food company and a head of the innovation lab in a public sector company formulated these respective issues as follows:

> To not start projects, we need to be more critical. We try a project pretty easy. So, I think we need to be more focused on that.

And

> We have put several projects on hold, but this cannot always remain the case. These projects should be removed completely over time if we don't continue with them.

When asked to evaluate their current NPD portfolio, almost all interviewees responded that their portfolio was as good as the individual projects in the portfolio. In other words, they based their evaluation of the portfolio on an evaluation of the projects. This is understandable as evaluating projects is easier and more tangible than an evaluation of the entire portfolio. Moreover, we already mentioned in Section 7.2 that many criteria mentioned in literature and practice can be used on the project level as well as the portfolio level. We found that criteria such as balance, strategic fit, and resource fit can and are used for project-level monitoring: Does a project disturb or strengthen a certain balance, does it fit the existing strategy, are resources still available for the project? Such criteria can also be used at the portfolio level, then managers consider the balance in the entire portfolio and examine which type of future projects must be stimulated and which ones must be phased out.

Also, managers take into account current or expected changes in the strategy, and review which type of projects to stimulate or phase out. With respect to resource fit, managers examine whether they have the right number of projects in their portfolio. The latter way of monitoring at the portfolio level is closely connected to the four NPPM objectives, mentioned in Section 7.1, maximizing value, balance, strategic alignment, and the right number of projects, which are often conflicting and whose relative priority may change over time, as formulated by a head of the innovation lab of a public sector company:

> Having a balance between long and short term projects has become more important. But also having a balance between easy and difficult projects.

7.4 Guidelines for Effective Portfolio Management

Based on both our literature review and field research, we formulate six guidelines helpful for managers who intend to improve the NPPM in their firm.

First, while the distinction between new product portfolio management (NPPM) and new product development (NPD) is clear cut in the academic literature, this distinction is more vague in practice. This was manifested in our field research when we had to interrupt managers and refocus the interview on NPPM. The academic literature defines NPD as the process of conceptualizing, developing, and launching a new product, while NPPM refers to the process of managing a coherent set of such projects. While NPD concerns projects with limited duration and can benefit from general project management lessons, NPPM is an ongoing process without start and finish. In practice, many firms do not clearly differentiate between NPPM and NPD, and sometimes manage their new product portfolio more or less as a number of separate projects. Although managing NPD seems to be more obvious as a response to daily challenges, it is important to evaluate the entire portfolio of NPD projects regularly to ensure the right balance. A simple yet powerful way for doing so is to create bubble diagrams in which each project is displayed as a circle, the X- and Y-axes show how a project scores on important factors, while the size of the circle reflects either expected profits or costs (see for example Cooper et al. 2001b).

Second, the observed confusion between NPPM and NPD decision-making stems partly from the fact that multiple NPPM criteria are also relevant for NPD decision-making. Examples include project progress, strategic fit, financial attractiveness, technical feasibility, and market acceptance. On the other hand, criteria such as balance, resource fit, and distribution of projects over the various NPD phases are unique to NPPM. In short, managers need to treat NPPM decision-making as a different though overlapping challenge compared to NPD; see Table 7.2.

Third, one of the aspects making NPPM decision-making complex is that most of the NPPM objectives and criteria are qualitative by nature. For example, this is true for all unique NPPM criteria mentioned above (balance, resource fit, and distribution of projects over the NPD phases). Not only are these criteria qualitative, their "optimal" level may also change over time. A good distribution of projects over the various NPD phases needs to ensure that a firm can launch new products regularly. However, if a firm has recently launched multiple new products successfully, not having many projects in the final stages of the NPD funnel may matter less. But if it has not introduced new products recently or if such launches failed, the situation would be different.

TABLE 7.2 IMPORTANT DECISION-MAKING CRITERIA FOR NPD AND NPPM.

Important NPD Decision-making Criteria	Important NPPM Decision-making Criteria
Financial attractiveness of NPD project	Financial attractiveness of the entire portfolio
Technical feasibility of NPD project	Technical feasibility of all projects in the portfolio
Market acceptance of NPD project	Market acceptance of all projects in the portfolio
Strategic fit of the NPD project	Alignment of the portfolio with strategy
Resources available for NPD project	Resource fit (right number of projects in the portfolio)
Progress of project in NPD (stage-gate) process	Progress of all projects in the portfolio
	Balance in the portfolio, e.g., short term versus long term, various product-market combinations, technologies, incremental versus radical
	Distribution of projects in the portfolio over NPD phases

Fourth, the process of NPPM decision-making in firms tends to be less structured than NPD decision-making is. An NPD project gets managed by a project team that meets regularly and considers more or less standardized criteria, and many firms use an NPD process with milestones that serve as natural decision moments, even when not using the classical stage-gate process. NPPM decision-making, on the other hand, is often less well structured, not always clearly separated from NPD decision-making, and sometimes conducted by the same managers as involved in NPD, which enlarges the confusion between both and may create loyalty conflicts. We recommend managers to clearly differentiate both decision-making processes. When making NPPM decisions, the focus needs to be on the entire set of ongoing NPD projects, not so much on individual projects.

Fifth, our research shows that managers recognize the value of effective NPPM. They realize that in order to prevent negative consequences, such as ending up with an unbalanced collection of too many and mostly incremental projects, active NPPM is required. We have observed that in firms currently not performing well-structured NPPM, managers expressed a need for a more structured process. Scheduling dedicated NPPM meetings regularly, inviting a balanced group of cross-functional, high-level managers, deciding about the most important NPPM criteria and providing the corresponding information about all ongoing projects in a systematic manner contribute to a structured NPPM process. As only a small minority of the interviewed firms had a dedicated NPPM manager or expressed the need for such a person, we recommend to focus first on creating a well-designed process and ensuring involvement of the right set of managers.

Sixth, firms find it much harder to assess NPPM performance than NPD performance. Well-known NPD success measures include NPD cycle time (versus planning), market acceptance, and financials (e.g., profit or revenue percentage from products launched in the past three years). Due to a lack of separate NPPM success measures, firms tend to either not measure NPPM performance or apply the same metrics as they do for NPD performance measurement. To ensure that

NPPM meetings do not deviate from their purpose, we recommend managers to define dedicated NPPM success criteria, and to have reports prepared before NPPM meetings that show how their current portfolio fares with regard to these criteria.

7.5 Conclusion

Despite the acknowledged importance of NPPM (e.g. Cooper et al. 2001a), the NPPM literature is much more limited compared to the attention that has been paid to the organization and setup of the NPD process. Exceptions are the classical papers of Cooper and colleagues around the turn of the century (Cooper et al. 1997a, 1997b, 1999, 2002b, 2004) and, for instance, the work of Kester and colleagues a decade later (Kester et al. 2011, 2014). This chapter summarizes the most important literature and presents NPPM as an ongoing decision process aimed at four objectives (maximizing value, balance, strategic alignment, and right number of projects), and that entails managing two related portfolios, the innovation portfolio and the NPD portfolio.

In our empirical research we find that NPPM in practice is less developed than NPD, and that both are often confused with each other. Multiple times we had to remind managers during an in-depth interview that our study focused on NPPM and not on NPD. Nevertheless, the interviewed managers recognize the value of NPPM. Our case studies show that the first part of NPPM, innovation portfolio management, is the most mature part. In the second part, NPD portfolio management, most firms emphasize monitoring of individual NPD projects. Such firms can benefit from a more mature monitoring of their entire portfolio. This is an important activity as successful individual NPD projects bring firms revenues and profits, but it is the joint set of such projects that determines whether a firm is well-prepared for the uncertain future.

Note

1 We thank Sigrid Breel, Nadi Dijksterhuis, Hélène Furon, Franziska Hansen, Lotte Hartstra, Anna van der Heide, Bart Themmen, Justin van der Veen, Marloes Wesseling, and Teun van der Zee for their crucial support in conducting the case study interviews.

References

Adams, M. and Boike, D. (2004). The PDMA foundation 2004 comparative performance assessment study. *Visions* 28 (3): 26–29.

Allen, B.J., Chandrasekaran, D., and Basuroy, S. (2018). Design crowdsourcing: the impact on new product performance of sourcing design solutions from the "crowd". *Journal of Marketing* 82 (2): 106–123.

Barczak, G., Griffin, A., and Kahn, K.B. (2009). Perspective: trends and drivers of success in NPD practices: results of the 2003 PDMA best practices study. *Journal of Product Innovation Management* 26 (1): 3–23.

Behrens, J. (2016). A lack of insight: an experimental analysis of R&D managers' decision making in innovation portfolio management. *Creativity and Innovation Management* 25 (2): 239–250.

Boyd, H.W. and Headen, R.S. (1978). Definition and management of the product-market portfolio. *Industrial Marketing Management* 7 (5): 337–346.

Brown, S.L. and Eisenhardt, K.M. (1997). The art of continuous change: linking complexity theory and time-paced evolution in relentlessly shifting organizations. *Administrative Science Quarterly* 42 (1): 1–34.

Carbonell-Foulquié, P., Munuera-Alemán, J.L., and Rodrıguez-Escudero, A.I. (2004). Criteria employed for go/no-go decisions when developing successful highly innovative products. *Industrial Marketing Management* 33 (4): 307–316.

Chao, R.O. and Kavadias, S. (2008). A theoretical framework for managing the new product development portfolio: when and how to use strategic buckets. *Management Science* 54 (5): 907–921.

Cooper, R.G. (1990). Stage-gate systems: a new tool for managing new products. *Business Horizons* 33 (3): 44–54.

Cooper, R.G. (2008). Perspective: the stage-gate idea-to-launch process – update, what's next, and NextGen systems. *Journal of Product Innovation Management* 25 (3): 213–232.

Cooper, R.G. (2013). Where are all the breakthrough new products?: using portfolio management to boost innovation. *Research Technology Management* 56 (5): 25–33.

Cooper, R.G., Edgett, S.J., and Kleinschmidt, E.J. (1997a). Portfolio management in new product development: lessons from the leaders—I. *Research Technology Management* 40 (5): 16–28.

Cooper, R.G., Edgett, S.J., and Kleinschmidt, E.J. (1997b). Portfolio management in new product development: lessons from the leaders—II. *Research Technology Management* 40 (6): 43–52.

Cooper, R.G., Edgett, S.J., and Kleinschmidt, E.J. (1999). New product portfolio management: practices and performance. *Journal of Product Innovation Management* 16 (4): 333–351.

Cooper, R.G., Edgett, S.J., and Kleinschmidt, E.J. (2001a). *Portfolio Management for New Products*. Cambridge, MA: Perseus Publishing.

Cooper, R.G., Edgett, S.J., and Kleinschmidt, E.J. (2001b). Portfolio management for new product development: results from an industry practices study. *R&D Management* 31 (4): 361–380.

Cooper, R.G., Edgett, S.J., and Kleinschmidt, E.J. (2002a). Optimizing the stage-gate process: what best-practice companies do–I. *Research Technology Management* 45 (5): 21–27.

Cooper, R.G., Edgett, S.J., and Kleinschmidt, E.J. (2002b). Portfolio management: fundamental to new product success. In: *The PDMA Toolbook for New Product Development* (ed. P. Belliveau, A. Griffin, and S. Sommermeyer), 331–364. New York: John Wiley & Sons.

Cooper, R.G., Edgett, S.J., and Kleinschmidt, E.J. (2004). Benchmarking best npd practices-II. *Research Technology Management* 47 (3): 50–59.

Cooper, R.G. and Sommer, A.F. (2020). New-product portfolio management with agile: challenges and solutions for manufacturers using agile development methods. *Research Technology Management* 63 (1): 29–38.

Day, G.S. (1977). Diagnosing the product portfolio. *Journal of Marketing* 41 (2): 29–38.

Eggers, J.P. (2012). All experience is not created equal: learning, adapting, and focusing in product portfolio management. *Strategic Management Journal* 33 (3): 315–335.

Hart, S., Hultink, E.J., Tzokas, N., and Commandeur, H.R. (2003). Industrial companies' evaluation criteria in new product development gates. *Journal of Product Innovation Management* 20 (1): 22–36.

Hrebiniak, L.G. (2006). Obstacles to effective strategy implementation. *Organization Dynamics* 35 (1): 12–31.

Hu, F., Bijmolt, T., and Huizingh, K.R.E. (2020 February-March). The impact of innovation contest briefs on the quality of solvers and solutions. *Technovation* 90–91: 1–12.

Jonas, D., Kock, A., and Gemünden, H.G. (2012). Predicting project portfolio success by measuring management quality – a longitudinal study. *IEEE Transactions on Engineering Management* 60 (2): 215–226.

Kester, L., Griffin, A., Hultink, E.J., and Lauche, K. (2011). Exploring portfolio decision-making processes. *Journal of Product Innovation Management* 28 (5): 641–661.

Kester, L., Hultink, E.J., and Griffin, A. (2014). An empirical investigation of the antecedents and outcomes of NPD portfolio success. *Journal of Product Innovation Management* 31 (6): 1199–1213.

Lifshitz-Assaf, H., Lebovitz, S., and Zalmanson, L. (2021). Minimal and adaptive coordination: how hackathons' projects accelerate innovation without killing IT. *Academy of Management Journal* 64 (3): 684–715.

Lockett, M., De Reyck, B., and Sloper, A. (2008). Managing project portfolios. *Business Strategy Review* 19 (2): 77–83.

Martinsuo, M. and Geraldi, J. (2020). Management of project portfolios: relationships of project portfolios with their contexts. *International Journal of Project Management* 38 (7): 441–453.

Mathews, S. (2010). Innovation portfolio architecture. *Research Technology Management* 53 (6): 30–40.

Mathews, S. (2011). Innovation portfolio architecture—Part 2: attribute selection and valuation. *Research Technology Management* 54 (5): 37–46.

McNally, R.C., Durmuşoğlu, S.S., and Calantone, R.J. (2013). New product portfolio management decisions: antecedents and consequences. *Journal of Product Innovation Management* 30 (2): 245–261.

Meskendahl, S. (2010). The influence of business strategy on project portfolio management and its success – a conceptual framework. *International Journal of Project Management* 28 (8): 807–817.

O'Connor, G.C. and Ayers, A.D. (2005). Building a radical innovation competency. *Research Technology Management* 48 (1): 23–31.

Paulson, A.S., O'Connor, G.C., and Robeson, D. (2007). Evaluating radical innovation portfolios. *Research Technology Management* 50 (5): 17–29.

Repenning, N.P. (2001). Understanding fire fighting in new product development. *Journal of Product Innovation Management* 18 (5): 285–300.

Wheelwright, S.C. and Clark, K.B. (1992). Creating project plans to focus product development. Harvard Business Review, March-April: 70–82.

J.D. (Hans) van der Bij is Associate Professor of Innovation Management at the Economics and Business Department of the University of Groningen. He gained his PhD from the Eindhoven University of Technology and had a visiting associate professorship at the University of Washington in Seattle. His current research interests include innovation and entrepreneurship, especially the cognitive facets of innovative and entrepreneurial behavior, new product portfolio management, cross-functional teams, and project flexibility. He is currently member of the editorial review board of the Journal of Product Innovation Management.

K.R.E. (Eelko) Huizingh is Associate Professor of Innovation Management at the University of Groningen, the Netherlands, and Vice President Scientific Affairs of ISPIM (International Society for Professional Innovation Management). He was director of the Innovation Centre of Expertise Vinci, and (co-)authored over 500 articles, which have appeared in a wide range of journals including Marketing Science, International Journal of Research in Marketing, Journal of Product Innovation Management, Organizational Behavior and Human Decision Processes, Technovation, R&D Management, Journal of Retailing, Marketing Letters, Decision Support Systems, and Information & Management. Dr. Huizingh is an experienced trainer in research design and academic writing (see www.hacademic.com).

CHAPTER EIGHT

THE POLITICS OF PROCESS: THE PORTFOLIO MANAGEMENT FRAMEWORK

Stephen K. Markham

8.1 Introduction

Perhaps the most effective political tool in modern management is the product development process. From a corporate politics point of view, management agrees to be bound to unknown future decisions made by a group of people that don't report to them and are often many levels below them. New product decisions are not trivial; they are among the most consequential decisions organizations make. They not only commit the organization to the direct cost of development, but they also constrain it with the opportunity cost of not doing something different. During development and for a significant time in the market, the organization is committed to a strategy that has profound impact on revenue, growth, talent acquisition, cost of capital, customer satisfaction, and competitive positioning.

Product development processes always define roles and responsibilities to ensure good product decisions (Ettlie and Elsenbach 2007; Farris and Cordero 2002; Roberts and Fusfeld 1980, 1982). This chapter expands on the product development process as a political framework in the organization. This framework has been utilized by numerous organizations in the author's consulting practice and fully implemented in a global 50 firm by the author while serving as the firm's Senior Vice President of Global Strategy and Portfolio.

Politics is not the content of decisions, rather it is an unnamed part of the decision-making process (Smith 2007). In this chapter politics refers to influencing individuals and the decision-making process around new products.

The PDMA Handbook of Innovation and New Product Development, Fourth Edition. Edited by Ludwig Bstieler and Charles H. Noble.

A host of influence tactics leveraging relationships, resources, positional power, opinions, and data are used to persuade others to take a preferred course of actions (Howell and Higgins 1990; Kipnis et al. 1980; Markham 1998, 2000; Yukl and Tracey 1992).

Politics is generally thought of in a negative light because it can interject destructive aspects into final outcomes (Frost and Egri 1990; Pfeffer 1981, 1991). For example, when empire building, ego, revenge, greed, or ambition enters the process, the decisions do not reflect what is best for the organization but what is best for a person or group. Many people and groups vie to influence organizational decisions with their own well-intentioned but often self-interested goals. Of course, politics is often used to further legitimate aims of the organization. It can be seen as good or bad depending on whether you get your way in relation to a decision. But whenever decisions are made through politics rather than accessible, transparent processes, some will feel aggrieved and the legitimacy of the organization, process, and decision-makers will be diminished, and future cooperation compromised. The aim of this chapter is to help establish accessible, transparent processes to alleviate the use and negative consequences of politics.

Our revulsion to politics is often related to our inability to influence decisions. Our perception of political behavior and the outcomes is heavily influenced by our own preferences for the outcomes. Politics is the tool used to make decisions when clear data, information, or experience is lacking. Nowhere is the lack of clarity more prevalent than when a company is trying to decide on its product portfolio. The very nature of developing a new product means that decisions must be made without the benefit of clear data and experience. The more radical a product, the more political the decisions become, because there is a greater lack of information and experience. When information is not available politics can be a substitute for knowledge in the decision process. Of course, some people and companies engage in political behavior with or without information, no matter the reason; however, an agreed upon decision-making framework helps to turn political decisions into analytical decisions. Understanding politics as a way decisions are made in the absence of convincing information is critical to managing politics and not letting it spiral into destructive behavior.

The Framework relies on three fundamental agreements. First, we must recognize the political gift senior managers bestow on the organization by embracing and being bound to a process that makes decisions whose outcome they do not know ahead of time. Second, the product process divides development into segmented tasks that must meet predefined criteria for the project to continue. Each task defines work that reduces the risk of a decision by performing the appropriate research and analyses to make decisions needed at that step. Third, decisions are binding on the whole organization. If the agreed upon decision-makers decide based on the agreed upon criteria, all parts of the organization must support the decisions whether they like them or not.

An unrecognized reason product development processes are so important is their help in managing politics. Given the high level of politics often present in product development, one may rightfully wonder if the processes produce political turmoil. When embroiled in constant political fighting over every tiny facet of a project, we can easily forget that the process reduces politics. No matter how consuming the politics, it is preferable to the political chaos that attends the inability to decide. Organizations can correct wrong decisions but being too paralyzed to make any decision is perhaps the most frustrating and destructive outcome possible – far worse than the politics surrounding product development decisions. Ultimately, the Framework is a political process to replace political behavior, whether it is perceived as good or bad, with analytical behavior.

Not only are senior managers bound by the process decisions but so are their parts of the organization. The willingness of all parts of the organization to honor its decisions is a critical measure of an effective new product development process. Therefore, at the highest level the political process of producing new products must politically bind all parts of the organization to a common course of action. No matter how much political acrimony arises in the development decision-making process, it results in a decision – which can be hard to come by in some organizations.

As the process progresses political behavior is focused on the task at hand as defined in the phase of development. This is as it should be; all parts of the organization should focus on that stage of development and not on disparate issues spread throughout the life of the project. All parts of the organization can participate in the decision-making and the decision-making should be transparent. Political behavior is directed at a particular task along the development continuum and then it is resolved – or at least a decision is made, and the project moves forward. Without a process, differences remain unresolved because no intermediate decisions are made; with a process, decisions are made, right or wrong, popular, or not.

The power of making decisions is great because the consequences of not making them are even greater. Therefore, a good decision-making process will have strong decision points with the tools in place at each point to provide the necessary information to settle questions and resolve arguments about the right course of action(Cooper 1983, 1994, 1997; Mugge and Markham 2012). The literature has hundreds of product development tools to gather, assess, and make decisions about every facet of projects at every step along the way (Cooper et al. 2004; Markham 2002; Pinto and Slevin 1989). The important point is that these tools also function as analytical antidotes to political behavior. To the degree they generate convincing information about the development path of a particular product, the more they also reduce politics. To help the company make decisions, the tools should be thoughtfully selected and rigorously and meticulously executed.

8.2 The Portfolio Management Framework

Selecting product development tools is critical, but more critical is the larger decision-making structure within the organization that uses the tools. Rather than sifting through the most politically able tools, this chapter focuses on the decision-making apparatus. Product decisions are part of a company's larger operational and/or capital budget. In addition to products, the budget must provide for company expansion, manufacturing, distribution, marketing, sales, quality, maintenance, replacement, and of course salaries, in addition to numerous expenses necessary to keep the doors open. We may feel the organization is shortsighted when funding product decisions, but this is often the myopic view. Fighting for project funding without understanding the full budget is a sure way to damage your political standing and your current and future ability to make budget requests.

Some development projects are funded through current operations, others through capital outlay. The funding mode depends on the type of project and on decisions made by the CEO and CFO, who in turn follow policies set by the board in response to investor and regulatory policies. The nature of the project and its funding has implications for project priority, amount of funding, timing of completion and other aspects that if not understood can contribute to needless and impossible political disagreements. Most mature portfolios will need to be revitalized, a need that brings technical, financial, and of course political challenges. Without a common process, corporate decision-making is as fractured and politically fraught as in the development of new products.

The purpose of the Framework is to drive better portfolio investment decisions that will strengthen future success. This objective requires a structure that can not only make strategically, financially, and technically good decisions, but also tame the politics that flourishes from paralyzed decision making. The joint strategic/technical/financial/political problem addressed by the Framework has multiple facets. 1) Products released by existing development programs – in the middle of the night, deep in the woods so nobody knows about them. Product releases must be part of an integral strategic plan with specified objectives rather than simply the culmination of a development program. 2) The "fuzzy" front end of innovation, activities that are not under management control. They may not be connected to a portfolio strategy and may therefore be rife with political behavior. Clarity in portfolio planning must start here rather than later, in development or even after launch. 3) Lack of a clear, transparent portfolio strategy that integrates the earliest ideation, the full development process, life-cycle management, and end-of-life planning. 4) Fractured decision-making with no clear decision rights or accountability for the outcomes. 5) Execution teams that are expected to develop a product with very positive commercial outcomes without a clear mandate, scope, or objectives and without the cooperation of other critical parts of the organization. 6) Not addressing all the critical parts of the portfolio. Some parts of the portfolio may be over or underfunded relative to the market opportunity and objectives for customer satisfaction and competitive positioning. 7) The compelling urge for an intense, one-time solution rather than a persistent process executed over time. This list of issues is overwhelming, but the surprising thing is they don't all have to be solved or any of them completely solved to greatly improve portfolio performance.

The goals of the Framework are to: 1) Simplify portfolio management decision-making. 2) Make fast, reliable decisions. 3) Optimize the portfolio instead of pushing out individual product releases. 4) Ensure the portfolio and every constituent part support the strategy of the business unit (BU), which entails influencing the strategy with proper portfolio performance expectations. 5) Actively manage the portfolio through on-going portfolio analysis, results reports, and roadmap updates. 6) Integrate portfolio offerings across the BU, such as products, support, services, maintenance, warranty, pricing, and delivery. 7) Actively manage the product life cycle of each part of the portfolio to ensure end-of-life decisions are timely and maximize customer satisfaction and future revenue.

> *Success Tip* – Build a structure inside the organization with the authority to make portfolio decisions that all parts of the organization participate in and support.

The Framework consists of three main components: 1) the Portfolio Management Structure, 2) the Animation, and 3) Portfolio Administration. The Portfolio Management Structure sets up the organization to actively manage a portfolio. Since portfolios are intensely cross-functional and most organizations are just as intensely functional in nature, smoothly managing the portfolio is not a natural act. The portfolio is often managed either by a staff function or by an isolated executive or in a haphazard or periodic fashion. But the organization needs clearly defined decision-making processes and criteria, well-established timing, clear roles, clearly documented responsibilities, and accepted performance measures. The Framework described in this chapter meets these needs by routinizing the portfolio function through interrelated structures and processes. Without this level of structure, the portfolio and the products within the portfolio will be weakly managed and continuously embroiled in contentious contests of opinions. This structure is just as important for small product development programs as for large organizations.

The Portfolio Management Framework consists of these interrelated parts:

- Portfolio Governance Structure
  - Governance Structure
  - Portfolio Decision-making Process
  - Investment Flowchart
- Animation
  - Portfolio Cadence
  - Portfolio Roles and Responsibilities
  - Interlocking Meetings and Agendas

- Portfolio Management
  - Framework Documents
  - Measures and Metrics

8.2.1 Governance Structure

Governing the portfolio is a complex political process. It is not a simple decision made by the portfolio manager about one project at a time that everyone else supports. Rather it involves all levels in the organization agreeing to what decisions they make and what decisions other levels make – and not crossing the lines. It establishes accountability for providing inputs and executing decisions. Critically, it also assigns resources and sets expectations for what work needs to be done by whom and by when.

Central to any governance model is defining the decision rights, or who makes what decision. At a minimum governance specifies the proposer and the disposer at each level and phase of portfolio management. These roles must be identified by the organization's preferred roles, titles, or assigned responsibilities. The critical element is that the person or group with control over the resources is different from the people who will use them. A person or group makes a proposal to the governing person or group who decides (disposes) if the resources will be made available to the proposer. If they are made available, a set of interlocking roles and responsibilities is created necessary to progress to fruition. The rights to make resource decisions at every level must be granted at the highest level in the organization. The disposer must make good on timely delivery of resources and must also set criteria and performance expectations. The proposer must make good use of the resources and agree to the criteria and expectations. Effective governance ensures accountability and reduces political wrangling about who has the resources and how they are applied.

> *Success Tip* – Separate the decision to approve a project from the people proposing the project.

Figure 8.1 portrays a five-level Portfolio Management Framework Governance Structure describing the levels of decision-making in a typical development organization. The figure identifies each level of decision-makers, what they require as inputs and what they deliver. In many instances a simple two-level structure of a proposer and disposer is best.

1. Level 1. The Corporate Governance level of decision-makers requires the Corporate Strategy to ensure that enough investment goes to the right places to achieve the corporate objectives. To assess how each portfolio element is performing, and to make funding continuation, reduction, and enhancement decisions, the Corporate Governance level requires timely portfolio performance results in a comparable format. This level of review is often done semiannually but can be done quarterly or annually depending on the nature of the environment such as customer acquisition and churn, sales stability, innovation initiatives, and new technologies.

FIGURE 8.1 PORTFOLIO MANAGEMENT FRAMEWORK GOVERNANCE STRUCTURE.

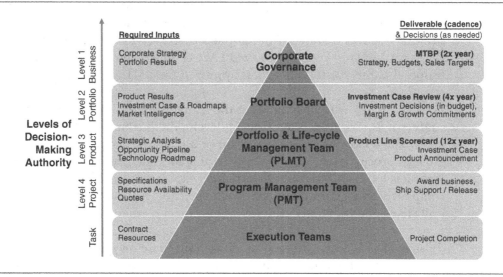

2. Level 2. The Portfolio Board (PB) is a standing interdisciplinary group, often department heads with direct interests in the portfolio such as sales, marketing, production, and development. The Board makes portfolio investment decisions based on optimization of revenue, resources, competition, new opportunities, and internal capacity. The Board needs individual product results, product roadmaps, Investment Cases, and market intelligence. In turn it delivers Investment Case reviews and decisions and makes margin and growth commitments. The Portfolio Board proposes a portfolio plan with the necessary budget and expected performance results to the Corporate Governance body. In turn, the Board considers proposals from the PLMT (see below) and makes investment decisions (disposes) about individual projects.

3. Level 3. The Portfolio and Life-cycle Management Team (PLMT) is also a standing interdisciplinary group, usually representatives of the department heads. To fulfill the portfolio strategy put forward by the Portfolio Board and approved by the Governance body, the PLMT proposes individual projects to the Board. The PLMT needs strategic analyses, a transparent opportunity pipeline, and technology roadmaps. In turn it delivers Product Line Scorecards to the Portfolio Board, proposes Investment Cases, and makes product announcements. The PLMT proposes plans for individual projects to the Board and considers proposals from individuals and departments from anywhere in the organization, as well as suppliers, partners, and customers.

4. Level 4. The Program Management Team (PMT) is another standing interdisciplinary group created to meet the development needs of the portfolio. The PMT creates and manages the actual work done by the Execution Teams. The PMT ensures resources are assigned to projects that deliver results to the PLMT, the Board, and the governance body. The PMT needs product

specifications, resources, and quotes from vendors for development components costs. In return the PMT delivers project contracts, awards business to vendors, and makes ship support and release commitments. The PMT may make proposals to the PLMT as a group or as individuals. If the portfolio strategy requires additional development streams to meet objectives, the PMT may initiate a proposal to the PLMT or join with the PLMT to make a proposal to the Portfolio Board. If approved the PMT applies the necessary resources.

5. Level 5. The Execution Team does the actual work of development. In this chapter it is portrayed as a technical development team, but it could just as easily be a services development team or a maintenance program team, an internal process team, or another type of team. The Execution Teams need contracts for resources and performance expectations. In turn they deliver completed projects on time that meet expectations. Team members are typically reassigned when the development project is completed.

In smaller programs or smaller organizations there may only be two levels – the proposer and the disposer. Their level in the organization does not matter. The important point is there is a person that makes decisions separate from the person that carries out the decision. For example, there may only be a single director of product development and one product development team.

8.2.2 Portfolio Decision-making Process

While the governance structure specifies the decision rights or who makes decisions with specific resources, the portfolio decision-making process describes the flow of information and decisions among the players involved in the portfolio. This further elaboration of decision rights is necessary to clarify how decisions are made in real time. Without a clear agreement from all players about who makes decisions about information and resources, projects can be starved for what they need. To avoid debilitating political fights over scarce resources, these decisions need to be crystal clear. The diagram in Figure 8.2 provides a clear view of decision-making rights at all levels in the organization. This clarity ensures accountability and transparency; it eliminates surprise decisions and the frustration of people wondering who decided or why they were not consulted or sought out for approval. Similarly, violations are immediately obvious.

- Corporate Governance provides guidance and budget to the Business Unit.
- Business Unit Governance provides performance reports and targets to Corporate Governance.
- Corporate Strategy provides Corporate Governance and the Portfolio Team with strategic direction and analysis.
- Portfolio Team provides portfolio strategy recommendations to the Portfolio Board.
- Portfolio Board provides product performance reports to Corporate Governance.

FIGURE 8.2 DECISIONS, RESOURCES, AND INFORMATION FLOWS.

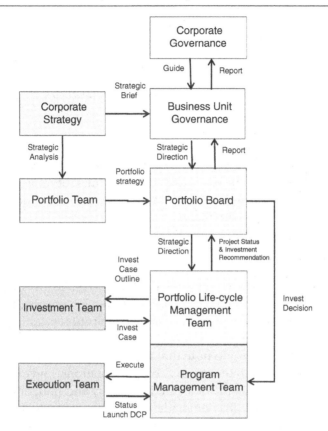

- Corporate Governance, with input from Corporate Portfolio Board and Corporate Strategy:
  - Assesses overall performance, targets, and inputs. Makes Level 1 Portfolio Decisions.
  - Allocates budget 2X per year during Mid-Term Business Plan checkpoints.
  - Provides budget guidelines and strategic direction to Portfolio Board.
- The Portfolio Board (PB) cascades strategic direction to the Portfolio Team and Portfolio Life-cycle Management Team (PLMT).
- The Portfolio Team analyzes the portfolio and brings forward portfolio recommendations to the Portfolio Board and PLMT. New investment recommendations start as "Project Charters."
- If charter request is approved, the PLMT assigns an Investment Team to create an Investment Case.
- The PLMT evaluates the Investment Case and makes a recommendation to the Portfolio Board.
- The Portfolio Board makes Go/No-Go investment decision.
- The Program Management Team charters an Execution Team to execute the project.

- The Execution Team reports progress to the Program Management Team.
- The Execution Team completes the project, Launch Decision Checkpoint (DCP) is approved by PLMT, the product becomes Generally Available (GA) and enters the Life-cycle phase.
- The Portfolio Team reports product performance to the PLMT and Portfolio Board.
- Investment recommendations from Life-cycle Reviews are brought forward as charters.

Obviously, how decision-making works in organizations will vary greatly; hopefully, this example provides enough detail to plan your own process. In smaller organizations all executive functions may be integrated into a single entity such as the Portfolio Board and all execution functions integrated into the Program Management Team.

8.2.3 Investment Flowchart

Further depth is necessary to understand how to make investment decisions for a specific proposal in the portfolio through the life of a product. As mentioned before the governance process requires a proposer and a decision-maker. Figure 8.3 sets up the proposer/decision-maker process to make investment decisions. It is important to note that this is for substantial decisions. Small projects such as simple enhancements, fixes, adaptations, and extensions are made by lower levels with specific delegated authority and budget. For example, bug fixes less than $50,000 may be made by the program managers without bringing it to the portfolio board or PLMT. Even though authority might be delegated to a lower level of governance, the total budget remains the responsibility of the delegating body. It is also important to note this is an investment decision flowchart, not a development gates and stages or phases model. The decision to invest in a portfolio and each element in the portfolio is not the same as review gates that check progress of a project. Investment decisions determine the amount and timing of funding while gates and stages are a project management tool used after an investment is made. The investment decision may be revisited depending on the results of a given gate review.

The Investment Flowchart in Figure 8.3 describes:

- Ideas for investment can come from anywhere in the organization.
- A formal request for project support is made by filling out a one-page project charter.
- All charters are reviewed by the PLMT at their next biweekly meeting where they can (Charter DCP PLMT):
  1. Provide feedback for improving the proposal but withhold support.
  2. Recommend the project for immediate consideration to the Portfolio Board through a fast trial.
  3. Form an Investment Team to build a complete Investment Case (proposal). The PLMT may adjust the Investment Case.

FIGURE 8.3 INVESTMENT FLOWCHART.

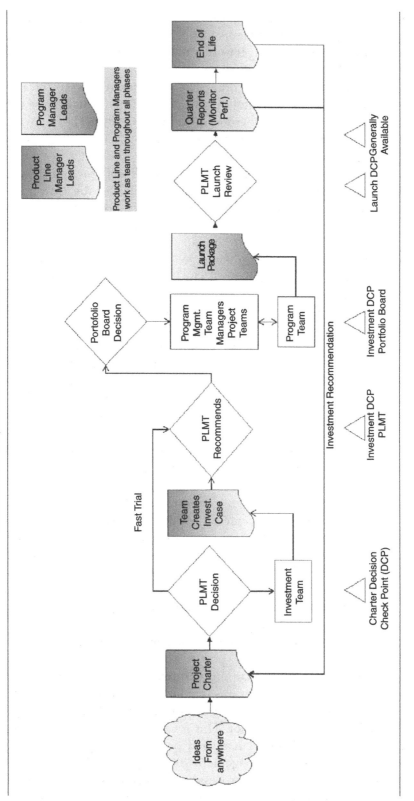

- The Investment Team begins to create an Investment Case. The PLMT has a set budget to support teams in this work. All work is in budget. Except for charters, no out-of-budget work is to be started or continued without approval of the PLMT.
- The Investment Team proposes an Investment Case to the PLMT. At their next biweekly meeting the PLMT decides if they will propose the Investment Case to the Portfolio Board (Investment DCP PLMT).
- The Portfolio Board at the next biweekly meeting (Investment DCP Portfolio Board) can:
  1. Provide feedback but no investment. All work on that project must stop.
  2. Ask for more information to be included in the Investment Case. The Investment Case has a template that if followed should provide enough information for the Portfolio Board to make an informed decision the first time the Case is presented.
  3. Make an investment and charge the Program Management Team with following the Investment Case. The Portfolio Board may adjust the investment amount, timing, or scope.
- The Execution Team does the work specified in the Investment Case.
- The Execution Team creates a Launch Package that includes all aspects of technical, production, service, maintenance, financial, sales, marketing, and distribution plans to optimize the investment (Launch DCP).
- The PLMT reviews the launch plans; if all parameters of the original Investment Case remain the same, the product is launched. Note: The Portfolio Board does not re-authorize the Investment Case or give a separate approval to launch the product – that was done at the investment decision checkpoint. Of course, it's not wise to surprise the executives. If the Investment Case materially changes, the Portfolio Board will need to reconvene. For example, if the cost or prices change, the unit volume decreases, or partners retreat, the Case should be reconsidered before launch.
- The product becomes Generally Available (GA).
- The Product Line Managers prepare quarterly reports for the executives' review. The executives may recommend further investment or redirection or make other recommendations.
- The Product Line Managers continue to monitor the product and when appropriate make an end-of-life recommendation to the PLMT.

Success Tip – Ensure the decision-making process is open and transparent.

Establishing and maintaining a rational investment process has many details and nuances. Ensuring that it is regular, predictable, and transparent and that it makes all parties accountable increases decision quality, reduces ambiguity, and lessens political behavior. This is especially true for smaller organizations that cannot afford poorly managed or delayed projects. While the Portfolio Board and Program Management Team may conduct all the steps in the Investment Flowchart, this process represents the minimal process any organization should follow. Decision shortcuts should not be taken here.

8.2.4 Portfolio Planning Cadence

Together the Governance Structure, Portfolio Decision-making Process, and Investment Flowchart make up the structure needed to manage portfolio content and reduce political behavior. Nevertheless, structure alone does not make a complete Framework. The actions must be coordinated through a unified method to ensure all parts function together at the right time. This unifying and coordinating function is accomplished by establishing a regular cadence of portfolio actions, defining roles and responsibilities, and setting regular meetings with regular agendas.

It may seem counterintuitive that such an innovative process as managing the portfolio relies on such a tight structure and process. Nevertheless, without binding processes the disparate parts of the organization will not be able to function cohesively, resulting in disagreements and high levels of political behavior.

The first step to put the Framework into action is to create a cadence of when key activities need to be completed. Figure 8.4 presents an annual schedule used by a major electronics division of a global conglomerate to plan portfolio activities. Key dates and deliverables for the business unit must be coordinated with the larger corporation's planning process as depicted in the top half of the figure. The bottom half of the figure presents a simplified "swim lane" chart. Each department is required to do certain work by a specified time. In this way all the departments know what others are doing. A transparent process increases accountability and decreases political behavior.

> *Success Tip* – Regularly publish the cadence and deadlines for all decisions. Repeat requests for input and feedback.

As the organization becomes familiar with a repeated process, its members learn their own role and others' roles. Clarity of roles decreases duplication and confusion about who is doing what and resolves when tasks must be done since dependencies become clear. It reduces anxiety and bounds the tasks both with timelines and the work preceding and following each part. Gaps in responsibilities become easier to identify and specific assignments to rectify problems are straightforward. As the cadence is repeated, handoffs become routine and performance expectations become clear.

In smaller organizations and for less complex projects, the planning process may be simplified if the project plan retains the cadence in terms of when different resources will be called upon to work on the project.

8.2.5 Portfolio Roles

The next step to animate the Framework is to define roles and responsibilities. Although numerous roles must be identified, two paired roles – product manager/product line manager and program manager – are critical. These roles can easily be confused if not carefully crafted. The examples below will likely be modified for your organization to account for already established job descriptions, as well as for the number and capability of staff and the actual workload.

FIGURE 8.4 PORTFOLIO PLANNING CADENCE.

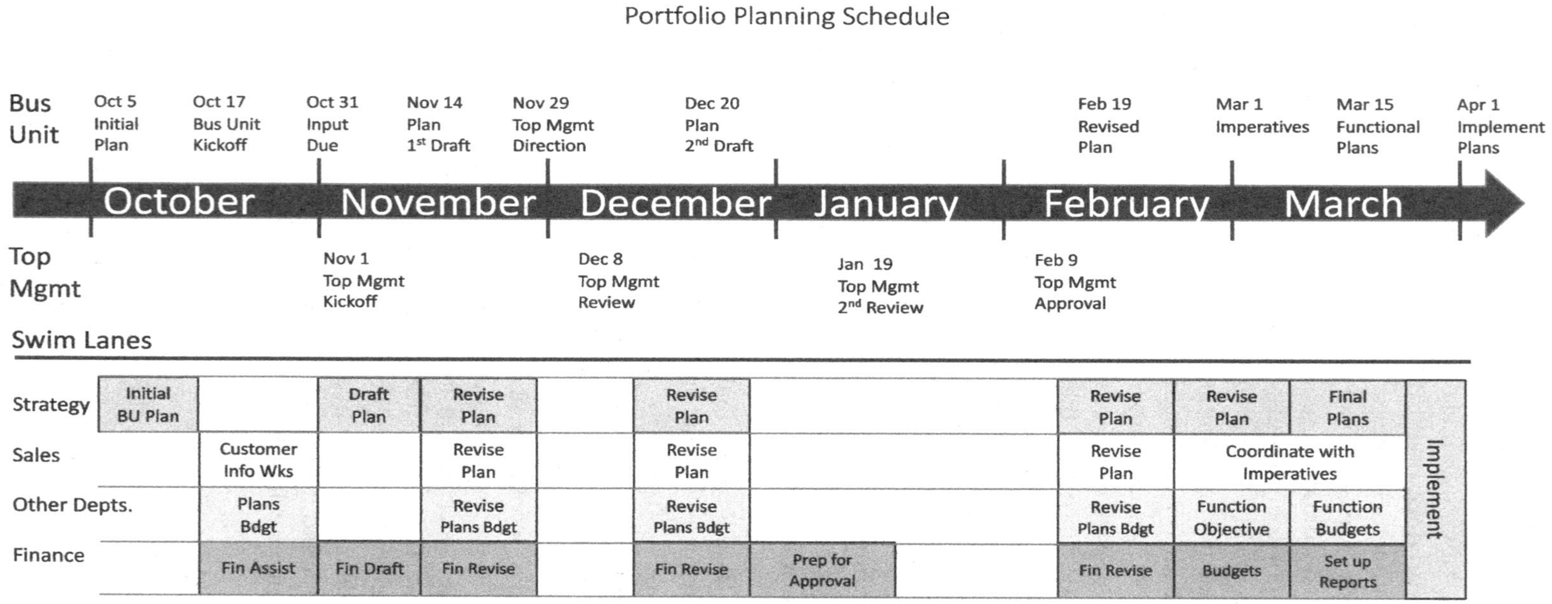

Portfolio Planning Schedule

	October	November			December	January	February		March	
Bus Unit	Oct 5 Initial Plan	Oct 17 Bus Unit Kickoff	Oct 31 Input Due	Nov 14 Plan 1st Draft	Nov 29 Top Mgmt Direction	Dec 20 Plan 2nd Draft		Feb 19 Revised Plan	Mar 1 Imperatives	Mar 15 Functional Plans	Apr 1 Implement Plans
Top Mgmt			Nov 1 Top Mgmt Kickoff			Dec 8 Top Mgmt Review	Jan 19 Top Mgmt 2nd Review	Feb 9 Top Mgmt Approval			

Swim Lanes

									Implement		
Strategy	Initial BU Plan		Draft Plan	Revise Plan		Revise Plan		Revise Plan	Revise Plan	Final Plans	
Sales		Customer Info Wks		Revise Plan		Revise Plan		Revise Plan	Coordinate with Imperatives		
Other Depts.		Plans Bdgt		Revise Plans Bdgt		Revise Plans Bdgt		Revise Plans Bdgt	Function Objective	Function Budgets	
Finance	Fin Assist		Fin Draft	Fin Revise		Fin Revise	Prep for Approval	Fin Revise	Budgets	Set up Reports	

Product Line Manager. Product Line Managers (PLM) operate as the overall, end-to-end business owner for their product line, including hardware, software, services, and maintenance. Their responsibilities include vision and strategy, profit and loss ownership, pricing strategy, setting product requirements, and cross-functional leadership to drive business results. Within the product management team, PLMs help align solution areas through collaboration to produce a more cohesive, customer solution-oriented portfolio. PLMs create project charters for the PLMT to review and approve and often take the lead to create the Investment Case. They work closely with the Program Manager under the direction of the PLMT and Portfolio Board. The PLM manages and tracks progress against financial targets with corrective actions through the quarterly review process.

Program Manager. The Program Manager has three primary responsibilities. The first is to work with the PLM to prepare Investment Cases for the PLMT. The second is to manage project execution from Investment DCP to General Availability (GA). This responsibility includes coordinating, tracking status, and leading a team of representatives from all functional areas required to develop, deliver, and launch a product. The third responsibility is quickly and efficiently to develop and launch cost competitive, high-quality products and provide expertise to the Product Line Manager on matters relating to development strategy and technology. For projects, the program manager is responsible for achieving development scope, expense, product cost, schedule, and quality commitments. The program manager reports consolidated status at Program Management Team meetings, escalates issues to the appropriate areas, and works with peers to resolve cross-product dependencies.

Success Tip – Make sure roles and responsibilities are assigned and understood by everyone.

To further elaborate on this critical pairing of roles, each role is further defined in Table 8.1 by what it is required to do at each Decision Checkpoint (DCP) in the Investment Flowchart.

8.2.6 Interlocking Meetings and Agendas

Finally, to bring the Framework to life and further animate the decision-making flowchart, the roles are set in motion through a regular series of meetings that integrate all levels of decision-makers through interlocking agendas as described in Table 8.2. Often, political behavior erupts because people do not know where in the process they can raise concerns or objection or even request basic information. Holding regular well-defined meetings allows people to contribute positively. It also allows information randomly injected into the process to be redirected and properly addressed rather than inadequately argued about in the moment.

To keep the process moving as well as ensure the proposer and disposer responsibilities are executed consistently and timely, the agendas for the different

TABLE 8.1 ROLES AND RESPONSIBILITIES AT EACH DECISION CHECKPOINT.

Product Line Manager	Program Manager
Charter DCP	**Charter DCP**
Responsible for identification and analysis of market/customer needs, evaluation and validation of business opportunity. Responsible for creating charter based upon external and internal inputs.	Provides budgetary sizings (cost, schedule, expense) inputs for the effort needed to do the work to achieve Investment Decision Check Point (DCP).
Charter to Investment DCP	**Charter to Investment DCP**
Responsible for driving the Investment team, along with strong support from the program manager and supporting teams to create an Investment Case for review at the PLMT and PB. The Product Line Manager is responsible for solution definition and the financial aspects of the Investment Case.	After charter, the Program Manager leads a team responsible for working with the product Line Manager to convert the product requirements and project objectives into a technical proposal for input into the Investment Case. This includes a detailed product specification to achieve the product requirements or user stories with development, manufacturing expenses and product cost, and interlocking schedule with all required product teams to deliver the solution as agreed with the Product Line Manager. The Program Manager is the person in Development with the authority to provide approved sizings (cost, schedule, expense) or make commitments on behalf of Development.
Investment DCP to Launch DCP/ GA	**Investment DCP to Launch DCP/GA**
Responsible for providing status to Program Managers for Brand, Finance, and Product Manager activities related launching the product.	Following an approved Investment DCP, the Program Manager leads the program team to develop and launch the product to general availability. This includes managing dependencies with peer program managers to deliver entire solution. They are responsible for achieving development scope, expense, product cost, schedule, quality commitment, and successful GA of the projects. The Product Line Manager is a member of the Program Team.
Launch DCP/GA through Portfolio Life Cycle	**Launch DCP/GA through Product Production Life-cycle End**
The Product Manager is responsible and accountable for achieving the P&L's commitments in the Investment Case. The program results are reviewed at the Quarterly Review and recommendations are given as needed to manage the portfolio product.	Responsible for leading a team to maintain the product after launch through the production life. This includes production continuity, customer technical issues, cost downs (HW only), and other sustaining activities. This may include product updates driven by the Portfolio Life-cycle activities.

levels of decision-making are interlocked (see Table 8.2). For example, if the Program Management Team proposes an idea to develop through a charter to the PLMT, the PLMT then decides if it should move forward. If the PLMT recommends investment, it in turn becomes the proposer of that investment to the

TABLE 8.2 INTERLOCKING MEETINGS AND AGENDAS.

Meeting	Frequency	Proposals to Make	Decisions to Make	Reports to Send Out
Quarterly Portfolio Review (QPR)	Quarterly	Proposals to Corporate	Decisions about Portfolio Board Proposals	Portfolio Results to Corporate
Portfolio Board	Biweekly	Proposals to BU Governance	Decisions about PLMT proposals	Portfolio Results to BU Governance
Portfolio Life-cycle Management Team	Biweekly – opposite week as the Portfolio Board	Proposals to the Portfolio Board	Decisions about PMT and other people's proposals	Development status of each project. Performance status of each product in the market
Program Management Team	Weekly	Proposals to the PLMT	Decisions about Execution Team proposals	Status of each development project and status of each product in the market
Execution Team	Weekly/daily as needed	Proposals to make to the PMT or PLMT	Decisions about member proposals	Status of each component of each development project

Portfolio Board that makes the final investment decision. Regular meetings with predictable agendas help all parties to know where and how to bring up concerns, when to have work completed, and what performance standards are expected.

It should be noted that the Portfolio Board makes a single investment decision at one point in time. It does not manage the development process. If the business case changes substantially during the development process, then the PLMT should bring the project back to the Portfolio Board for reconsideration. If the business case remains intact the PLMT is preauthorized to launch the product. Providing accurate and complete information in the business case is therefore critical. Before making a proposal, managers should utilize rigorous but not make-work processes to ensure the Portfolio Board is well informed.

- Quarterly Portfolio Review (QPR)
  - Chair: Portfolio Management Team Executive.
  - Description: The QPR is informational. The Portfolio Management Team presents to upper management the status of the overall portfolio as well as the performance of each element and may make recommendations to modify the portfolio.
  - Attendees: Portfolio Board, Portfolio Management Team, PLMT.
  - Frequency: Once a quarter.
- Portfolio Board Meetings
  - Chair: Portfolio Management Team Executive.

- – Description: Review Portfolio Status presented by the Portfolio Management Team. Decide on recommendations brought forward by PLMT. Integrate Corporate Strategy into building the portfolio plan.
- – Attendees: Portfolio Board and Portfolio, Sales, Services, and Development Executive directors.
- – Frequency: Biweekly.
- Portfolio Life-cycle Management Team (PLMT) Meetings
  - – Chair: Hardware, Maintenance, Services, and Software Portfolio Managers chair their respective sections.
  - – Description: Review Portfolio Status presented by the Portfolio Management Team. Decide on recommendations brought forward by Investment Teams and on new charters, recommend fast trials, ensure QPR recommendations are brought forward as charters. Evaluate end-of-life recommendations. Implement product launch plans.
  - – Attendees: Portfolio Management Team, PLMT, Product Line Managers, Program Managers, supporting functions' representatives (finance, supply chain, etc.).
  - – Frequency: Monitor product performance and propose corrective action if needed post product launch.
- Program Management Team Meetings
  - – Chair: Product Line or Program Manager respectively.
  - – Description: Review each element of the portfolio for timelines and quality performance. Solve problems, assign, and adjust resources, coordinate with other parts of the organization to accomplish the full range of tasks prior to product launch.
  - – Attendees: Investment and Execution Team members.
  - – Frequency: As deemed necessary by individual teams.
- Execution Team Meetings
  - – Chair: Individual assigned by the Execution Team leaders.
  - – Description: Plan and execute the plan approved by the Portfolio Board under the direction of the PLMT and Program Managers.
  - – Attendees: Individuals assigned to the team by their functional managers and the PLMT and Program Managers. Though highly interdisciplinary, the team may focus on a single discipline for some time as a technical or other requirement is addressed.

Every product should be reviewed on a periodic basis. Of course, smaller portfolios, incremental improvements, and smaller organizations may need simplified agendas and meetings that follow the changes that smaller operations will need to make to the structure in Figure 8.1.

8.2.7 Framework Documents

Much of the information necessary to run the Framework comes from the agendas and minutes in the meetings. This is as it should be. Meetings before meetings and long reports run counter to making timely decisions and are often a symptom of highly political environments. Many product development systems have unwieldly documentation requirements, but managers and executives should start by requesting the minimum viable documentation and add to it only information required by multiple later projects. Don't start with requesting everything that could possibly be used and then try to eliminate parts that are not used. It is likely that nothing will be used, resulting in poor information sharing and charges of political behavior.

The information to run the Framework should be available to everyone concerned. The information on which decisions are made and how performance is evaluated should be transparent. Transparency reduces political behavior since everyone is dealing with the same set of facts rather than contests of opinions.

Nevertheless, not all information can be presented in a meeting format. For example, market research reports may contain more information than can be reported in a standing meeting. In these cases, documents should be prepared to make sure all parties are equipped to make well-informed, high-quality decisions. To ensure portfolio elements are fairly and consistently decided upon and managed, use a few standard documents to collect and present information in a consistent way. This standard not only helps to make good comparative decisions but also helps make it clear why decisions were made.

Success Tip – Keep paperwork and reports to a minimum – but not below the minimum.

The Framework requires a consistent set of deliverables contained in a standard set of documents. These documents not only trace the work of the project but also contain the necessary reasoning for the investment, the market problem, the solution, the budget, timelines, customer information, etc. A few documents that should be used:

- Project Charters. This one-page document has two parts. 1) Outline the potential opportunity by clearly describing the idea that has commercial merit, the market segment, the market problem, and the proposed solution (this can be applied to internal projects as well). The charter should describe what is unique, beneficial, and believable about the proposed idea. 2) Outline how the opportunity will be delivered, providing some sense of project duration, the amount and type of resources needed, the path to market, and of course a rough estimate of impact on the company – health, safety, environmental, revenue potential, reputational, cost reduction, etc.
- Investment Cases, also called business plans or business cases. They should contain the information the executives believe necessary to make well-informed investment decisions. Typically, they include detailed market analyses and competitor analyses. They may include ties to strategy, development plans, launch plans, sales, market, development, partnerships, intellectual property, the team, timelines, roadmaps, and cost and revenue projections. Plans should be prepared by the PLM with input from the Program Manager. They should contain forward-looking revenue estimates that can later be

used to evaluate performance. Estimates will likely be wildly inaccurate at the start but will soon start to be much more accurate as they become part of the open record and people are held accountable for delivery. Public accountability helps eliminate inflated estimates and missed deadlines as well as sandbagging and passive-aggressive foot-dragging by parts of the organization.

- Launch Plans. As indicated earlier, launch plans are a near universally ignored part of planning, budgets, and execution. Because launch is at the crossroads of so many handoffs, such as development and operation, it is often left to ill-equipped, uninformed, and unidentified people to pick it up and make something out of the new product. Launch Plans should have a clear template and execution timeline with clearly identified resources. Launches should be a top line item in quarterly reviews by senior executives.

- Quarterly Performance Reports. Quarterly reports will change over time as the portfolio matures or is reinvigorated. Different products or portfolio segments may have very different reporting formats. For example, a global product trying to break into a foreign market with entrenched competitors may look different than an existing cash cow. Nevertheless, the portfolio management team must make the reports comparable by reporting on what the executives determine to be crucial for each portfolio element. Often the portfolio or product manager calls attention to what is critical to monitor, usually with both activity and performance measures. Always there are detailed financial analyses, including sales data analysis, price analysis, competitor analysis, life-cycle estimates. Even if a project has not entered the market, it is not exempt from cost analysis and continued rigorous price and volume estimates as development continues.

- Portfolio on a Page. Across the top is a column heading for each phase that your organization uses to describe the stage of development or life-cycle stage of a project/product. For example, the first column might be "Charter," meaning a project may be chartered. Next might be each stage of development, followed by launch, followed by "In Market," and finally by "End of Life." Each horizontal line represents a different project/product. In the cell is the date on which each project entered the phase identified in the column heading. In this one-page format, the entire portfolio can easily be grasped by even the most cursory executive examination.

- Roadmaps. Each product plan in the Investment Case should include a product roadmap. Integrated roadmaps for each product line and for the entire portfolio are also helpful. While the portfolio on a page looks at the existing portfolio, the integrated roadmap shows direction, that is, the impact of the company's forward-looking investment decisions. This is important for talking with customers so they can rely on your organization's vision for the future. It also galvanizes the view inside the company about the direction the company is taking, which reduces political controversy over direction.

Many product and portfolio programs fall down under the weight of their own required paperwork. Every effort must be taken to keep paperwork at a minimum. For example, although many other requirement documents are often generated and sometimes used, critical requirements might become a section in the Investment Case rather than separate documents. Market reports, detailed cost analyses, partner agreements, and others might be necessary. Quality systems might require your organization to create and maintain excessive documentation that weighs down development. But remember that these requirements result from your own organization's decision-making. Even with a regulated quality system, the organization specifies what is needed to manage systems in accordance with objectives. While the Framework strongly encourages quality systems, organizations should scrub unnecessary and unused documents. For either small or large companies a "light-touch" process ensures that time is not wasted documenting obvious, trivial, or non-consequential information. The appropriate level of detail must be determined by experienced managers giving direction to their staff, reviewing outcomes, and over time applying the discipline of continuous improvement.

8.2.8 Portfolio Measures

Profit and loss (P&L) responsibility entails measuring the performance of the portfolio and each product/service, at all stages of life, and being accountable for that performance. Like the Investment Case, the measures are driven by what the executives want to know to make well-informed decisions. The measures will change as the portfolio matures and is revitalized. Rather than reviewing typical measures and metrics for reporting numbers, this section describes how executives measure the portfolio. Portfolio measurement is a political process. Both the method and criteria are socially derived by executives with many objectives and motivations.

Success Tip – Establish measures ahead of time, but be prepared to be flexible.

Executive interests in the Quarterly Portfolio Review (QPR) meetings drive the use of measures. Portfolio review meetings last a long time, typically one to two days in a sequestered format with senior executives teasing apart every aspect of the portfolio. The meeting usually starts with an overview of the portfolio by the portfolio executive who calls attention to strengths and weaknesses in last quarter's performance and, for the coming quarter, reviews major drivers for and against performance in specific areas. This meeting is the portfolio manager's chance to direct the attention of other executives to important items, but in the absence of a standard, agreed approach, it never works; executives wander all over the map as they touch on one seemingly unrelated point after another. Make no mistake, however, there are always reasons for the questions.

Portfolio measurement is a social, political agreement among the executives about what they consider important to the organization's success. This is true for both the portfolio overview and for each portfolio element. To reduce wandering questions, understand what executives want and present it in the same format quarter after quarter. A few cycles may be needed for executives to agree on the content and format of the information. At the start of using a Framework, the executives will have intuitive insights and ask insightful but disjointed questions as they begin to articulate their interests and concerns. As they coalesce around what is presented and how it is adjudicated, the rest of the organization will quickly adopt what the executives judge as important. This process can be painful at first, but it is a powerful tool to reduce disagreements about what is important through the organization.

Success Tip – Use the meetings and agendas to focus executives on the most important success factors.

After the portfolio executive's overview, the product line managers present their product lines to the executives. When all the product presentations contain common information in the same format, measuring the portfolio and each product becomes much easier and it also becomes easier to call attention to areas that are different or that need special attention from one or more of the executives.

Typically, measuring portfolio performance begins with an in-depth sales analysis for each product in the market. A sales analysis starts with volume and price. Typically, each geographical region is then considered for volume growth and price and margin performance, followed by segment analysis by region. Volume and price growth or declines are noted and explained in detail. Since executives focus on problem areas, explanations for decreasing volume or price should be detailed and authoritative. This is no place for suppositions. Further analysis will examine planned vs actual performance and, where there are deficits, remediation plans with the probability of success. In addition, forward-looking revenue forecasts will be evaluated in terms of present performance problems. These numbers will be used by the finance department to provide reports to investors; executives do not want to provide inaccurate guidance based on inaccurate or out-of-date estimates. These numbers are crucial to the company's stock price and in turn to the executives' stock option plans. While executives want to show growth, they do not want to raise unrealistic expectations in the market. Not meeting published performance marks does more harm than inflated numbers do good. Finally, executives will want to examine areas of upside potential. Be ready for questions about the marginal effect on sales of changes in approach or product mix or additional investments.

Another critical measure is cost. The focus here will be to maintain costs and reduce the total cost of delivery. As a product matures, many opportunities arise to reduce component costs and warranty and maintenance costs. Typically, given the ongoing expectation that progress will be made on this front every quarter, be ready with cost reduction plans. Similarly, as repeat customers start to make up a larger part of the volume, attention may shift to customer retention or getting a larger share of customer spending. This may spark ideas about derivative product offerings or bundling of other products and services. Executives love these kinds of innovations because development cost and time to market are low and potential for additional revenue comes with low risk.

The cost of delivery methods will also be scrutinized at the QPR. If repeat customers are growing in volume, contracts with distributors and even the company's own sales force will be examined for ways to reduce sales and distribution costs. Marketing may also be an area of cost reduction. But if sales are declining, be ready with an analysis of how increasing spending in the sales channel might stimulate additional sales. Asking to spend more in the face of declining sales will likely be a high-stress conversation; a track record of making the numbers is required to have the credibility to successfully make that pitch. Your analysis should also include a harvesting strategy where the company makes no or limited investments and just collects revenue on existing sales.

Detailed sales and operational plans for the coming quarter will typically be presented last. If executives give directions contrary to your assumptions, the forward-looking plan will look terribly out of touch. Rigorous analysis will help avoid this outcome. Avoiding the urge to make the future sound bright is important because next quarter accountability will be in full force for what is said this quarter. Bad performance in one quarter is not solved by promising short-term solutions. The best approach is to own the performance and present only realistic views of future performance and not be pressured into making overly optimistic forecasts. Of course, without plans for increasing performance you may not be at the next portfolio review. It is important to note that no matter how well a product is performing there will be pressure to improve performance in some other way.

8.3 Summary and Conclusion

The purpose of the Framework is to increase decision-making quality and align the organization's decision-making apparatus to efficiently and effectively deploy limited resources to make the most impact on the organization and to reduce political behavior as a deliberate and necessary outcome. Nevertheless, every action can be subject to political influence. Membership on committees and teams is routinely used to advantageously position oneself or department. Executives must work together to set the example for the organization to put politics aside, or any process will lapse into politics as usual.

The attempt here is to give well-meaning people the tools to make good decisions. Nevertheless, there are additional facets of portfolio management not included here such as the use of analytical tools, tying corporate and portfolio strategies, generating requirements, business case development, launch strategies, as well as software tools and staffing.

Not all decisions can be cleanly made even using a structured approach such as this Framework. There will be close calls for which manager intuition or preferences are the only final arbiter. Of course, skilled political players may be successful at influencing the project no matter how rigorously a process is followed. Adequate information may simply be unavailable to make well-informed decisions. Even in these situations, a process is helpful to identify the known, the unknown, and the unknowable. Managers who make decisions under imperfect information conditions will likely be subject to political behavior and a broad array of influence attempts. In these cases, political decision-making is still preferable to no decision.

This chapter reports highlights of years of experience implementing decision-making structures in many organizations by many hundreds of people. Significant thought and work must be taken before using any of these tools or structures in your organization.

References

Cooper, R.G. (1983). The new product process: an empirically based classification scheme. *R&D Management* 13 (1): 1–14.

Cooper, R.G. (1994). Third-generation new product processes. *Journal of Product Innovation Management* 11 (1): 3–14.

Cooper, R.G. (1997). Fixing the fuzzy front end of the new product process: building the business case. *CMA Magazine* 71 (8): 21–23.

Cooper, R.G., Edgett, S.J., and Kleinschmidt, E.J. (2004). Benchmarking best NPD practices. *Research Technology Management* 47 (1): 31–43.

Ettlie, J.E. and Elsenbach, J.M. (2007). The changing role of R&D gatekeepers. *Research Technology Management* 50 (5): 59–66.

Farris, G. and Cordero, R. (2002). Leading your scientists and engineers 2002. *Research Technology Management* 45 (5): 13–25.

Frost, P.J. and Egri, C.P. (1990). Influence of political action on innovation: part I. *Leadership & Organization Development Journal* 11 (1): 17–25.

Howell, J.M. and Higgins, C.A. (1990). Leadership behaviors, influence tactics, and career experiences of champions of technological innovation. *The Leadership Quarterly* 1 (4): 249–264.

Kipnis, D., Schmidt, S.M., and Wilkinson, I. (1980). Intraorganizational influence tactics: explorations in getting one's way. *Journal of Applied Psychology* 65 (4): 440–452.

Markham, S.K. (1998). A longitudinal examination of how champions influence others to support their projects. *The Journal of Product Innovation Management* 15 (6): 490–504.

Markham, S.K. (2000). Corporate championing and antagonism as forms of political behavior: an R&D perspective. *Organization Science* 11 (4): 429–447.

Markham, S.K. (2002). Moving technology from lab to market. *Research Technology Management* 45 (6): 31–42.

Mugge, P. and Markham, S.K. (2012). An innovation management framework: competencies and dimensions. In: *PDMA Handbook of Product Development*, 3e (ed. K. Khan and S. Uban). New York, NY: Wiley and Sons.

Pfeffer, J. (1981). *Power in Organizations.* Marshfield, Mass: Pitman Pub.

Pfeffer, J. (1991). Organization theory and structural perspectives on management. *Journal of Management* 17 (4): 789–803.

Pinto, J.K. and Slevin, D.P. (1989). Critical success factors in R&D projects. *Research Technology Management* 32 (1): 31–35.

Roberts, E.B. and Fusfeld, A.R. (1980). Critical functions: needed roles in the innovation process. Working Paper: Sloan School of Management.

Roberts, E.B. and Fusfeld, A.R. (1982). Staffing the innovation technology-based organization. *Sloan Management Review* 22 (3): 19–33.

Smith, D. (2007). The politics of innovation: why innovations need a godfather. *Technovation* 27 (3): 95–104.

Yukl, G. and Bruce Tracey, J. (1992). Consequences of influence tactics used with subordinates, peers, and the boss. *Journal of Applied Psychology* 77 (4): 525–535.

Stephen K. Markham is the Goodnight Distinguished Professor of Innovation and Entrepreneurship and Executive Director of Innovation and Entrepreneurship at North Carolina State University. Dr Markham served as Toshiba's Senior Vice President of Global Strategy and Portfolio in the Global Commerce Solutions division. He co-founded BP's Innovation Board and co-founded numerous high-tech companies and served as Chair, CEO, CFO, and VP Marketing and Product Development. His research focuses on Champions of Innovation. He was director of the Center for Innovation Management Studies and the Director of the Technology Entrepreneurship and Commercialization program at NCSU.

CHAPTER NINE

INTEGRATING IP ACTIONS INTO NPD PROCESSES: BEST PRACTICES FOR PROTECTING (*AND PROMOTING*) NEW PRODUCT INNOVATION

Joshua L. Cohen, Esq.[1]

IP-savvy development teams strive to create products that not only delight consumers, but that also enjoy strong IP protection and avoid conflicts with others' IP rights. But how is this best accomplished? And how can firms trust their processes to reliably deliver this result time and time again? The answer lies in firms adopting integrated IP programs that complete IP-related actions as a natural byproduct of their NPD processes.

9.1 Introduction

Firms that innovate share a common mandate – they need a robust program of intellectual property (IP) practices to safeguard their IP, avoid conflicts with others' IP rights, and catalyze further innovation. In short, they need to *protect*, *respect*, and *promote* IP.

Firms know that strong IP rights can protect their core technologies and help maximize the ROI of their product development efforts. They also know that IP risks should be identified and avoided proactively. Yet they often fail to execute the *right IP actions*, at the *right times*; instead, too many firms rely on *ad hoc* or reactive IP actions as opposed to systemic IP awareness and integrated processes.

The PDMA Handbook of Innovation and New Product Development, Fourth Edition. Edited by Ludwig Bstieler and Charles H. Noble.
© 2023 John Wiley & Sons, Inc. Published 2023 by John Wiley & Sons, Inc.

FIGURE 9.1 FOUR-STEP FRAMEWORK.

By adopting a robust IP program – a set of coordinated, sustainable, and consistent measures and behaviors – firms can ensure that their innovations are well-positioned from the IP perspective. This chapter provides a clear framework for doing so.

Many resources provide a practical reference for basic IP principles.[2] This chapter builds on that substantial body of knowledge. It provides product development professionals with an actionable framework for executing important IP actions at the right junctures in NPD efforts, so that strong IP positions become a natural byproduct of their NPD processes.

This chapter is organized according to a four-step framework (Figure 9.1) by which stakeholders in successful new product innovation can:

- In STEP I, adopt an IP directive – a top-down mandate for effective IP practices;
- In STEP II, commit to IP activities consistent with achieving the IP directive;
- In STEP III, select IP deliverables that ensure completion of the right IP activities; and
- In STEP IV, integrate the selected IP deliverables into their NPD processes so that they are completed at the right times.

Thus, the audience for this chapter includes the community of product development professionals that seek to optimize their firms' NPD processes, the executives and managers responsible for ensuring successful product development and IP oversight, and scholars interested in strategies for generating IP and innovation.

9.2 Adopt an IP Directive (STEP I)

No matter their level of sophistication around IP management, all firms can benefit from a top-down directive for their IP programs. The *Case Study* illustrates this point – it shows how NPD efforts can be compromised or fail if IP review of new product offerings is delayed or bypassed (Cohen V, 2006). Specifically, the *Case Study* is a reminder that late IP review can:

- Disrupt an NPD effort with eleventh-hour design changes that waste resources and introduce inefficiencies (e.g., scrapped inventory and mold tooling, unanticipated redesigns, and missed launch dates);
- Expose the firm to unanticipated IP risks and deprive the NPD team the ability to proactively avoid those risks (e.g., by missing the opportunity to design around a competitor's patent early in the original concept generation effort); and
- Forfeit valuable IP rights and opportunities (e.g., by prematurely disclosing inventions without first ensuring timely patent filings and confidentiality agreements).

Whether or not firms have experienced these pitfalls of delayed IP review first hand, a clear IP directive – their North Star – will guide them toward improved IP management.

Case Study 9.1

Imagine a ten-month NPD effort culminating in an improved surgical instrument for laparoscopic procedures. The instrument has a new handle design with superior performance. The NPD team readies for launch on schedule after completing a staged process – much like Robert Cooper's Stage-Gate^{TM} process – including idea generation, scoping, building a business case, development, and testing/validation. The instrument's design is now frozen: component inventory has been ordered, prototypes have been tested to satisfaction, custom mold tooling for the handle has been produced, and marketing has announced the launch date.

But at this late stage, the NPD team informs IP counsel about the project for the first time and asks counsel to confirm that the instrument is patentable and clear of infringement risks. In an urgent study, counsel discovers a competitor's patent that would be infringed by sales of the instrument's handle. Counsel also learns that the NPD team had openly disclosed instrument prototypes during VoC review and presented them at a trade show.

Because of the infringement risks posed by the competitor's patent rights, management decides to pull the proposed instrument to redesign it. In an all-hands effort, the instrument is redesigned – this time with input from IP counsel – to avoid the infringement risk. In its "design around" effort, the development team changes the instrument's handle design to clear the competitor's patent claims while satisfying performance and manufacturability requirements.

Though the redesigned instrument proved to be even better than the original design – requiring fewer components and having improved performance – the urgent redesign required scrapping the mold tooling and inventory, postponing the launch date, and repeating much of the development effort. It also burdened management with uncertainty and both tangible and intangible costs.

9.2.1 Proposed Directive

As alluded to earlier, this author believes that an IP directive ideally includes *protecting*, *respecting*, and *promoting* IP rights – keystone objectives of a robust IP program. As illustrated in Figure 9.2, this directive safeguards the firm's technology and design innovations (and the commercial advantages they confer), avoids conflicts with others' IP rights (and the risks and distraction associated with such conflicts), and accelerates the generation of new innovations and IP assets.

The *protect-respect-promote* directive forms a strong foundation for a robust IP program. It ensures *balanced IP management*, powers a *virtuous cycle of IP creation*, and establishes an *IP ecosystem* supportive of new product innovation.

- *Balanced IP Management*: The three objectives of protecting, respecting, and promoting IP do not reside in separate silos; instead, they harmoniously cooperate to ensure balanced IP management. If one is overlooked, the IP program is left incomplete. And if one is overemphasized to the detriment of others, the IP program is unbalanced. This "three-legged stool" of IP management won't stand if a leg is missing, and it will topple if a leg is under- or overdeveloped.
- *Virtuous Cycle of IP*: Each of these three objectives enables the others, and together they power a virtuous cycle of IP-catalyzed innovation. By *protecting* IP rights, the firm in turn gains leverage when *respecting* (or being challenged by) IP rights of others. And when design around efforts are needed to *respect* others' IP rights, those efforts *promote* (and even drive) new innovations and IP. Completing the cycle, *promoting* IP generates new *protectable* IP assets.
- *IP Ecosystem*: By advancing these three objectives, firms establish an ecosystem within which stakeholders are both "willing and able" to innovate (Hill et al.

FIGURE 9.2 IP DIRECTIVE.

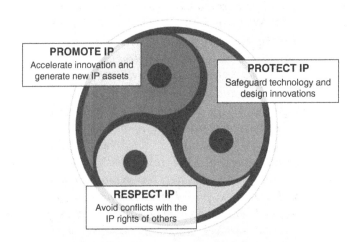

2014) – it provides the resources and environment that enable firms to innovate and generate valuable IP rights as a natural response to business challenges and opportunities.

9.3 Commit to IP Activities (STEP II)

Once an IP directive is adopted in STEP I, firms must commit to completing the specific IP-related activities that advance that directive. Those IP activities are summarized below, organized according to the *protect-respect-promote* objectives of the IP directive.

9.3.1 Activities to *Protect IP*

Protecting IP requires active harvesting of IP from within the firm, securing comprehensive IP rights for the harvested IP using all IP protection regimes, safeguarding proprietary information against inadvertent or early disclosure, and policing and enforcing IP rights as needed to maintain a competitive advantage. Two primary activities – *securing IP rights* and *safeguarding proprietary information* – are addressed here.

- *Securing IP Rights:* IP protection grants its owner the right to exclude others from unauthorized use of an intellectual asset. It therefore represents valuable leverage firms can use to safeguard and sustain a competitive advantage, including by deterring competitors from adopting the firm's technology and design innovations, by demanding royalty income if the firm elects to license its IP rights to others, or by asserting the firm's IP rights as needed to resolve disputes (Cohen VI, 2008). Especially in the case of consumer products, a full "cocktail" of IP rights can be secured, including utility patent, design patent, trade secret, copyright, trade dress, and trademark rights (Cohen I, 2002). The *IP Overview* provides a primer on IP rights, using Crocs™ Classic Clogs for illustration.
- *Safeguarding Proprietary Information:* Until inventions are published or otherwise shared openly outside the firm, they remain proprietary information – or secrets – of the firm. Even when patent protection is sought with the understanding that the patent (and the patent application in most cases) will be published with full details of an invention, the invention remains secret proprietary information of the firm until that publication occurs. NPD efforts, especially those developing consumer products, involve many market-facing activities when collaborating with customers for idea generation, VoC research, preliminary market assessments, concept testing or test marketing, and of course market launch. Steps should therefore be taken to protect proprietary information and avoid inadvertent or early external disclosures. Additionally, some inventions may be selected for protection by way of trade secret rights – in lieu of patent protection – if they cannot be reverse engineered, can be kept secret, and meet other criteria for trade secret protection.

IP Overview

Crocs™ Classic Clogs – originally conceptualized as boat shoes by three college friends sailing the Caribbean – illustrate various forms of IP rights. Crocs secured IP rights to protect its global sales, now exceeding 850 MM pairs sold in over 85 countries. With reference to Crocs™ Classic Clogs, here is an overview of IP rights and how they can protect a product's function, form, and source identifiers.

FUNCTION
Utility patent rights *protect new, non-obvious, and functional aspects of an invention for a limited term. US patents grant patent owners the right to exclude others from making, using, selling, offering to sell, and importing the claimed invention in the US for 20 years from the patent filing date. For example, US Patent No. 6,993,858 for "Breathable Footwear Pieces" allows Crocs to exclude others from selling footwear having, among other features, friction between a strap and a footwear base to maintain the strap in place in position after it is repositioned.*
Trade secret rights *protect proprietary information if the owner has taken reasonable measures to keep it secret and if the information derives value from not being generally known by others. Trade secret rights grant owners the right to bring an action if the trade secret is misappropriated, for as long as the trade secret remains secret. For example, Crocs' proprietary closed-cell resin material, Croslite™, is considered a substantial innovation in footwear comfort and functionality that enables Crocs to produce soft and lightweight, nonmarking, slip- and odor-resistant footwear.*
FORM
In contrast to utility patents, ***design patents*** *protect new, non-obvious, and ornamental aspects of an invention for a limited term. US design patents grant their owners the right to exclude others from making, using, selling, offering to sell, and importing the claimed invention for 15 years from the patent grant date. For example, Crocs' US Patent No. D 517,789 for "Footwear" allowed Crocs to exclude others from selling footwear having an ornamental design for footwear, as shown and described in the patent's specification and figures.*

(Continued)

Copyright rights protect original works of authorship – including pictorial, graphic, and sculptural works – that are fixed in a tangible form of expression. Copyright provides the owner with the right to exclude others from various acts – such as reproducing and displaying the protected work – for a limited time generally based on the life of the author plus seventy years after the author's death. For example, Crocs' subsidiary Jibbitz has obtained numerous copyright registrations for its decorative shoe charms for personalizing and customizing Crocs™ Classic Clogs.

SOURCE IDENTIFICATION

 Trademark rights protect indicators – any word, phrase, symbol, design, or combination of these things – that identify a source of goods or services. They grant trademark owners the right to exclude others from unauthorized use of a trademark or service mark with goods or services in a manner that is likely to cause confusion, deception, or mistake about the source of the goods and services. Unlike utility patents and design patents, trademark rights can last as long as the trademark is used in interstate commerce. For example, Crocs' US Trademark Registration No. 3,836,415 protects the word mark CROCS for use in connection with "lightweight slip-resistant footwear." Similarly, US Trademark Registration No. 4,179,090 protects the Crocodile Logo.

Trade dress rights protect a product's total image or overall appearance, including the shape, color, or any other nonfunctional product features. Like traditional trademarks, trade dress rights grant trade dress owners the right to exclude others from unauthorized use of the trade dress with goods or services in a manner that is likely to cause confusion, deception, or mistake about the source of the goods and services, and can last as long as it is used in interstate commerce. For example, Crocs' US Trademark Registration No. 5,149,328 protects the "three-dimensional configuration of the outside design of an upper for a shoe," the trade dress of the shoe featuring holes placed across the horizontal portion of the upper.

9.3.2 Activities to *Respect IP*

The commercialization of every new product or service innovation represents a new risk of infringing IP rights of others. This risk is especially acute when firms expand their offerings by launching products or services in new product categories or industries and encounter new competitors and technologies. Put simply, the development of new products not only generates important business opportunities – like the protection of IP assets – but also raises new IP risks. Two primary activities needed for respecting IP – *identifying IP risks* and *mitigating IP risks* – are addressed here.

- *Identifying IP Risks*: Firms are accustomed to planning for *known* business risks. But unanticipated risks can cause costly business disruptions like the eleventh-hour design changes, scrapped mold tooling, and wasted development resources illustrated in the *Case Study*. Early IP review in NPD efforts identifies risks and affords firms the opportunity to accept, avoid, or manage them proactively.

- *Mitigating IP Risks*: Once a risk is identified, the NPD team should communicate it to management so that steps can be taken to reduce the chances (uncertainty) that the risk will "hit" and/or reduce the amounts at stake (impact) if the risk actually does hit. Mitigation efforts can include "designing around" IP rights of others, electing to choose product concepts having reduced risk profiles, limiting commercialization strategically such as by jurisdiction or market, or by licensing rights to use others' IP.

9.3.3 Activities to *Promote IP*

A robust IP program will promote further innovation within the firm and should do so in a purposeful way. Firms can use patent literature, including issued patents and published patent applications, as it was originally intended – to accelerate their innovation of new product and service offerings. Patent literature can also be researched for gathering competitive intelligence. And an active IP program itself can increase IP awareness and foster a culture of innovation. Two primary activities for promoting IP – *inspiring ideation* and *collecting competitive intelligence* – are addressed here.

- *Inspiring Ideation*: Like many such systems worldwide, the US patent system encourages early publication of inventions – via patents and published applications – for others to build on. Patent publications spark further ideation when firms look at, and learn from, the work others have done. And by educating the firm's engineering and business corps about IP assets and the leverage they provide, development teams are encouraged and incented to innovate more.
- *Collecting Competitive Intelligence*: By monitoring patent filings of others, firms and their NPD teams learn about their current competitors and prospective competitors. Patent filings often reveal the trajectory of competitive development efforts and provide a window into competitors' R&D efforts and where they may be heading with new product or service offerings. Patent filings also reveal emerging competitors and industry newcomers that may later become significant rivals. Firms can also use competitive intelligence from patent filings to discover emerging technologies in their industry, identify prospective partners and acquisition targets, and generally inform product mapping, product positioning, and development efforts.

Once a firm commits to IP activities aligned with its IP directive in STEP II, firms can then select the specific IP deliverables required to complete those IP activities (and thus achieve the directive) in STEP III.

9.4 Select IP Deliverables for Each NPD Stage (STEP III)

Once the activities (of STEP II) needed to advance the firm's IP directive (of STEP I) have been identified, firms are poised to select the specific IP deliverables required of their NPD teams. Because virtually all effective NPD processes progress through defined stages from the inception of a development effort to launch, these IP-related deliverables are best described generally and in the context of the stages when they are best completed. Later, in STEP IV, guidance is provided for formally integrating specific IP deliverables into the NPD process itself.

To illustrate IP deliverables and generally where they properly fit in NPD processes, Cooper's Stage-Gate™ process (Cooper 2017) is simplified into an integrated NPD process – the STEPWISE™ process illustrated in Figure 9.3 – which incorporates IP activities and deliverables (Cohen III, 2004).

Like typical NPD processes, each stage includes defined activities and deliverables, and each gate includes defined criteria and output, as well as a decision and action plan for the next stage. But unlike typical NPD processes – aptly drawn in a left-to-right progression, the STEPWISE™ process allows us to think differently in an upward journey from a wide foundation of research from which ideas and concepts emerge, through defined stages into which IP-related activities are integrated, and to a pinnacle where a fully vetted product is introduced to the consuming public. It has five sequential stages – ideation, concept generation, feasibility, development, and commercialization – with which IP-related deliverables are aligned, thus ensuring that they are all completed as an organic outcome of the NPD process. IP deliverables are described for each sequential stage below.

9.4.1 Ideation Stage

Whether a new product development effort is considered incremental (such as a cost reduction or customer-requested modification), platform (next-generation product or line extension), or breakthrough (new to the firm or the world), NPD processes typically begin with a "fuzzy front end" of ideation (Koen 2005)

FIGURE 9.3 INTEGRATED NPD PROCESS.

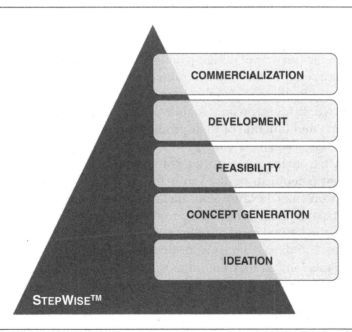

in which the NPD team defines the challenge and a path forward for innovative solutions. As IP deliverables, the NPD team should also (1) outline an IP strategy (*Outline IP Strategy* below) and (2) plan to learn from patent literature (*Conduct Search (Initial)* below).

Outline IP Strategy: Because the commercial success of new products depends not only on a first-to-market advantage, but also the ability to prevent knockoffs and fast followers from eroding that advantage (Cohen IX, 2014), it is important to set strategies to secure comprehensive IP rights and erect barriers for competitors, thus transforming a first-to-market edge into a sustained business asset. Design briefs provide the first and best opportunity to set a design project on the right path – they help ensure that the resulting design meets the NPD team's key criteria for a successful product design, including a strong IP position (Cohen VII, 2012). Among other IP strategies, the NPD team should consider (1) securing a *"cocktail" of IP rights*, (2) protecting both the *form and function* of innovative product designs, (3) protecting the entire *lifecycle of the product*, and (4) *strategic patenting.*

- *Cocktail of IP Rights*: Various IP protection regimes are available, and different forms of IP protection can protect different aspects of the same product. Consider Crocs™ Classic Clogs – described in the *IP Overview* above – as an example of a single product protected by utility patent, design patent, trademark, trade dress, trade secret, and copyright rights. To maximize the leverage of IP protection, an IP strategy should plan to secure a full "cocktail" of IP rights and provide a roadmap for IP to be generated by the NPD team and protected by the firm.
- *Form and Function*: Though designers blend form with function in the realm of consumer products (and then advertise products as "*The Fusion of Design & Technology*"), IP regimes for protecting such products separate form from function (Cohen II, 2004). A product can enjoy utility patent or trade secret protection for its functional attributes, but nonfunctional ornamentation is instead the purview of design patents, and trademark and trade dress protections are reserved for those nonfunctional features that can link a product to its source in the minds of consumers. If this form-function separation is respected, firms can plan – in their NPD team's IP strategy – to protect both the form and function of their product innovations.
- *Product Lifecycle*: The IP strategy should contemplate the entire lifecycle of the product, from launch to sunset. For instance, the exclusive rights afforded by a design patent can be used to help secure "secondary meaning" for trade dress protection, because a product's form can enjoy dual protection through design patent and long-lasting trade dress protections (Cohen II, 2004). Unlike the limited term of utility patents and design patents – 20 years from filing and 15 years from issue, respectively – trademark and trade dress rights can last as long as they continue to identify a product's source (Cohen IV, 2005). Honeywell's Round® thermostat exemplifies this strategy. It still enjoys trademark protection for its dome shape despite the expiration of the corresponding design patent long ago in 1970 (*In re Honeywell* 1988).

- *Strategic Patenting*: Strategic patenting seeks protection of a firm's competitive position in the context of the supply chain it occupies. For example, patent rights can safeguard the firm's downstream product sales, upstream component supplies, and help secure the firm's exclusive position as both a customer and a supplier up and down the supply chain. Additionally, IP protection is sought not only to cover the firm's core products but also for alternative concepts that may later be preferred by the firm or that may present viable alternatives for competitors. By obtaining exclusive rights to all of the best solutions to a product development challenge, commercial advantages over competitors are expanded.

In sum, an *Outline IP Strategy* deliverable should (1) specify the types of IP to be generated and protected, (2) anticipate the likely IP-related risks to be identified and mitigated, and (3) target the specific way(s) – by performance, aesthetics, or otherwise – the end product should differentiate from competitive offerings, so that those differentiators can be protected.

Conduct Search (Initial): By studying the "prior art" related to a new development effort, the NPD team can learn valuable information from the patent literature. It will help catalyze innovation, prevent "reinvention of the wheel," and allow the NPD team to improve upon the efforts of others. To date, over 11 million issued utility patents describe existing technical innovations, and nearly a million issued design patents illustrate ornamental product configurations. Pending utility patent applications are published every Thursday, typically eighteen months after they are filed. The result is a massive collection of patent literature – a vast (and underutilized) resource for inspiration, ideation, and competitive intelligence. NPD teams and managers can conduct or commission a "state-of-the-art" or "landscape" search of this patent literature, using a detailed classification system – which categorizes patents in searchable "classes" and "subclasses" based on technology for utility patents and ornamentation for design patents – as well as key words (USPTO 2004 and www.uspto.gov). Where a prolific inventor or current (or prospective) competitor is known, their patents and published applications are of particular interest to the NPD team. The results of this initial patent literature search are synthesized – with the assistance of IP counsel (Cohen IV, 2005) – and presented to the NPD team early in the NPD process, well before the development process is in full swing, to (1) *inspire ideation*, (2) *identify general risks*, and (3) *identify IP-related opportunities*.

- *Inspire Ideation*: Patent literature helps inspire ideation and guide later concept development efforts. Just as Apple's iPod® designers would have looked to Sony's 1979 Walkman® and related patent literature as a catalyst for ideation, NPD teams benefit from studying past product solutions. This is in fact a fundamental tenet of the US constitutional mandate "to promote the progress of science and useful arts." By encouraging inventors to promptly and thoroughly reveal their inventions (and the best way to produce them) in patents and published patent applications, patent systems in the US and worldwide make technologies available for review and improvement by others. While a patent search grants the NPD team access to efforts of others, a firm's own patent portfolio also provides a starting place when developing new products or expanding into new business opportunities (Schoppe and Pekar 2005).

- *Identify General Risks:* In addition to inspiring ideation, early patent searches reveal general areas of IP-related risk. Proactive identification of patent infringement risks (and the early elimination of risky product concepts during the concept generation and feasibility stages) streamlines the NPD process. Though specific risks will be identified and managed later in the feasibility and development stages after a targeted search is completed to evaluate specific product features, this early identification of general risk areas helps the NPD team to navigate IP minefields.
- *Identify Opportunities:* Patent searches often uncover valuable competitive intelligence and fruitful IP-related opportunities. Partners or merger targets may be revealed, and the patent literature will help the NPD team to develop strategies for securing IP rights from others as needed (by license or acquisition), explore areas for prospective patent protection, find technologies that are already available in the public domain in expired patents or abandoned patent applications, and learn about the trajectory of industry and competitive technologies.

In sum, a *Conduct Search (Initial)* deliverable should include a report to the NPD team with (1) patent literature that poses general infringement risks, (2) intelligence about specific and emerging competitors, (3) prior technologies that are available for use or that can spark ideation, and (4) potential IP-related opportunities such as licensors or acquisition targets.

9.4.2 Concept Generation Stage

In this stage, NPD teams generate product concepts based on criteria from technology, business, and operational disciplines. IP-related activities integrated in this stage not only facilitate concept review and selection, but actually foster continued ideation and differentiation (Cohen and Donnelly 2015). NPD teams should also (1) take measures to maintain confidentiality (*Establish Confidentiality* below) and (2) document their inventions (*Document Inventions* below).

Establish Confidentiality: Under US law, patent rights are generally barred if an invention is made public before a patent application is filed. Prefiling events – like trade shows, sales offers, and nonconfidential disclosures to customers – can frustrate prospective patent rights. While there are exceptions to this bar under US law (for certain events within a year before an application is filed), they are limited and do not apply in most other countries. If a product concept must be disclosed (for Voice of the Customer (VoC) review or to communicate with vendors for example), Non-Disclosure Agreements (NDAs) can help the NPD team avoid a bar to patenting. But ideally, IP filings should be completed before concepts are disclosed outside of the firm. Additionally, harvested inventions can be selected for trade secret protection if they cannot be reverse engineered, can be kept secret, and meet other criteria for trade secret protection. An *Establish Confidentiality* deliverable should therefore result in the execution of NDAs by vendors, customers, and collaborators outside the firm to ensure (1) the confidentiality of proprietary information, (2) the firm's exclusive ownership of IP rights, and (3) the avoidance of premature invention disclosures.

Document Inventions: The fruits of NPD team efforts need to be actively harvested and documented. Otherwise, they represent missed opportunities and lost assets left to die on the vine. Ideally, all inventive developments should be documented and presented to a management body – such as a patent committee formed from technology, business, and operations stakeholders – so that commercially valuable inventions can be selected for protection.

- Firms should adopt templates – whether in conventional form or a workflow environment – for documenting inventions. Templates should be tailored to align with the firm's technology, industry, and culture.
- In use, completed templates should provide details of the invention (its objective, differences from the closest known prior art, and a thorough description); information about any current or planned commercial activities relevant to the invention (whether it is likely to be embodied in a commercial product, the likelihood it will be of interest to the firm's competitors); possible secrecy of the invention (whether it can be reverse-engineered by competitors or kept as a secret); possible events that may bar patenting of the invention; and information impacting ownership of the invention (names of contributors inside or outside the firm).
- Once completed, this template should be reviewed by IP management to determine whether there is a business case for protecting the invention. And if so, what forms of protection are appropriate.

A *Document Inventions* deliverable should therefore result in completed Disclosure of Invention Templates for consideration by a patent committee or IP management for possible utility patent or design patent application filings or trade secret protection.

9.4.3 Feasibility Stage

In this stage, NPD teams must consider the feasibility of product concepts generated in the concept generation stage from perspectives of all involved disciplines. Marketing professionals, for example, ask whether it is feasible to promote selected product concepts and penetrate market opportunities, while engineers and designers consider whether it is feasible to develop the concepts within cost, performance, and aesthetic constraints. NPD teams must also consider whether product concepts are *legally* feasible – capable of protection and free of intolerable infringement risks. In this stage, NPD teams should (1) perform initial IP clearance (*IP Clearance (Screening)* below) and (2) update its review of the patent literature (*Conduct Search (Targeted)* below).

IP Clearance (Screening): Various competing concepts are typically generated upstream of the feasibility stage of the NPD process – during the earlier concept generation stage – and a product concept is chosen to be developed downstream in the later development stage. The feasibility stage therefore provides the best opportunity for the NPD team to screen out concepts that pose intolerable risks of infringing IP rights of others. For example, functional features of product concepts may come dangerously close to exclusive rights claimed in third-party utility patents, or an ornamental feature of a product concept may resemble a design claimed in another's design patent. Early screening of product concepts allows the NPD team to pivot away from risky concepts and focus resources on surviving product concepts. Also, the NPD team should select product concepts that offer greater differentiation from competitive products and prior patent literature – these are the product concepts that may give rise to stronger IP rights or even enjoy longer lasting trade dress or trademark rights. An *IP Clearance (Screening)* deliverable should therefore summarize a review of patents and published patent applications, identify any product concepts that should be screened out because of a high risk profile, and direct the NPD team's attention to product concepts having the greatest potential for differentiation and strong IP rights.

Conduct Search (Targeted): In contrast to (and yet supplementing) the initial state-of-the-art search performed earlier in the NPD process, a "targeted" search of the patent literature identifies patent documents specifically relevant to the product concept ultimately selected by the NPD team for further development. Now that features of front-running product concepts are known, yet well before a product design is "frozen," those features can be searched to identify any specific infringement risks. A *Conduct Search (Targeted)* deliverable should therefore report to the NPD team with the results of a "clearance" or "freedom-to-operate" search and identify patent documents that pose specific infringement risks to be mitigated by the firm.

9.4.4 Development Stage

In this stage, the NPD team develops the selected product concept. Specific IP risks and opportunities associated with features of that product concept must now be addressed. It is in this stage that the NPD team should (1) take actions to clear IP risks of concern (*IP Clearance (Design Around)* below) and (2) file patent applications (*File Applications* below).

IP Clearance (Design Around): If specific infringement risks have been identified by the NPD team, a "design around" effort evaluates the scope of protection provided by a patent of concern and helps the NPD team to navigate around that scope, thus moving the final product design outside the scope of others' IP rights. As illustrated in the *Case Study*, such design around efforts often lead to new innovations and IP opportunities; like other constraints in NPD efforts (such as aggressive price points, challenging performance requirements, and stringent specifications) that drive innovation through problem-solving exercises, the need to avoid infringement can also spark ideation. If a relevant published patent application has not yet been granted, the NPD team can monitor its progress as it is examined by the Patent Office in case it matures into a new infringement risk. An *IP Clearance (Design Around)* deliverable should therefore report to the NPD team regarding (1) the scope of any third-party IP rights of concern, (2) the steps taken to modify the proposed product to mitigate infringement risks, and (3) the results of any "design-around" efforts.

File Applications: By this stage, inventive aspects of the product design are developed sufficiently to be described and claimed in a provisional or nonprovisional patent application. It is important to uniformly apply clear, objective criteria when selecting which inventions should be protected by a patent filing. For example:

- Will the invention be embodied in a core product of the firm?
- Is the inventive technology likely to be platformed into a standard product offering?
- Will the firm's competitors need (or wish) to adopt the technology to compete with the firm?
- Will patent protection provide important business leverage to improve a particular business situation?

If an invention meets one or more of these criteria, then the effort and cost of patent protection may be justified. If not, then resources should be redirected to better-justified endeavors. A *File Applications* deliverable should therefore report on (1) the IP rights to be protected by filings, (2) the IP-related actions, including filing patent applications and trademark applications, to be completed consistent with the NPD team's IP strategy, and (3) the completion of those filings.

9.4.5 Commercialization Stage

In the commercialization stage, the now-completed product design is readied for launch by the NPD team. To memorialize assessments conducted in earlier stages and ensure that IP protections have been pursued, the NPD team should (1) document clearance positions (*IP Clearance (Documentation)* below), (2) confirm that IP protections are in place (*Confirm IP Protections* below), and (3) initiate an IP monitoring program (*Establish IP Monitoring* below).

IP Clearance (Documentation): Integrated NPD processes require formal IP clearance and confirmation that any risks associated with the launch of a product are tolerable. It may in some cases be prudent to memorialize steps taken to mitigate IP risks or the reasons why the product does not infringe any identified patents of concern. An *IP Clearance (Documentation)* deliverable should therefore (1) report on the completion of any risk mitigation steps, (2) communicate any substantial risks to management for consideration, and (3) memorialize the bases for clearance positions, such as by non-infringement or patent invalidity analyses.

Confirm IP Protections: During commercialization efforts leading up to launch, NPD teams should ensure that IP protections are in place, including trademark, patent, trade secret, and copyright protections, as appropriate. This includes possible foreign protection as well as ensuring coverage of the final product design.

- Foreign protection may be important when a product will be manufactured or sold overseas, or when value can be derived from licensing the product innovation to foreign companies. Criteria for foreign patenting filings should be clear and objective – inquiring where competitors are located, where products would be sold or imported, where products would be produced, and what costs will be incurred – so that business-oriented decisions can be made.

- Throughout NPD processes, product designs will inevitably evolve. So will the scope of prospective patent rights as patent claims are amended and patent applications progress through examination. It is therefore important to ensure that the claims granted in patents adequately cover the final product design. If the patent claims and product design diverge, then resulting patent rights will have sharply diminished value.

A *Confirm IP Protections* deliverable should therefore include a report on (1) the status of IP filings and (2) alignment between the product design and the description and claims of pending patent application filings.

Establish IP Monitoring: When approaching product launch, NPD teams can establish procedures for monitoring competitive activities – including competitors' new patent filings and relevant product releases. Competitive information is also collected from sales, distributors, and other customer-facing stakeholders. If competitive activities call for the enforcement of IP rights, action can be taken to license the firm's IP for royalty revenue, cross-license in exchange for another's rights, or otherwise use IP rights for leverage to challenge a competitive threat or infringement. An *Establish IP Monitoring* deliverable should therefore initiate monitoring of third-party product offerings and IP filings, including the monitoring of trademark usage, to identify infringing acts.

By selecting from among the foregoing IP deliverables to be expected from their NPD teams in STEP III, firms set in motion the activities (identified in STEP II) needed to advance the firm's IP directive (of STEP I). As illustrated in Figure 9.4, each NPD stage includes timed deliverables that enable firms to *protect*, *respect*, and *promote* IP rights.

FIGURE 9.4 MATRIX OF NPD STAGES AND IP DELIVERABLES.

Stage	Deliverable	Directive		
		Protect IP	Respect IP	Promote IP
Ideation	Outline IP Strategy	√	√	√
	Conduct Search (Initial)		√	√
Concept Generation	Document Inventions	√		√
	Establish Confidentiality	√		
Feasibility	IP Clearance (Screening)		√	
	Conduct Search (Targeted)		√	√
Development	IP Clearance (Design Around)		√	√
	File Applications	√		
Commercialization	IP Clearance (Documentation)		√	
	Confirm IP Protections	√		
	Establish IP Monitoring		√	√

9.5 Integrate IP into the NPD Process (STEP IV)

Integrated NPD formally infuses IP activities into NPD processes and ensures that end products meet not only the requirements of interdisciplinary NPD teams (Correa 2004), but also pass legal muster before they go to market. Now that the IP deliverables that firms expect from their NPD teams are selected (in STEP III), firms can integrate those deliverables into their NPD efforts. This means defining the IP deliverables succinctly and formally incorporating them into NPD efforts – into the NPD process itself and into the NPD team.

9.5.1 Define IP Deliverables

IP deliverables are *generally* identified above in STEP III, but they should be *precisely* defined in STEP IV; that way, NPD teams understand exactly what is expected of them. The definitions should be concise yet complete. Concise so that they won't bog down the NPD process documentation. Yet complete in that they provide the NPD team with the specific guidance it needs to properly complete each deliverable.

An example of deliverables and their respective descriptions are provided in Figure 9.5. Of course, one size does not fit all; instead, the definitions can and should be tailored – and the deliverables themselves should be selected – based on each firm's IP directive as well as its culture, NPD process, industry, product category, and other factors influencing the firm's IP management.

FIGURE 9.5 DESCRIPTIONS OF IP DELIVERABLES.

IP Deliverable	Description
Outline IP Strategy	Report to NPD team regarding (1) types of IP to be sought/protected, (2) IP-related risks likely to be encountered, and (3) way(s) the end product should differentiate from competitive offerings.
Conduct Search (Initial)	Conduct "state-of-the-art" or "landscape" search and report to NPD team regarding (1) patent literature posing general infringement risks, (2) IP intelligence about competitors, (3) prior technologies available for use and ideation, and (4) IP-related opportunities such as licensors or acquisition targets.
Document Inventions	Complete Disclosure of Invention Templates for consideration by IP management for utility or design patent application filings or trade secret protection measures.
Establish Confidentiality	Execute Non-Disclosure Agreement by vendors, customers, and others to ensure (1) confidentiality of proprietary information, (2) firm ownership of IP rights, and (3) avoidance of premature disclosures.
IP Clearance (Screening)	Report to NPD team regarding the review of patents and published patent applications and product concepts screened out because of their higher risk profile.
Conduct Search (Targeted)	Conduct "clearance" or "freedom-to-operate" search and report to NPD team regarding patent literature that (1) relates to features of proposed product and (2) poses specific infringement risks.
IP Clearance (Design Around)	Report to NPD team regarding (1) scope of third-party IP rights of concern, (2) steps taken to modify the proposed product to mitigate infringement risks, and (3) results of the "design-around" effort.
File Applications	Report to NPD team regarding (1) IP rights to be protected by filings, (2) IP-related applications, to be filed consistent with IP strategy, and (3) completion of those filings.
IP Clearance (Documentation)	Report to NPD team regarding (1) completion of risk mitigation steps, (2) substantial risks communicated to management, and (3) documentation of bases for clearance positions.
Confirm IP Protections	Report to NPD team regarding (1) status of IP filings and (2) confirmation of alignment between the product design and application filings.
Establish IP Monitoring	Initiate monitoring of third-party product offerings and IP filings to identify infringing acts.

9.5.2 Incorporate IP Deliverables into NPD Efforts

Every NPD program defines a social system in which design and development work is carried out (Krishnan and Ulrich 2001). That social system includes a cross-functional NPD *team* of individuals from various disciplines that collaborate throughout a staged NPD *process*. This structured NPD ecosystem presents an opportunity to integrate defined IP deliverables into the NPD *process* and into the NPD *team*.

Incorporate IP Deliverables into NPD Process: About 70 percent of business units utilize formal NPD processes (Boike and Adams-Bigelow 2005), and IP rights are optimized when actions (and precautions) are taken at the right times in such NPD processes. We saw from the *Case Study* that learning just prior to product launch that a product would infringe IP rights of others is inefficient at best and disastrous at worst (Cohen IV, 2005). Conversely, certain IP actions can be handled too early. Completing a targeted patent search well before a product design is "frozen," for example, means that later-developed features will not be searched and could introduce unknown infringement risks.

In STEP III, we aligned a number of IP deliverables with five general NPD stages for illustration. In practice, it is necessary to precisely time the firm's specific IP deliverables in actual stages of the firm's NPD process. That timing is based in part on when certain *risks* arise and when certain *milestones* occur.

- Regarding *risks*, NPD processes are generally designed to increase the likelihood of the success of the resulting product (and decrease the chances of failure). This includes decreasing the risk of a compromised IP position with weak (or no) IP protection or with intolerable infringement risks. IP-related activities are therefore best positioned in the NPD process to steadily and methodically reduce unknown IP risks.
- Regarding *milestones*, IP-related activities should be timed ahead of milestones when the stakes are increased. Timing is therefore dictated by milestones such as capital investments (for example expenses for mold tooling, inventory, and manufacturing equipment), product launch announcements, and significant commitments of resources for technology development.

Also, the deliverables should appear on checklists at the gates between NPD stages. These checklists ensure timely completion of deliverables within the respective stages and before the NPD team can progress to the next stage. IP oversight of the timely and proper completion of IP deliverables ensures accountability and a strong IP position.

Incorporate IP Oversight into NPD Teams: New product development is aptly considered the transformation of market opportunities into commercial products (Krishnan and Ulrich, p. 1), and effective NPD processes draw input from various disciplines to effect that transformation. Of best performing business units, about 80 percent of NPD teams are cross-functional, and such teams have been embraced by a great majority of businesses (Cooper 2005). The establishment of interdisciplinary NPD in industry is also reflected in academic programs like the Integrated Product Development (IPD) program at Lehigh University implemented in 1996 (Watkins et al. 1996), and in later programs (Melamed et al. 2004).

Though NPD teams may include members from a variety of functions including marketing, design/engineering, manufacturing/operations, finance, purchasing, customer support, and quality depending on the industry and product category (Kahn 2005), cross-functional teams generally include key representatives from design, engineering, and marketing disciplines (Cagan and Vogel 2002). Figure 9.6 illustrates that a target product design should meet overlapping criteria of technology, business, and operational disciplines. NPD teams should also include a representative responsible for IP management, to layer legal criteria with those set by the other disciplines. The IP representative should be made part of the NPD team from the start and should be fully supported by (or specifically include) IP counsel.

IP counsel is ideally engaged to facilitate the NPD effort and to promote innovation that can be protected. When fully integrated into NPD processes, IP counsel fosters ideation and concept generation and spots general IP risks and opportunities early on. Collaborative input from IP counsel thus supports the ideation and filtration functions that ready concepts for feasibility review. Additionally, IP counsel should be expected to help the NPD team to complete IP deliverables at each NPD process stage. Doing so ensures that IP tasks are completed and that the NPD process will not progress otherwise.

Once the IP deliverables have been specifically defined and thoughtfully incorporated into the NPD process, and once an IP representative or stakeholder is incorporated into the NPD team, firms can execute the updated process and deploy IP integration via procedures, documents, and education. As firms roll out updated NPD processes, IP awareness will improve and IP-related opportunities and risks will become more top-of-mind for the firm and its management.

The process of creating or updating an IP-integrated NPD process requires coordination among project managers that oversee the NPD process, IP counsel

FIGURE 9.6 INTEGRATED NPD TEAM.

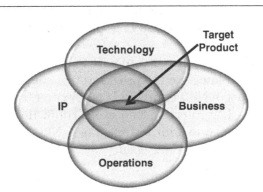

FIGURE 9.7 IP DIRECTIVE, ACTIVITIES, AND DELIVERABLES.

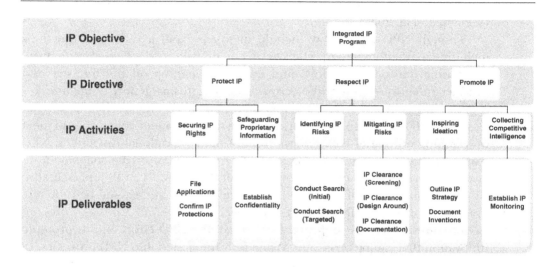

(whether in-house counsel or outside counsel), and executives responsible for successful NPD efforts. When done right, compliance with the NPD process ensures that IP deliverables are completed, IP activities are performed, and the IP directive is fulfilled (see Figure 9.7, illustrating the hierarchical arrangement of the IP directive, activities, and deliverables).

9.6 Conclusion

From this chapter, product development professionals are equipped with a pragmatic framework for integrating IP practices into their NPD processes, enabling their firms to protect valuable intellectual assets and catalyze further innovation. With this framework, NPD professionals can answer the *Who, What, When, How,* and *Why* of integrated IP: Who does the work of overseeing the IP actions? What are the IP activities and deliverables needed for a robust IP Program? When should those IP deliverables be completed? How are they best integrated and systematized? And, importantly, this chapter answered the question why we should care.

As NPD professionals strive to optimize their firms' NPD processes, they can (and should) adopt integrated IP programs that complete IP-related actions as a natural byproduct of NPD efforts and empower NPD teams to create products that enjoy strong IP protection and avoid conflicting with others' IP rights (Cohen VIII, 2014). And with an integrated NPD process they can reliably deliver this result time and time again.

Notes

1 Always consult counsel with your legal questions. This chapter represents the author's personal views and shouldn't be taken as legal advice; instead, it is provided solely for informational purposes.

2 In the Second Edition of this Handbook (Chapter 19), Laura Schoppe and Nancy Pekar explained strategies for extracting value from patent portfolios (Schoppe and Pekar 2005). In the Third Edition of this Handbook (Chapter 23), Rel Ambrozy explained the types of IP, the patent application process, ownership of IP, and trade-offs between trade secrets and patent rights (Ambrozy 2013). And in Chapter 10 of *The PDMA ToolBook 3*, Sharad Rastogi, Ari Shinozaki, and Matthew Kaness explained the patent lifecycle process, IP creation strategies, and IP deployment strategies (Rastogi et al. 2007).

References

Ambrozy, R. (2013). Understanding the most common types of intellectual property rights and applying them to the product development process. In: *The PDMA Handbook of New Product Development*, 3e, 368–384. John Wiley & Sons, Inc.

Boike, D.G. and Adams-Bigelow, M. (2005). *Trends and drivers of success in NPD practices: findings from the 2004 PDMA Foundation CPAS Survey*. Philadelphia, PA: PDMA Philadelphia.

Cagan, J. and Vogel, C.M. (2002). *Creating Breakthrough Products: Innovation from Product Planning to Program Approval*. Prentice Hall, 138.

Cohen, J.L. (2002). Coordinating enhanced IP for consumer products. *The Legal Intelligencer*, December (Cohen I).

Cohen, J.L. (2004). Managing design for market advantage: protecting both form and function of innovative designs. *Design Management Review* 15 (1) (Cohen II).

Cohen, J.L. (2004). *Integrating intellectual property review into product development processes: strategies for seizing IP opportunities and managing IP risks*. Philadelphia, PA: PDMA Philadelphia (Cohen III).

Cohen, J.L. (2005). Integrated new product development: what's law got to do with it? The role of legal counsel in new product development. In: *Proceedings of the 2005 Eastman Industrial Design Society of America (IDSA) National Education Conference* (Cohen IV).

Cohen, J.L. (2006). Strategies for integrating a strong intellectual property review into new product development processes. *Visions Magazine*, September (Cohen V).

Cohen, J.L. (2008). Sustaining the competitive edge of design innovations: strategies for protecting the fruits of design thinking in postmodern organizations. In: *Proceedings of the International Design Management Institute (DMI) Education Conference – Design Thinking: New Challenges for Designers, Managers and Organizations*. ESSEC Business School, Cergy-Pointoise, France (Cohen VI).

Cohen, J.L. (2012). Strategies for creating a strong intellectual property position. In: *Creating the Perfect Design Brief: How to Manage Design for Strategic Advantage* (ed. P.L. Phillips). Allworth Press (Cohen VII).

Cohen, J.L. (2014). Integrating intellectual property strategy into the design cycle: reinforced brand as a natural byproduct of the design process. In: *Proceedings of the 2014 Annual Conference of the American Intellectual Property Law Association (AIPLA)* (Cohen VIII).

Cohen, J.L. (2014). Integrated IP: a key ingredient of successful new product innovation. *INSIGHT Newsletter* 24 (1) (Cohen IX).

Cohen, J.L. and Donnelly, R. (2015). Deliberate differentiation: strategies for creating and protecting iconic designs (how planning trumps serendipity in pursuit of the real thing and other true-life stories of design protection). *Trademark Reporter of the International Trademark Association (INTA)* 105 (6).

Cooper, R.G. (2005). New products–what separates the winners from the losers and what drives success. In: *PDMA Handbook of New Product Development*, 2e, 3–28, at p. 25. John Wiley & Sons, Inc.

Cooper, R.G. (2017). *Winning at New Products: Creating Value through Innovation*. Basic Books.

Correa, A. (2004). Interdisciplinary academic projects within the industrial design, business, and manufacturing environments. In: *Proceedings, Eastman National IDSA Education Conference*, 57–61.

Hill, L.A., Brandeau, G., Truelove, E., and Lineback, K. (2014). *Collective Genius: The Art and Practice of Leading Innovation*. Harvard Business Review Press.

In re Honeywell, Inc., 8 U.S.P.Q.2d (BNA) 1600 (Trademark Trial & App. Bd. 1988).

Kahn, K.B. (2005). The PDMA glossary for new product development. In: *The PDMA Handbook of New Product Development*, 2e, 582. John Wiley & Sons, Inc.

Koen, P.A. (2005). The fuzzy front end for incremental, platform, and breakthrough products. In: *PDMA Handbook of New Product Development*, 2e, 81–91, at p. 82. John Wiley & Sons, Inc.

Krishnan, V. and Ulrich, K.T. (2001). Product development decisions: a review of the literature. *Management Science* 47 (1): 1–21, at p. 11.

Melamed, S., Page, A.L., and Scott, M.J. (2004). Lessons learned year two: teaching interdisciplinary product development at the University of Illinois at Chicago. In: *Proceedings, Eastman National IDSA Education Conference*, 171–175.

Overview of the Classification System, United States Patent and Trademark Office (USPTO), December 2002 Edition, Rev. 3 (2004).

Rastogi, S., Shinozaki, A., and Kaness, M. (2007). Intellectual property and NPD. In: *The PDMA Toolbook for New Product Development*, 3e, 275–313. John Wiley & Sons, Inc.

Schoppe, L.A. and Pekar, N. (2005). Extracting value from your patent portfolio. In: *The PDMA Handbook of New Product Development*, 2e, 302–318, at p. 303. John Wiley & Sons, Inc.

Watkins, T.A., Ochs, J.B., Boothe, B.W., and Beam, H. (1996). *Learning across functional silos: Lehigh University's integrated product development program*. Orlando, FLA: Education Innovation in Economics and Business, Orlando.

Joshua L. Cohen, Esq. is an IP counselor, and a shareholder and director of the IP law firm RatnerPrestia. He founded and chairs RatnerPrestia's Design Rights Group, which protects products ranging from luxury automobiles and time-pieces to consumer electronics for some of the world's most innovative firms. Joshua has authored numerous articles and book chapters about IP law and strategy. He is an Adjunct Professor in Lehigh University's graduate Technical Entrepreneurship program, and has been a frequent Lecturer in graduate programs of the University of Pennsylvania, New York University, and Thomas Jefferson University. He also lectures regularly on topics of IP strategy, patent practice, and innovation at conferences of AIPLA, DMI, IDSA, INTA, LES, and PDMA. Joshua is President Emeritus of the Product Development and Management Association (Greater Philadelphia Chapter), and President Emeritus of the Benjamin Franklin American Inn of Court, an association formed to advance ethics, professionalism, and civility in IP law.

CHAPTER TEN

MANAGING THE FRONT END OF INNOVATION (FEI): GOING BEYOND PROCESS

Jelena Spanjol[1] and Lisa Welzenbach

10.1 What Is the Front End of Innovation (FEI) and Why Is It Important?

In the 2019 PwC annual global CEO survey, 62% of CEOs reported pursuing new product development as a key instrument to drive revenue growth (PwC 2019). Yet, 55% of those same CEOs indicated they were extremely concerned with their ability to innovate effectively – a clear disconnect between what is needed and what is possible. Indeed, business leaders have been consistently reporting innovation *both* as a strategic priority *and* a critical challenge in their organizations.

One central challenge to innovating effectively is populating the innovation portfolio with the right mix of breakthrough and incremental innovation. A study of over 22,000 new product introductions over almost 20 years in the consumer packaged goods sector (Sorescu and Spanjol 2008) demonstrates that breakthrough innovation (i.e., products with new to the world consumer benefits) delivers organizational growth, while incremental innovation (i.e., products that include variations of existing benefits) merely ensures a company's survival. What firms ultimately introduce into the market is determined early in the innovation process (i.e., during the front end of innovation), not at the launch stage. As a result, comprehensively and effectively managing the front end of innovation (FEI) is essential to organizational vitality and new product success (Markham 2013).

The PDMA Handbook of Innovation and New Product Development, Fourth Edition. Edited by Ludwig Bstieler and Charles H. Noble.
© 2023 John Wiley & Sons, Inc. Published 2023 by John Wiley & Sons, Inc.

Formally, the FEI represents "work that is done toward developing a product before that project enters into the formal product development system" (Markham 2013, p. 77). It encompasses the *identification and assessment of problems to solve and solution alternatives.* In contrast to the formal new product development (NPD) process, innovation managers are often challenged with variable funding models in the FEI, having to "bootleg" projects, all while being unable to plan out major insights and turning points and deliver clear revenue expectations (Koen et al. 2001). In other words, the FEI is a balancing act along two dimensions: flexibility↔stability and creativity↔discipline (Gassmann and Schweitzer 2014). As a result, an effective FEI requires a comprehensive management approach across four aspects – process, ownership, people, and metrics – enabling companies to ask four key questions:

Process: *How should we conduct the FEI?*
Ownership: *Where in our organization does the FEI belong?*
People: *Who do we involve in the FEI and what kind of leadership is needed?*
Metrics: *How do we assess our FEI effectiveness in delivering on objectives?*

The remainder of this chapter will briefly characterize each FEI management domain and provide evidence-based insights as well as short illustrations. Along the way, we also highlight selected sustainability considerations that may arise during the FEI, based on interviews we conducted with over 50 small- and medium-sized (SME) companies in Germany (Welzenbach and Spanjol 2022). We note that the chapter only provides an overview, not an exhaustive account of FEI management practices – more detailed discussions are delivered in other chapters. Our main objective is to argue for a comprehensive approach to FEI management, going beyond the typical process view (i.e., how do we conduct FEI activities) and addressing the interdependence among FEI management domains (i.e., process, ownership, people, and metrics).

10.2 Managing the Process: How Should We Conduct the FEI?

Typically, FEI management in companies centers around *how* to conduct various FEI activities, i.e., what the FEI process should look like. While most companies implement some variation of a formal NPD process built sequentially around activity stages and decision gates (Cooper 1998, 2014) for the back end of innovation, an effective FEI process demands moving away from a sequential toward a nonlinear, iterative, and concurrent set of activities (Koen et al. 2001). Figure 10.1 illustrates how key activities feed into the formal back end of innovation. Those key FEI activities (elaborated on in the next sections) are:

1. Understanding the problem to be solved (i.e., research),
2. Generating possible solution alternatives (i.e., ideation), and
3. Testing as well as refining selected solution alternatives (i.e., concept testing and prototyping).

FIGURE 10.1 FRONT END OF INNOVATION VS BACK END OF INNOVATION.

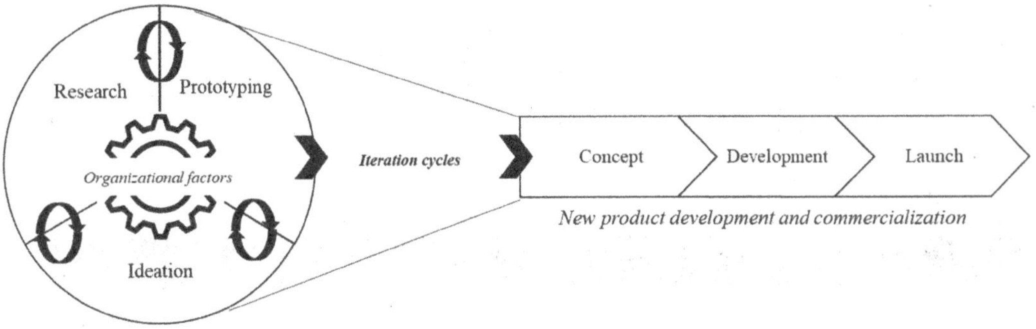

Source: Adapted from Koen et al. (2001) and Gaubinger and Rabl (2014).

10.2.1 FEI Process Building Block: Research

Research is a discovery process and provides the foundation for hypotheses regarding both the problem to be solved and the solutions that might be implied by those hypotheses. Because the FEI is problem-driven, a detailed understanding of the context, the stakeholders (users, buyers, sellers, intermediaries, etc.), and the motivations of those stakeholders is required. Managing the research component of the FEI process requires openness to hear and refute assumptions, revise interpretations, and update expectations.

Companies can deploy many different research tools and approaches, spanning primary research (conducted for the specific purpose of the NPD project through surveys, interviews, focus groups, and observation, for example) and secondary research (relying on data collected for other purposes, such as sales reports, customer feedback, government data, discussion forums, and published studies, among others). Both are needed to lay a strong foundation for understanding the problem space. Yet, most companies (over) rely on what customers say, using predominantly interviews and focus groups *both* when getting started with exploring the problem space *and* when moving closer to a final NPD concept (see Figure 10.2). This means that many companies are using the same tools when they are trying to understand the problem space (i.e., early FEI) as well as when they are refining and testing concepts (i.e., late FEI). Put differently, interviews and focus groups may be used without properly differentiating the underlying objectives (i.e., understanding problems versus testing concepts). One possible explanation for this is that innovation teams choose research methods based on familiarity, rather than on how well the research methods fit with the objectives in the early versus late FEI (Creusen et al. 2013; see Figure 10.2).

Yet, discovering customer and market insights that lead to robust problem definitions in the FEI and enable breakthrough innovation concepts to emerge is unlikely to happen by simply listening to (potential) customers

FIGURE 10.2 HOW COMPANIES LEARN ABOUT PROBLEMS TO SOLVE AND DEVELOP ALTERNATIVE SOLUTIONS.

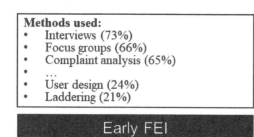

Methods used:
- Interviews (73%)
- Focus groups (66%)
- Complaint analysis (65%)
- …
- User design (24%)
- Laddering (21%)

Early FEI

Methods used:
- Interviews (64%)
- Focus groups (55%)
- Brainstorming with customers (50%)
- …
- Co-design (28%)
- User design (24%)

Late FEI

process

Source: Adapted from Creusen et al. (2013).

through interviews and focus groups. As demonstrated by Griffin and Hauser (1993) in their seminal Voice of the Customer (VOC) study, conducting just ten interviews and focus groups enables the discovery of about 80–90% of customer attitudes and opinions as well as usage and buying situations and processes. They do not, however, provide unique insights into latent needs and complex socio-cognitive, emotional, and culturally embedded factors that shape human behaviors – insights needed for generating breakthrough innovation.

In contrast, rich ethnographic accounts (Wasson 2000) enable deep market learning (Cayla and Arnould 2013; see Spotlight 10.1) as a foundation for more radical new product ideas. Thus, organizational practices that allow for a deep understanding of the problem space are critical for innovation success. This is why organizations such as Virgin Atlantic "observe and shadow customers, rather than involve them in formal surveys or focus groups" (Micheli and Perks 2015). As cited in Micheli and Perks (2015, p. 211), the head of Design at Virgin Atlantic explains why this is so important: "It wasn't necessarily asking people what they want; it's about looking at someone and seeing how they behave, seeing the things that irritate them, not necessarily asking them, because often they don't know what they want, until you show them what they could have." Virgin Atlantic's "Departure Beach" exemplifies how a better understanding of (latent) customer needs and problems through observing flight-related behaviors enabled the company to question existing offerings and create a space for customers that serves the needs identified through this ethnographic practice.[1] (see Spotlight 10.1). The "jobs to be done" approach (Christensen et al. 2016; Ulwick and Bettencourt 2008) is another method that probes customer needs and generates deep understanding.

In sum, the discovery of a more complete picture of the problem space (and particularly the users inside that space) requires multiple research methods. Each method uncovers a different facet of the problem space that is subsequently deployed toward more effective ideation.

SPOTLIGHT 10.1   ETHNOGRAPHIC MARKET LEARNING IN THE AIR TRAVEL INDUSTRY

Understanding **how, why, and when** customers seek relief and comfort at the airport allowed Virgin Atlantic to identify a unique concept called the "Departure Beach" for their Barbados destination.

What is the status quo? Ethnographic inquiries of the **problem space** led to key insights:

- Possibly long waits at the airport (e.g., because of early hotel checkout times or overnight flights)
- Wait is neither comfortable nor enjoyable; passengers sleep twisted on the seats, or even lie on the floor trying to relax and prevent health issues (e.g., legs and feet swelling)
- Potential delays make things even worse

What might the solution look like? Virgin Atlantic introduces the Departure Beach:

- Virgin picks up travelers from hotel (taking care of their luggage) and transfers them to the Departure Beach
- Travelers can enjoy time on the beach or in air-conditioned lounges while waiting
- Full shower and changing facilities are provided so enjoyment of Departure Beach does not come with reduced travel comfort (e.g., sand in shoes)
- Reasonable prices while at Departure Beach; lunch and non-alcoholic drinks included in ticket
- Transfer to airport synchronized with actual departure time

(For more on Virgin Atlantic's Departure Beach, see *https://www.gq-magazine. co.uk/article/virgin-departure-beach*, *https://hungrybecky.com/the-departure-beach-barbados.html*)

10.2.2 FEI Process Building Block: Ideation

Ideation is a structured and generative process that leverages iteration. To effectively ideate, a deep understanding of the problem space is needed (generated through effective research). Structure in the ideation process comes from tools and techniques that allow for maximizing options for the solution space (Cooper & Edgett 2008). However, according to a Booz & Company survey, companies are woefully ineffective with their ideation efforts: "Perhaps the most surprising result of our study of the up-front innovation process is how many companies say they simply aren't very good at it. Just 43% of participants said their efforts to generate new ideas were highly effective, and only 36 percent felt the same way about their efforts to convert ideas to product development projects. Altogether, only a quarter of all respondents indicated that their organizations were highly effective at both" (Jaruzelski et al. 2012). The same survey also showed a strong correlation between companies' ideation effectiveness and financial performance.

Companies can leverage a wide array of ideation methods, either seeking to counteract mental blocks (i.e., intuitive methods, such as brainstorming, fishbone diagrams, and method 635) or to systematically decompose and analyze problems (i.e., logical methods, such as TRIZ and analogical ideation; see Spotlight 10.2) (Shah et al. 2003). Importantly, ideation methods need to be aligned with innovation objectives in order to be effective.

10.2.3 FEI Process Building Block: Prototyping

Once ideas are more refined, the probing of form, function, and appearance parameters begins through the use of prototyping. Prototyping is an iterative process to test these alternative solution ideas. Different forms of prototyping

SPOTLIGHT 10.2   FROM IMPROVING A VEGETABLE PEELER TO CLEANING UP A MAJOR OIL SPILL WITH ANALOGICAL IDEATION.

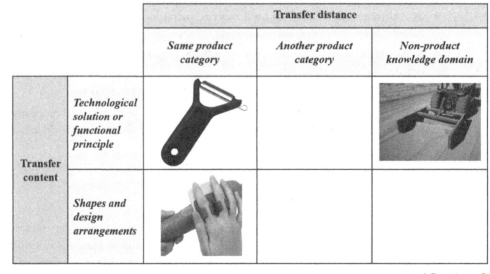

		Transfer distance		
		Same product category	*Another product category*	*Non-product knowledge domain*
Transfer content	*Technological solution or functional principle*			
	Shapes and design arrangements			

(Continued)

SPOTLIGHT 10.2 (CONTINUED)

Status quo: "Unwieldy", traditional potato peeler
Not easy or comfortable to hold; makes peeling potatoes difficult and lengthy
How can we modify or improve the peeler to address these problems?

Source domain(s)

FROM WHERE to transfer solution and/or design aspects/elements to the "unwieldy" peeler?

- **Same product category**: E.g., another peeler (whose use is more comfortable etc.)
- **Another product category**: E.g., peeler for other (food) products, razors
- **Non-product knowledge-related domain**: E.g., Other areas where something has to be removed evenly and thinly, such as stripping soil

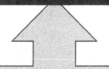 ANALOGICAL TRANSFER

Transfer content

WHAT can be transferred from other product categories or knowledge domains to the "unwieldy" peeler?

- **Technological solution** or functional principles
- Shapes and **design** arrangements

Target domain
(analogical ideation target)

"Unwieldy", traditional potato peeler

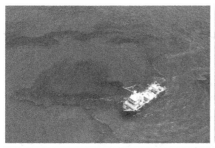

Original problem: Cleaning up a large oil spill	**Analogical problem: Tagging whales for scientific studies**
Structural elements of the problem:	*Structural elements of the problem:*
▪ Oil pools are moving across a large body of water and are difficult to identify without a large fleet of specialized boats and staff ▪ Dispatching cleanup teams to identified oil pools takes time and oil pools migrate by the time crew arrives	▪ Whales are moving across a large body of water and are difficult to identify without a large fleet of specialized boats and staff ▪ Dispatching science teams to identified whales takes time and whales migrate by the time crew arrives

Transfer of solution elements from analogical problem (whale tagging) to original problem (oil spill cleanup)

Unmanned water monitoring craft with solar energy panels and integrates sensor and communication electronics

allow for various types of solution testing (see Figure 10.3). As prototyping continues, the focus shifts from assessing the appropriateness of a potential solution to refining of ideas and concepts. Most importantly, prototypes are created to learn something specific, at the minimum effort required (Frascara 2002).

Managing the prototyping process during the FEI effectively requires innovators to recognize that prototypes represent "a kind of material conversation" between the FEI team and user or customer; prototyping is thus critical as a learning instrument since "objects mediate our experience and understanding" (Poggenpohl 2002, p. 70). Yet, prototyping can also lead to wasted time and effort during the FEI if the most important hypothesis engendered in the concept or the most challenging technical component is not tackled first. The "singing monkey on the pedestal" problem[2] illustrates this challenge: rather than prototyping the pedestal first, innovation teams need to tackle the "singing monkey" as it is the most challenging component of the concept.

In sum, the first FEI building block to be managed (i.e., the process itself) consists of three key activities (research, ideation, and prototyping). Who is in charge of this process, however, differs across companies. We outline two key considerations for FEI ownership next: governance structure and functional engagement.

FIGURE 10.3 HOW DIFFERENT PROTOTYPES MOVE IDEAS TOWARD CONCEPTS.

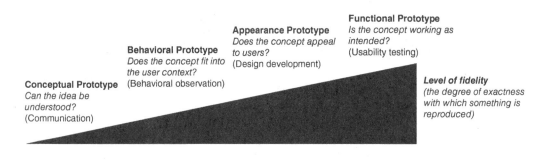

What about sustainability and the FEI process?

How (effectively) sustainability considerations are integrated in innovation processes and outcomes is already shaped in the FEI.

- *FEI research: Innovating (more) sustainably requires the consideration and integration of a firm's stakeholders' interests during NPD. The earlier firms adopt a stakeholder perspective, the more likely it becomes to address stakeholder interests in innovation projects (Juntunen et al. 2019; Stilgoe et al. 2013). Especially early FEI research plays a critical role in innovating more*

(Continued)

> *sustainably and should not only focus on the customer as one stakeholder group but also gather insights and knowledge from other groups, such as suppliers and other value chain partners.*
>
> - *FEI ideation: In the ideation phase, it is important to explore potential (intended and un-intended) societal and environmental consequences arising from innovation. A useful approach are scenario analyses or anticipatory life cycle assessments, which require individuals to think in desirable and/or undesirable futures related to NPD (Bocken et al. 2014; Wender et al. 2014).*
> - *FEI prototyping: Given that potential societal and environmental consequences connected to NPD are uncertain, continuously testing underlying hypotheses is important to be able to quickly adapt to new knowledge as it emerges (Stilgoe et al. 2013). This is especially important in the FEI as firms are in a better position to avoid negative societal and environmental impacts before these emerge in later NPD phases.*
>
> *In sum, a good FEI process management helps firms in thinking ahead of possible (un)intended societal and environmental impacts connected to innovation, especially through listening to stakeholders and iteratively exploring these potential impacts (and responding to them) in early phases of NPD.*

10.3 Managing Ownership: Where in Our Organization Does the FEI Belong?

One key outcome of effective FEI management is the population of the innovation pipeline with breakthrough innovation concepts. This is important as breakthrough innovations are needed for the organic growth of firms (Jelinek and Schoonhoven 1990; Sorescu and Spanjol 2008). Yet, breakthrough innovation is more difficult (compared to incremental innovation) because of the associated higher level of uncertainty and perceived risk stemming from longer development times (Leifer et al. 2000), conformist decision-making cultures in companies (Benner and Tushman 2003; Pech 2001), and the confusion of roles and responsibilities at project and organizational levels (Jones and Butler 1992). In addition, breakthrough innovation projects often face a hostile environment in ongoing operations, which seek to minimize variation, exploit the known, and develop repeatable processes. As a result, gaining clarity on appropriate *governance* of FEI processes and activities and *functional ownership* is critical (see Figure 10.4).

10.3.1 FEI Organizational Governance: Centralized vs Decentralized

According to the 2017 BCG Global Innovation Survey, companies find themselves in one of three roughly equally sized innovation governance segments: a third of companies' innovation function governance is fully centralized, a third is fully decentralized (owned by business units), and a third has some business unit and some centralized governance elements. Yet, the same survey also found

FIGURE 10.4 FEI OWNERSHIP AT ORGANIZATIONAL AND PROJECT LEVELS.

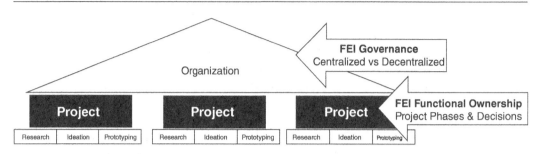

that over 70% of companies with strong innovation performance follow a centralized innovation governance model (BCG 2017). This makes good sense as centralized FEI ownership allows for coordinated strategic investment and leadership. To ensure an effective centralized FEI governance model, transparent investment, and decision-making during innovation project selection are required, along with decision-makers representing functional and business unit stakeholders (Robeson and O'Connor 2007). Importantly, such centralized FEI governance needs to be focused on "achieving objectives" rather than simply moving through the stages of the FEI (Griffin et al. 2007).

Yet, there are limitations to FEI centralization, especially when organizations are absorbing innovation projects through acquisitions. Moving too quickly to a centralized governance model can have particularly detrimental effects on new product introductions if the firms being acquired do not have any experience commercializing new products themselves (Puranam et al. 2006). Thus, any decision regarding whether and how to centralize FEI management must consider the circumstances and characteristics of the organization.

10.3.2 FEI Project Ownership: Roles and Responsibilities

At the project level, a myriad of ownership-related issues usually arise. We highlight two in this section: psychological ownership and functional ownership.

10.3.2.1 Psychological Ownership. Not surprisingly, as individuals and teams work on identifying problems and developing new product ideas, attachment to those ideas also grows. Such possessiveness can be both cognitive (thoughts) and affective (emotions). Formally known as psychological ownership, it refers to "a relationship between a person and an object (material or nonmaterial) in which the object of ownership is experienced as being closely connected to the self, that is being part of the extended self" (Pierce et al. 2003, p. 86). On the upside, psychological ownership can increase stewardship, responsibility, and investment. On the other hand, psychological ownership also increases territoriality, resistance, and control.

In the FEI, such psychological ownership dynamics can wreak havoc, especially during idea development and project transitions. Imagine an innovation team member ideating around a particular problem. By investing effort and time into idea development, feelings of psychological ownership naturally arise and strengthen over time. Ownership of ideas could either encourage acceptance or lead to rejection of input and feedback from the remainder of the team. Feelings of ownership might lead to an "I own it, so I want to make it better" logic. Although it might also spur feelings of "I own it, so I want to protect it." It turns out that idea owners with strong psychological ownership feelings will welcome additive feedback (i.e., input that expands or builds on an idea) but reject subtractive feedback (i.e., input that reduces an idea) (Baer and Brown 2012). In sum, both psychological ownership tendencies and feedback formats need to be managed carefully during the FEI.

10.3.2.2 Functional Ownership. Psychological ownership can also play out across functional lines. Many times, technical development teams (e.g., R&D or design staff) and business managers (e.g., product or brand managers) will be at odds regarding the ownership of innovation projects as they move toward the later stages of the FEI and toward entering the formal NPD process. What might have been a bootlegged FEI project "owned" by technical or design staff, will become "owned" by business managers for market viability evaluation decisions, for example (Alexander and Barkema 2015). Naturally, the psychological ownership of the technical or design staff will collide with the decision ownership of the management team. The FEI therefore represents "contested territory" with organizational, technological, and process dynamics preventing a full "resolution" of ownership struggles (Alexander 2014). Thus, a key management aspect in the FEI is ensuring clarity in roles. Innovation teams should be clear on who is in charge of the project management, who owns the research or technical input, and how these (and other) roles and ownership parts interact and integrate. A clear organizational chart, transparency, and consistent and open communication are essential.

A great way to zero into functional and psychological ownership issues is to examine hand-off dynamics during the FEI. Hand-offs represent a shift in formal responsibility (i.e., organizational accountability) for continuing the development of an idea or concept as those move between an originator and receiver (Rouse 2016). If difficulty in letting an idea or concept go arises during hand-offs, if the receiver exhibits resentment in having to take on an idea, or if there is ambivalence after the hand-off, a better-managed ownership transfer process between stages of the FEI is likely necessary. Other behaviors can tip off innovation managers that ownership struggles exist and must be addressed. For example, R&D staff might be reluctant to participate in meetings with product managers, unwilling to disclose work schedules to delay scheduling meetings (Alexander 2014; Alexander and Barkema 2015). This reluctance could be a reflection of a psychological and functional ownership collision and lack of ownership clarity in the team.

> **What about sustainability and FEI ownership?**
>
> *Our insights from a qualitative study among innovation and sustainability managers from Germany show that questions related to FEI ownership can have an impact on how sustainability aspects are considered during NPD.*

- *Innovation governance: Dedicated sustainability positions or departments can play an important role for a more sustainable FEI and therefore NPD processes and outcomes. Instead of operating as an isolated function (as is often the case), sustainability departments need to serve as sparring partners and "knowledge islands" for NPD teams and assist with specialized societal and/or environmental knowledge during the FEI, in the sense of cross-functional collaboration. Sustainability departments can secure a dedicated space for in-depth exploration of societal and environmental NPD considerations for which the innovation function often lacks time and expertise. Ideally, they also have voting power in NPD decisions.*

- *Employee sustainability commitment: Similar to the concept of psychological ownership, raising the sustainability commitment among NPD teams can foster the consideration and integration of sustainability aspects in the FEI. This can happen, for example, through dedicated sustainability trainings and workshops or through the establishment of recruiting criteria that explicitly require a certain degree of sustainability knowledge.*

In sum, flexible and collaborative organizational structures, combined with high levels of sustainability commitments, represent important catalysts for a sustainable FEI.

10.4 Managing People: Who Do We Involve in the FEI and What Kind of Leadership Is Needed?

A successful FEI requires management of people in two key domains: leadership and teams. Since companies typically have multiple innovation projects going on concurrently, a coordinated portfolio approach to ideation, for example, demands both leadership and team dynamics to be accounted for. There are three key aspects in a portfolio approach to ideation (one of the key FEI activities, see Section 10.2.2): (a) organizational conditions under which a sufficient number of creative and valuable ideas can be generated; (b) processes and routines that need to be in place to effectively evaluate, select, and further elaborate the most promising ideas and concepts; and (c) how to ensure that ideas and opportunities relevant to the firm's future business are addressed (Kock et al. 2015). These three factors (creative encouragement, ideation strategy, and FEI process formalization) depend on effective FEI leadership putting in place relevant practices and innovation teams carrying them out.

10.4.1 FEI Leadership

Interestingly, only about a third of Fortune 500 firms have a dedicated senior innovation executive (Lovric and Schneider 2019), such as a Chief Innovation Officer (referred to as CIO or CINO). While popular business press offers some insight into the typical tasks of a CIO (Di Fiore 2014; Kaplan 2019), the position profile and scope have to fit with the organization's strategic orientation and structure. Thus, one company might need an "investor" type CIO while a close competitor might benefit from an "advocate" type CIO (see Lovric and Schneider 2019 for other CIO types). Creating a CIO position, however, is not without controversy, with some arguing that such a position relieves the rest of the organization from the responsibility to innovation (Swoboda 2020).

Whether a company decides to create a CIO position or not, effective innovation requires strong leadership. Yet, even with a CIO or other form of innovation leadership, a differentiated approach to the FEI might not be in place. FEI leadership requires a different approach and distinct skills from those necessary for the formal NPD (i.e., the back end of innovation). While the formal NPD process (often a stage-gated one) requires a focus on execution, control, and discipline, an effective FEI requires a more flexible leadership approach centered on organizational creativity and motivation of engaged employees (Deschamps 2005; see Table 10.1).

Effective FEI leadership also requires a shift toward embracing "failure" as learning opportunities (Cannon and Edmondson 2005), as most "failure" in the FEI context is essentially an unexpected outcome during experimentation, hypothesis testing, assumption probing, and uncertainty reduction activities (Edmondson 2011; see Table 10.2).

TABLE 10.1 FEI VS FORMAL NPD LEADERSHIP.

	Front End	Back End
Aims	Explore opportunities, generat and select great ideas on the right customer problem, turn ideas into viable/desirable/feasible concepts	Convert concepts into winning new products, bring them to market quickly and cost effectively
Key Requirement	Organizational creativity	Organizational discipline
Leader characteristics	• Extreme openness to and curiosity about the external world	• Implementation focus
	• Out-of-the-box thinking	• Operational knowledge
	• Patience regarding results	• Speed in decision and action
	• Acceptance of risks	• Pragmatic risk management
	• Willingness to experiment and tolerance of failures	• Coordination skills
		• Urge to win in the market
People management	• Perceived as inspiring visionaries, great motivators and good coaches	• Willingness to lead their troops from within a group and to engage in product battles
	• Effective at attracting, retaining, and motivating innovators and entrepreneurs	• Demanding, particularly in terms of time-to-market
	• Create an exciting environment of adventure and challenge	• Tend to be available, accessible, and emotionally committed
Leadership	• Transformational leadership style	• Transactional leadership style

Source: Adapted from Deschamps (2005).

TABLE 10.2 THE NECESSITY FOR LEARNING-ORIENTED FEI FRAMES.

	Traditional Frame	Learning-oriented Reframe
Expectations about failure	Failure is not acceptable	Failure is a natural byproduct of a healthy process of experimentation and learning
Beliefs about effective performance	Involves avoiding failure	Involves learning from intelligent failure and communicating the lessons broadly in the organization
Psychological and interpersonal responses to failure	Self-protective	Curiosity, humor, and a belief that being the first to capture learning creates personal and organizational advantage
Approach to leading	Manage day-to-day operations efficiently	Recognizing the need for spare organizational capacity to learn, grow, and adapt for the future
Managerial focus	Control costs	Promote investment in future success

Source: Reproduced from Cannon & Edmondson (2005, p. 316).

10.4.2 FEI Team Dynamics

10.4.2.1 Internal Perspective. Innovation, of course, is a team sport. While leadership is important, understanding and managing team dynamics is critical for a successful FEI process. Recall the challenges that can arise around psychological ownership in teams (see Section 10.3). The question therefore arises: what do we know about what makes teams more successful at innovating? A meta-analysis examining pooled evidence collected over 30 years, from more than 100 studies and 50,000 observations, provides some answers (Hülsheger et al. 2009). Figure 10.5 summarizes the factors that have been proven to enhance team innovation performance, regardless of contextual differences, such as industry and organizational maturity.

These meta-analytic findings show that both leadership (support for innovation, vision) and shared team principles (task orientation, cohesion) are needed. A study of R&D teams across companies (33 team leaders with 188 employees), for example, showed that supportive leadership enhances team innovation only when whole team is concerned with excellence (task orientation). Supportive leadership falls on barren ground without the team's commitment to a shared set of principles and accompanying practices to pursue high performance in the FEI. One example of such team practices is decision-making comprehensiveness, reflecting how extensive and in-depth all available strategic options are being evaluated during the FEI. Benefits from decision-making comprehensiveness accrue not only during FEI but carry over into greater new product advantage of ultimately launched new products (Slotegraaf and Atuahene-Gima 2011). Teams should therefore engage in reflection and debate regarding objectives (are market/user needs clear among members of the team?), strategies (what are the ways in which concepts address needs?), and processes (how is the team working and do processes need to be adjusted?).

FIGURE 10.5 FACTORS THAT ENHANCE TEAM LEVEL INNOVATION ACROSS CONTEXTS.

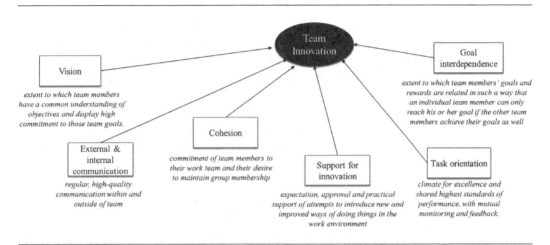

Source: Adapted from Hülsheger et al. (2009).

Such reflection and debate will be particularly helpful during concept generation (greater exploration and insights) but less productive during concept selection (reduced efficiency) (Seidel and Fixson 2013).

10.4.2.2 External Perspective. How firms leverage external sources plays an important role in the FEI. Through collaborations with external stakeholders, firms can access valuable knowledge from outside the firm – which is the core idea underlying the concept of open innovation (Chesbrough 2003). Sources of external knowledge include suppliers, customers, competitors, or scientific partners such as universities. In general, companies can decide to collaborate with specialist sources that possess a dedicated body of knowledge in a specific domain, or they can choose to work with a broad variety of external stakeholders (West and Bogers 2014). One especially popular and important external source of innovation are a firm's customers. Engaging customers may enable the firm to better understand (latent) customer needs during the FEI to develop superior new products and gain a competitive advantage (Gassmann et al. 2010).

In addition to the question of whom to involve, firms also need to consider the depth of collaboration or, more precisely, the power they are willing to transfer to the external party. As a general rule, the greater the involvement of an external actor in the FEI, the more power and influence is given to that actor. Following Cui and Wu (2016), customer engagement[3] can be classified into three types: customers as information sources, as co-developers, or as innovators. When involved as information sources, companies still dominate the innovation process and focus on gathering information and feedback from customers (or other external actors) and then apply these in the FEI. This corresponds to the manufacturer-active paradigm (MAP) (von Hippel 2005). As co-developers, power is equally distributed among the company and customers, meaning that they share the responsibility for the FEI as partners. In line with the customer-active paradigm (CAP), they already get actively involved in early stages of the innovation process and carry out activities in the FEI such as idea generation (von Hippel 2005). A good example from practice is the "Co-create IKEA" initiative. Launched in 2018, it represents a digital platform where customers (but also other actors such as start-ups) can become partners in new product development, through submitting new ideas, experimenting with and further developing them (with IKEA and other partners), and ultimately creating new products (Ting 2018). Finally, as innovators, customers will dominate the innovation process, which corresponds with the highest level of power transferred to them. In such cases, the company is often only responsible for the production and distribution of the resulting products, which were designed and developed by the customer (or any other external actor) (Cui and Wu 2016). Developing new products without direct involvement of the manufacturing firms is also called the customer-dominated paradigm (CDP) (von Hippel 2005). The question of what form of customer involvement firms should pursue depends on several factors such as nature and heterogeneity of customer needs, or firm strategic orientation (e.g., exploration vs exploitation). For example, when customer needs are highly tacit, involving them as information sources or co-developers may be less effective as these approaches are less helpful in understanding difficult-to-articulate problems (Cui and Wu 2016).

Deciding on whom to involve in innovation and to what extent is not sufficient to benefit from external innovation sources. It is equally important that companies integrate external knowledge into their FEI, which is dependent on factors such as a culture of openness (especially toward external input) or the technical capability to assimilate information from outside the firm. In sum, for any case of external FEI involvement, firms need to carefully design the involvement initiative and compare expected benefits with costs (West and Bogers 2014).

What about sustainability and managing people in the FEI?

How both internal and external stakeholders of a firm are managed can make an important contribution toward more sustainable FEIs.

- *Leadership: Given the uncertainty in the FEI and relating to societal and environmental NPD considerations (Stilgoe et al. 2013), it is particularly important that FEI leadership encourages experimentation, learning from failure, openness, continuous feedback, and reflection. Actively involving and empowering employees to integrate sustainability considerations into the FEI can also mean that there is no "fixed" or permanent leader, but that leadership is project-based. A popular management tool or system in that context is provided by "Objectives and Key Results" (OKR).*

- *External collaborations: Exploring, identifying, or evaluating non-economic consequences arising from innovation is knowledge-intensive and complex. Therefore, it is important for firms to create, maintain, and develop strong external networks with a broad variety of stakeholders to better access the required knowledge or skills. When engaging in external relationships during*

(Continued)

> the FEI, firms should pay attention to several relationship management aspects, such as transparency, extensive information sharing and communication, or mutual support. Additionally, relationships should be based on an eye-level partnership, with both parties having a similar mindset or shared values when it comes to the importance of sustainability in innovation.
>
> In sum, involving employees and external stakeholders is only half the battle. It is equally important <u>how</u> they are involved.

10.5 Managing Metrics: How Do We Assess Our FEI Effectiveness in Delivering on Objectives?

We spent the chapter so far outlining a comprehensive FEI management approach, presenting arguments and evidence that such an integrated FEI management model (linking process, ownership, and people) will lead to a more successful FEI. This section addresses how to tell whether your FEI is more (or less) successful – i.e., how to metricize the FEI. To do so, it is helpful to start with a few remarks on metrics generally and innovation metrics specifically.

10.5.1 Organizational Metrics

A myriad of metrics is deployed in companies, with big data and digital transformation further increasing the number and variety of performance measures. As the saying "what gets measured gets managed" indicates, however, just because something is being measured (and therefore directs efforts toward that metric) does not mean that the right thing is being measured. About 65% of companies that invest 15% or more of revenue in innovation struggle to bridge the gap between innovation strategy and business strategy (PwC 2017). Asked about the most important indicators for measuring innovation, companies overwhelmingly rely on revenue metrics to tell whether they are doing a good job – 69% use sales growth as the key indicator for a healthy innovation function. This is followed by customer satisfaction ratings (43%), number of new ideas in the pipeline (40%), and market share (36%) (PwC 2017). Yet, sales growth, customer satisfaction, and market share are influenced by more than launching new products and the number of new ideas does not track whether ideas have potential or are a good fit for the company. Thus, a more thoughtful and nuanced approach to innovation and FEI metrics is going to be more helpful to steer activities and outcomes.

Given that the most glaring disconnect between measuring FEI performance and broader firm-level long-term strategic goal performance lies in the distance between them, a helpful way to think about FEI metrics is as an innovation "thermostat" (Hauser and Katz 1998; see Figure 10.6). The idea is that metrics determine actions and decisions and are good only if they also improve long-term desired outcomes. By recognizing the interplay between organizational culture and ownership

FIGURE 10.6 FEI METRICS: A "THERMOSTAT" APPROACH.

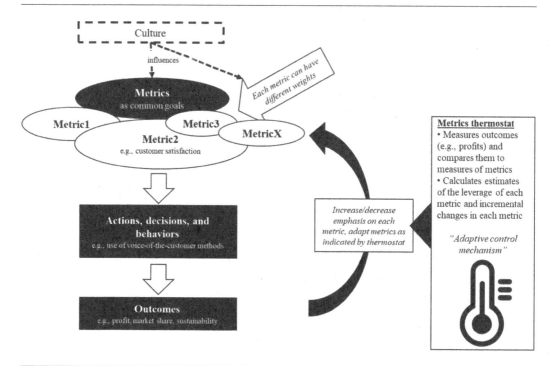

Source: Adapted from Hauser & Katz (1998) and Hauser (2001).

dynamics (as discussed earlier in this chapter) representing "soft" metrics, and such "soft" metrics shaping everyday behaviors, a firm can engage in continuous monitoring and recalibration of metrics to keep the FEI team on course.

10.5.2 FEI and General Innovation Metrics

A great start into defining appropriate innovation metrics is Tom Kuczmarski's (2001) list of key mistakes companies and innovation managers typically make when designing and introducing metrics[4]:

- Too many metrics – implementing too many metrics either because the innovation function is trying to do too much too quickly and/or not letting go of existing measures that are not relevant anymore.
- Too focused on outcomes – project-focused metrics can orient the team toward short-term financial outcomes. Therefore, metrics that reflect the process of continuous innovation are more suitable to measuring, assuming, and managing risk.
- Too infrequent – keeping track of continuous innovation efforts helps avoid P&L-driven decisions.
- Too focused on cutting costs – ultimate goals that are not tied to the market (i.e., customer needs and opportunities) create P&L (i.e., short term) focus.

- Too focused on the past – metrics that punish and assign blame lead to risk aversion. Tracking learning (particularly from failure) alleviates risk aversion (see also Section 10.4, Table 10.2).

While the above list is helpful to consider what to measure, the reason behind measuring something (Kuczmarski 2001) is foundational and requires clarity and consensus across innovation stakeholders (team, leadership, etc.). Is the metric trying to incentivize (i.e., reward) the right behaviors (see Spotlight 10.3)?

SPOTLIGHT 10.3   FEI REWARD METRICS IN R&D TEAMS

FEI activities can be months and years away from a new product launch. Therefore, FEI metrics that track performance and provide appropriate rewards (i.e., incentives) need to consider both early and later success (Manso 2017). First, the best FEI results may involve initial failure and thus require rewards for early failures and learnings. Second, substantial time will pass before benefits become apparent thus requiring rewards for outcomes later in the project. However, metrics for such early and delayed rewards are not equally effective across organizations. A key variable for effectiveness is centralization (see also Section 10.3.1). According to a study by Lerner and Wulf (2007), firms with centralized R&D organizations and long-term incentives also produce more heavily-cited, award-winning patents with greater originality. Short-term incentives don't work, and effects are weaker in decentralized R&D organizations, where the authority of corporate R&D over R&D decisions is more limited.

Is it trying to flag (i.e., diagnose) what does not vs what works well? Is the metric targeted at comparing the teams' performance against others in the same or competitor organizations (i.e., benchmarking)? Is the metric assessing resource flows (i.e., allocation)? A multitude of reasons for metricizing some aspects of innovation projects and functions exist. Starting with gaining clarity what the most important questions and decisions are is imperative to avoid meaningless and sprawling metrics.[5]

What about sustainability and FEI metrics?

Metrics play a considerable role in the relationship between FEI activities and sustainability.

- *Having metrics that do not exclusively focus on economic or financial measures (e.g., cutting costs, revenues) opens the discussion on how the FEI can create social and environmental value. Hence, firms need to integrate non-economic measures or key performance indicators (KPIs) into their FEI metrics systems to formally incentivize and direct behaviors and decisions toward greater sustainability in innovation*

(Björklund and Forslund 2018). Moreover, it is especially critical that introduced metrics are characterized by a (long-term) future orientation, representing a core principle of responsible innovation (Stilgoe et al. 2013). Examples for potential sustainability KPIs from an environmental perspective are carbon footprint rates, resource consumption rates (e.g., water), or the use of recycled materials in potential solutions explored during the FEI. Social value KPIs could relate to the quality of work conditions, employee satisfaction, or workplace diversity and equality, child labor, supplier and consumer safety rates.

- *Lastly, given the interplay between organizational culture and "soft" metrics (Hauser and Katz 1998), a culture of continuous reflection and feedback is important to support close monitoring of non-economic metrics as well as being able to quickly adapt and respond to new knowledge as it emerges during the FEI (Stilgoe et al. 2013).*

10.6 Conclusion

In this chapter, we have taken a broader view of FEI management, proposing four key management areas: process, people, ownership, and metrics. We outlined each FEI management aspect and highlighted interconnections and interplays among them throughout the chapter. Importantly, a comprehensive FEI management approach allows companies to more effectively manage the entire innovation as different demands for the front vs back end of innovation are clarified.

Acknowledgments

Over the years, the first author collaborated with colleagues across various teaching formats that address the front end of innovation. Those collaborations have inspired the development of this chapter. Special thanks go to Stephen Melamed, Donald Bergh, and Michael Scott from the Innovation Center at the University of Illinois at Chicago (UIC). For more information on how students from engineering, design, business, and medicine work together with company partners through the front end of innovation at UIC, see the chapter in a prior PDMA edited volume (Spanjol et al. 2014).

Notes

1   For more information on how Virgin Atlantic makes use of ethnographic approaches to better understand their customers' needs and problems for innovation, please see Chapter 14 (Micheli and Perks 2015) in "Design Thinking: New Product Development Essentials from the PDMA" on "Strategically Embedding Design Thinking in the Firm."

2   See blog post on this at https://blog.x.company/tackle-the-monkey-first-90fd6223e04d.
3   These levels of involvement can similarly be applied to other external actors involved in a firm's innovation effort.
4   For more innovation metric pitfalls, see Hauser and Katz (1998).
5   For a humorous view of this problem, see cartoons by Tom Fishburne at https://marketoonist.com/2019/11/kpi-overload.html.

References and Further Reading

Alexander, L. (2014). *People, Politics, and Innovation: A Process Perspective* (No. EPS-2014-331-S&E). ERIM Ph.D. Series Research in Management. Erasmus Research Institute of Management. http://hdl.handle.net/1765/77209.

Alexander, L. and Barkema, H.G. (2015). Innovation ownership struggles. *Academy of Management Proceedings* 2015 (1): 13590.

Baer, M. and Brown, G. (2012). Blind in one eye: how psychological ownership of ideas affects the types of suggestions people adopt. *Organizational Behavior and Human Decision Processes* 118 (1): 60–71.

BCG (2017). *The Most Innovative Companies 2018.* https://web-assets.bcg.com/img-src/BCG-Most-Innovative-Companies-Jan-2018_tcm9-207939.pdf.

Benner, M.J. and Tushman, M.L. (2003). Exploitation, exploration, and process management: the productivity dilemma revisited. *Academy of Management Review* 28 (2): 238–256.

Björklund, M. and Forslund, H. (2018). Exploring the sustainable logistics innovation process. *Industrial Management & Data Systems* 118 (1): 204–217.

Bocken, N.M.P., Farracho, M., Bosworth, R., and Kemp, R. (2014). The front-end of eco-innovation for eco-innovative small and medium sized companies. *Journal of Engineering and Technology Management* 31 (1): 43–57.

Cannon, M.D. and Edmondson, A.C. (2005). Failing to learn and learning to fail (intelligently): how great organizations put failure to work to innovate and improve. *Long Range Planning* 38 (3): 299–319.

Cayla, J. and Arnould, E. (2013). Ethnographic stories for market learning. *Journal of Marketing* 77 (4): 1–16.

Chesbrough, H.W. (2003). *Open Innovation: The New Imperative for Creating and Profiting from Technology.* Boston, MA: Harvard Business Press.

Christensen, C.M., Hall, T., Dillon, K., and Duncan, D.S. (2016). Know your customers' "jobs to be done." *Harvard Business Review* 94 (9): 54–62.

Cooper & Edgett (2008). Ideation for product innovation: what are the best methods? *PDMA Visions Magazine,* 32.

Cooper, J.R. (1998). A multidimensional approach to the adoption of innovation. *Management Decision* 36 (8): 493–502.

Cooper, R.G. (2014). What's next?: after stage-gate. *Research-Technology Management* 57 (1): 20–31.

Creusen, M., Hultink, E., and Eling, K. (2013). Choice of consumer research methods in the front end of new product development. *International Journal of Market Research* 55 (1): 81–104.

Cui, A.S. and Wu, F. (2016). Utilizing customer knowledge in innovation: antecedents and impact of customer involvement on new product performance. *Journal of the Academy of Marketing Science* 44 (4): 516–538.

Dahl, D.W. and Moreau, P. (2002). The influence and value of analogical thinking during new product ideation. *Journal of Marketing Research* 39 (1): 47–60.

Deschamps, J. (2005). Different leadership skills for different innovation strategies. *Strategy & Leadership* 33 (5): 31–38.

Di Fiore, A. (2014). A chief innovation officer's actual responsibilities. *Harvard Business Review.* https://hbr.org/2014/11/a-chief-innovation-officers-actual-responsibilities.

Edmondson, A.C. (2011). Strategies for learning from failure. *Harvard Business Review* 89 (4): 48–55.

Frascara, J. (2002). *Design and the Social Sciences: Making Connections.* London, UK: Taylor & Francis.

Gassmann, O., Enkel, E., and Chesbrough, H. (2010). The future of open innovation. *R&D Management* 40 (3): 213–221.

Gassmann, O. and Schweitzer, F. (2014). Managing the unmanageable: the fuzzy front end of innovation. In: *Management of the Fuzzy Front End of Innovation* (ed. O. Gassmann and F. Schweitzer), 3–14. Springer, Cham.

Gaubinger, K. and Rabl, M. (2014). Structuring the front end of innovation. In: *Management of the Fuzzy Front End of Innovation* (ed. O. Gassmann and F. Schweitzer), 15–30. Springer, Cham.

Griffin, A. and Hauser, J.R. (1993). The voice of the customer. *Marketing Science* 12 (1): 1–27.

Griffin, A., Hoffmann, L.N., Price, R.L., and Vojak, B.A. (2007): How serial innovators navigate the fuzzy front end of new product development. *Institute for the Study of Business Markets*, Pennsylvania State University, 1–35.

Hauser, J. and Katz, G. (1998). Metrics: you are what you measure! *European Management Journal* 16 (5): 517–528.

Hauser, J.R. (2001). Metrics thermostat. *Journal of Product Innovation Management* 18 (3): 134–153.

Hülsheger, U.R., Anderson, N., and Salgado, J.F. (2009). Team-level predictors of innovation at work: a comprehensive meta-analysis spanning three decades of research. *Journal of Applied Psychology* 94 (5): 1128–1145.

Jaruzelski, B., Loehr, J., and Holman, R. (2012). Making ideas work. *PwC Strategy& Global Innovation*, 1000. https://www.strategy-business.com/article/00140.

Jelinek, M. and Schoonhoven, C.B. (1990). *The Innovation Marathon: Lessons from High Technology Firms.* Jossey-Bass Publishers.

Jones, G.R. and Butler, J.E. (1992). Managing internal corporate entrepreneurship: an agency theory perspective. *Journal of Management* 18 (4): 733–749.

Juntunen, J.K., Halme, M., Korsunova, A., and Rajala, R. (2019). Strategies for integrating stakeholders into sustainability innovation: a configurational perspective. *Journal of Product Innovation Management* 36 (3): 331–355.

Kaplan, S. (2019). The Chief Innovation Officer. Inc. https://www.inc.com/soren-kaplan/the-chief-innovation-officer-job-description.html.

Kock, A., Heising, W., and Gemünden, H.G. (2015). How ideation portfolio management influences front-end success. *Journal of Product Innovation Management* 32 (4): 539–555.

Koen, P., Ajamian, G., Burkart, R. et al. (2001). Providing clarity and a common language to the "fuzzy front end." *Research-Technology Management* 44 (2): 46–55.

Kuczmarski, T.D. (2001). Five fatal flaws of innovation metrics. *Marketing Management* 10 (1): 34.

Leifer, R., McDermott, C.M., O'Connor, G.C. et al. (2000). *Radical Innovation: How Mature Companies Can Outsmart Upstarts.* Harvard Business Press.

Lerner, J. and Wulf, J. (2007). Innovation and incentives: evidence from corporate R&D. *Review of Economics & Statistics* 89 (4): 634–644.

Lovric, D.S.G. and Schneider, G. (2019). What kind of chief innovation officer does your company need. Harvard Business Review. https://hbr.org/2019/11/what-kind-of-chief-innovation-officer-does-your-company-need.

Manso, G. (2017). Creating incentives for innovation. *California Management Review* 60 (1): 18–32.

Markham, S.K. (2013). The impact of front-end innovation activities on product performance. *Journal of Product Innovation Management* 30 (1): 77–92.

Micheli, P. and Perks, H. (2015). Strategically embedding design thinking in the firm. In: *Design Thinking: New Product Development Essentials from the PDMA* (ed. M.G. Luchs, K.S. Swan, and A. Griffin), 205–220.

Pech, R.J. (2001). Reflections: termites, group behaviour, and the loss of innovation: conformity rules! *Journal of Managerial Psychology* 16 (7): 559–574.

Pierce, J.L., Kostova, T., and Dirks, K.T. (2003). The state of psychological ownership: integrating and extending a century of research. *Review of General Psychology* 7 (1): 84.

Poggenpohl, S.H. (2002). Design moves. In: *Design and Social Sciences: Making Connections* (ed. J. Frascara), 66–81. New York: Taylor & Francis.

Puranam, P., Singh, H., and Zollo, M. (2006). Organizing for innovation: managing the coordination-autonomy dilemma in technology acquisitions. *Academy of Management Journal* 49 (2): 263–280.

PwC 2017. *Reinventing innovation – Five findings to guide strategy through execution.* https://www.pwc.com/gr/en/publications/assets/innovation-benchmark-report.pdf.

PwC (2019). *22nd Annual Global CEO Survey - CEO's curbed confidence spells caution.* https://www.pwc.com/gx/en/ceo-survey/2019/report/pwc-22nd-annual-global-ceo-survey.pdf.

Robeson, D. and O'Connor, G. (2007). The governance of innovation centers in large established companies. *Journal of Engineering and Technology Management* 24 (1–2): 121–147.

Rouse, E.D. (2016). In the space between: creative workers' psychological ownership in idea handoffs. *Academy of Management Proceedings* 2016 (1): 10715.

Seidel, V.P. and Fixson, S.K. (2013). Adopting design thinking in novice multidisciplinary teams: the application and limits of design methods and reflexive practices. *Journal of Product Innovation Management* 30 (1): 19–33.

Shah, J.J., Smith, S.M., and Vargas-Hernandez, N. (2003). Metrics for measuring ideation effectiveness. *Design Studies* 24 (2): 111–134.

Slotegraaf, R.J. and Atuahene-Gima, K. (2011). Product development team stability and new product advantage: the role of decision-making processes. *Journal of Marketing* 75 (1): 96–108.

Sorescu, A.B. and Spanjol, J. (2008). Innovation's effect on firm value and risk: insights from consumer packaged goods. *Journal of Marketing* 72 (2): 114–132.

Spanjol, J., Scott, M.J., Melamed, S. et al. (2014). Collaborative innovation across industry-academy and functional boundaries: how companies innovate with interdisciplinary faculty and student teams. In: *Open Innovation: New Product Development Essentials from the PDMA* (ed. A. Griffin, S. Durmusoglu, and C. Noble), 175–222. Hoboken, NJ: John Wiley & Sons Inc.

Stilgoe, J., Owen, R., and Macnaghten, P. (2013). Developing a framework for responsible innovation. *Research Policy* 42 (9): 1568–1580.

Swoboda, C. (2020). Hiring a chief innovation officer is a bad idea. *Forbes.* https://www.forbes.com/sites/chuckswoboda/2020/02/24/hiring-a-chief-innovation-officer-is-a-bad-idea/?sh=2c843d214138.

Ting (2018). *IKEA: Crowdsourcing ideas to co-create a better everyday life.* Harvard Business School. https://digital.hbs.edu/platform-digit/submission/ikea-crowdsourcing-ideas-to-co-create-a-better-everyday-life.

Ulwick, A.W. and Bettencourt, L.A. (2008). Giving customers a fair hearing. *MIT Sloan Management Review* 49 (3): 62.

von Hippel, E. (2005). *Democratizing Innovation.* Cambridge, MA: MIT Press.

Wasson, C. (2000). Ethnography in the field of design. *Human Organization* 59 (4): 377–388.

Welzenbach, L. and Spanjol, J. (2022). *Enacting Responsible Innovation: Insights from German Small and Mid-sized Firms.* Ludwig-Maximilians-Universität München, Working paper.

Wender, B.A., Foley, R.W., Hottle, T.A. et al. (2014). Anticipatory life-cycle assessment for responsible research and innovation. *Journal of Responsible Innovation* 1 (2): 200–207.

West, J. and Bogers, M. (2014). Leveraging external sources of innovation: a review of research on open innovation. *Journal of Product Innovation Management* 31 (4): 814–831.

Prof. Dr. Jelena Spanjol, PhD, is head of the Institute for Innovation Management (IIM) at the Munich School of Management, Ludwig-Maximilians-Universität (LMU) in Munich, Germany. She is co-founder and Chair of the Board of Directors of the LMU Innovation & Entrepreneurship Center (IEC). Prior to joining LMU, she held faculty positions at the University of Illinois at Chicago (UIC) and Texas A&M University, receiving her PhD from the University of Illinois at Urbana-Champaign. Her research examines innovation dynamics across micro-, meso-, and macro-levels.

Lisa Welzenbach is a doctoral candidate at the Institute for Innovation Management, Ludwig-Maximilians-Universität (LMU) Munich. Before starting her doctoral studies she earned a Master's degree in Corporate Management & Economics at Zeppelin University (Friedrichshafen) and a Bachelor's degree in Business Administration at LMU Munich. She also holds a Master in Business Research from LMU Munich. In her doctoral dissertation, she is investigating how firms can become more responsible innovators, i.e., also account for societal and environmental consequences when innovating. Other research interests are related to the field of sustainable consumption.

CHAPTER ELEVEN

OPPORTUNISTIC NEW PRODUCT DEVELOPMENT

Floor Blindenbach-Driessen and Jan van den Ende

11.1 Introduction

This chapter focuses on what it takes to manage new product development (NPD) when firms innovate infrequently and (re)act to opportunities as they arise, instead of as part of a well-defined growth strategy and portfolio of NPD projects. We label the first as Opportunistic NPD and the latter as Strategic NPD.

Strategic NPD is the norm in the innovation and strategy management literature (Moore 2007; Teece 2007). However, there are many organizations that don't have an innovation strategy, especially firms that get away with introducing a new offering once every couple of years, such as for example accounting, law, and construction firms. Also in agriculture (van den Ende 2021), we find companies in which innovation projects are one-off opportunities instead of part of a portfolio of ongoing NPD projects aligned with the organization's innovation strategy. There are also indications in the literature, that this opportunistic approach to NPD is rather prevalent. For example, the Belgian Community Innovation Survey shows that at the beginning of this century, 25% of companies introduced one new product or service in a five-year period (Geerts 2019). It is safe to conclude that these companies applied an opportunistic approach to NPD.

Based on our research and practical experience, this chapter provides insights into what Opportunistic NPD entails and offers guidance to innovators and managers of organizations which innovate infrequently and bring new

The PDMA Handbook of Innovation and New Product Development, Fourth Edition. Edited by Ludwig Bstieler and Charles H. Noble.
© 2023 John Wiley & Sons, Inc. Published 2023 by John Wiley & Sons, Inc.

products or services to the market in an opportunistic manner. As we will explain in this chapter, these one-off new product development endeavors are often executed with fewer resources and fewer capabilities, while they require more discipline, because when acting opportunistically there is just one shot to get it right.

11.2 Opportunistic versus Strategic NPD

Opportunistic and Strategic NPD are fundamentally different; see Figure 11.1. Strategic NPD is based on the same logic and power law that venture capitalist use. A handful of startups take of exponentially, however, most end up being worth zero. The world's first Venture Capitalist, Arthur Rock, had great difficulty in evaluating startups. Nevertheless, it was not fatal to his investment strategy. The subjective nature of his methods meant he would be wrong much of the time, with a tiny number of winners making up for the losers (Mallaby 2022). In the same way, Strategic NPD is a numbers game – in which the high odds of each bet are offset by a few big wins.

Firms that innovate in an opportunistic manner do not benefit from this power law, simply because they don't have the quantity of projects to make it work. Instead, they have just one shot to get each project right. Furthermore, because opportunistic NPD is most common among mid-size and smaller firms there is an additional challenge that successes rarely come at a scale that is sufficient to make up for any losses on other NPD projects. That means, that Opportunistic NPD is about placing one bet in a game with relatively modest prize money.

Wheelwright and Clark (1992) mention that some firms may prefer to use a few big bets funnel over a regular funnel. While this may sound similar to the Opportunistic approach, it is different because a few big bets funnel assumes that there are a few big bets to make. In our experience, firms that apply opportunistic NPD don't have a few big bets but just one single opportunity to innovate at a time. These firms (re)act to opportunities as they arise or are created. Take for example a hospital that created a crowdfunding platform for its pediatric research projects or a solar installer that developed a software solution to better keep track of its clients. In both cases, these were new product development

FIGURE 11.1 STRATEGIC VERSUS OPPORTUNISTIC NEW PRODUCT DEVELOPMENT.

Strategic New Product Development	**Portfolio of projects** A numbers game – The low odds of each bet are offset by a few big wins
Opportunistic New Product Development	**One-off projects:** Sharp shooting - Placing one bet in a game with relatively modest prize money

endeavors driven by the opportunity that existed because of the circumstances and people available. These projects were not part of a well-thought-out strategy or part of a portfolio of other initiatives. And for the people on the team, undertaking the project was not part of their job description. Opportunistic NPD is thereby about one-off innovation projects.

In the agricultural sector, you can find many examples of such one-off opportunities. Innovation in varieties, innovation in pest control, innovation in energy systems, etc., are all driven by the suppliers that operate in this sector, not by the farmers themselves. Farmers do not have the capital, expertise, or experience to develop any of these solutions. Acting on such opportunities enables farmers to upgrade their offerings and crop yields with minimum risks and product development costs. However, these supplier-driven new products and services are rarely plug and play solutions. Instead, as we will illustrate later in this chapter, it requires NPD efforts to implement these new solutions successfully. Making a vendor solution work in their specific context provides a farmer with a competitive advantage, just plugging in a solution won't, if anyone can do that. There are also disadvantages to such supplier-driven innovations, because it will make the strategic direction of the firm rather ad hoc. Depending on the dynamics in very different sectors the farmer may identify opportunities that require investments in energy upgrades, new seedlings, etc.

In the rest of the chapter, we will use vignettes with examples of opportunistic NPD projects to illustrate how firms develop new offerings and solutions when they are one-offs. From these vignettes and our experience with facilitating such teams, we provide best practices for the different phases of the innovation process. In the conclusion, we summarize what the implications are of the differences between strategic and opportunistic NPD for practitioners and researchers.

11.2.1 What Types of Opportunistic NPD Can We Distinguish?

Figure 11.2 provides an overview of the typical NPD activities, with on the top row new product, new service, new business model types of innovation activities and on the bottom row process improvements.

In the left column we have one-off innovation projects and we contrast these initiatives with innovation projects in the right column, that are executed in the context of a larger portfolio of innovation initiatives.

11.2.2 The NPD Process

For Strategic and Opportunistic NPD, the innovation process itself is no different and consists of the same steps of scouting, validating, transforming, and scaling. Scouting is the phase where and when the new opportunity gets defined. Validating is the phase where the opportunity is examined closer and its feasibility, viability, and desirability get tested. Transforming is the phase where the solution gets developed. Scaling is the phase where the solution gets implemented

FIGURE 11.2 OPPORTUNISTIC–STRATEGIC NPD MATRIX.

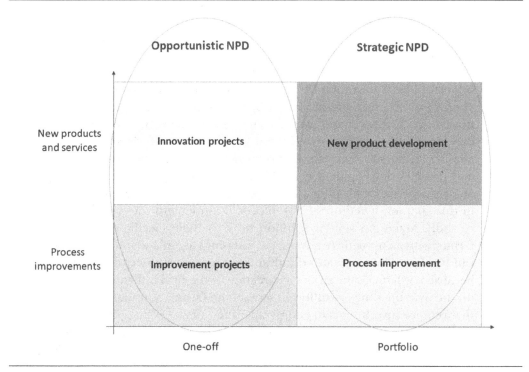

and delivered to as many users or customers as possible. While the phases are universal, the questions that need to be addressed in each phase are different when comparing strategic and opportunistic NPD; see Figure 11.3.

In the sections below we will use vignettes with examples of Opportunistic NPD to point out challenges and best practices for each of the phases. We will thereby differentiate between best practices at the organizational and at the project level to bring Opportunistic NPD projects about. The organizational level is the level of the firm or business unit as a whole. What should top management or business unit management do to make the activities in each phase run smoothly? The project level is the level of the individual projects. What should project leaders and project teams do to execute their project successfully?

11.2.2.1 Scouting. In the case of Strategic NPD, the scouting phase is a hunt for novel ideas to feed the pipeline. With many options to choose from, the question is what are the best opportunities for the firm to pursue? Whereas in the case of opportunistic NPD, the scouting phase is more a matter of being attuned to possibilities that arise. When the occasional, too good to miss opportunity arises, the question will be, can the firm pull it off? It is like the difference between a real estate developer looking for a house as an invest opportunity, versus an individual looking to buy a new home. For the first it is their job and part of the routine, for the latter it is a major event in their lifetime.

FIGURE 11.3 DIFFERENT QUESTIONS THAT NEED TO BE ADDRESSED AT THE DIFFERENT PHASES.

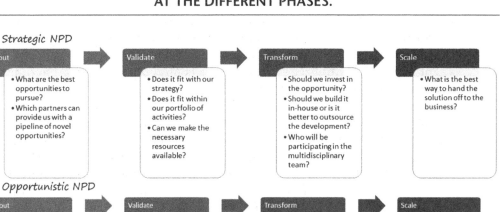

11.2.2.1.1 Scouting at the organizational level Firms pursuing an opportunistic innovation approach, should scout relatively intensively. Since the opportunities can come from different directions and fields, the CEO and the business unit manager themselves, are the ones who often are most active in the scouting activities. They must spend time and effort to connect to learn about new developments. Vendors and suppliers are a great source of ideas for organizations that apply opportunistic NPD.

Since scouting is a lot of work, Lex Mundi, a Global Network of 150 law firms, offers technology scouting as one of their services to member firms. The innovation team of this network scouts and tests technology with interested member firms. And when proven useful, the pioneering member firms are then asked to share their experiences with others.

Greenhouse farmers in the Netherlands visit trade fairs, have so called study clubs, and read news about university research in their professional journals. Since this sector has a strong cluster character, with many growers, suppliers, and trading companies in the same region, even private contacts can be important sources of information. Discussions at birthday parties are characterized by an exchange of a mix of professional and private experiences. In other words, greenhouse farmers dedicate a lot of effort on scouting, both during work and in their personal time.

Best practices

- Use your network to scout for available options – so you know what is out there and can explore more opportunities in collaboration with others, while reducing risks at the same time.
- For larger (decentralized) firms – have a central repository to avoid duplications elsewhere in the organization.

11.2.2.1.2 Scouting at the project level Scouting at the project level is about identifying the opportunity and defining what needs to happen to act upon it.

In a midsize law firm, each partner scouted for their own legal technology solutions. That did not work well, since it led to the situation in which the firm had more software licenses to maintain than there were lawyers in the firm. As mentioned in the best practices above, when everyone in the firm is engaged in scouting activities, it becomes a necessity to keep a central repository of projects that have been undertaken and that are being executed to avoid duplications. The law firm above reduced the redundancy by installing a simple rule. They would only allow further exploration and testing, if at least two practice groups vouched that they would pay and use the technology.

Physicians at a Children's Hospital were scouting for new funding opportunities for their research projects. When the Obama administration made crowdfunding possible, they were intrigued by the idea and started exploring if they could create a crowdfunding platform for their pediatric research projects. The goal of the project was to see if crowdfunding could be a viable funding alternative to grant funding and philanthropy. Thanks to the availability of a group of enthusiastic summer interns the project came about. The interns were the ones who took it upon them to research the opportunity, investigate available platforms, find suitable projects, talk with stakeholders in the organization, etc. Without this team of summer interns, this crowdfunding platform would not have come about.

Because of the scale at which opportunistic NPD takes place, the one-off projects are typically undertaken by the business unit. With limited resources, these teams need to be creative in how they scout, where to find information, and how to validate the assumptions they are making with regards to the opportunity.

Best practices

- Provide the opportunity to explore novel ideas by encouraging people to seek out new knowledge and providing support to going to events, etc.
- Accept that this phase may take some time, since the individuals or team that elaborate the idea may have to acquire knowledge from fields are new to them.
- Strengthen the idea and outline the project by having many conversations with people inside and outside of the organization.

- Get creative in who you can engage to help with the data-gathering activities that follow the scouting; summer students and interns are very suitable for this job.
- Scouting can be done by individuals but before investing, a decentralized organization may want to put rules in place such as that at least one other colleague and/or practice group is willing to support the efforts.

11.2.2.2 Validating. In the case of Strategic NPD, initiatives must fit within the firm's strategy, within the portfolio of ongoing projects, and the resources need to be available to form the team and execute the project. That means, that even when the idea is great, but it does not hit all these check boxes, the project won't continue beyond the validation phase. Under the Opportunistic NPD regime, none of this applies or is relevant during the validation phase, as there is no overall innovation strategy that guides project selection. In addition, there are no dedicated resources to take on the identified opportunities. The organization will need to make people and resources available for the occasion.

11.2.2.2.1 Validating at the organizational level In the case of Strategic NPD, there is funding available, but whether a project receives funding is a matter of priorities. In the case of Opportunistic NPD, there are no available funds. Instead, the funding will need to be made available or be found somewhere.

Deciding to invest in an innovation project is a task for management. Because the innovation projects are opportunity driven, an innovation strategy typically does not exist. And because of the ad hoc nature and low frequency of projects, portfolio management does not make much sense. As a result, projects must be evaluated one by one as they come in, which begs the question, what criteria should be used to give a project a go?

Take a new startup service project of a law firm. The small-business practice-group of the firm wanted to automate their offering with the goal to attract more startups. The IT department was reluctant about the project. They reasoned that the practice group first had to find startup clients before automation would be worthwhile. An impasse that could only be solved by allowing the practice group to undertake the project and prove there was a need for the self-service solution. It was the task of the team to engage in customer discovery and collect the necessary evidence, so that management could decide whether to invest.

When the task to validate the project is delegated to the team, then management must learn to listen to evidence the team collected. Unfortunately, that does not always happen (Blindenbach 2018). Firms engaged in Strategic NPD will have an innovation strategy to assist in evaluation of projects. In a firm engaged in Opportunistic NPD such a strategy will not be available. In this void, managers tend to rely on their gut feelings to evaluate projects. This may have detrimental effects. In a construction firm, senior management was notorious for saying "but my gut says otherwise." It killed all innovation efforts because there is no arguing against a decision-maker's gut feeling. Worse, such gut feelings are often misplaced. In this case, the firm missed out on several opportunities. Nobel Prize laureate Kahneman discovered that gut feeling do not work well in cases of innovation because it is precisely here that familiar paths and routines are broken. What you would expect, suddenly no longer applies due to new technology, a different revenue model, and changing customer expectations, etc. (Kahneman and Klein 2009). In short, gut feelings are often wrong when it comes to decisions that have to do with new situations. That is why it is so important for senior management to listen to the evidence collected by project teams.

Another way to select projects is by looking at the effort, teamwork, and learning ability of teams while they worked on putting their project proposal together. The dedication and commitment of a team in the early phases is a good predictor of how vested they are in getting it right. As Ed Catmull said "If you give a good idea to a mediocre team, they will screw it up. If you give a mediocre idea to a brilliant team, they will either fix it or throw it away and come up with something better."

With that in mind, Lex Mundi, the Global Law Firm Network, set up an innovation challenge, in which participating teams of the member firms were given a task to complete their project description on an online learning platform. Instead of evaluating the ideas, they looked at the team composition, the commitment of the team during the program, and the conciseness of the project descriptions to select the winners. From the ten teams that showed interest in the program, the five most committed teams could be easily identified. What more, six months later, four of these five teams delivered working prototypes while only one team dropped out due to unforeseen and unrelated circumstances.

Best practices

- Use data-driven decision making.
- Focus on the quality of the team when making go/no-go decisions.
- Have the commitment of the team outweigh the perception of the value of the solution and trust that high performing teams won't waste their time on things that have a low chance of success.

11.2.2.2.2 Validating at the project level One of the tasks at hand during the validation phase is the creation of the business case. That is not a simple task under any circumstances.

In the case of Opportunistic NPD, the organization and the team are likely unfamiliar with this task. It is the reason why a law firm had been holding off for nine years to invest in an expensive software package to digitize their physical minute books – minute books are used to contain the corporate history of a firm and capture data on shareholders, outstanding debts, etc. Transitioning to virtual minute books required a significant investment, that some in the partnership favored and others opposed. Many heated debates and outspoken opinions had not brought these opposing factions any closer. It took a team that was willing to put effort in collecting factual data about the current problem – for example how many unbillable hours were lost each month by searching for misplaced books, and interviews with clients – to get confirmation about their willingness to pay for virtual access to get a unanimous decision from leadership in the end.

The Minute Book team is an example of a team that had to validate their own idea. When this team asked clients about their willingness to pay for virtual access to their minute book, they received a negative response at first. It took effort and guts to dig deeper and learn that clients were willing to pay for third-party access. It is an example of a high-performing team, that is committed, where the members worked well together, and that learned quickly. Such a team can be trusted to deliver the best possible results (Blindenbach & van de Ende 2010). High performing teams are not going to waste time on a project that doesn't have a high chance of success.

Ipsun Solar, a solar panel installation company could not convince their software provider to make changes that would help them to keep better track of their customers. After their requests for what they considered must-have features were denied, they took on the opportunity and developed the complete software solution in-house with the goal to market it to their peers and other installers. During the validation phase, they had to figure out all that would be entailed in developing the software solution and verify with other installers if they had similar needs. Developing their software solution Sunvoy cost-effectively, meant that they had to engage their peer network and engage in customer discovery at the early stages of the project, to make sure that the solution they had in mind solved not only their problem, but also that of the other solar installers in their network. As a result of the customer discovery work the firm undertook under the guidance of the Virginia Small Business Development Center, the firm quickly learned that creating a solution even just for the members of the peer network would be more complex and labor intensive than anticipated. They realized that to execute the project, they had to secure a government research grant or find funding elsewhere. Alternatively, which they end up doing, they could bootstrap the development of Sunvoy and accept that it would take longer.

It pays off to invest in the validation phase. The eData platform team of a law firm had won the in-house idea challenge and was given the green light to create online courses to help the lawyers of the firm with common e-discovery questions and needs. While the team was ready to start building the solution, they were asked to first verify the need by the innovation program they signed up for. What specific problem should these online courses solve? During the customer discovery conversations with potential users (Blank and Dorf 2012), it turned out that the envisioned eData platform did not solve a user need. The team was overworked and did not have time to spend on higher value-added tasks because they provided individualized technical e-discovery training to anyone who asked for assistance. Clients and colleagues who knew about the team were happy with this dedicated service, but it could not always be provided as quickly as they wanted. For the e-discovery group it was not cost effective and there were also a number of newer clients and colleagues who were not aware of the team's capabilities and services. So instead of launching the online eData platform with basic e-discovery courses and resources as initially envisioned, the team started by creating a list with the most frequently asked questions and links to existing materials and asking everyone in the group to begin using it when responding to requests. That list alone solved a significant part of the problem. They did build the eData platform, but it ended up having a different focus than originally expected. The eData platform is another example of a team that was tasked to validate their own idea and project. The team was able to do so successfully, using customer discovery which is considered a best practice for founders to validate their startup (Blank and Dorf 2012).

In other words, even when you know your clients well, it pays off to spend time on fact finding and customer discovery. During the validation phase, you want to make sure that you know exactly which problem to solve. As Albert Einstein said, "If I had an hour to solve a problem, I'd spend 55 minutes thinking about the problem and five minutes thinking about solutions." However, instead of thinking as Einstein said, we would like to suggest that you engage in conversations with your end-users and clients to understand their current workflows, user ecosystem, and to validate the assumptions you are making – as the lean startup methodology suggests.

We mentioned before that many of the new product and service development efforts are driven by suppliers. In that case, the task of the firm is to adjust and implement an existing solution, and not so much developing it. That can lead to the impression that vendors can solve any problem, but beware. First, when the vendor is selling a standardized solution, make sure it fits your specific problem. Second, when adopting such generic solutions, know that it will be difficult to create a competitive advantage from adopting such solutions. After all, anyone else can do the same.

Best practices

- Make sure you know what it takes to create the business case for an innovation project. These business cases cannot be derived from historic data, instead they are built on assumptions.
- Use an outsider – an external coach or facilitator – to help with identifying the most critical assumptions that must come true for the project to be a success.
- Use customer discovery practices to learn more about your client's current workflows, ecosystems, willingness to pay, etc.
- Investigate the availability of government grants, if possible, to off-set the development costs and reduce risks.

11.2.2.3 Transforming. For Opportunistic NPD, outsourcing is nearly always a must during the Transformation phase. The question is, how much of the work will be outsourced. Whereas on the strategic side, the organization may have the capabilities and resources to execute the entire project in-house.

11.2.2.3.1 Transforming at the organizational level During the transformation phase, most of the activities takes place at the project level. While one-off and executed in an opportunistic manner, it still pays-off to track the progress and performance of these innovation initiatives. For example, in a hospital, patentability of the idea was a requirement to get internal funding. The hospital's residents and fellows – physicians in training – brought many ideas to the IP committee. However, nearly all their ideas were declined because these physicians in training were unlikely to stay with the hospital long enough to bring their idea to fruition. It took the hospital leadership years, before realizing this was a flaw. They only started to notice, when they realized they lost innovative talent at a higher rate than other hospitals in the area. Tracking the progress of NPD initiatives over time is the only way to identify bottlenecks and roadblocks in the innovation process.

In the Opportunistic NPD model, you cannot afford much overhead to manage the innovation function, because it is challenging to create a positive return on investment from the sum of innovation activities at such a small scale. First, as mentioned above, the numbers game does not play out. The frequency is too low to have big winners make up for the losses. Secondly, big winners are rarely that big. In part that is because many of these projects are vendor-based innovations. A vendor solution will enable the firm to do something new or better. However, adopting such an off-the-shelf solution will rarely enable these firm to venture into new markets and scale. Service firms, which seems to be another segment where Opportunistic NPD prevails, run into scaling limitations for different reasons. These firms usually have a regional presence only and their scaling is limited to how fast they can train other providers to adopt the new offering. Altogether, in the Opportunistic NPD model it is a challenge for an innovation manager to create a return on investment that covers his or her salary. Assume that such a person earns $100k and that the firm has a 20% profit margin on its offerings. That means that $500,000 in new revenue need to be generated from new offerings just to pay for the salary of this one innovation manager. That is a steep challenge, when you just undertake just one to two projects each year that have relatively small impact.

Without a structured innovation process and innovation function, it can be tempting to make giving the green light to a team a one-time and done decision. However, also for Opportunistic NPD, discovery-driven planning (McGrath and MacMillan 2009) is a best practice that should be used. It requires senior management to evaluate at each milestone, if and whether a team should proceed based on the data they have collected and the progress they have made.

Best practices

- Watch your overhead costs, because engaging in innovation projects only every so often does not justify a full-time innovation function.
- In the absence of a stage-gate model, using a milestone-based investment approach becomes critical to avoid overcommitting resources.

- Taking advantage that things are less formalized means that discipline in the execution should be combined with flexibility in direction.
- Track everything that you can, so that over time, you can identify weaknesses in your innovation value chain.

11.2.2.3.2 Transforming at the project level While the Transformation phase is all about developing and testing the idea, customer discovery does not stop after the Scouting and Validation phases. Take an accounting firm that created a client portal. They built a solution that standardized a commonly used form and automated the follow up for a client. However, to the dismay of the team, when the prototype of the solution was offered to this client for beta testing, the client never logged in nor used the form. Continued customer discovery revealed that the client of the accounting firm and the users of these forms were different people. Human resources (HR) was to fill out these forms, not the accountants who were their clients. Once HR was made aware of the available forms, they were grateful for the new automated solution.

Under the Opportunistic NPD model, there are nearly always tasks that are outsourced to vendors. While vendors and contractors are invaluable to bring novel ideas about, you cannot ask them to perform miracles. For example, an emergency medicine physician noticed that nosebleeds took up a disproportionate amount of attention of the nurses. Patients who came with a nosebleed to the ER typically were covered in blood and for that reason often in a panicked state. Nevertheless, their situation was rarely life threatening. Stopping a nosebleed can be difficult when on blood thinners, but with the right amount of pressure, the appropriate ice pack, medication, and a lot of patience, even the most severe nosebleed can easily be solved. So, this physician thought, why not create a device that enables patients to do this themselves so they don't have to come to the ER. Now, more than five years later, there is finally a prototype of this device called NasaClip, that will be going to market soon. While the problem was clear, the solution was not as simple as the medicated nasal clip with an optional ice pack that this emergency physician had envisioned. Solving the problem turned out to require quite an engineering feat and investment.

When engaging vendors in the transformation phase, be careful that you don't create your own competition. For example, a small firm contracted a software developer to help with their digital transformation efforts. The first solution was barely in use, when the firm learned that the developer had offered the same solution also to their main competitor. The small firm had taken on all the risk and spend the time to make it work, but it did not gain a competitive advantage over their main competitor.

Risk is a big issue in Opportunistic NPD. Firms cannot share risk between different projects. In addition, since they are often inexperienced in the field of innovation, even estimating the size of risk may be difficult. A greenhouse owner who created a geothermic energy source took an insurance against the case that no appropriate hot water source would be found in the deep layers. The whole project cost several millions; the insurance policy itself cost him over a million. But he took it, because it was impossible for him to estimate the size of the risks while the potential downside would affect his financial situation substantially.

Another issue we would like to raise, is that the examples used in this chapter are of projects executed by teams for whom the innovation project was a task that came on top of their regular duties. Given the inexperience of these teams and the leaders they report to, it really helps these teams to get coaching or sign up for a course for instance at a local university that teaches how to validate innovative ideas, how to create the business case for an innovation project, and how to bring innovative ideas about. There is a wealth of information available on the internet on how to do these tasks. However, it is a steep learning curve if you must figure it out while also pursuing your project.

What more, most of these projects took place within the operational context. The inability to spent significant time on the innovation project becomes especially problematic during the development phase, as that is a time the NPD project will start to become time consuming, significant investments are being made, and when development speed starts to matter.

Best practices

- Follow a proven process so you can focus on making your project happen instead of figuring out what to do when.
- Get support for your team from experienced innovators so you don't make beginners' mistakes during the execution.
- Continue customer discovery during the development – get as much feedback as you can from actual clients.
- Use contractors for skills you don't have in-house, but make sure that you have something unique to add and do not give your intellectual property away – as you want to create a competitive advantage.
- Apply for insurance if failure of the development is too big of a financial risk.

11.2.2.4 Scaling. Scaling consists of implementing the solution and scaling it. Strategic NPD projects are typically executed independent of operations. That makes that during the scaling phase, the key concern is how to integrate the new solutions within the existing operations. Once the new solution becomes the responsibility of operations, scaling will be taken care of. In contrast, Opportunistic NPD projects typically arose because of a unique combination of the right people at the right place at the right time and these projects are typically executed by operations. The question of adoption is therefore less relevant. However, as we will learn from the vignettes below, that does not mean that scaling is self-evident. Instead, it is something that requires deliberate effort.

11.2.2.4.1 Scaling at the organizational level While we discuss the innovation process here sequentially, in practice, projects rarely proceed orderly through the Scouting, Validating, Transforming, and Scaling phases. Not in the least, because there is no formalized stage-gate process. For example, one project team was doing customer discovery interviews in the Validation phase. During one of these interviews, the interviewee got curious and eagerly asked the team, "Can you solve that problem for me?" While far from ready to serve clients with the automated solution the team had in mind, they decided to use the client as a test case. They ended up delivering and developing the new service on the go, offering this interviewee a manual version. While not profitable, the learning experience was invaluable for the development of the actual solution. The example above is rather rare, but it certainly happens that minimum viable products are sold to clients during the Transformation phase.

The Scaling phase is a time that leadership needs to become more actively involved, to ensure that solutions get scaled. Unfortunately, when innovating opportunistically, the opportunities to scale often get missed. That is how it happened, that a client of a construction firm got offered a very innovative approach on the first project they contracted out, but not on the next job they granted to this same construction firm. To the surprise of the client, the second project team was unaware of the novel approach used by the preceding team. The engineers on the first project had moved on to the next new thing and those in the field reverted to their old routines.

Unless incentivized, scaling generally benefits the organization and not necessarily the individuals who took it upon them to take advantage of the opportunity in the first place. Often for the team members there is little, if any, motivation to share the learnings and invest time to teach others in the organization what it takes to adopt or deliver the newly developed solution. That means that incentives are needed to ensure scaling happens.

Creating a culture for innovation is also part of Scaling at the organizational level. While creating a culture for innovation may sound vague and fuzzy, it is in fact a deliberate effort by management to enable individuals and teams to act upon novel opportunities and to put in place the processes, parameters, and incentives to make it happen in a cost-effective manner.

Often, management thinks that showing commitment starts by hiring a full-time innovation manager. In the Strategic NPD model, an innovation manager would be hired and given the appropriate budget to execute the innovation strategy and support several innovation projects. In the Opportunistic NPD model, without a budget and innovation strategy, many of these new hires operate in a vacuum and are set up to fail.

It may be the reason that the most successful innovation functions that used the Opportunistic NPD approach, started at the other end. Take for instance the Center for Innovation at the Mayo clinic. This center got started by helping the dermatology department, which was struggling at that time. After helping this department reinvent their practice and turn it around financially, other departments stood in line to send a team to the Center for Innovation.

In a similar fashion, a consultancy for the Dutch government, saw that blockchain technology, big data, cyber, and a variety of other technologies perplexed several parts of the Dutch government – their primary client. One of their consultants envisioned how their firm could help Dutch government agencies by creating an inspiration lab – a collaborative environment to test out these new technologies. How to make that happen was the question. Transparency about metrics, a successful use case, and clarity about the outcomes, helped the consultant to set up this inspiration lab. This lab got spun off as DigiCampus, which was possible because by then the lab had proven its impact and created a positive return on investment.

Scaling at the organization level requires management to create a culture for innovation with a clear process, the necessary support, and appropriate incentives, so that when an opportunity arises, the firm can take action. What is different, is that for Opportunistic NPD, management needs to be even more diligent and transparent, in order to ensure that each one-off effort pays off.

Best practices

- Put incentives in place that motivate teams to scale their solution.
- When building innovation infrastructure, start small and get results first as overhead costs are difficult to recuperate when there are just a few innovation projects and when these new products or services are not very scalable.
- Invest in creating a culture for innovation by making sure people who engage in innovation projects have the time and proper support to act upon ideas.

11.2.2.4.2 Scaling at the project level If possible, avoid trial and error during the scaling phase because such iterations become more costly. Take for example a law firm that had developed a client portal. The first offering on the portal was a template that was frequently requested by one of their high-profile clients. Despite the high demand for the template, the version on the portal did not see any uptake. Only when the team rethought the entire project, they realized that their high-profile clients currently could get this template tailored and filled out for free, by simply calling the paralegal of the firm that was working on their account. Placing the call was for them simpler, more effective, efficient, and came at no cost, while they had to pay a fee to download the new do-it-yourself template. Creative thinking, a new business model, and a new workflow solution resulted in a major overhaul of the client portal that in the end was successful. Ideally, many of these issues should have been picked up and be addressed during the validation and transformation phases, not during implementation.

In another law firm, the IT department had been trying to convince their lawyers to use Microsoft Teams since the start of the COVID-19 pandemic. Nevertheless, uptake had been virtually non-existence despite their attempts to tell lawyers about all the benefits. For the lawyers, it was a novel software solution of which they failed to see the upside. That changed, when one of the lawyers showed her colleagues the benefits. Note, this first project did not come about because this lawyer needed Microsoft Teams. She set out to develop an intranet solution for her team, to help her team communicate better, as virtual collaboration had become a significant challenge when the pandemic kept them all home for more than a year. When interviewing the IT department about the possible tools, she learned about Microsoft Teams, which seemed the perfect solution for her problem. Next, she ran an experiment, forcing her team to use Microsoft Teams for 3 weeks instead of WhatsApp, which they had been using thus far as the tool to communicate. This experiment went so well, that her team decided to stick to the new solution. Seeing the positive impact Microsoft Teams had on her team and using this lawyer as champion, it took the IT department less than a year to convince nearly all the other lawyers in the firm to switch to Microsoft Teams. This example is to show that even for an elsewhere widely used solution like Microsoft Teams, adoption and scaling are not a given in the Opportunistic approach.

Unfortunately, creating a successful first version is not always sufficient to guarantee successful scaling. Take for example the new patent service for startups – an early-stage startup program – developed by a law firm. One of their lawyers put a lot of time and effort in developing the program, engaging seasoned entrepreneurs in the vetting committee, and getting the program out to the local entrepreneurial community. Despite the program attracting many applications and its initial success, the law firm decided to cease accepting further applicants to the program at a time when it was too early to see the first impactful results. That is, before any of the participating startups had become successful. A loss for everyone involved. Without continued support, commitment, and protection – the program could not continue. The latter is an example to show once more, that initial success does not guarantee long-term success or firmwide scaling when innovation is approached opportunistically.

Scaling at the project-level is on the one hand easier, because teams are embedded and close to the day-to-day operations of the firm. That makes adoption easier. However, don't be fooled, because without senior management support to protect the project when the going gets tough, projects may be pulled prematurely. So, even when using Opportunistic NPD, it is essential to ensure that the commitment of an appropriate project champion exists from the start.

Best practices

- Set clear expectations that scaling is also part of their job, when forming the team and giving them the opportunity to develop the new solution.
- Avoid trial and error learning during the scaling phase as it is costly and time consuming at this stage.
- To encourage adoption, show the benefits instead of telling what these are.
- Make sure you secure from the start project champions that support your project and who have close relationships with the business unit or the department that will be responsible for the implementation and roll out of the new solution.

11.3 Discussion and Conclusion

When reading the best-practice tips, these may sound like best practices that would apply under any circumstance. So, how different is Opportunistic NPD? If we go back to the grid of Figure 11.2 and use the vignettes discussed above, (see Table 11.1), we can start to put more detail into each box and clear similarities and differences emerge; see Figure 11.4.

TABLE 11.1 OVERVIEW OF THE USED VIGNETTES.

Vignettes	Process Improvements	NPD
Scouting	Law firm legal tech	Crowdfunding
Validating	Law firm startup offer	Lex Mundi challenge
	eData online platform	Virtual minute books
		Solar
Transforming	Digital transformation	Hospital fellows
	Green-house farmer	Client portal
		Nose bleeds
Scaling	Teams RMS	Laboral
	Mayo clinic	Construction
	Digicampus	Early-stage startup program

FIGURE 11.4 OPPORTUNISTIC–STRATEGIC NPD MATRIX (PROCESS, TECHNOLOGY, TEAM).

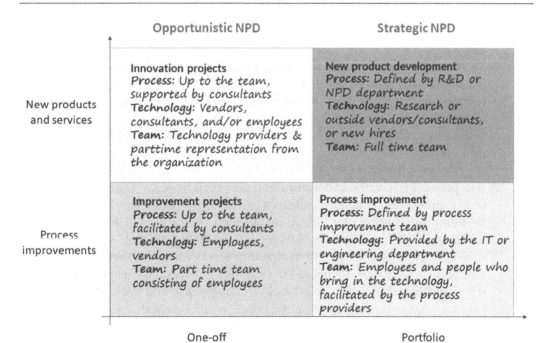

Opportunistic NPD   Strategic NPD

New products and services

Innovation projects
Process: Up to the team, supported by consultants
Technology: Vendors, consultants, and/or employees
Team: Technology providers & parttime representation from the organization

New product development
Process: Defined by R&D or NPD department
Technology: Research or outside vendors/consultants, or new hires
Team: Full time team

Process improvements

Improvement projects
Process: Up to the team, facilitated by consultants
Technology: Employees, vendors
Team: Part time team consisting of employees

Process improvement
Process: Defined by process improvement team
Technology: Provided by the IT or engineering department
Team: Employees and people who bring in the technology, facilitated by the process providers

One-off   Portfolio

The innovation process needs structure. However, what this structure consists of is different. In Figure 11.4, we highlight for each of the four boxes in the matrix the differences when it comes to the process used, who provides or develops the technology, and who forms the team that gets the work done.

On the right side, the overall direction is provided by the firm's strategy and a portfolio of activities gets developed to act on the strategy and spread risks. On the left side, each project is a one-off, with no opportunity for portfolio management to balance risks across projects.

In the Strategic NPD model, there is a distinct difference in the process, technology, and team between product and process innovation activities. New products and services are developed and supported by the innovation unit and paid from the corporate budget, whereas continuous improvement activities are supported and paid for by the business units and supported by the engineering and/or IT groups.

On the Opportunistic side, innovation projects can originate anywhere in the organization, and it is in most cases the size of the investment which will determine whether support comes from the business unit or from senior management. When comparing the vignettes that entailed more process innovations with the projects that concerned new products or services, the process, technology, and team were also very similar. Probably, because in both cases, the execution fell on teams of employees who were willing to take on the innovation project as an additional task.

The innovation process itself from discovery to scaling still consists of the same tasks and processes, albeit the phases are less distinct on the left side. Validating the need and testing early in the development phases are equally important as when innovating at a large scale. The difference is that when innovating at a small scale, there is less room to make up for (costly) failures.

The innovation process remains a value chain in both models. That means, that for Opportunistic NPD, the organization must have capabilities in Scouting, Validating, Transforming, and Scaling processes in order to bring ideas to practice. When this knowledge is not present in the firm, outside consultants or programs are a must to provide innovation teams with the structure and support they need to be successful. Leaving it up to a team to learn how to innovate and bring a new offering to market is a steep learning curve and likely sets them up for failure.

Because Opportunistic NPD typically is used when the frequency of innovation projects is low, the innovation function must be lean. That makes the role of the innovation manager different. First, in the Opportunistic Model, the innovation managers often came up through the ranks and had a functional background. This innovation manager can decide to get deeply involved him or herself in each project. However, that only works when the breadth of capabilities of the firm is rather narrow. For example, in a specialized law firm that only does intellectual property cases, such an approach could work. In a generalist law firm, that covers areas of law that concern mergers and acquisition, finance, corporate, and human resources, such an approach would require the innovation manager to have state-of-the art knowledge in all these domains, which is sheer impossible. That may be the reason, why more often these innovation managers take on the role of facilitator. That way, they can rely on the business units for the technical side while making sure that innovation teams know what they are doing when it comes to the process.

For smaller firms using the Opportunistic NPD model, it is not uncommon that the innovation function resides with the chief operating officer. While for Strategic NPD that would be a no-go, for Opportunistic NPD it seems to work well. The chief operating officer has oversight over all the operational activities and can make sure that the initiatives undertaken are aligned with the firm's strategy and make sure that there are no duplications. One of the reasons this approach works, is that these smaller organizations are less focused on efficiency and have often more flexibility in their operational organizations, which makes it easier to integrate innovation activities in this context.

In the Strategic NPD model, innovation teams consist of employees who are skilled and trained in new product development. In the Opportunistic NPD model, the teams consist of individuals who are passionate about the cause, willing to give their time and risking their careers to make change happen. These innovators need to be provided with the appropriate support, training, and guidance, so they know how to go about the NPD tasks at hand.

Working with employees, and within the operational context, means NPD incentives are different too. Bringing ideas forward is not the challenging part in the Opportunistic Model, so that does not need to be incentivized. Instead, it is better to incentivize people to risk their (personal) time and participate in innovation activities. Since each project is a one-off, it is most likely that completely different teams will undertake subsequent tries, when the first attempt(s) were not successful. These tries can be years apart. It is therefore even more important to capture lessons learned and incentivize knowledge sharing, so the knowledge gained is available for later reuse.

For those who study NPD, making a difference between Opportunistic and Strategic NPD is important too. Among others, when doing research into best practices at the project team level – it will be important to take into account if the team members are NPD professionals or employees from the business units who drive the project. In addition, when studying innovation strategies, researchers must realize that in the Opportunistic NPD model, the projects that get executed are not the result of a well-crafted strategy. Instead, the sum of projects that got executed successfully, will make up the firm's innovation strategy.

References

Blank, S. and Dorf, B. (2012). The startup owner's manual: the step-by-step guide for building a K&S Ranch.

Blindenbach-Driessen, F. (2018). Leveraging constraints for innovation: new product development essentials from the PDMA. In: *Time Commitment as the Scarcest Resource*, PDMA (ed. S. Gurtner, J. Spanjol, A. Griffin). Wiley Publishing.

Blindenbach-Driessen, F. and van den Ende, J. (2010 September). Innovation management practices compared: the example of project-based firms. *Journal of Product Innovation Management* 27 (5): 705–724.

Geerts (2019). Effective innovation strategies for incumbent firms. Doctoral Thesis, University of Twente and the KU Leuven. https://ris.utwente.nl/ws/portalfiles/portal/162072478/Proefschrift_Annelies_Geerts_20191128.pdf.

Kahneman, D. and Klein, G. (2009). Conditions for intuitive expertise: a failure to disagree. *American Psychologist* 64 (6): 515–526.

Mallaby, S. (2022). *Silicon Valley Invention of Venture Capital.* Washington Post. https://www.washingtonpost.com/opinions/2022/01/27/silicon-valley-invention-venture-capital-sebastian-mallaby-book.

McGrath, R.G. and MacMillan, I.C. (2009). *Discovery Driven Growth.* Harvard Business School Press.

Moore, G.A. (2007). To succeed in the long term, focus on the middle term. *Harvard Business Review* (July–August): 84–90.

Teece, D.J. (2007 December). Explicating dynamic capabilities: the nature and microfoundations of (sustainable) enterprise performance. *Strategic Management Journal* 28: 1319–1350.

van den Ende, J. (2021). *Innovation Management.* Red Globe Press, MacMillan/Bloomsbury.

Wheelwright, S.C. and Clark, K.B. (1992). *Revolutionizing Product Development.* The Free Press.

Dr. Floor Blindenbach-Driessen is the founder of Organizing4innovation. She has a PhD in management from the Erasmus University in the Netherlands. She has published articles in *Research Policy, the Journal of Product Innovation Management, IEEE Transactions on Engineering Management,* and *the Journal of Medical Practice Management.* She is the inventor of the O4I Collaborative Learning Platform for Innovation Teams. This platform turns innovation into an organizational learning habit. Innovation teams on the platform, deliver concrete results within six months. Over the years, she has guided hundreds of innovators leading to millions of dollars in new revenues. She is determined to create pathways to affordable and sustainable innovation cultures in all types of organizations.

Prof. Dr. Jan van den Ende is a professor of management of technology and innovation at Rotterdam School of Management, Erasmus University. He is also professor of Horticulture Innovation. His current research interests include firm-internal and -external idea management, ecosystem innovation, and sustainable innovation. Jan van den Ende has published in numerous journals such as *Organization Science, Harvard Business Review, Journal of Product Innovation Management,* and *Research Policy.* He recently published the textbook *Innovation Management,* targeted at both students and practitioners (Bloomsbury/Macmillan). He is founder of the New Business Roundtable, in which innovation managers of large companies meet. He teaches frequently in executive courses, for companies such as Philips, ING, and ASML.

CHAPTER TWELVE

REALLY NEW PRODUCT LAUNCH STRATEGIES: PRESCRIPTIVE ADVICE TO MANAGERS FROM CONSUMER RESEARCH INSIGHTS

Sven Feurer[1], Steve Hoeffler, Min Zhao, and Michal Herzenstein

12.1 Introduction

Consider the following fictional scenario in which two consumers, Kaitlin and Jacob, are contemplating space travel as several commercial spaceships have announced they will soon start accepting reservations. At first the couple is excited about the possibility of flying to outer space like astronauts, orbiting around the earth weightlessly, and seeing earth from space with their own eyes. But when the flight date nears, they face many difficult questions as they think through adopting this cool new experience. Will it be safe, how would they feel during the flight, will they breathe normally and adjust well to the loss of gravity? How might their family and friends perceive them if they take this expensive and risky trip? Ultimately, the key question for them is, "Should we do it?"

Offerings such as space tourism are considered really new products (RNPs). RNPs revolutionize existing product categories, define new ones, or defy classification within existing categories (e.g., Gregan-Paxton and John 1997; Lehmann 1994). This characteristic distinguishes RNPs from INPs (incrementally new products) that represent incremental improvement to existing products. In this sense, consumers are, by definition, unfamiliar with RNPs (e.g., Veryzer 1998), giving rise to the necessity to learn about the RNPs before the

The PDMA Handbook of Innovation and New Product Development, Fourth Edition. Edited by Ludwig Bstieler and Charles H. Noble.
© 2023 John Wiley & Sons, Inc. Published 2023 by John Wiley & Sons, Inc.

adoption decision (Lehmann 1994). It is evident from the opening scenario that some of the key issues consumers face when deciding whether to adopt an RNP stem from the difficulties associated with learning about the RNP and evaluating its novel benefits in light of the behavioral changes required. As the literature has consistently shown, since RNPs enable consumers to do things they have never done before (e.g., Zhao et al. 2012), RNPs often require consumers to alter their behavior in order to take advantage of the novel benefits (e.g., Veryzer 1998). Unfortunately, as a result of such unresolved behavioral uncertainties, many RNPs introduced in the marketplace are characterized by slow adoption and diffusion, and even outright failure (TiVo, Google Glass, 3D TV, and Segway are some prominent examples). These problems are devastating from a managerial stance because development and introduction of RNPs are costly, and key for firms' long-term success (e.g., Sorescu and Spanjol 2008). From a consumer standpoint, the difficulty of learning about RNPs and making sense of their utility may lead to suboptimal adoption/rejection decisions.

In this chapter, we categorize and advance existing knowledge to identify prescriptive advice for managers through an interesting theoretical lens. We aim to guide firms to leverage existing knowledge and insights from three decades of published behavioral research, such that managers can communicate the benefits of the RNP to consumers in a more effective way to facilitate the adoption decision process. Our synthesized findings could also help consumers like Kaitlin and Jacob estimate the potential usefulness of an RNP and their preferences more accurately.

12.2 Organizing Prior RNP Research Based on Adoption Process

Over the past three decades, research on consumer response to RNPs has yielded a staggering number of published papers, nearly 600, across various academic journals (Feurer et al. 2021). Extracting relevant insights and ideas from such a wide network of publications spanning many different domains and theories is no easy feat. Therefore, we decided to concentrate on domains that affect consumers' initial response to RNPs – the response that will ultimately determine whether they adopt the product. For this stream of extant research identified in top marketing and innovation journals, we extract the theoretical concepts tested in the papers and group similar concepts together. We then follow the framework in Feurer et al. (2021) to present these synthesized insights that marketers of an RNP should consider during consumers' adoption process.

In accordance with Hoeffler (2003), we break down the adoption process into three stages: The first stage "learning about RNPs and its attributes" implies that consumers have difficulties applying existing category knowledge to an RNP that, by definition, revolutionizes a product category or defines a new one. In the second stage "evaluating the RNP's utility," consumers seek to understand whether and in which situations the RNP's functionality will be of value to them. The final stage, "initial response to the RNP," captures the potential outcome we focus on (e.g., adoption, adoption intention, rejection). Thus, after having learned about what an RNP is or does, consumers will go on to evaluate

FIGURE 12.1 CONCEPTUAL FRAMEWORK.

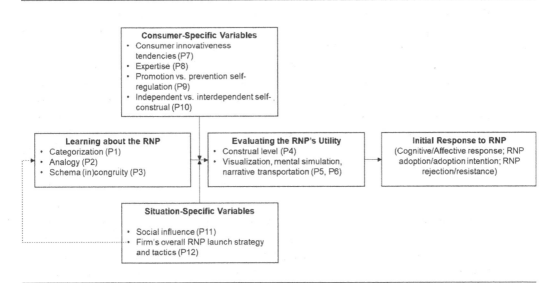

Notes: P = prescription.
--- There may also be a direct effect of launch strategies on how consumers learn about the RNP
Source: Adapted from Feurer et al. (2021).

whether the RNP's functions provide benefits valuable to them, which in turn leads to the initial adoption/rejection decision.

Consumers' decisions are never made in a vacuum; hence in the last two sections we identify processes that are not directly related with the product itself but may affect the stages described above. These moderating effects broadly fall into two categories: (a) consumer-specific moderators that affect the learning or evaluation of RNPs such as expertise or self-construal, and (b) situation-specific moderators such as the social influence and firm's overall tactical and strategic launch decisions. The resulting organizing framework appears in Figure 12.1.

In the following sections we review insights based on the concepts and theories regarding RNP learning, RNP evaluation, and these moderating factors, all of which play a critical role in consumer's initial reaction to the RNP and ultimately their adoption decision. For each theoretical concept, we briefly discuss the most influential published articles and elaborate on their insights, and then provide recommendations for managers.

12.3 Learning about the RNP

Existing work in RNP research has suggested that learning what an RNP is or can do might occur through categorizing and classifying it into either well-known or completely new categories, or by associating it with familiar products through analogies. These processes might also be facilitated by having a congruent schema between the RNP and prior knowledge consumers have (or be impeded if the schemas are incongruent). We discuss these three ideas in depth below.

12.3.1 Categorization

In an exploratory study on the consumer innovation evaluation process, Olshavsky and Spreng (1996) note that "when presented with an innovative concept, consumers first attempt to categorize the product" (p. 512). Thinking about consumers adopting space travel experience (i.e., becoming "space travelers") one can categorize them simply as tourists (albeit going to a completely new destination), or they can be categorized as astronauts, a title that until recently was used only by professional spaceship operators. These options point to the fact that consumers often find it difficult to attach a category label to a particular RNP. Consequently, consumers must arduously construct new knowledge structures rather than adopting existing ones (Moreau, Markman et al. 2001).

Researchers illuminate the process by which consumers may draw information from more than one domain to categorize an RNP: a digital camera is both a camera and a scanner. However, while consumers are able, they are typically unwilling, to draw information from multiple categories, unless significant contextual support is provided. Instead, "the first plausible category label provided to the consumer significantly influences their categorizations, expectations, and preferences" (p. 489), suggesting that it is important to positioning the RNP along one clear product category to aid learning. This suggestion, however, also depends on the type of product. For hybrid RNPs which result from a combination of features and functionality of two or more existing products, it is beneficial for categorization to remain vague (Noseworthy and Goode 2011). Consider Apple's introduction of the iPhone in 2007, at the time a unique combination of the features of a mobile phone, the iPod, and an internet communication device. Here, a simple positioning as belonging to one of these categories would have hindered the ability of consumers to recognize (and appreciate) the innovative benefits created by the features taken from the respective secondary categories.

Clearly, if consumers attempt to place the RNP in an existing category, the novel benefits may be overlooked (or discounted) because those novel features and benefits are not usually associated with the existing category. On the other hand, placing the RNP within a familiar category can facilitate learning. This tradeoff should be considered by marketers, as well as the predominant categorization scheme most likely to be used by consumers. Accordingly, we conclude the following prescription for marketing managers:

Prescription 1: Managers should have a good understanding of the tradeoffs associated with how RNPs are positioned:

(a) Positioning within an existing category may lead to faster consumer learning and lower uncertainty about the product; however, consumers may discount the novel features and product benefits as the RNPs do not fit with the existing category.

(b) Creating an entirely new category may generate excitement and curiosity among consumers eager to learn more; however, this may be hard and even frustrating for some customers who may consequently withdraw from the learning process entirely.

12.3.2 Analogy

In the early days of cloud computing, when consumers did not fully understand the concept, SunGard ran an ad campaign comparing their offering to surviving a zombie attack, and explained that in both instances, the attackers are "after your brain" and the only way to successfully survive is to be prepared (Herzenstein and Hoeffler 2016). The campaign tripled traffic to the company's website. Advocates of using analogical learning argue that RNPs are often so new and distinct that closely related product categories from which knowledge could potentially be transferred do not exist at the time of launch, hence category-based learning is not feasible. However, the problem is that the path to analogical enlightenment is paved with many necessary and sequential steps (Gregan-Paxton and John 1997). Specifically, when consumers encounter an analogy, they first create a map that aligns the two bases of comparison in the analogy (e.g., cloud computing and a zombie attack). Then consumers must transfer their knowledge from one base to the other: how will knowledge about a zombie attack help me understand what is cloud computing?

Obviously, effective analogical learning only occurs if no mistakes happen in the process. When consumers select (or managers offer) an inappropriate knowledge base (e.g., cloud computing is like a zombie attack in that both are fictional), learning would be impaired. As such, research on analogies identifies boundary conditions for effective learning by analogy to occur. Gregan-Paxton and John (1997) argue that base domain experts (novices) are more (less) likely to perceive common relations between the base domain and the RNP target domain. Herzenstein and Hoeffler (2016) show that divulging multiple aspects of an analogy can help the persuasive impact of an analogy, but only if the analogy is difficult to understand (such as the case with cloud computing and a zombie attack). If the analogy is easy to understand, consumers prefer to "solve" it themselves as they experience a positive feeling with a simple aspect, which is then translated into higher RNP evaluations.

> **Prescription 2:** To help consumers understand an RNP, managers can relate the product to a relevant base domain by using an analogy. The more conceptually distant the base domain is from the RNP (due to product nature or consumer expertise), the more the analogy needs to be fully described to consumers.

12.3.3 Schema (In)Congruity

This literature stream examines how the congruence of a product with its activated product schema influences learning and evaluation. Meyers-Levy and Tybout (1989) define schema congruity as a situation in which the activated schema represents all the fundamental attributes of a product, and schema incongruity as a situation in which the activated schema does not represent the total configuration of a product's attributes. Jhang et al. (2012) characterize levels of incongruity in terms of the type of associations the consumer must make to resolve the incongruity to understand a benefit of the new product.

RNPs that offer an extremely incongruent attribute completely lack links to shared associations with the activated category.

Several moderators related with product positioning, promotion, and product offering affect the relationship between the level of incongruity and RNP learning. For example, it has been shown that positioning along functional dimensions (versus experiential) is preferred when the product form is incongruent with consumers' normative expectations (Noseworthy and Trudel 2011). For example, if space travel seems incongruent with the tourism category, the positioning should stress how safe and easy it is. But if space travel is congruent then positioning should stress how breathtaking the experience will be (seeing earth from space, experiencing weightlessness, etc.). In terms of promotions, marketers can incorporate "enablers" (e.g. the color green to make sense of a semantically related feature such as vitamin enriched) in their product design to help resolve extreme incongruity (Noseworthy et al. 2018). Inducing calmness in consumers during the RNP learning stage can also help improve consumers perceptions of highly incongruent products, as Noseworthy et al. (2014) show that consumers prefer such products more when they are relaxed rather than excited, but prefer moderately incongruent products more when they are excited rather than relaxed. A relaxed state, the authors note, helps consumers overcome the anxiety that a highly incongruent product may elicit. As for the product offering, Ma et al. (2015) advise that when the RNP has an incongruent feature, marketers should offer the incongruent feature as an optional add-on (peripheral) rather than building it into the product (core). For example, a mind control feature that allows video gamers to move objects with their mind can be positioned as a core product construction if it is built into a new virtual reality (VR) console, but it can also be positioned as a peripheral product construction offered as an optional accessory for a new VR console. The VR was better received when the mind control feature was offered as peripheral.

In many cases RNPs are extremely incongruent with the activated schema, and as Meyers-Levy and Tybout (1989) note, resolving extreme incongruity requires complex strategies because "there is no ready associative pathway to traverse" (p. 42). Thus, managers responsible for the launch of RNPs should take measures that facilitate the resolution of extreme incongruity to help consumers learn (Jhang et al. 2012). Space travel, for example, may activate the product category schema of tourism, but the space aspect may not associate with the activated tourism schema or category expectations, thus it is likely that positioning it as a safe and enjoyable experience, adding cues that remind consumers of travel (like checking-in), all while inducing a relaxed state in consumers (perhaps with music or a calm voice) will facilitate learning and thus evaluations.

Prescription 3: Managers seeking to position an RNP as a creator of a new category should help consumers resolve perceived extreme incongruity by incorporating some familiar designs in the RNP or offering the novel aspects as peripheral components separate from the core product, positioning properly on functionality and adding enablers in the product design or promotion, and inducing a more mellow emotional state.

12.4 Evaluating the RNP's Utility

When consumers evaluate the RNP's utility and the value they will derive from adopting it, they consider reasons for and against adoption (Claudy et al. 2015; Talke and Heidenreich 2014). However, evaluating the RNP could prove difficult because essentially consumers must understand their usage process (Heidenreich and Kraemer 2015; Hoeffler 2003) and evaluate the benefits they will enjoy. Hence, managers need to draw consumers' attention to the novel functions of the RNP. Insights from research on construal level and visualization can help guide the marketing efforts of RNPs. We elaborate on these concepts next.

12.4.1 Construal Level

In our opening scenario, Kaitlin and Jacob were excited about space travel when it was first announced, but months later, they reflect on more concrete issues associated with adoption such as the travel uncertainty and high price tag. This apparent change in perspective can be explained by construal level theory which links psychological distance to the extent to which consumers' thinking is abstract or concrete (Trope and Liberman 2003). In the context of RNPs, the more temporally distant the product is from the consumer, the more abstract it will be thought of (high level considerations, or the "why" aspects of the desirability of adopting it) and the greater focus is put on product benefit. But as the product becomes temporally closer, it will be thought of more concretely (low level considerations, or the "how" aspects of the RNP's usage feasibility; Alexander et al. 2008) and consumers' focus is shifted to behavioral change required to adopt the product. As a result, consumers may be "all bark but no bite" as they intend to adopt when product launch is far, but that intention is dramatically reduced when the product is actually launched. Since it is beneficial for the marketers to make accurate forecast of consumers' RNP preference that will be consistent with their actual adoption behavior, managers should adjust their communication strategy between product announcement and launch such that consumers consider both types of information (abstract/desirability and concrete/feasibility). More specifically, during concept testing or market forecasting (where consumers' adoption decision is for the far future), they should remind consumers of the usage process of the RNP to facilitate a realistic forecast of consumers' preference; however, at the time when consumers are making the adoption decision, they should remind consumers of the benefit of the RNP in their communication (Zhao et al. 2007).

> **Prescription 4:** Managers communicating RNPs should help consumers appreciate the naturally ignored product aspect to increase preference stability over time and firm's forecasting success rate: usage process information should be highlighted during concept testing/market forecasting, and product benefit information should be highlighted for imminent adoption decisions.

12.4.2 Visualization, Mental Simulation, Narrative Transportation

One tool that can help focus consumers on the usage process is mental simulation, or mental imagery, which is defined as an imitative mental representation of events (Taylor and Schneider 1989). The idea is that mental simulation enables consumers to evoke relevant personal experiences and more effectively align the RNP with patterns of usage (Taylor et al. 1998). A similar concept is visualization or "thinking in pictures" (Dahl and Hoeffler 2004) which MacInnis and Price (1987) define as a process by which sensory information is represented in the working memory. As a result, "by having people simulate using the product in current usage scenarios, respondents are better able to think through the subtle effects that adopting the product would have on their day-to-day routine, which improves their ability to predict the personal usefulness of the new benefits" (Hoeffler 2003, p. 416). However, consumers may find it difficult to imagine how they are going to use an RNP.

Continuing with the example of space travel, consumers may wonder if they would be comfortable during the journey. Since most of the population has not experienced zero gravity, it might be hard for consumers to envision their experience in that environment. Indeed, recent research finds that imagination difficulty decreases evaluation of RNPs because consumers perceive imagination difficulty as diagnostic in their RNP assessment (Zhao et al. 2012) – if it is hard to imagine using the product, the product must be difficult to use or not that good. However, with the right framing of (or guidance for) the mental simulation exercise, managers can shift consumers' focus from the usage process toward the outcome of recognizing the benefits of the RNP (Zhao et al. 2011). Therefore, in the space travel example, marketers should encourage consumers to think about how amazing the view of the earth from space could be, rather than the overwhelming process of getting the spacecraft to that view. Similarly, Zhao et al. (2009) emphasize the importance of imagination-focused (versus memory-focused) visualization for higher evaluation of RNPs. Their findings suggest marketers should encourage consumers to push their imaginations while simulating a brand-new usage scenario (of space traveling) rather than relying on their past memories and being constrained by their experience and existing behavioral patterns (of traveling). Another possible way of easing visualization is to supply "visualization aids" that are either concrete or abstract, consistent with consumers' natural mindset for past- vs. future-focused visualization (Zhao et al. 2014). Advances in virtual reality (VR) may also become a useful visualization tool as it further facilicates consumers to simulate the novel benefits provided by a RNP.

Affect also plays a role in the effectiveness of mental simulation as a tool for the evaluation of an RNP. Since these products are often beleaguered by uncertainty, thinking about them can elicit positive (hope, optimism) or negative (anxiety, fear) emotions, which subsequently impacts product evaluation (Lin et al. 2020). Marketers, obviously, would like consumers to experience positive emotions as they evaluate the RNP, which is more likely to be achieved if consumers think about the benefits they will gain when using the product in the distant future (rather than thinking about the product usage process which is naturally salient in the near future; Castaño et al. 2008; Zhao et al. 2011). Encouraging an affective rather than cognitive focus during mental simulation is also important as relying on feelings usually results in less critical analysis of the arguments, fewer negative thoughts and a more positive overall product attitude (Green and Brock 2000; West et al. 2004).

> **Prescription 5:** Managers should actively reduce the difficulty of mental simulation and encourage consumers to focus on the benefits of the RNP (rather than the process of using it) or rely on their feelings to enhance RNP evaluations.

A related concept, *narrative transportation*, is defined as "a mixture of attention, imagery, and feelings that people experience when they watch a movie or read a narrative" (Van den Hende and Schoormans 2012). Being transported means that consumers are immersed in what they are watching or reading and have vivid images in their minds; consumers see themselves in the action while forgetting about the world around them. Similar to watching a movie, a narrative represents a surrogate experience by "transporting" the consumer into a story line with related events that typically involve a protagonist, a setting, and a plot with a series of actions (van den Hende et al. 2012).

Consumers may spontaneously generate mental simulations in an attempt to assess the value of RNPs, and when these self-generated simulations are sufficiently engrossing and vivid narratives, they evoke transportation. The affect associated with becoming transported during evaluation positively impacts the evaluation of the RNP (Nielsen et al. 2018). Consumers with extensive experience in a related product category are more likely to be transported during the evaluation, as it is easier for them to engage in an engrossing mental simulation. Further, as previously mentioned, advances in virtual reality may increase the likelihood that consumers will be transported when viewing information about an RNP via a VR headset.

> **Prescription 6:** Managers should strive to transport consumers into a story about an RNP to facilitate consumers' understanding of the RNP and improve their evaluations.

12.5 The Role of Consumer-specific Variables

So far, we discussed ideas based on psychological concepts examined in the RNP literature that can guide managers in helping consumers learn and assess the RNP from the product perspective. However, as mentioned earlier, these learning and evaluation processes are affected by a myriad of other things, some of them related to consumers themselves and others related to the broader situational contexts. In this section we discuss the moderating factors related to consumers, including their innovativeness tendencies, expertise, promotion versus prevention self-regulation focus, and independent versus interdependent self-construal.

12.5.1 Consumer Innovativeness Tendencies

Consumer innovativeness is a latent concept that ultimately explains why some consumers adopt an RNP earlier than others (Goldsmith and Hofacker 1991; Midgley and Dowling 1978; Rogers 2003; Rogers and Shoemaker 1971). Many RNPs are surrounded with uncertainties regarding the benefits and deterrents of adoption (Castaño et al. 2008), and individuals with high consumer innovativeness deal with these uncertainties better than individuals low in consumer innovativeness (Rogers 2003). As such, consumer innovativeness is an important segmentation criterion and consumers with different levels of innovativeness should be targeted with different messages. For instance, consumers high in innovativeness should be targeted by signaling exclusive innovativeness or originality (Schuhmacher et al. 2018), which can be achieved with a message that focuses on the product's unique features, an exclusive distribution system, and a very high launch price. However, if the RNP is conducive for mass market introduction, these consumers are usually low on innovativeness and therefore marketers should signal security. This suggests that managers should adjust their marketing strategies and focus on exclusive originality versus security for different types of customers ranging from innovators, early adopters, early majority to late majority and laggards during the product life cycle, especially once the influence of innovators on the remaining segments decreases (Mahajan and Muller 1998).

Despite the obvious effect of consumer innovativeness on the speed of adoption, it actually cannot fully account for the likelihood of adoption. Individuals low in consumer innovativeness might not be the quickest ones to adopt an innovation, but they eventually do (Heidenreich and Handrich 2015; Heidenreich and Kraemer 2015; Talke and Heidenreich 2014). Accordingly, recent researchers proposed a more nuanced concept: people's general inclination to resist innovations, or "passive innovation resistance." They showed it is a dominant barrier, which must be overcome before the RNP adoption can take place. Further, these authors prescribe the use of analogies and the provision of detailed information about the RNP as methods that can minimize the passive innovation resistance (Heidenreich and Kraemer 2015).

> **Prescription 7:** Managers need to carefully align their launch tactics to the requirements of the targeted segments based on consumer innovativeness and a predisposition to resist innovations.

12.5.2 Expertise

Expertise, a correlated concept of consumer innovativeness, plays a crucial role in consumer learning about RNPs. Expertise is a function of the amount of domain-specific knowledge acquired through experience or training (Wood and Lynch 2002), and it helps in the learning stage by providing a "framework explicitly designed for organizing incoming information about that stimulus" (Gregan-Paxton et al. 2002, p. 536). But while these features of expertise certainly seem positive, Moreau, Lehmann et al. (2001) show that experts are better able than novices to recognize a lack of understanding for an RNP which leads to a decrease in comprehension and perceived net benefits. In a similar vein, Wood and Lynch (2002) argue that high levels of product expertise is often accompanied by overconfidence, motivational deficits and a "feeling of knowing" that ultimately hinders experts' ability to learn about and fully appreciate a RNP's benefits. Zhao et al. (2012) note that for an RNP which constitutes a new product category, expertise in the domain of the RNP is generally lower, making consumers more likely to draw on contextual factors rather than product information itself as diagnostic cues in forming their RNP evaluation.

An interesting tactic comes in the form of bundling the RNP together with a familiar product when applicable. For example, a digital pen (at the time an RNP) could be sold separately or as an accessory for a laptop (bundled together). The latter strategy is most effective for novices who evaluate the RNP more positively when it is bundled with a familiar product that has a good fit with the RNP, although interestingly the bundle does not increase comprehension of the RNP, in comparison to marketing it as a stand-alone product (Reinders et al. 2010).

> **Prescription 8:** Marketers should strive to determine the level of expertise of adopters in order to choose marketing strategies (communication, bundling) that would enable these consumers to use their expertise level appropriately to fully appreciate the novel features of the RNP.

12.5.3 Promotion versus Prevention Self-Regulation

Regulatory focus theory posits the presence of two self-regulation systems for decision-makers: the promotion system and the prevention system. The former is derived from nurturance needs such as advancement and growth, and uses approach strategies when regulating toward desirable ends. The latter is derived from security needs and uses avoidance strategies when regulating away from undesirable ends. Herzenstein et al. (2007) theorize and find that self-regulation affects the adoption of RNPs because consumers weigh the benefit provided by a new product against its costs differently, depending on their self-regulation focus. Further, they note that the associated risk/uncertainty perception (which is inherent in RNPs due to their newness) may vary as a function of self-regulation focus. Consumers with a chronic disposition to be promotion-focused own more novel high-technology products than prevention-focused consumers. As such, regulatory focus may be another effective basis for segmentation.

Regulatory focus can also be momentarily induced, therefore "marketers can frame communications to encourage promotion focus and shift consumer focus toward positive outcomes, which may favorably affect evaluations and purchase likelihood" (Herzenstein et al. 2007, p. 259). More recently, Luo et al. (2016) showed that exposure to an advertisement for an RNP could trigger a prevention goal among consumers, and therefore they advise marketers to frame product benefits "in terms of negative outcomes avoided by adopting the products." In our space travel example, one can elicit promotional communications that focus on the benefits of space travel and achieving a goal that was beyond the reach of prior generations, or work around consumers' natural prevention focus while encountering RNPs by warning consumers not to be left behind (e.g., "Don't be the only one who doesn't get to see earth from space!"). Thus, existing research suggests that marketers encourage a promotion focus, but also communicate with the default prevention focus to reduce uncertainty during the evaluation process.

> **Prescription 9:** Managers should attempt to create a promotion-focused mind-set in consumers by highlighting what an RNP allows them to accomplish. For prevention-focused consumers, managers can frame RNP benefits as avoiding negative outcomes.

12.5.4 Independent versus Interdependent Self-construal

Self-construal theory refers to the bases of self-definition in relation to the extent to which the self is defined independently of others or interdependently with others (Markus and Kitayama 1991). An independent self-perspective views the self as autonomous and separate from others, and focuses on the person's abilities, preferences, and wishes. In this view, the individual's goals supersede those of the in-group. In contrast, an interdependent self-perspective is characterized by a sense of connectedness with others and a focus on one's role in their in-groups. Therefore, in this view, the group's goals supersede those of the individual. While some cultures have values that are more in line with the independent self (usually Western) and others with the interdependent self (usually Eastern), self-construal can also be situationally activated (Aaker and Lee 2001).

Because RNPs are often seen as novel, unique, and different, they are more likely to be adopted by consumers with an independent self-construal as these people desire a high level of distinctiveness (Ma et al. 2014). This finding suggests that managers should temporarily induce a more independent self-construal while presenting the RNP. For example, presenting a digital pen (which was a new product at the time) as an apparatus for "the art of expressing yourself: spell out your inspiration" (inducing an independent self) versus "the art of sharing: write to your loved ones" (inducing an interdependent self) increased adoption significantly. Coming back to our opening example, promoting the space travel experience as "fulfilling your ultimate personal dream," would highlight independent self-construal and potentially enhance adoption intentions!

> **Prescription 10:** Managers should attempt to create an independent self-perspective that highlights the uniqueness of consumers who adopt the RNP.

12.6 The Role of Situation-specific Variables

The second category of factors moderating consumers' learning and evaluation processes we identified has to do with the broader context surrounding the RNP, such as social influence and the company's overall launch strategies.

12.6.1 Social Influence

Social influence is a critical driver of RNP adoption and subsequent diffusion (Bass 1969), meaning that at the time of adoption, consumers may be affected by the decisions of others in the social system. Furthermore, potential adopters such as Kaitlin and Jacob would think about how their family/friends would perceive them for taking the risky and expensive space trip. Indeed, superordinate group influence affects the consumption for early adopters seeking social approval (Fisher and Price 1992). Anecdotally, social influence and the visibility of consumption were arguably one major reason for the failure of the Google Glass (because a small camera was mounted on the frame and continuously took pictures, as what Wartzman (2013) called "the creep factor").

Moreover, consumers low in innovativeness might be negatively influenced by those who have already adopted the RNP such that the early adopters' uniqueness is preserved (Moldovan et al. 2015). For instance, negative online RNP reviews might stop other consumers from adopting the product (Xiao et al. 2018). Conversely, positive ratings might serve as a trustworthiness signal and thus increase adoption likelihood (Konya-Baumbach et al. 2019). Morvinski et al. (2017) examine the effect of adoption stock ("Ten Million Readers Can't Be Wrong!") and homophily (the tendency to act like similar others) on subsequent adoption. Interestingly, they found that in high uncertainty situations, which is typically the case for RNPs, the effect to be null or even negative in low-homophily situations.

On the other hand, there are RNPs that depend on herd adoption to succeed, for example the first Messenger application – a consumer would only adopt it if many of their friends and family members adopt it too (otherwise, who will they message?). Similarly, innovations related to the shared economy such as Uber and Airbnb need to be adopted by many at once to create sufficient supply and demand. In situations like these, the effect of social influence and word-of-mouth is even more critical.

> **Prescription 11:** Marketers should strive to understand the nuanced role of potential social influence when introducing RNPs to ensure social desirability while maintaining a good balance between wide-scale adoption and uniqueness.

12.6.2 Firm's Overall RNP Launch Strategy and Tactics

Besides firms' launch decisions specifically designed for the target product such as customer segmentation efforts (by consumer innovativeness, expertise, etc.), positioning efforts (new category, level of uniqueness etc.) and product offering (bundling it with familiar product, or offering it as an optional accessory versus integrating it into the core product), a vast number of other factors pertaining to the firm's overall marketing strategy also influence the learning, evaluation, and adoption decision process. This broader firm context includes the overarching RNP entry strategy (skimming vs. penetration), and the corresponding tactical launch decisions regarding pricing, distribution, and promotion (e.g., Montaguti et al. 2002).

For example, research related to the price of an RNP and its effect on adoption shows that the negative effect of a high launch price on perceived price fairness is less pronounced for really innovative products (Kuester et al. 2015), suggesting acceptance of higher prices for such products. When consumers expect the price of an RNP to decline shortly after launch (as in a skimming strategy), they perceive the price as unfair, independent of whether it was originally priced high or low, suggesting that firms might want to manage such expectations. As we mentioned earlier, considering consumer innovativeness as an important segmentation criterion, marketers can use distribution and pricing strategies to reinforce the product positioning along the lines of uniqueness and novelty (Schuhmacher et al. 2018).

Launch tactics that aim to increase consumer trust, especially for RNPs based on new technologies and unfamiliar suppliers, can substantially increase adoption (Konya-Baumbach et al. 2019). These authors document three launch tactics that foster trust: providing customer reviews and ratings, using benefit-based promotional messages, and clear prices that do not include any obfuscated information, hidden fees, or monetizing opportunities (such as selling customer data to a third party).

Finally, a promising new tool to facilitate learning about the RNP and ultimately its adoption is to convey product information in the form of a game (Müller-Stewens et al. 2017). It enhances adoption through stimulating curiosity and enhancing perceived vividness of information presentation. For example, instead of telling consumers about the benefits of a new type of bicycle tires (in which the new feature is that it allows them to ride on very uneven terrain), consumers who click on the ad would see a game in which they virtually ride a bike with regular tires and then with the new tires, and see how far they can get in the allotted time. The game would "teach" consumers about the traction features of the really new tires with a more direct product experience.

> **Prescription 12:** Firms should consider different aspects of its overall marketing strategy and latest technology to optimize the adoption process (such as using gamification or virtual reality to teach consumers novel benefits).

12.6.3 Business-to-Business Insights

While our chapter has been focused on consumer response to RNPs – many of the insights carry over to business-to-business (B2B) situations. In fact, extant research suggests that aspects play a role in organizational buying "that go beyond the core purchasing criteria" (Bornemann et al. 2020, p. 448). First, purchase decisions in those situations are being made by managers who – even though they are not acting in the role of an individual consumer but that of a businessperson (Heide and Wathne 2006) – might be influenced by the same conceptual issues. For example, when considering the adoption of an RNP product or service, business managers may visualize how the new service adds value to their offering (and the degree to which final consumers will understand and appreciate those novel benefits), or use an analogy to better understand the RNP. Second, when businesses consider the purchase of a completely new solution, they are faced with the same types of tradeoffs consumers face: the efficiency the new solution allows and the risk it bears. The difference between businesses and consumers adoption is scale and accountability (Doney and Armstrong 1996). For example, if a new solution is adopted and it is found to be harmful, it may influence that business's employees or customers.

12.7 Concluding Thoughts

Space travel, as amazing as it may be, is also risky, just like many RNPs on the market. This example, that we have woven into the fabric of our chapter, illustrates how a large variety of behavioral aspects may facilitate or hinder its adoption. Starting from the learning process, through evaluations, personal characteristics of potential customers, and the broader situational context surrounding them, many factors could play a role to influence RNP adoption. Our literature review demonstrates that findings from well-tested theoretical lenses can be applied to illuminate the steps in the adoption process, identifying a variety of influencing variables as summarized in the 12 prescriptions for managers. We also provide managers with a "scorecard" of the 12 identified critical factors that need to be considered as they introduce RNPs in their industries and expect a successful diffusion (see Figure 12.2). We do not believe that there is one particular path to market for RNPs that is the right path for all products. Instead, our scorecard focuses on the tradeoffs and relevant issues that need to be considered when designing the optimal launch strategy for a specific RNP.

FIGURE 12.2 RNP LAUNCH STRATEGY SCORECARD.

RNP LAUNCH STRATEGY SCORECARD

Factor in how consumers learn about the RNP	Consider how consumers evaluate the RNP's utility	Pay attention to consumer-specific variables	Acknowledge situation-specific variables
☐ Evaluate the tradeoff between positioning as a member of an existing category versus creating an entirely new one	☐ Evaluate the type of information that is used to convey product benefits or usage process during testing and launch	☐ Evaluate the innovativeness tendencies of the targeting segment and their predispositions to resist the innovation	☐ Evaluate the nuanced role of social influence toward a goal of increasing social desirability
☐ Evaluate the potential use of an analogy and the degree to which the analogy is defined when communicating benefits	☐ Evaluate the degree to which consumers can simulate the usage of RNP and whether VR headsets may aid	☐ Evaluate the expertise level of the targeted segment to aid launch strategies	☐ Evaluate fit with overall firm strategy and novel approaches (use of VR applications, gamification) to optimize adoption process
☐ Evaluate ways to reduce extreme incongruity via product design or positioning decisions	☐ Evaluate ways in which consumers could be transported when evaluating RNP	☐ Evaluate degree of promotion/ prevention focus to help craft positioning strategy	
		☐ Evaluate methods for creating an independent self-perspective	

Note

1   Authorship is random and all authors contributed equally to this research. This research was supported by Boston College, University of Delaware, University of Toronto, the Social Sciences and Humanities Research Council of Canada (SSHRC), and a postdoc fellowship of the German Academic Exchange Service (DAAD).

References

Aaker, J.L. and Lee, A.Y. (2001). "I" seek pleasures and "we" avoid pains: the role of self-regulatory goals in information processing and persuasion. *Journal of Consumer Research* 28 (1): 33–49.

Alexander, D.L., Lynch, J.G., Jr., and Wang, Q. (2008). As time goes by: do cold feet follow warm intentions for really new versus incrementally new products? *Journal of Marketing Research* 45 (3): 307–319.

Bass, F.M. (1969). A new product growth for model consumer durables. *Management Science* 15 (5): 215–227.

Bornemann, T., Klarmann, M., and Moosbrugger, M. (2020 December). Verhaltenswissenschaftliche Forschung zum organisationalen Einkaufsverhalten: Überblick über die Marketingliteratur. *Schmalenbachs Zeitschrift für Betriebswirtschaftliche Forschung* 72: 447–478.

Castaño, R., Sujan, M., Kacker, M., and Sujan, H. (2008). Managing uncertainty in the adoption of new products: temporal distance and mental simulation. *Journal of Marketing Research* 45 (3): 320–336.

Claudy, M.C., Garcia, R., and O'Driscoll, A. (2015). Consumer resistance to innovation—a behavioral reasoning perspective. *Journal of the Academy of Marketing Science* 43 (4): 528–544.

Dahl, D.W. and Hoeffler, S. (2004). Visualizing the self: exploring the potential benefits and drawbacks for new product evaluation. *Journal of Product Innovation Management* 21 (4): 259–267.

Doney, P.M. and Armstrong, G.M. (1996). Effects of accountability on symbolic information search and information analysis by organizational buyers. *Journal of the Academy of Marketing Science* 24 (1): 57–65.

Feurer, S., Hoeffler, S., Zhao, M., and Herzenstein, M. (2021). Consumers' response to really new products: a cohesive synthesis of current research and future research directions. *International Journal of Innovation Management* 25 (08): 2150092.

Fisher, R. and Price, L. (1992). An investigation into the social context of early adoption behavior. *Journal of Consumer Research* 19 (8): 477–486.

Goldsmith, R.E. and Hofacker, C.F. (1991). Measuring consumer innovativeness. *Journal of the Academy of Marketing Science* 19 (3): 209–221.

Green, M. and Brock, T. (2000). The role of transportation in the persuasiveness of public narratives. *Journal of Personality and Social Psychology* 79 (5): 701–721.

Gregan-Paxton, J., Hibbard, J.D., Brunel, F.F., and Azar, P. (2002). "So that's what that is": examining the impact of analogy on consumers' knowledge development for really new products. *Psychology and Marketing* 19 (6): 533–550.

Gregan-Paxton, J. and John, D.R. (1997). Consumer learning by analogy: a model of internal knowledge transfer. *Journal of Consumer Research* 24 (3): 266–284.

Heide, J.B. and Wathne, K.H. (2006). Friends, businesspeople, and relationship roles: a conceptual framework and a research agenda. *Journal of Marketing* 70 (3): 90–103.

Heidenreich, S. and Handrich, M. (2015). What about passive innovation resistance? Investigating adoption-related behavior from a resistance perspective. *Journal of Product Innovation Management* 32 (6): 878–903.

Heidenreich, S. and Kraemer, T. (2015). Innovations-doomed to fail? Investigating strategies to overcome passive innovation resistance. *Journal of Product Innovation Management* 33 (3): 277–297.

Herzenstein, M. and Hoeffler, S. (2016). Of clouds and zombies: how and when analogical learning improves evaluations of really new products. *Journal of Consumer Psychology* 26 (4): 550–557.

Herzenstein, M., Posavac, S.S., and Brakus, J.J. (2007). Adoption of new and really new products: the effects of self-regulation systems and risk salience. *Journal of Marketing Research* 44 (2): 251–260.

Hoeffler, S. (2003). Measuring preferences for really new products. *Journal of Marketing Research* 40 (4): 406–421.

Jhang, J.H., Grant, S.J., and Campbell, M.C. (2012). Get it? Got it. Good! enhancing new product acceptance by facilitating resolution of extreme incongruity. *Journal of Marketing Research* 49 (2): 247–259.

Konya-Baumbach, E., Schuhmacher, M.C., Kuester, S., and Kuharev, V. (2019). Making a first impression as a start-up: strategies to overcome low initial trust perceptions in digital innovation adoption. *International Journal of Research in Marketing* 36 (3): 385–399.

Kuester, S., Feurer, S., Schuhmacher, M.C., and Reinartz, D. (2015). Comparing the incomparable? How consumers judge the price fairness of new products. *International Journal of Research in Marketing* 32 (3): 272–283.

Lehmann, D.R. (1994). Characteristics of 'really' new products. Boston (September 29–30).

Lin, Y.-T., MacInnis, D.J., and Eisingerich, A.B. (2020). Strong anxiety boosts new product adoption when hope is also strong. *Journal of Marketing* 84 (5): 60–78.

Luo, Y., Wong, V., and Chou, T.J. (2016). The role of product newness in activating consumer regulatory goals. *International Journal of Research in Marketing* 33 (3): 600–611.

Ma, Z., Gill, T., and Jiang, Y. (2015). Core versus peripheral innovations: the effect of innovation locus on consumer adoption of new products. *Journal of Marketing Research* 52 (3): 309–324.

Ma, Z., Yang, Z., and Mourali, M. (2014). Consumer adoption of new products: independent versus interdependent self-perspectives. *Journal of Marketing* 78 (2): 101–117.

MacInnis, D.J. and Price, L.L. (1987). The role of imagery in information processing: review and extensions. *Journal of Consumer Research* 13 (4): 473–491.

Mahajan, V. and Muller, E. (1998). When is it worthwhile targeting the majority instead of the innovators in a new product launch? *Journal of Marketing Research* 35 (4): 488–495.

Markus, H.R. and Kitayama, S. (1991). Culture and the self: implications for cognition, emotion, and motivation. *Psychological Review* 98 (2): 224–253.

Meyers-Levy, J. and Tybout, A.M. (1989). Schema congruity as a basis for product evaluation. *Journal of Consumer Research* 16 (1): 39–55.

Midgley, D. and Dowling, G. (1978). Innovativeness: the concept and its measurement. *Journal of Consumer Research* 4 (4): 229–242.

Moldovan, S., Steinhart, Y., and Ofen, S. (2015). "Share and scare": solving the communication dilemma of early adopters with a high need for uniqueness. *Journal of Consumer Psychology* 25 (1): 1–14.

Montaguti, E., Kuester, S., and Robertson, T.S. (2002). Entry strategy for radical product innovations: a conceptual model and propositional inventory. *International Journal of Research in Marketing* 19 (1): 21–42.

Moreau, C.P., Lehmann, D.R., and Markman, A.B. (2001). Entrenched knowledge structures and consumer response to new products. *Journal of Marketing Research* 38 (1): 14–29.

Moreau, C.P., Markman, A.B., and Lehmann, D.R. (2001). What is it? Categorization flexibility and consumers' responses to really new products. *Journal of Consumer Research* 27 (4): 489–498.

Morvinski, C., Amir, O. and Muller, E. (2017). "Ten Million Readers Can't Be Wrong!," or can they? On the role of information about adoption stock in new product trial. *Marketing Science* 36 (2): 290–300.

Müller-Stewens, J., Schlager, T., Häubl, G. and Herrmann, A. (2017). Gamified information presentation and consumer adoption of product innovations. *Journal of Marketing* 81 (2): 8–24.

Nielsen, J.H., Escalas, J.E., and Hoeffler, S. (2018). Mental simulation and category knowledge affect really new product evaluation through transportation. *Journal of Experimental Psychology: Applied* 24 (2): 145–158.

Noseworthy, T.J., Di Muro, F., and Murray, K.B. (2014). The role of arousal in congruity-based product evaluation. *Journal of Consumer Research* 41 (4): 1108–1126.

Noseworthy, T.J. and Goode, M.R. (2011). Contrasting rule-based and similarity-based category learning: the effects of mood and prior knowledge on ambiguous categorization. *Journal of Consumer Psychology* 21 (3): 362–371.

Noseworthy, T.J., Murray, K.B., and Di Muro, F. (2018). When two wrongs make a right: using conjunctive enablers to enhance evaluations for extremely incongruent new products. *Journal of Consumer Research* 44 (6): 1379–1396.

Noseworthy, T.J. and Trudel, R. (2011). Looks interesting, but what does it do? Evaluation of incongruent product form depends on positioning. *Journal of Marketing Research* 48 (6): 1008–1019.

Olshavsky, R.W. and Spreng, R.A. (1996). An exploratory study of the innovation evaluation process. *Journal of Product Innovation Management* 13 (6): 512–529.

Reinders, M.J., Frambach, R.T., and Schoormans, J.P.L. (2010). Using product bundling to facilitate the adoption process of radical innovations. *Journal of Product Innovation Management* 27 (7): 1127–1140.

Rogers, E.M. (2003). *Diffusion of Innovations*, 5e. New York, NY: Free Press.

Rogers, E.M. and Shoemaker, F.F. (1971). *Communication of Innovations*. New York, NY: Free Press.

Schuhmacher, M.C., Kuester, S., and Hultink, E.J. (2018). Appetizer or main course: early market vs. majority market go-to-market strategies for radical innovations. *Journal of Product Innovation Management* 35 (1): 106–124.

Sorescu, A.B. and Spanjol, J. (2008). Innovation's effect on firm value and risk: insights from consumer packaged goods. *Journal of Marketing* 72 (2): 114–132.

Talke, K. and Heidenreich, S. (2014). How to overcome pro-change bias: incorporating passive and active innovation resistance in innovation decision models. *Journal of Product Innovation Management* 31 (5): 894–907.

Taylor, S.E., Pham, L.B., Rivkin, I.D. and Armor, D.A. (1998). Harnessing the imagination: mental simulation, self-regulation, and coping. *American psychologist* 53 (4): 429–439.

Taylor, S.E. and Schneider, S.K. (1989). Coping and the simulation of events. *Social Cognition* 7 (2): 174–194.

Trope, Y. and Liberman, N. (2003). Temporal construal. *Psychological Review* 110 (3): 403–421.

Van den Hende, E.A., Dahl, D.W., Schoormans, J.P. and Snelders, D. (2012). Narrative transportation in concept tests for really new products: the moderating effect of reader–protagonist similarity. *Journal of Product Innovation Management* 29: 157–170.

Van den Hende, E.A. and Schoormans, J.P.L. (2012). The story is as good as the real thing: early customer input on product applications of radically new technologies. *Journal of Product Innovation Management* 29 (4): 655–666.

Veryzer, R.W. (1998). Key factors affecting customer evaluation of discontinuous new products. *Journal of Product Innovation Management* 15 (2): 136–150.

Wartzman, R. (2013). Would Peter Drucker wear Google Glass? https://forbes.com. https://www.forbes.com/sites/drucker/2013/05/21/drucker-wear-google-glass.

West, P., Huber, J., and Min, K. (2004). Altering experienced utility: the impact of story writing and self-referencing on preferences. *Journal of Consumer Research* 31 (3): 61–68.

Wood, S.L. and Lynch, J.G., Jr. (2002). Prior knowledge and complacency in new product learning. *Journal of Consumer Research* 29 (3): 416–426.

Xiao, Y., Zhang, H., and Cervone, D. (2018). Social functions of anger: a competitive mediation model of new product reviews. *Journal of Product Innovation Management* 35 (3): 367–388.

Zhao, M., Dahl, D.W., and Hoeffler, S. (2014). Optimal visualization aids and temporal framing for new products. *Journal of Consumer Research* 41 (4): 1137–1151.

Zhao, M., Hoeffler, S., and Dahl, D. (2009). The role of imagination-focused visualization on new product evaluation. *Journal of Marketing Research* 46 (1): 46–55.

Zhao, M., Hoeffler, S., and Dahl, D. (2012). Imagination difficulty and new product evaluation. *Journal of Product Innovation Management* 29 (S1): 76–90.

Zhao, M., Hoeffler, S. and Zauberman, G. (2007). Mental simulation and preference consistency over time: the role of process-versus outcome-focused thoughts. *Journal of Marketing Research* 44 (3): 379–388.

Zhao, M., Hoeffler, S., and Zauberman, G. (2011). Mental simulation and product evaluation: the affective and cognitive dimensions of process versus outcome simulation. *Journal of Marketing Research* 48 (5): 827–839.

Sven Feurer is Professor of Marketing at the Business School, Institute Marketing & Global Management, Bern University of Applied Sciences (Switzerland). His work focuses on technology acceptance, really new products, pricing, and sustainable/healthy consumption. His work has appeared in such journals as the *Journal of Consumer Research*, the *Journal of the Academy of Marketing Science*, the *Journal of Product Innovation Management*, and the *International Journal of Research in Marketing*.

Steve Hoeffler is Professor of Marketing at Vanderbilt University's Owen Graduate School of Management, Nashville (USA). As an expert in consumer products marketing, brand management and consumer behavior, he focuses on how radically new products are marketed. His work has appeared in such journals as the *Journal of Consumer Psychology*, the *Journal of Product Innovation Management*, and the *Journal of Marketing Research*.

Min Zhao is an Associate Professor of Marketing at Boston College. Her primary research focuses on decision over time that pertains to consumers' financial decisions, everyday task completion, new product adoption, and hedonic consumptions. She has published in leading marketing and psychological journals such as the *Journal of Marketing Research, Journal of Consumer Research, and Psychological Science*.

Michal Herzenstein is an Associate Professor of Marketing at the Lerner College of Business and Economics, University of Delaware. Her research focuses on really new products and natural language processing in a variety of domains, and was published in leading journals such as the Journal of Marketing Research, Journal of Consumer Psychology, and Organizational Behavior and Human Decision Processes. She won multiple awards including her department's research and service awards, and her university's highest teaching award.

CHAPTER THIRTEEN

MANAGING THE SUPPLY CHAIN IMPLICATIONS OF LAUNCH

C. Anthony Di Benedetto and Roger J. Calantone

13.1 Introduction

The launch activity for new consumer products is a risky endeavor and usually is the most expensive stage in the new product development process. Typically, launch involves the efforts of brand management and the supply chain (logistics and operations). Accordingly, launch represents a major stumbling block in coordinating brand management, logistics, and operations, and the financial amounts at stake are often a critical factor in determining the success of new consumer products. A particular launch consideration is the clean handoff from the development team to the team that will manage the product during and following launch. This handoff provides considerable opportunity to turn a successful product development into a commercial failure. By *handoff* we mean that tactical decisions made at launch must align with the strategy that justified the product's development. For example, distribution logistics must be in place; a reliable demand forecast for the new product must be made to guide manufacturing ramp-up; and promotional activities aimed at both the consumer and the trade must be appropriately timed. A large electronics manufacturer, for example, may run a marketing/logistics process in parallel with their innovation stage-gate process to ensure that both marketing and production are ready for global launch when the product comes out of the development stage. While many of these launch aspects have been well studied and understood by product development professionals, the importance of supply chain and distribution logistics to product launch success has been largely overlooked.

The PDMA Handbook of Innovation and New Product Development, Fourth Edition. Edited by Ludwig Bstieler and Charles H. Noble.

There is much to be gained by a more thorough consideration of supply chain capabilities to support successful launches and minimize losses from unsuccessful ones. As we will see later, integrating supply chain capability issues explicitly into the product launch strategy is central to *lean launch* methods. Lean launch methods involve the use of a flexible supply chain system to enable the firm to react quickly to emerging customer needs and market demands. Lean launch uses the principle of postponement, that is, making product form decisions late in the process when customer needs are better articulated, while gradually increasing resource commitments and ramping up manufacturing capability (Bowersox, Stank et al. 1999; Calantone et al. 2005; Zinn 2019). Companies have been adopting lean launch methods for some time, and lean launch has become a requirement in a variety of consumer product launch scenarios.

One must also consider a more nuanced view of speed as a strategic lever with respect to technological content and competitive advantage. Speed is usually treated as a macro- or, less often, a meso-conditional in strategic new product development (NPD) theory-driven models (e.g., McCardle et al. 2018). This view has been reinforced notably by Charles Fine in his book on clockspeed as a competitive necessity (Fine 2010); he documents the nature of technological innovation as, at best, a temporary and transitory source of competitive advantage.

Yet the micro-aspects of speed as a barrier to optimal team dynamics seems to elude NPD team leaders. These dynamics occur within and between expert teams in both commercial and academic settings. Factually, those significant differences exist between teams based on the nature of their expertise, the emergent personality characteristics of the teams, and the compensation and control mechanisms of the individual teams.

It is clear that micro-inefficiencies and even conflicts occur between teams universally, as the culture of teams conforms to their compensated response to challenges. This is because specific expertise, in critical amounts necessary to market competitive success and, ultimately, dominance tends toward complementary rather than supplementary staffing within teams and consequent increased differences. Between-teams overall project management becomes necessary around the natural clockspeed of each team, and the individuals within each team. The emergent culture of each team creates differences in outputs, turning the cooperation paradigm on its head.

Overall, the goals of this chapter are as follows:

- To describe the potential pitfalls at the launch stage
- To outline the development of a launch strategy to manage these pitfalls
- To present the advantages of lean launch and flexible supply chain processes in launch strategy development
- To illustrate the successful application of lean launch methods
- To draw managerial insights and conclusions regarding the benefits of a lean launch

13.2 Pitfalls at the Launch Stage

Traditionally, product launch is managed using anticipatory methods; that is, manufacturing, marketing, and supply chain/distribution decisions are made in advance of the launch based on early forecasts of demand, assumptions of competitors' actions, and assumptions of resellers' costs that must be absorbed to gain efficacious market access. Yet, it is possible for the firm to time the product launch poorly using these methods. If the marketing programs are carried out well but the launch is slow, the product may never achieve its marketplace potential or the market window could be totally missed (Calantone and Di Benedetto 2012). Similarly, if the launch is timed too early, promotion strategy or distribution channels may not be in place, or key marketplace information may be missing or unavailable (such as information about changing product technology or customer requirements) (Calantone and Di Benedetto 2012). Firms that accelerated product development using lean launch principles were able to establish an option to launch early or late, depending on market conditions; this option was lost if the firm was unable to successfully deploy lean launch and accelerate time to market. In the extreme case, the market window might be missed altogether.

Market orientation, leadership style of the project manager, and extent of cross-functional integration are all antecedents of timing, speed to market, and proficiency of launch activities (Calantone et al. 2012). Thus a firm may improve new product performance and build competitive advantage through organizational resources as well as through market orientation (assessing customers and competitors, and choosing an appropriate response) (Saeed et al. 2015). There is also evidence that market orientation has a significant impact on speed to market, which ultimately affects new product performance (Zamani et al. 2016).

As an additional complication, there may be disagreement on the time to market itself: top managers might want to launch immediately, customers might be impatient (and ready to switch to a competitor with advanced technology), marketing may need more time to get promotional materials ready, and production is having difficulty ramping up to full scale production (Calantone et al. 2012).

Three negative, and avoidable, outcomes of a traditional anticipatory launch are possible. When a product is both technically and financially successful across a broad range of market segments, unplanned out-of-stock problems are likely to materialize. Even when a product has widespread success, its popularity and its adoption rate are likely to vary among market segments. Replenishment inventory needed for markets experiencing rapid penetration may not be available due to pre-introduction inventory commitment to other segments. When products are highly successful, the manufacturing and logistics capacity may not be able to keep up with the demand because of scheduling lead-time and material procurement inflexibility. If inventory is available in the aggregate, the product may still be out of stock on retail shelves in some markets while being overstocked in other markets. For at least the time it takes to reposition inventory to where it is needed and to ramp up manufacturing support, the launch success may be in jeopardy. For products that are neither technical nor financial successes, pre-allocation of inventory results in overstock. In this case, inventory is positioned forward in the channel, resulting in excess reclamation expense.

Realize that new product introductions are seldom clear-cut successes or failures. Products may initially appeal to only a narrow segment of the target market, such as a specific geographic region or usage group, as contrasted to the broader market to which they are presented. Financial success depends on sufficient penetration to cover manufacturing, inventory, and promotional startup costs. Therefore, products that experience limited technical success but have a potential for achieving broader appeal over time may fail at launch due to the inability to focus resources (including logistical support), generate sufficient segmental revenue, and cover market rollout costs (McCardle et al. 2018).

A launch of a new cracker product provides an example of the pitfalls of anticipatory launch strategies. Two variants of a new thin cracker were introduced prior to the year-end holiday season. One variant was flavored similarly to the established cracker brand, while the other was onion-flavored. The manufacturing process for the products involved production of the regular-flavored cracker with an additional flavoring process for the onion-flavored variant. Significant inventory of each variant was sold to retailers and forward deployed for the expected holiday sales. The market enthusiastically received the regular-flavored cracker. The onion-flavored version, however, was not well received and sales lagged behind those of the regular variant by a considerable margin. Unfortunately, the supply chain was unable to fully replenish the regular-flavored version, resulting in out-of-stock situations, while high levels of onion-flavored stock remained on retail shelves until after the holiday season. The combination of reclamation costs for unused onion-flavored inventory and out-of-stock costs for the regular-flavored cracker resulted in limited financial success of the overall launch.

13.3 Launch Strategy

At a (more) macro level, the launch strategy is simply the decision to launch or not launch the product. More specifically, launch strategy decisions are concerned with both product and market issues: the innovativeness of the new product, the targeted market, the competitive positioning, and so forth (Cooper 2019; Slater et al. 2014). As noted earlier, careful timing of the launch is also an important part of the launch strategy. On the engineering side, the launch strategy is supported by market tests that confirm the adequacy of the product prototypes, as evidenced in internal alpha testing or beta testing with select customers. On the marketing side, the launch strategy requires knowledge of the product's ability to satisfy the customer's value proposition, in requisite quantities, at a price with sufficient margin over the cost to provide an adequate financial return to justify the production and marketing investments the commercialization stage requires. This stage of the new product development process requires actual financial returns rather than just the promise of returns.

The launch strategy needs another component in order to calibrate the cost basis of the decision: The scale of the launch with regard to the size of the potential addressable market is required. As shown in Figure 13.1, the challenge is to get close to the right size of the market, to properly scale both the size of the marketing investment and the size of the production and distribution facility.

As shown in Figure 13.1, there are risks inherent in inaccurate forecasts of market size. In the case of a large actual market, planning for a small launch leads to opportunity costs – missed sales as well as creation of an opportunity window for competitors. If the actual market size is low, planning for a large launch leads to cost overruns in production and inventory holding costs, which will not be recovered. The figure proposes lean launch strategy as an intermediate between low and high launch size. If a lean launch is enacted, the risks of over- or under-estimating actual market size are minimized, as supply chain flexibility allows for exploiting of information of demand as it becomes available. The effects of a flexible supply chain will be further examined in the next section.

Often, demand and profit assessment, and the decision to develop and launch the product, are supported by teardown analysis. For example, in traditional teardown analysis of a new sport-utility vehicle (SUV) aimed at consumers, a carmaker will buy several competitive models, move them to a central location, and disassemble them to examine their individual components. Each part is cost estimated, leading to a very good projection of material costs per unit or even the bill of materials. The type of labor involved is also assessed (i.e., whether human labor or robotics is used), and a labor cost per unit is estimated. A usual accompanying step is to take the public plant tour to confirm exactly what kinds of robots and other equipment are being used, the number of workstations and

FIGURE 13.1 MATCHING CONSEQUENCES BETWEEN LAUNCH SIZE AND MARKET SIZE.

		Actual Market Size	
		Low	High
Launch Size	Low	Commercial Failure	Opportunity Cost
	Intermediate (Lean Launch)	Reduced Chance of Commercial Failure	Reduced Opportunity Cost
	High	Cost Overrun, Commercial Failure	Commercial Success

inspection points, and so on. Yet, in a complete teardown analysis, the carmaker would project the total size of the SUV market and then assess the total unit costs at various levels of production (keeping in mind that average total costs will decrease as production increases). The carmaker can then make an intelligent decision as to whether the SUV should be launched.

For example, if the firm believes it can get 10% of the SUV market with this new product, how many units does that translate to? What would be the average total cost incurred? And, given the going selling price, could the company make a large enough profit to generate target net present values or to pay back the development costs in the desired period? The forecasting and management challenge is to properly size the launch for the market demand. When this is impossible, that is, when the firm cannot know the market reality in advance of the launch sizing decision, they must try to increase the flexibility of the production response tactics for marketing and distribution resource allocation. This would permit the firm to rapidly respond to early sales success without overcommitting to inventory during the introductory rollout phase.

Closely monitoring sales trends, through the use of point-of-sale (POS) information, can assist the firm in responding in a timely manner to sales fluctuations. These efforts can be further facilitated by regional rollouts that build a response to demand slowly, and slowly ramp up productive and distributive capacity, again while avoiding overcommitment. Furthermore, political elements come into play when sizing the production facility.

The opportunity cost scenario as well as the overcommitment scenario are both addressed by a variety of supply chain strategies, discussed later. For now, suffice it to say that flexibility and staged market commitment are necessary to a right-sized launch strategy. The next section describes how lean launch methods can help firms achieve the required level of supply chain flexibility described here.

13.4 The Flexible Supply Chain and Lean Launch

Advanced supply chain capabilities offer an alternative way to support a successful new product launch as well as contain losses when products fail to meet expectations. The lean launch method involves development of a flexible supply chain system capable of rapidly responding to early sales success in order to limit commitment of inventory during introductory rollout. Flexible supply chain logistics systems are characterized by coordinated source, make, and deliver operations that drastically cut raw material to consumer cycle times and enable the firm to respond to actual market needs rather than anticipate demand with inventory.

Postponement is the basic principle driving the development of lean launch strategies. Leading-edge firms increasingly use postponement as the logic for flexible operations that enable quick reaction to customer needs and actual market demand. Postponement delays finalization of product form and identity to the latest possible point in the marketing flow and postpones commitment of inventory to specific locations to the latest possible point in time. Cutting lead times can reduce uncertainty and increase operational flexibility so that products can be produced to order or at least manufactured at a time closer to when demand materializes. The volatility of demand for new products can be managed by reducing lead times, which shortens the forecasting horizon and lowers the risk of error (Bowersox, Stank et al. 1999; Kou and Lee 2015; Kou et al. 2015).

Postponement of time and form can be employed (Zinn 2019). In time postponement, the key differential is the timing of inventory deployment to the next location in the distribution process. In contrast to anticipatory shipment to distribution warehouses based on forecasts, the goal of time postponement is to ship exact product quantities from a central location to satisfy specific customer requirements. The practice of shipping exact quantities to specific destinations greatly reduces the risk of improper inventory deployment and eliminates duplicate inventory safety stocks throughout the channel. Time postponement provides inventory-positioning flexibility by alleviating the need for forward deployment of inventory to cover total forecasted sales. Positioning flexibility allows firms to strategically position only limited inventory in the market and selectively replenish stock based on closely monitored sales information. Benefits from time postponement may be realized regardless of whether one or multiple new product variants are launched (Bowersox, Stank et al. 1999; Kou and Lee 2015; Kou et al. 2015).

Form postponement provides product variation flexibility by alleviating the need to lock in feature design prior to gaining some understanding of a product's market appeal. Assembly, packaging, and labeling postponement are options in which firms initially manufacture products to an intermediate or neutral form with the intent to delay customization until specific customer orders are received. Benefits from form postponement become significant when introducing multiple product variants. As a classic example of form postponement, clothing manufacturer Benetton innovated its production process to make garments from bleached, undyed yarn (traditionally the dyeing process comes first). By delaying dyeing until market information on color preferences is available, overruns and stockouts are minimized (Gawas 2021).

Postponement of product differentiation reduces the need to stock inventory of all product variations. For example, computers are often assembled, packaged, and labeled to meet specific configurations during customer order processing. Demand variations from forecasted volumes for each product variant following launch can be accommodated without the out-of-stock or overstock risk associated with traditional anticipatory launch strategies.

Form postponement may also involve forward deployment of materials or components to support final customized manufacturing to specific customer requirements. The shipment of house paint to retailers as a neutral base with subsequent mixing to customer-specified colors provides the classic example of postponing form until end-consumer purchase. International shipments that necessitate language-specific labels and support materials, such as instruction manuals, also frequently utilize form postponement. Such products are shipped in bulk quantities to a regional distribution center where labeling and packaging are completed as customer orders are processed.

The application of lean launch strategies is driven by key competence in five areas of supply chain management (Bowersox, Closs et al. 1999). These include collaborative relationships, information systems, measurement systems, internal operations, and external operations – all representing critical elements of a firm's supply chain strategies, structures, and processes. Competence in collaborative relationships requires a willingness on the part of supply chain partners to create structures, frameworks, and metrics that encourage cross-organizational behavior. This consists of sharing strategic planning and operational information as well as creating financial linkages that make firms dependent upon mutual performance. Suppliers, manufacturers, third-party providers, and customers are encouraged to identify and partner with firms that share a common vision and are pursuing parallel objectives pertaining to partnership interdependence and the principles of collaboration. Efforts must focus on providing the best end-customer value, regardless of where along the supply chain the necessary competencies exist. This collaborative relationship perspective is key to developing effective supply chain structures that align the functional operations of multiple firms into an integrated system.

Supply chains capable of supporting lean launch also depend upon the availability of sophisticated and economical information technology that allows businesses to quantify sales, define requirements, and trigger production and inventory replenishment 24 hours a day, seven days a week. Such systems provide the input needed for short-, mid-, and long-term plans, which translate strategic goals and objectives into action and work to guide each operating area. Effective information systems provide thorough, accurate, and timely information from customers, material and service suppliers, and internal functional areas regarding current and expected conditions. Managers with access to data throughout the supply chain, and with the hardware and software needed to process them, are better positioned to gain rapid insight into demand patterns and trends.

Accessibility allows integrated operational decisions to be made in complex global supply chains. Rather than relying upon forecast sales, inventory replenishments are driven by precise sales information regarding specific stock items in the market. The success of such technology and planning integration rests upon a firm's ability to manage information on supply chain resource allocation through seamless transactions across the total order-to-delivery cycle. It requires adaptation of technological systems to exchange information across functional boundaries in a timely, responsive, and usable format and to extend such internal communications capabilities to external supply chain partners.

Measurement system integration is also required to manage coordinated supply chain lean launches. These systems must track performance across the borders of internal functional areas and external supply chain partners, measuring both the operations of the overall supply chain and the financial performance of individual firms. Measurement systems must also reflect the operational performance of the overall supply chain and the financial performance of individual firms. Integrated performance measurement provides the basis for calibrating the many parts of the supply chain. Good metrics and strong measurement systems serve to provide timely feedback so that management can take corrective action and drive integrated operations.

Greater coordination of internal source, production, and delivery operations also enables lean launch applications. Integration of internal operations provides a firm with the ability to seamlessly link activities across internal functional areas in order to achieve synergies that lead to better performance in meeting customer requirements. Internal integration is achieved by linking operations into a seamless, synchronized operational flow, encouraging frontline managers and employees to use their own discretion, within policy guidelines, to make timely decisions.

Empowered employees have the authority and information necessary to do a job and they are trusted to perform work without intense over-the-shoulder supervision, enabling them to focus resources on providing unique and customer-valued product/service offerings that competitors cannot effectively match.

Coordination of procurement and production techniques such as concurrent engineering and design, supplier partnerships, agile manufacturing, and improved transportation performance has the potential to create flexible processes that enable firms to accommodate actual market needs rather than rely on anticipatory forecasts.

The need to reduce redundancies and achieve greater economies of scale in launch operations is not limited to internal activities alone. External integration synchronizes the core competencies of selected supply chain participants to jointly achieve improved service capabilities at lower total supply chain cost. The goal is to outsource specialized activities that previously were developed and performed internally. After outsourcing activities are identified and appropriate suppliers are chosen, systems and operational interfaces between firms must be synchronized to reduce duplication, redundancy, and dwell time (the ratio of days inventory sits idle in the supply chain relative to the days it is being used productively). Synchronization requires extensive information sharing between firms to standardize processes and procedures.

Additionally, synchronization ensures that all activities are conducted by the supply chain entity that best creates the service and cost configuration to meet customer requirements. That is, the timing of the launch is in line with supply chain partner concerns and customer needs, and allows the firm to capitalize optimally on the marketplace opportunity (Calantone and Di Benedetto 2012; Calantone et al. 2012). Innovative firms have utilized the principles of response-based logistics to customize product and service offerings without increasing manufacturing capacity or stock levels. The following example illustrates how these principles have been applied in industry (from Cheng et al. 2010, pp. 125–132).

13.5 An Illustration in a Manufacturing Setting

Cheng et al. (2010, pp. 125–132) report on a Hong Kong-based manufacturer of electric toasters that successfully deployed a form postponement strategy in its production process. The company's products are sold under 11 different brand names throughout the world; it produces about 12,000 toasters per day. A typical order is completed in 15 days, which can be shortened to nine days if the company maintains work-in-process (WIP) inventory. Thus, it is advantageous to convert raw materials to WIP and have this inventory on hand. In reality this is challenging, however, since customer orders are highly customized (product's outer shells and gift boxes, for example) and repeat orders are not common. As a result, the company strictly relied on make-to-order production, converting raw material into components, and ultimately finished goods, only when the customer orders were received. Consequently, the production schedule was highly variable, prone to wide swings when large customers placed orders.

To overcome this problem, the company used a form postponement strategy by standardizing internal components. Applying this strategy, heaters, power cords, and other component parts, are now produced on a make-to-stock basis, and on a predictable production schedule (since make-to-stock production was based on forecasted demand). The standardized components become WIP inventory, ready to use when customer orders were placed. Upon receiving the orders, the company manufactures the outer shells, gift boxes, and other customized parts, and adds these to the standardized components in WIP inventory, thus producing finished customized products in a timely manner and at reduced overall cost.

The company reported several advantages from adopting the postponement strategy: lead time was reduced from 15 to nine days; raw material costs for standardized component products was reduced by about 5% since high-quantity purchases were based on predictable production volumes; raw materials holding costs were reduced as these were converted to WIP components; and production of make-to-stock components led to a 15% reduction in setup time. Further, production machines were running at 85% capacity – a level that was not previously possible due to highly fluctuating demand. In sum, the production process is smoother, any kind of customer demand can be more easily met more quickly, and labor costs are more predictable as these are based on forecasted fluctuations in demand. Customers also directly benefit from the postponement strategy. Since the manufacturer can more easily adjust orders in a timely fashion, they can place smaller and more frequent orders, allowing them to make adjustments based on end user demand. This effect minimizes costs associated with unsold inventory of unwanted styles, as well as opportunity costs due to shortage of popular styles (Cheng et al. 2010, pp. 125–132).

13.6 Summary

The pressure is increasingly on firms to meet customer needs and marketplace demands more quickly and completely than the competition. Many firms see the development and launch of successful new products as their lifeblood, and their ability to identify and meet emerging customer needs and demands quickly as a key component of their competitive strategy. Until relatively recently, however, new product launch had been business as usual in many firms: Marketing, manufacturing, and distribution channel decisions pertaining to launch had been made in anticipatory fashion based on early forecasts.

Launches which are designed on inaccurate forecasts of actual market size result in cost overruns or opportunity costs, raising the likelihood of commercial failure. At the same time, the importance of clockspeed is increasingly recognized; "business as usual" technological innovation provides temporary competitive advantage at best. Further, the rate of technological change and obsolescence will be industry-specific; thus, the competitive advantage obtained by technological innovation may be highly transitory. The launch decision becomes even more complex in cases where multiple new products may be launched on a planned schedule. For example, companies such as 3M, Hewlett-Packard, or Samsung may launch new products on a weekly basis. This reality can place further strain on the launch process, which can cause supply chain disruptions and challenge launch timing and clockspeed objectives.

By including distribution and logistics employees more fully on the launch team, firms can become more adept at increasing supply chain flexibility and improve the effectiveness and efficiency of the new product launch. Those firms employing lean launch methods have been able to accelerate the time to market and cut lead times drastically, thereby enabling them to match emerging customer needs more rapidly and minimizing the risks inherent in inaccurate market size forecasts.

By postponing major decisions as long as possible, even large firms can seem to turn on a dime; match product features and production to customer demand much more effectively than before; and reduce costs through cheaper distribution and reduced manufacturing change orders gained by postponement. External integration synchronization ensures that the timing of the launch meets the requirements of supply chain members and end customers. The discussion of lean launch and principles of postponement described above provide a starting point for analysis of one's own company in search of ways to obtain lean launch advantages. In sum, launch is a key stage in the new product development process and deserves a much more strategic view.

References

Bowersox, D.B., Stank, T., and Daugherty, P. (1999). Lean launch: managing product introduction risk through response-based logistics. *Journal of Product Innovation Management* 16: 557–568.

Bowersox, D.J., Closs, D.J., and Stank, T.P. (1999). *21st Century Logistics: Making Supply Chain Integration a Reality*. Oak Brook, IL: Council of Logistics Management.

Calantone, R.J. and Di Benedetto, C.A. (2012). The role of lean launch execution and launch timing on new product performance. *Journal of the Academy of Marketing Science* 40: 526–538.

Calantone, R.J., Di Benedetto, C.A., and Rubera, G. (2012). Launch timing and launch activities proficiency as antecedents to new product performance. *Journal of Global Scholars of Marketing Science* 22: 290–309.

Calantone, R.J., Di Benedetto, C.A., and Stank, T.P. (2005). Managing the supply chain implications of launch. In: *The PDMA Handbook of New Product Development*, 2e (ed. K.B. Kahn, G. Castellion, and A. Griffin), 466–478. Hoboken, NJ: Wiley.

Cheng, T.C.E., Li, J., Wen, C.L.J. et al. (2010). *Postponement Strategies in Supply Chain Management*. New York: Springer.

Cooper, R.G. (2019). The drivers of success in new-product development. *Industrial Marketing Management* 76: 36–47.

Fine, C. (2010). *Clockspeed: Winning Industry Control in the Age of Temporary Advantage*. New York: Basic Books.

Gawas, T. (2021). Benetton supply chain: differentiating the brand. Thestrategystory.com, March 24, 2021.

Kou, T.-C. and Lee, B.C.Y. (2015). The influence of supply chain architecture on new product launch and performance in the high-tech industry. *Journal of Business & Industrial Marketing* 30 (5): 677–687.

Kou, T.-C., Lee, B.C.Y., and Wei, C.-F. (2015). The role of product lean launch in customer relationships and performance in the high-tech manufacturing industry. *International Journal of Operations & Production Management* 35 (8): 1207–1223.

McCardle, M., White, J.C., and Calantone, R. (2018). Market foresight and new product outcomes innovation and strategy. *Review of Marketing Research* 15: 169–203.

Saeed, S., Yousafzai, S., Paladino, A., and De Luca, L.M. (2015). Inside-out and outside-in orientations: a meta-analysis of orientation's effects on innovation and firm performance. *Industrial Marketing Management* 47: 121–133.

Slater, S.F., Mohr, J., and Sengupta, S. (2014). Radical product innovation capability: literature review, synthesis, and illustrative research propositions. *Journal of Product Innovation Management* 31 (3): 552–566.

Zamani, S.N.M., Abdul-Talib, A.-N., and Ashari, H. (2016). Strategic orientations and new product success: the mediating impact of innovation speed. *Information* 19: 2785–2790.

Zinn, W. (2019). A historical review of postponement research. *Journal of Business Logistics* 40 (1): 66–72.

C. Anthony Di Benedetto is Professor of Marketing and Senior Washburn Research Fellow at Temple University and Co-Editor-in-Chief of *Industrial Marketing Management*. He received his B.Sc., MBA, and PhD from McGill University, Montreal, Canada. His work has been published in the *Journal of Product Innovation Management, Industrial Marketing Management, Management Science, Strategic Management Journal, Journal of International Business Studies*, and elsewhere.

Roger J. Calantone is University Distinguished Professor Emeritus at the Eli Broad Graduate School of Management at Michigan State University. He received his PhD from the University of Massachusetts, Amherst, and has published in numerous journals, including the *Journal of Marketing Research, Journal of Marketing, Marketing Science*, and *Management Science*.

CHAPTER FOURTEEN

NEW PRODUCT DEVELOPMENT IN EAST ASIA: BEST PRACTICES AND LESSONS TO BE LEARNED

Martin Hemmert

14.1 Introduction

While modern new product development (NPD) systems and practices first evolved in Western countries, since the 1980s, the NPD performance of East Asian companies has often exceeded that of their Western rivals in major industries. This strong performance has been enabled by organizing NPD in ways that are different from the West. Japanese automobile manufacturers have shown far superior NPD performance over their American and European competitors in terms of lead time, quality, and productivity (Clark and Fujimoto 1991). Similarly, Japanese electronics firms achieved NPD excellence based on organizational processes focused on organizational knowledge creation and knowledge transfer, which fundamentally differ from those in Western firms (Nonaka and Takeuchi 1995). NPD tools such as quality function deployment (QFD) and total quality management (TQM) have been intensively discussed globally (Chan and Wu 2002; Hackman and Wageman 1995) since they have been implemented widely in Japan, and Japanese companies have continuously applied these and other NPD practices effectively. Western companies have attempted to learn from their Japanese rivals in a variety of ways.

More recently, competitors from other East Asian countries, most notably South Korea (hereafter, Korea) and China, have also challenged incumbents

The PDMA Handbook of Innovation and New Product Development, Fourth Edition. Edited by Ludwig Bstieler and Charles H. Noble.

with impressive NPD performance across knowledge-intensive manufacturing and technology industries. Korean conglomerates, such as Samsung and Hyundai, have achieved competitive advantage through their short NPD lead time, cost leadership, and excellent product quality (Hemmert 2018; Song and Lee 2014). In addition, Chinese companies, including large manufacturing and technology groups such as Haier, Huawei, and Alibaba, as well as relatively smaller and more specialized firms, have achieved strong global competitiveness through their NPD excellence (Yip and McKern 2016). In other words, new East Asian rivals that outperform their competitors through their effective NPD are continuously challenging Western multinational firms.

Importantly, the NPD management of these firms is not only different from Western competitors but is also highly diverse across East Asian countries. Japanese, Korean, and Chinese companies have different and distinct ways of organizing and managing NPD. As a result, it has become ever more difficult to understand from a global perspective how East Asian firms achieve competitive advantage in NPD and how these competitive challenges can be effectively addressed.

Even after deciphering the NPD management of East Asian firms, managers of non-East Asian firms often question what they can learn from these rivals. Japanese, Korean, or Chinese NPD systems may appear so strongly embedded in the cultural, economic, and institutional context of the firms' home countries that Western managers might conclude that it is not possible to emulate their managerial practices. Alternatively, they may try to implement East Asian NPD practices and find that they do not work well in their firms because they are not well understood or accepted by their managerial or NPD staff. In other words, adopting East Asian NPD tools appears to be a challenging task.

Despite these challenges, however, global competitors can greatly benefit from studying the effective NPD practices of their East Asian counterparts. Not all of these practices are equally difficult to understand, and Western and other firms can certainly enhance their NPD performance by pragmatically adopting East Asian managerial tools that are effective in their own business environments. Furthermore, a good understanding of Japanese, Korean, and Chinese NPD practices is highly valuable for effective innovation and competition in East Asian countries, which are among the largest and most attractive markets in the world.

This chapter first discusses the specific features of Japanese, Korean, and Chinese firms' NPD management and how they differ from those of the West. Specifically, the themes of NPD strategy, technology sourcing, NPD project management, and new product introduction are reviewed. Thereafter, the country-specific embeddedness of East Asian NPD practices and their implications for adopting these practices are discussed. The chapter concludes with takeaways on what global competitors can learn from the best NPD management practices of East Asian firms.

14.2 East Asian NPD Practices

Subsequently, the NPD practices of Japanese, Korean, and Chinese companies across different types of activities will be identified and discussed. The key observations are summarized in Table 14.1.

TABLE 14.1 NPD PRACTICES OF EAST ASIAN FIRMS.

Field of Activity	Japan	Korea	China
Overall NPD investment and strategy	Very strong NPD investment; technology-based differentiation strategy	Very strong NPD investment; combination of cost leadership strategy based on process innovation and quality-based differentiation strategy	Strong NPD investment; cost leadership strategy based on low cost product and business model innovation
Technology sourcing	Stepwise, cumulative in-house development of core technologies; NPD in close collaboration with *keiretsu* suppliers	In-house core technology development, supported by *chaebol* member firms; pragmatic acquisition of external knowledge by hiring foreign experts and via acquisitions	External technology sourcing via international acquisitions; in-house development of new products and business models for emerging markets
NPD project management	Fast NPD through cross-functional integration with heavyweight project managers and extensive informal communication	Very fast NPD through cross-functional integration, task-force teams, internal NPD competition, strong executive leadership and champion behavior	Very fast NPD through flexible, informal task processing, quick top-level decision-making, and entrepreneurial bottom-up new product concept development
New product introduction	Rigorous pre-launch product testing in collaboration with customers; sequential market introduction (first in Japan, followed by other markets); limited product customization for international markets	Rigorous, but fast-tracked pre-launch product testing; parallel market introduction across countries; extensive product customization for international markets	Limited pre-launch product testing; "quick-and-dirty" market introduction with subsequent quality improvement based on customer feedback; focus on home market and other emerging markets

14.2.1 Overall NPD Investment and Strategy

On the surface, it is widely believed that East Asian companies apply conservative NPD strategies by following the technological trends established by Western companies and refraining from radical product innovation. A view exists that East Asian firms are "copycats" that engage in minimal genuine innovation (Luo et al. 2011). Furthermore, Western observers tend to perceive that companies from Japan, Korea, and China follow broadly similar strategies that rely mainly on in-house NPD and have little inclination to work with external partners when developing new products.

However, these blanket assessments are inaccurate and overlook extraordinarily strong NPD investments by East Asian firms. Furthermore, such views do not consider the significant differences among Japanese, Korean, and Chinese companies.

In 2020, the overall R&D expenditures of Japanese and Korean firms amounted to 2.58% and 3.81% of their countries' gross domestic product (GDP), respectively, and were among the highest of all Organisation for Economic Cooperation and Development (OECD) member countries, far above the OECD average of 1.92%. The R&D expenditures of Chinese firms were 1.84% of China's GDP in the same year (OECD 2022). While the R&D spending level of Chinese firms is lower than that of Japanese and Korean firms, it is extraordinarily high for emerging market firms. Such firms tend to focus on catching up with their developed country counterparts by relying on existing technologies, and therefore spend much less on R&D than developed country firms.

Furthermore, East Asian firms have outstanding innovation outcomes. For example, the number of international patents in the information and communication technology (ICT) industries is much higher in Japan and Korea than in leading Western countries, such as the USA and Germany (Hemmert 2020). Japanese, Korean, and Chinese firms also hold strong global export market shares in high-tech sectors including the computer, electronic, and optical industries (OECD 2022). Japanese and Korean brands, such as Sony and Samsung, have become globally renowned for their highly innovative products. Overall, there is no doubt that East Asian firms, far from imitators, invest strongly and effectively in NPD.

However, regardless of their generally strong NPD investment and performance, Japanese, Korean, and Chinese companies tend to apply different NPD strategies, which are connected to their business strategies. Japanese firms, which often reached the technological forefront in the late twentieth century, frequently apply technology-based differentiation strategies. Based on their strength in core technologies (Hu 2012), they develop and incorporate technological features in their products that exceed those of their competitors, thereby seeking competitive advantage. For example, Japanese automobiles and IT devices often feature advanced user functionalities that competitors' products lack (Kusunoki 2006).

Korean firms have been rapidly catching up with leading Western and Japanese competitors and have taken technologically leading positions in some knowledge-intensive manufacturing industries since the 1990s, while still catching up in other industries. This hybrid status of partial technological leadership and followership has induced many Korean firms to apply diverse NPD strategies (Hobday et al. 2004). Supported by relentless process innovation, they have focused on cost leadership strategies in major industries, such as memory chips, and have sought to leverage first-mover advantages. However, in the development of other products, such as automobiles and smartphones, Korean firms have primarily focused on seeking technology-based quality advantages (Hemmert 2018).

Finally, from a global perspective, while rapidly accumulating technological knowledge, Chinese firms remain mostly in a technological followership position. Accordingly, they mostly seek cost leadership by developing innovative new products at low cost (Yip and McKern 2016). Furthermore, they apply business model innovations to support their cost leadership. For example, they may lower their costs by outsourcing activities such as manufacturing, which are conducted in-house by most other firms.

14.2.2 Technology Sourcing

Japanese, Korean, and Chinese firms have applied distinct technology sourcing strategies. Japanese firms have focused on stepwise, cumulative in-house development of core technologies while emphasizing collaboration with their suppliers in NPD. Such collaboration is embedded in *keiretsu* ties between manufacturers and suppliers, which feature close alliances based on long-term, trust-based business relationships (Aoki and Lennerfors 2013). Suppliers are part of a manufacturer's NPD project from the outset, and develop customized key parts and components for the new product, which are approved by the manufacturer. Consequently, Japanese suppliers develop relationship-specific skills, including NPD skills, in collaboration with manufacturers (Choi and Hara 2018). Strong trust between manufacturers and suppliers is based on predictability and reliable support (Dyer and Chu 2011) and results in low transaction cost (Dyer and Chu 2003). Such practices in close collaboration with suppliers have resulted in a superior NPD performance of Japanese manufacturers such as Toyota (see Box 14.1) in comparison to their Western counterparts, including shorter lead time, higher quality and productivity, and faster production ramp-up (Aoki and Staeblein 2018; Clark and Fujimoto 1991).

Owing to their latecomer status in innovation, Korean firms have applied NPD and technology sourcing strategies that differ from their Japanese counterparts. After initially acquiring technologies from foreign companies via licensing, reverse engineering, and the hiring of international experts, they internalized tacit knowledge from these sources and subsequently enhanced it by developing customized products for the markets they targeted (Kim 1997). In recent decades, based on the advanced capabilities of their researchers, designers, and engineers, various Korean firms have assumed global NPD leadership in industries such as semiconductors and smartphones (Hemmert 2018; Song and Lee 2014).

Box 14.1   NPD at Toyota Motor

Founded in 1937, the Toyota Motor company became famous for the introduction of a system of operations and supply chain management practices known as the "Toyota Production System," which later won global fame as "lean production." This system includes principles such as just-in-time delivery across the entire supply chain, producing only what is needed, when it is needed, and in the amount needed, and quality management that focuses on avoiding defects instead of post-hoc debugging. These pioneering and highly effective practices gave Toyota a strong competitive edge, resulting in rapid growth and enabling the company to become the largest automobile producer in the world. Toyota's management principles also strengthen the company's NPD performance, as Toyota relies on long-term collaboration with a network of *keiretsu* suppliers that are closely involved in the development of new products from the outset. Furthermore, Toyota has a strong internal engineering culture that focuses on cross-functional collaboration in achieving product excellence, while rigorously avoiding wasteful spending. As a result, Toyota's car models do not only top global quality ranks, but are also developed more quickly and at lower cost compared with competitors. Toyota's strong NPD capabilities further enabled the company to launch its first hybrid gasoline-electric engine model in 1997 and to become a global pioneer and leader in the hybrid car market. While Toyota's practices initially evolved from its Japanese home base, it has extended its *keiretsu* networks to many non-Japanese suppliers, which have further strengthened Toyota's NPD capabilities through their technical excellence. Collaboration with suppliers has further deepened at the early stage of product development, incorporating suppliers' insights and ideas systematically and effectively.

Overall, the NPD and technology sourcing of Korean firms is focused on in-house activities and support by member firms within business groups (*chaebols*). In contrast to manufacturing *keiretsu* in Japan, which is based on close, long-term relationships between independent firms, *chaebol* member firms are strictly controlled by owner-managers, who are either the groups' founders or their successors from the same families (Hemmert 2018). However, regardless of their in-house and in-group NPD focus, Korean firms retain flexibility in acquiring external technological knowledge. For example, the automobile manufacturer Hyundai-Kia enhanced its NPD performance by scouting leading developers and designers from European competitors and giving them strong authority to apply major new concepts in NPD projects. Similarly, Samsung Electronics acquired the American car mobility provider Harman to strengthen its NPD capabilities in audio and sensor technologies, while Hyundai-Kia acquired the Massachusetts-based robotics company Boston Dynamics. Fundamentally, Korean firms emphasize the swift acquisition of new resources and knowledge for NPD, which is often accompanied by discarding resources that are no longer useful. Such swiftness in acquiring new knowledge enhances their NPD performance (Park and Kim 2013).

Chinese firms entered global competition in knowledge-intensive industries even later than their Korean counterparts. Thus, they tend to rely strongly on external technology to enhance their NPD performance. Some aspiring Chinese firms have acquired major Western competitors to fast-track their technology sourcing by absorbing their NPD knowledge. Major examples of this technology sourcing strategy include the acquisition of IBM's PC division by Lenovo and the acquisition of Swedish car manufacturer Volvo by Geely. Other Chinese companies rely on collaboration with foreign multinational enterprises (MNEs) (Steinfeld and Beltoft 2014), which have often established R&D centers in China (von Zedtwitz et al. 2007). However, various Chinese technology companies, such as Huawei and Alibaba, have emphasized in-house NPD more strongly by focusing on emerging markets and new business models that are not utilized by competitors in developed countries (Yip and McKern 2016; Zhang 2014).

14.2.3 NPD Project Management

Companies from East Asia have introduced a variety of new effective tools in NPD project management, which have resulted in shorter lead times, lower cost, and higher product quality. Japanese, Korean, and Chinese companies have made distinct contributions to the introduction of effective new NPD project management practices.

Japanese companies have achieved strong NPD advantages over their Western rivals in knowledge-intensive industries such as automobile manufacturing by developing superior new products faster and at a lower cost (Clark and Fujimoto 1991). A key driver of this strong performance is their integrated NPD process, which has been epitomized by concepts such as TQM and QFD. While TQM aims to comprehensively incorporate all relevant customer requirements into NPD project management (Hackman and Wageman 1995), QFD focuses on translating these customer requirements into technical requirements in the NPD process (Chan and Wu 2002). Such concepts have not originated exclusively in Japan; however, they appear to have been implemented more effectively by Japanese companies than their Western counterparts. An important dimension behind the establishment of an integrated NPD process has been cross-functional integration through the appointment of "heavyweight" NPD project managers who hold strong authority and intensively engage in hands-on communication and coordination with all staff involved in an NPD project (Nobeoka 2006). Furthermore, cross-functional teams do not serve as mere NPD steering committees. Their members and all other technical and managerial staff frequently engage in informal communication across functions and departments to coordinate and synchronize their efforts. Due to this strong NPD integration, activities such as product engineering and process engineering can strongly overlap, resulting in shorter development times, lower cost, and higher customer satisfaction (Clark and Fujimoto 1991).

Korean firms have achieved strong NPD performance, particularly through their capability to quickly develop new products. Their NPD lead times are much shorter than those of both Western companies and Japanese competitors

(Song and Lee 2014). Korean firms' short NPD lead times have enabled them to achieve first-mover advantages and market dominance in some product lines in high-tech industries (Shin and Jang 2005). Firms such as Samsung Electronics (see Box 14.2) have shortened their NPD cycles using various project management techniques. In addition to strong cross-functional integration through task force teams, multiple teams are formed simultaneously for high-priority NPD activities and compete for who can solve critical technical problems first and best (Song and Lee 2014). Furthermore, Korean firms integrate process innovation activities into NPD projects, instead of conducting them separately (Choi et al. 2016). The focus of NPD teams in finalizing NPD projects successfully and in a timely manner is reinforced through the strong leadership of senior executives (Hemmert 2018) and the champion behavior of project managers (Shim and Kim 2018). With regard to quality management, Korean firms achieve not only technical excellence but also superior design quality in NPD, which often turns out to be an important competitive advantage. A strong and effective emphasis on design thinking in NPD has been applied by Samsung Electronics and LG Electronics as well as by tire manufacturers such as Hankook Tire and cosmetics firms such as Amore Pacific (De Mozota and Kim 2009).

Box 14.2   NPD at Samsung Electronics

Established in 1969 as a subsidiary of the Samsung conglomerate, Samsung Electronics entered the semiconductor industry in the late 1970s. By 1992, it achieved global market leadership in memory chips by dethroning Japanese competitors who had dominated the industry in the 1980s. Since that time, Samsung has maintained its global market leadership position. The company's impressive performance has been enabled by its NPD practices, which include close collaboration with *chaebol*-affiliated suppliers, cross-functional integration via task force teams, the parallel pursuit of various NPD processes, the simultaneous setup of multiple development teams that compete on task completion schedules, and rapid decision-making. These practices enabled Samsung to drastically cut its lead time and establish market leadership in the memory chip industry where first mover advantages are decisive. Thereafter, Samsung also became a leading global producer of smartphones through intensive in-house NPD efforts in collaboration with group-internal suppliers that focused on new, innovative designs and superior product quality. Short lead times also contributed to Samsung's competitive advantage in smartphones. Samsung initiates NPD with overall product planning and then combines all necessary technologies from internal and external sources, resulting in a highly focused NPD process. The company invests strongly in R&D and employs a higher number of engineers than competitors, while also engaging extensively in external technology sourcing. Samsung's emphasis on innovativeness and leadership in product designs has helped it to become a global pioneer in developing foldable smartphones.

Similar to Korean firms, Chinese firms strongly emphasize speed when developing new products (Hout and Michael 2014). They routinely pursue different development tasks and processes in parallel and in an informal manner to cut NPD lead times (Yip and McKern 2016). Furthermore, they apply various other practices that further increase NPD speed. From a top-down perspective, senior company executives hold strong authority and can make NPD-related key decisions quickly, without the need for a time-consuming discussion. From a bottom-up perspective, developers and engineers are encouraged to develop an entrepreneurial attitude toward NPD by freely proposing new ideas and concepts. Such ideas are then quickly evaluated and transformed into new NPD projects when assessed as promising (Yip and McKern 2016). The flexible, informal organization of NPD activities, in which engineers and designers are members of development teams rather than functional departments, is supported by organizational decentralization in some Chinese firms. The electrical appliance manufacturer Haier has reorganized itself into numerous micro-enterprises that were created based on employees' preferences, and that design, develop, manufacture, and distribute products and services autonomously (Frynas et al. 2018). The various organizational arrangements made by Chinese firms to speed up NPD processes have often afforded them a strong competitive advantage in fast-moving markets (Steinfeld and Beltoft 2014).

14.2.4 New Product Introduction

When introducing new products, Japanese firms tend to interact closely with their customers, engaging in a cycle of trial production, fault detection, and resolving defects before new product release (Marukawa 2009). These close customer interactions in product introduction are part of the generally strong collaboration among value chain participants in Japan that also features in technology sourcing, as discussed earlier. The intense testing of new products before their release aligns with differentiation-based strategies applied by Japanese firms. Furthermore, because customer orientation is generally strong in Japan, companies tend to introduce new products only when they have strong confidence that the products will work smoothly and that there will be no quality issues. NPD activities of Japanese firms are mostly conducted at their R&D facilities in Japan, and most customer interactions in this process primarily involve Japanese customers. Consequently, new products tend to be initially developed for the Japanese market and introduced to overseas markets only after a successful new product launch in Japan (Beise 2006). Product customization for overseas markets tends to be relatively limited as companies seek competitive advantage based on globally applicable core technologies and new product features.

Korean companies generally emphasize the rapid development and market introduction of new products and, therefore, tend not to engage in pre-release production trials and quality controls to the same extent as their Japanese counterparts. However, they generally feature strong market orientation, which is an important antecedent of NPD performance (Hong et al. 2013). As their products are often focused on high-end markets, companies engage in thorough quality testing before introducing new products while seeking the fastest possible completion of testing activities. Furthermore, in contrast to many Japanese companies, Korean manufacturers highly value international markets. Therefore, while most of the companies' NPD activities are conducted in Korea, they often develop customized product versions for international markets in parallel to ensure timely product introduction both in their home market and internationally (Hemmert 2018).

Chinese companies have a clearly different approach to introducing new products than their Japanese and Korean counterparts. Because they mostly apply cost leadership strategies, they tend to invest less in pre-release testing and quality control. Instead, companies such as Xiaomi (see Box 14.3) use post-release customer feedback to swiftly address quality problems (Steinfeld and Beltoft 2014). Thus, a substantial part of new product beta testing is conducted by early customers.

For example, the internet multimedia platform company Tencent releases early versions of its products to parts of its user base, and products are updated on a weekly basis based on user feedback (Yip and McKern 2016). This new product market introduction approach is often effective in China, where most customers are highly price sensitive and tolerate initial quality problems, as long

---

Box 14.3   NPD at Xiaomi

Founded in 2011, Xiaomi became the largest smartphone vendor in China only three years later, and the third largest competitor worldwide. The company initially developed smartphone software customized to the needs of Chinese users, combining ease of use with low cost. As its software applications became highly popular in China, Xiaomi quickly integrated forward into the smartphone industry. It did so by focusing only on R&D and customer service, while outsourcing manufacturing and producing only on demand. This lean structure has enabled Xiaomi to offer high quality products at a much lower price than global competitors. The high product quality has been established through an in-house team of highly capable developers who previously worked for leading Chinese and global companies. Additionally, Xiaomi strongly and regularly engages with its user community. Product managers closely monitor user forums for comments and ideas and present them to engineers. When evaluated positively, these ideas are quickly incorporated into weekly software updates for smartphone users. As a result of the company's very high perceived customer responsiveness, Xiaomi has gained a strong reputation in China. Furthermore, it has built an extensive ecosystem for its user base, which includes not only its own products, but thousands of consumer devices from hundreds of partner companies. Xiaomi invests in partner companies and takes a hands-on approach in supporting them. Consequently, the company reinforces its NPD capabilities with a relatively lean organization. Furthermore, Xiaomi has expanded its market reach to India and Southeast Asia, where user preferences are relatively similar to China.

as they are swiftly addressed by the exchange of faulty parts or software dispatches (Steinfeld and Beltoft 2014). Chinese firms mostly focus on their sizable home market when developing and introducing new products, and they often offer new functionalities tailored to the needs of Chinese customers. For example, the electronic appliance manufacturer Haier has adapted its washing machines so that farmers can use them to wash potatoes and vegetables (Yip and McKern 2016). However, successful companies also quickly introduced new products in other emerging markets, such as India, with similar customer preferences.

14.3 Challenges and Effective Means of Adopting Best East Asian NPD Practices

As discussed in the previous section, Japanese, Korean, and Chinese companies apply highly effective NPD practices across a wide range of activities. Therefore, there is potential for firms from other countries to enhance their NPD performance by learning from their East Asian competitors.

However, adopting East Asian NPD practices is not straightforward from an international perspective, as important aspects of the business environment strongly differ across countries. One such factor is the institutional environment. For example, Japan has a long history of close linkages between manufacturers and their suppliers in a vertical *keiretsu*. Their collaboration links can be traced back to Japan's industrialization since the late 19th century, when state-of-the-art technological knowledge from Western countries was initially absorbed by large manufacturers (Hemmert 1999). Subsequently, these manufacturers shared their knowledge with suppliers, forging close and sometimes proprietary supply chain relationships, which extended into collaborative NPD over time. Conversely, Korean manufacturers rely firmly on support from member firms in the same business group (*chaebol*) when developing new products. *Chaebols* emerged as highly diversified family-owned conglomerates when Korea industrialized. Consequently, valuable knowledge, including NPD knowledge, is freely shared within these conglomerates (Chang 2003). Furthermore, *chaebol* member firms can be effectively co-opted for joint NPD activities whenever needed.

Another important contextual aspect is the cultural environment. From a cross-cultural perspective, Japan, Korea, and China are recognized as the core countries of the "Confucian Asia Cluster," which features hierarchical relationships, family modeled organizations, and emphasis on principles such as diligence, self-sacrifice, and delayed gratification (Gupta and Hanges 2004). In East Asian business, there is also a general preference for relationship-based long-term collaboration over short-term transactional exchange (Chen and Miller 2011). Consequently, relational governance is more effective than contractual governance for effective NPD collaboration (Bstieler and Hemmert 2015). The importance of personal relationships for the effectiveness of long-term collaboration is especially emphasized in the Chinese context, where personal connections that follow social norms, such as maintaining long-term relationships, mutual commitment, loyalty, and obligation, are referred to as *guanxi* (Chen and Chen 2004). Similar social norms have been widely observed in Japan and Korea. For example, Japan has a strong cultural tradition of trust and extensive collaboration within communities, which supports collaboration and knowledge transfer in *keiretsu* business networks even in the absence of vertical control. These cultural features support NPD practices, such as long-term collaborative technology sourcing within stable business networks and business groups, and senior executives exerting strong pressure on developers to finalize NPD projects in a short period of time.

These considerations suggest that NPD practices from East Asian countries cannot be easily duplicated in other regions of the world, including Western countries. Institutionally or culturally embedded practices are notoriously difficult to transfer as they encounter lack of understanding or acceptance when applied in different environments. It is highly challenging to understand and internalize managerial practices that originate in distant institutional and cultural contexts. For example, it took General Motors a long time to absorb key aspects of Toyota's famous "lean production" system, even after a joint factory with Toyota was established in California (Inkpen 2008).

However, such hurdles do not mean that it is not possible or practical for companies from other parts of the world to learn effective NPD practices from East Asian firms. Instead, careful consideration is needed to effectively adopt best practices from Japan, Korea, and China. It is also important to note that not all Japanese, Korean, and Chinese firms necessarily excel in NPD and that their NPD practices have potential downsides. For example, the Japanese tendency to extensively involve a wide range of internal departments and external partners in NPD projects may result in technical and design compromises that deter companies from breakthrough innovations. Conversely, Korean and Chinese firms' emphasis on top-down decision-making and minimizing NPD lead times may potentially engender NPD staff burnout and quality problems that could negatively affect companies' performance in the long term. Therefore, it is prudent to focus on the best practices of leading Japanese, Korean, and Chinese firms, and to consider the fit of these practices with a given non-East Asian company's own market environment and organizational culture.

The best NPD practices can be effectively learned from leading East Asian counterparts by directly engaging with them via long-term collaborations such as strategic alliances, joint ventures, or supply chain relationships. While Japanese, Korean, and Chinese companies routinely aim to learn from their Western partners in such collaborations, foreign partners can also learn from their East Asian counterparts if they are fundamentally willing to learn and deploy sufficient resources for collaboration, which allow effective observation and absorption of best practices.

Another effective means to learn effective NPD practices from East Asian companies is the establishment of R&D units in East Asia, which can be leveraged for both technological collaboration with local partners and in-house R&D. Establishing R&D units in close geographical proximity to partner firms is a highly effective organizational arrangement for MNEs (Phene and Almeida 2008) and helps with absorbing local knowledge. Eventually, the effective absorption of such local knowledge can augment multinational firms' NPD both locally and globally (Zhao et al. 2020). Furthermore, East Asian countries have emerged as global lead markets in various ICT industries (Hemmert 2020). Multinational firms can learn about new technological trends and market developments in such lead markets through a local NPD presence while also being able to closely observe how East Asian firms shape and react to these trends in their home markets.

14.4 Learning from East Asian NPD Practices: Takeaways from a Global Perspective

Japanese, Korean, and Chinese companies have established a wide range of highly effective NPD practices, and their counterparts in other parts of the world can learn substantially from their best practices if motivated and willing to do so. Specifically, the following takeaways can be drawn from a global perspective.

14.4.1 No Pain, No Gain: Invest Strongly into NPD

While Japanese, Korean, and Chinese firms could initially catch up to some extent with their Western rivals through cost-effective means, such as licensing and reverse engineering, highly innovative and successful firms from the three countries have all strongly invested in NPD. Japan and Korea's business R&D intensity far exceeds that of most Western countries, and China's business R&D intensity far exceeds that of other emerging economies. The fundamental lesson from East Asia is that a strong, sustained NPD investment is needed to innovate successfully. There are no shortcuts.

14.4.2 Getting the Most out of Your Developers: Motivation is Everything

Successful innovators from East Asian countries have leveraged the capabilities of their NPD knowledge workers, including researchers, development engineers, and designers, to the greatest extent possible by effectively motivating them. While monetary incentives are undeniably important to enhance the motivation of these professionals, East Asian firms have additionally motivated them through various measures, such as exciting organizational goals and visions, a strong collaboration culture, extensive socialization efforts, tight project timelines, and internal contests between development teams. The measures taken by Japanese, Korean, and Chinese firms are highly diverse, and their effectiveness depends on the specific context in which the companies and their developers operate. Their overarching commonality is that firms see their knowledge workers as their most precious resource and effectively motivate and incentivize them by any possible means. The motivation of these professionals strongly determines the NPD performance of companies.

14.4.3 The Early Bird Catches the Worm: Speed, Speed, Speed

Another common feature of successful East Asian companies is their rapid NPD. East Asian firms tend to develop new products much more rapidly than their American and European counterparts (Markham and Lee 2013). The drastic shortening of NPD lead times has frequently given Japanese, Korean, and Chinese firms a competitive edge, particularly in high-tech industries, where first-mover advantages are highly important. Firms have achieved these short lead times through a range of organizational practices, such as creating cross-functional teams, prioritizing flexible, project-based organizations over functional organizations, speed-based team competitions, and quick decision-making. The effectiveness of each of these practices depends on the organizational and cultural context in which they are applied, but is definitely not limited to East Asian firms and business environments. Flexible and adaptive organizational arrangements that rely on rapid development efforts by cross-functional teams have recently been discussed in the Western context, using the concept of agile NPD (Cooper and Sommer 2016). While the specific practices of Japanese, Korean, and Chinese firms differ, companies from these three countries generally demonstrate high agility, as they understand that the rapid adaptation of new products to changing needs is crucial for successful innovation. Consequently, they do everything that helps shorten their NPD lead times. Pragmatism and open-mindedness toward new and unconventional practices are key to developing new products quickly and in an adaptive manner.

14.4.4 Partnering Effectively: Get Any Help You Need

East Asian companies also offer lessons for effective external partnering. While East Asian firms strongly invest in in-house R&D to develop core technologies, they also use external partners extensively in their NPD projects. Many of their partners, such as members of the same *keiretsu* networks in Japan and the same *chaebol* in Korea, are tightly connected to the focal firms via deep and stable long-term business relationships, which include extensive knowledge transfer and knowledge co-production. Therefore, *keiretsu* and *chaebol* suppliers can be engaged by East Asian firms in NPD projects from the outset, and focal firms can effectively absorb their suppliers' technology and knowledge. Alternatively, external knowledge can be effectively internalized via targeted acquisitions of firms that hold particularly valuable knowledge or scouting of leading NPD experts from competitors. Knowledge transfer among Chinese firms is facilitated by the *guanxi* networks of their representatives. While non-East Asian firms often find it challenging to develop long-term relationships with East Asian counterparts, the overall takeaway is to partner extensively and pragmatically with whoever can strengthen your NPD and consider all available collaboration modes flexibly. Furthermore, East Asian firms have demonstrated that deep and long-term supply chain collaboration can be effectively leveraged for the sourcing and co-production of valuable NPD knowledge.

14.4.5 Not All Customers Are the Same: Consider Market Needs in New Product Introduction

While it is widely known that a good fit with customer needs is crucial for the successful market introduction of new products, East Asian companies offer valuable lessons on how to introduce new products effectively, in consideration of different customer needs across countries. Korean firms, which tend to rely strongly on international markets, routinely develop customized product versions for different countries and regions in parallel with new products for their home market. This enables them not only to launch their products globally without delay but to also be more successful in international markets. Furthermore, Chinese firms have demonstrated that new products can be launched quickly and at affordable prices for customers in China and other emerging markets by saving cost and time on pre-launch quality testing and relying on customer feedback for post-launch quality improvement. The key point is that customers in different countries have different needs, and not all customers expect top-notch quality from a new product. Specifically, emerging market customers are highly price-sensitive. Consequently, firms can significantly enhance their market response to new products by seriously considering different customer needs from the outset in their NPD activities. For example, firms from advanced countries may develop scaled down product versions that are catered to the needs of emerging market customers.

14.4.6 Leverage Business Ties and Go Local to Learn the Most from East Asia

Because East Asia offers many valuable lessons for effective and successful NPD, companies from other parts of the world may gain substantially from understanding and absorbing the best NPD practices in this region. Whenever possible, companies should partner directly with Japanese, Korean, and Chinese firms to observe and potentially internalize their NPD practices. For effective partnering and learning, in-depth and long-term engagement is necessary considering the strong relational orientation of East Asian companies and managers. Furthermore, the establishment of local subsidiaries, including R&D centers, in East Asian countries can be highly effective in gaining a deep understanding of local NPD practices, as well as customer needs and market trends in the long term.

14.5 Conclusion

Over the past few decades, East Asian firms have excelled in NPD by developing new products more quickly, with lower costs and higher quality than global competitors. Companies from Japan, Korea, and China apply distinct NPD strategies, technology sourcing, project management, and new product introduction practices, which greatly enhance their NPD performance. These practices are embedded in the institutional and cultural environments of firms' home countries, and therefore cannot be easily transferred to other parts of the world. Nonetheless, there is great learning potential if companies and managers are sufficiently open-minded and willing to study effective East Asian NPD practices. Key managerial implications from East Asia are: (1) investing strongly in NPD activities; (2) effectively motivating and incentivizing developers; (3) speeding up NPD processes through cross-functional teams and the flexible organization of core activities; (4) absorbing knowledge from NPD partners by collaborating pragmatically in response to case-specific priorities; (5) adjusting the market introduction of new products according to differential customer needs; and (6) engaging directly with proficient Japanese, Korean, and Chinese partners to understand and absorb their best practices.

References

Aoki, K. and Lennerfors, T.T. (2013). The new, improved *keiretsu*. *Harvard Business Review* 91 (9): 109–113.

Aoki, K. and Staeblein, T. (2018). Monozukuri capability and dynamic product variety: an analysis of the design-manufacturing interface at Japanese and German automakers. *Technovation* 70–71: 33–45.

Beise, M. (2006). The domestic shaping of Japanese innovations. In: *Management of Technology and Innovation in Japan* (ed. C. Herstatt, C. Stockstrom, H. Tschirky, and A. Nagahira), 112–141. Berlin: Springer.

Bstieler, L. and Hemmert, M. (2015). The effectiveness of relational and contractual governance in new product development collaborations: evidence from Korea. *Technovation* 45-46: 29–39.

Chan, L.-K. and Wu, M.-L. (2002). Quality function deployment: a literature review. *European Journal of Operational Research* 143: 463–497.

Chang, S.J. (2003). *Financial Crisis and Transformation of Korean Business Groups: The Rise and Fall of Chaebols.* Cambridge: Cambridge University Press.

Chen, M.J. and Miller, D. (2011). The relational perspective as a business mindset: managerial implications for East and West. *Academy of Management Perspectives* 25 (3): 6–18.

Chen, X.-P. and Chen, C.C. (2004). On the intricacies of the Chinese guanxi: a process model of guanxi development. *Asia Pacific Journal of Management* 21 (3): 305–324.

Choi, K., Narasimhan, R., and Kim, S.W. (2016). Opening the technological innovation black box: the case of the electronics industry in Korea. *European Journal of Operational Research* 250: 192–203.

Choi, Y. and Hara, Y. (2018). The performance effect of inter-firm adaptation in channel relationships: the roles of relationship-specific resources and tailored activities. *Industrial Marketing Management* 70: 46–57.

Clark, K.B. and Fujimoto, T. (1991). *Product Development Performance: Strategy, Organization, and Management in the World Auto Industry.* Boston: Harvard Business School Press.

Cooper, R.G. and Sommer, A.F. (2016). Agile-Stage-Gate: new idea-to-launch method for manufactured new products is faster, more responsive. *Industrial Marketing Management* 59: 167–180.

De Mozota, B.B. and Kim, B.J. (2009). Managing design as a core competency: lessons from Korea. *Design Management Review* 20 (2): 66–76.

Dyer, J.H. and Chu, W. (2003). The role of trustworthiness in reducing transaction costs and improving performance: empirical evidence from the United States, Japan, and Korea. *Organization Science* 14 (1): 57–68.

Dyer, J.H. and Chu, W. (2011). The determinants of trust in supplier–automaker relations in the US, Japan, and Korea: a retrospective. *Journal of International Business Studies* 42 (1): 28–34.

Frynas, J.G., Mol, M.J., and Mellahi, K. (2018). Management innovation made in China: Haier's Rendaheyi. *California Management Review* 61 (1): 71–93.

Gupta, V. and Hanges, P.J. (2004). Regional and climate clustering of societal cultures. In: *Culture, Leadership, and Organizations: The GLOBE Study of 62 Societies* (ed. R.J. House, P.L. Hanges, M. Javidan, P.W. Dorfman, and V. Gupta), 178–218. Thousand Oaks: Sage.

Hackman, J.R. and Wageman, R. (1995). Total quality management: empirical, conceptual, and practical issues. *Administrative Science Quarterly* 40 (2): 309–342.

Hemmert, M. (1999). 'Intermediate organization' revisited: a framework for the vertical division of labor in manufacturing and the case of the Japanese assembly industries. *Industrial and Corporate Change* 8 (3): 487–517.

Hemmert, M. (2018). *The Evolution of Tiger Management: Korean Companies in Global Competition.* London and New York: Routledge.

Hemmert, M. (2020). Learning from Korea. In: *Doing Business in Korea* (ed. F.J. Froese), 118–134. London and New York: Routledge.

Hobday, M., Rush, H., and Bessant, J. (2004). Approaching the innovation frontier in Korea: the transition phase to leadership. *Research Policy* 33 (10): 1433–1457.

Hong, J., Song, T.H., and Yoo, S. (2013). Paths to success: how do market orientation and entrepreneurship orientation produce new product success? *Journal of Product Innovation Management* 30 (1): 44–55.

Hout, D. and Michael, D. (2014). A Chinese approach to management. *Harvard Business Review* 92 (9): 103–107.

Hu, M.-C. (2012). Technological innovation capabilities in the thin film transistor-liquid crystal display industries of Japan, Korea, and Taiwan. *Research Policy* 41 (3): 541–555.

Inkpen, A. (2008). Knowledge transfer and international joint ventures: the case of NUMMI and general motors. *Strategic Management Journal* 29 (4): 447–453.

Kim, L. (1997). *Imitation to Innovation: The Dynamics of Korea's Technological Learning.* Boston: Harvard Business School Press.

Kusunoki, K. (2006). Invisible dimensions of innovation: strategy for de-commoditization in the Japanese electronics industry. In: *Management of Technology and Innovation in Japan* (ed. C. Herstatt, C. Stockstrom, H. Tschirky, and A. Nagahira), 49–71. Berlin: Springer.

Luo, Y., Sun, J., and Wang, S.L. (2011). Emerging economy copycats: capability, environment, and strategy. *Academy of Management Perspectives* 25 (2): 37–56.

Markham, S.K. and Lee, H. (2013). Product Development and Management Association's 2012 comparative performance assessment study. *Journal of Product Innovation Management* 30 (3): 408–429.

Marukawa, T. (2009). Why Japanese multinationals failed in the Chinese mobile phone market: a comparative study of new product development in Japan and China. *Asia Pacific Business Review* 15 (3): 411–431.

Nobeoka, K. (2006). Reorientation in product development for multiproject management: the Toyota case. In: *Management of Technology and Innovation in Japan* (ed. C. Herstatt, C. Stockstrom, H. Tschirky, and A. Nagahira), 207–234. Berlin: Springer.

Nonaka, I. and Takeuchi, H. (1995). *The Knowledge-Creating Company: How Japanese Companies Create the Dynamics of Innovation.* New York and Oxford: Oxford University Press.

OECD – Organisation for Economic Co-operation and Development (2022). *Main Science and Technology Indicators.* https://stats.oecd.org/Index.aspx?DataSetCode=MSTI_PUB (accessed June 21, 2022).

Park, K. and Kim, B.-K. (2013). Dynamic capabilities and new product development performance: Korean SMEs. *Asian Journal of Technology Innovation* 21 (2): 202–219.

Phene, A. and Almeida, P. (2008). Innovation in multinational subsidiaries: the role of knowledge assimilation and subsidiary capabilities. *Journal of International Business Studies* 39 (5): 901–919.

Shim, D. and Kim, Y. (2018). Champion behaviour and product innovation performance in Korea. *Asian Journal of Technology Innovation* 26 (2): 172–201.

Shin, J.S. and Jang, S.-W. (2005). Creating first-mover advantages: the case of Samsung Electronics. *SCAPE Working Paper No. 2005/13*, Department of Economics, National University of Singapore.

Song, J. and Lee, K. (2014). *The Samsung Way: Transformational Management Strategies from the World Leader in Innovation and Design.* New York: McGraw-Hill.

Steinfeld, E.S. and Beltoft, T. (2014). Innovation lessons from China. *MIT Sloan Management Review* 55 (4): 49–55.

von Zedtwitz, M., Ikeda, T., Li, G. et al. (2007). Managing foreign R&D in China. *Research-Technology Management* 50 (3): 19–27.

Yip, G.S. and McKern, B. (2016). *China's Next Strategic Advantage: From Imitation to Innovation.* Cambridge and London: MIT Press.

Zhang, M.Y. (2014). Innovation management in China. In: *The Oxford Handbook of Innovation Management* (ed. M. Dodgson, D.M. Gann, and N. Phillips), 355–374. Oxford: Oxford University Press.

Zhao, S., Tan, H., Papanastassiou, M., and Harzing, A.-W. (2020). The internationalization of innovation towards the South: a historical case study of a global pharmaceutical corporation in China (1993–2017). *Asia Pacific Journal of Management* 37 (2): 553–585.

Martin Hemmert is Professor of International Business at Korea University. His research interests include international comparative studies of management systems, innovation systems, and entrepreneurial ecosystems, organizational boundaries of firms and inter-organizational research collaborations, with a focus on East Asian countries. He has published several books, including most recently *Entrepreneurship in Korea: From Chaebols to Start-ups* (Routledge, 2021), and many articles in peer-reviewed international journals such as *Journal of Product Innovation Management, Journal of World Business, Research Policy,* and *Technovation.*

SECTION THREE

USER PARTICIPATION AND VALUE CREATION IN NEW PRODUCT DEVELOPMENT

CHAPTER FIFTEEN

NAVIGATING OPEN INNOVATION

Rebecca J. Slotegraaf and Girish Mallapragada

15.1 Introduction

Innovation is an essential engine for corporate growth. By investing in new ideas, companies can explore new markets and create opportunities for increasing shareholder value. Historically, much of corporate innovation has occurred within the company's boundaries, and a vast amount of literature has documented the various processes and practices that enable company-driven innovation (e.g., Cooper and Kleinschmidt 1995; Griffin and Hauser 1996; Moorman and Slotegraaf 1999). As a form of innovation that has garnered much attention, open innovation (Chesbrough 2003) refers to a hybrid innovation model wherein the company cedes ownership of some innovation processes to an outside stakeholder, such as a channel partner or a customer. While it generates coordination challenges compared to the traditional company-driven model, open innovation can resolve other challenges that a company-centric model cannot. Furthermore, by democratizing innovation through an open approach, companies can better navigate an uncertain future by farming out the innovation process to important stakeholders (Slotegraaf 2012).

In this chapter, we begin by briefly explaining the evolution of open innovation before providing a framework that delineates the value of engaging with customers and competitors in open innovation across different stages of the innovation process. We then offer an overview of various open innovation approaches and discuss some of their advantages and disadvantages. We conclude with actionable insights for managers to use open innovation to better prepare for the future.

The PDMA Handbook of Innovation and New Product Development, Fourth Edition. Edited by Ludwig Bstieler and Charles H. Noble.

15.2 Tracing the Roots of Open Innovation

While mainstream attention to open innovation has increased after the seminal work of Chesbrough (2003), open innovation contexts have dramatically impacted the business landscape since the 1970s. The free software movement, which later transformed into the open-source movement (Stallman 2002), is the first example of open innovation. In his book, *The Cathedral and The Bazaar,* the technologist Eric Raymond equates the traditional company-centric model of developing software products to the "cathedral," a manifestation of a hierarchical approach to innovation. In contrast, he likens the open-source movement to the "bazaar," a manifestation of an arm's length market. The open-source movement, which has its roots in MIT's media lab in the early 1970s (Raymond 1997), is an astounding success story of an open innovation paradigm that changed the world. The open-source movement kickstarted a technology revolution that ceded control of the design, development, and improvement of products to users.

The lead-user research paradigm evolved in the 1980s (Lilien et al. 2002; Urban and von Hippel 1988) and centered initially on advantages associated with engaging with organizational customers to develop innovative products and solutions. In addition to customers as a source for ideation and design, engagements with other stakeholders have increasingly been documented and investigated, including, for example, channel partner contributions (Mallapragada and Srinivasan 2017), collaborations with academic institutions (Nerkar and Shane 2007), and formation of patent pools (e.g., Lerner and Tirole 2004).

Since the 1990s, broader recognition of open innovation practices and their applicability to other contexts beyond software and technology has taken shape (Chesbrough 2003). It is now accepted that open innovation offers companies an alternative way of governing innovation that is distinct from the traditional hierarchical structure. As different types of open innovation proliferate, companies often face a dizzying array of options. While the concept of open innovation has been refined over time, there is no integrative framework that offers tangible, managerial direction. We seek to address this gap in the current chapter.

15.3 Using Open Innovation at Different Stages in the Innovation Process

Companies use open innovation in different stages of their innovation process. Some early literature distinguished between inbound vs. outbound innovation, based on whether the information flow was from the firm to external stakeholders (i.e., outbound) or from external stakeholders to the firm (i.e., inbound) (Chesbrough 2003). We build on this typology and direct attention to the occurrence of open innovations across stages of the innovation process. We classify the innovation process into three stages; namely, ideation (i.e., generating and refining new product ideas), design and development (i.e., product design, prototype development), and commercialization (i.e., market testing, product launch). The use of open innovation at each stage generates advantages and disadvantages to consider when looking at external sources of innovation. We recognize that engagement with customers (i.e., channel partners, customers, or end consumers) is distinct from that with competitors and therefore differentiate advantages and disadvantages for competitors and customers; see Table 15.1 for a summary.

15.3.1 Ideation

Two key advantages to using open innovation in the idea generation stage of the innovation process are the sheer number of ideas and their variability, whether received from customers or competitors. If the net is cast widely to access a wide array of ideas, then "the quality of the average idea may fall, but the best one is more likely to be spectacular" (King and Lakhani 2013, p. 42). Customer participation in the ideation stage generates more novel ideas, delivers higher customer benefits (Poetz and Schreier 2012), and increases speed to market and financial performance (Chang and Taylor 2016). New product alliances with competitors enable higher knowledge overlap, stimulating creativity and market speed (Rindfleisch and Moorman 2001). Innovations that originated from channel partners can also lead to blockbuster products and efficient processes (e.g., Mallapragada and Srinivasan 2017). Indeed, research shows that companies can gain higher stock returns by working with competitors during ideation, especially for incremental products (Wu et al. 2015). Procter & Gamble's Connect + Develop initiative, launched in the early 2000s, has become the company's innovation model. It not only elevated P&G's engagement with consumers, business partners, and other external stakeholders but also decreased its innovation costs and increased its R&D productivity and innovation success.

Although involving customers or competitors in idea generation can be beneficial, there are some downsides. Foremost, consumers' ideas may be less feasible for a company to develop (Poetz and Schreier 2012). In addition, when assessing crowd-based ideas, consumers tend to undervalue feasibility and overvalue moderate originality, in contrast to companies that prefer feasible and highly original ideas (Hofstetter et al. 2018). Therefore, rather than using consumer voting, a panel that combines company interests, expert assessments, and consumer interest would provide a more effective way to evaluate crowd-generated ideas. When working with competitors, it is important to recognize that hesitation may occur with sharing the best ideas, especially due to concerns over intellectual property (IP). Address IP rights in advance can help to alleviate competitive collaboration concerns.

15.3.2 Design and Development

Engaging end-consumers or other customers in design offers advantages and disadvantages, including the potential to increase commercial success. When

TABLE 15.1 ENGAGEMENT TRADEOFFS WITH COMPETITORS AND CUSTOMERS ACROSS THE INNOVATION PROCESS.

	Ideation	Design and Development	Commercialization
Competitors	*Advantages:* Greater quantity and diversity of ideas Stimulates creative ideas Faster speed to market Higher stock returns, especially for incremental products	*Advantages:* Common frames of reference foster greater knowledge exchange and faster development speed Reduced R&D costs Access to new supply materials	*Advantages:* Sharing market knowledge across competitors can boost diffusion
	Disadvantages: Hesitation with sharing best ideas	*Disadvantages:* IP-related concerns may amplify opportunistic behavior Lower stock returns, though attenuated if competitor has strong technological capabilities	*Disadvantages:* Lower priority on customer interests Lower stock returns, though radical products attenuate this negative effect
Customers	*Advantages:* Greater quantity and diversity of ideas Ideas with more novelty and higher customer benefit Faster speed to market Higher financial new product performance	*Advantages:* Higher consumer willingness to pay for self-designed products Higher consumer purchase intention for self-customized products Reduced R&D costs and access to new supply materials from collaborations with channel partners	*Advantages:* Help refining market launch and product positioning Higher financial performance of new product
	Disadvantages: Ideas with potential for lower development feasibility Consumer assessment of crowd-based ideas may undervalue feasibility and overvalue only moderate levels of originality	*Disadvantages:* Slower speed to market When comparing self-designed product to professional designs, it reduces satisfaction	*Disadvantages:* Efforts may need to focus on Standards-based design Avoid proprietary lock-ins and high switching costs

customers self-design products, they have much more control over the characteristics that matter most to them. Customers are also more likely to think about eclectic product attributes, thereby expanding the scope of the product's design (e.g., Timoshenko and Hauser 2019). When end-consumers design a product, it increases their purchase intentions and generates a higher willingness to pay because they play an active role in the design process (e.g., Franke et al. 2009, 2010). However, consumers compare their self-designed products to those by professional designers, reducing their satisfaction with their self-design experience and the designed product (Moreau and Herd 2010). With product development, collaboration with channel partners or other business partners offers the opportunity to reduce costs and gain access to new supply materials. Business customers can help design and engineer products by offering useful solution-related knowledge (Coviello and Joseph 2012). However, it may not be leveraged in such a way to maximize its value, and it may slow speed to market (Chang and Taylor 2016).

When engaging with competitors during the development stage, common frames of reference foster greater knowledge understanding that facilitates development speed (Rindfleisch and Moorman 2001). For example, Pfizer and Merck collaborate to bring new cancer treatments to the market as soon as possible. In addition, several automotive manufacturers have collaborated in pursuit of lithium-ion and solid-state batteries. However, companies will likely face IP-related concerns during innovation that may amplify opportunistic behavior (Luo et al. 2007). Therefore, companies must develop reputations for fair dealings, especially with smaller companies. Regarding outcomes due to collaboration with competitors during design or development, research shows companies attain lower stock returns, yet this is attenuated when a company partners with a competitor with strong technological capabilities (Wu et al. 2015). This suggests that the stock market perceives collaboration with a competitor that has more technological power to signal specific strengths during design or development collaboration efforts.

15.3.3 Commercialization

Although the role of open innovation has garnered wide attention for earlier stages in the innovation process, where rules of engagement and investment are more flexible and companies can operate in an experimental mode, less is known about the impact of open innovation in the commercialization stage. Leveraging customer participation in the commercialization stage can be beneficial, with customers offering first-hand information on product usability, prototype testing, and market launch. Research does show that customer participation during commercialization can increase new product financial performance (Chang and Taylor 2016). When developing proprietary designs with lead users or other customers, higher switching costs are likely as firms face wide-scale implementation for commercialization. Thus, proprietary lock-ins might hinder open innovation practices in the long run. When engaging customers in open innovation during commercialization, companies must invest in standards-driven innovation to quickly reconfigure their implementation plans (Chen and Forman 2006).

Product launch is a central element in the commercialization stage, and marketing is critical. Companies that engage with competitors have the potential to share important market knowledge and capabilities to create a strong force in the marketplace. However, they should heed caution in selecting competitors, where power imbalance can limit sharing of core distribution channels, and lack of experience with the competitor can dampen knowledge exchange (Bucklin and Sengupta 1993). Engaging with competitors can also substantially reduce customer orientation (Rindfleisch and Moorman 2003), which may dampen success during commercialization because companies benefit when connecting with end consumers more closely. Research also shows lower stock returns for companies engaging with competitors during the commercialization stage, though this effect attenuates when the companies are getting ready to launch a radical product (Wu et al. 2015).

15.4 Tools for Open Innovation

Companies can use various approaches (or tools) to engage in open innovation. Some companies have pursued multiple open innovation tools, whereas others have found value in a single method. There is no single best approach for all companies, and each has specific advantages and disadvantages. Furthermore, these tools provide companies with different levels of control and speed (see Figure 15.1 for a map of these distinctions). For those tools that offer high control, retaining control means that the company determines much of the direction of the innovation endeavor, but it cannot control the input or quality of the external engagement. For those with low control, companies must be willing to hand over much of the innovation process to external stakeholders. Of note, companies that have not yet used open innovation are often not willing to give up control, which is necessary in open-sourcing and collaborative communities. With respect to speed, some tools facilitate faster engagement with partners and stakeholders, and might hasten the process of innovation. For a particular situation, brand, or strategic goal, companies should select the tool with the optimal combination of control and speed to achieve their innovation objectives. Although many tools are available, we focus on eight common tools across companies and industries (discussed in alphabetical order): collaborative community, corporate co-creation platform, crowdfunding, external innovation networking, innovation contests, innovation intermediaries, open-sourcing, and prediction markets. These tools enable companies to leverage knowledge outside of the company and use it within their innovation process.

15.4.1 Collaborative Community

A collaborative community is a group or community comprised of passionate users that cooperatively work toward a specific innovation objective. Companies can build their own community or tap into an existing one. For example, in

FIGURE 15.1 INNOVATION TOOLS: A FRAMEWORK BASED ON DEGREE OF CONTROL AND SPEED.

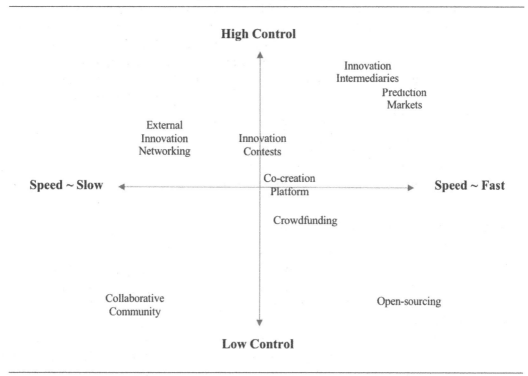

2012, BMW tapped the community of engineers, designers, and car enthusiasts that another company (Local Motors) had developed for ideas, concepts, and designs for a more urbanized future city. Many other companies build their own communities. For example, GE Appliances created its own community, FirstBuild, which involves a community of engineers, designers, home enthusiasts, and other passionate members to generate new product ideas and design new concepts. The goal is to allow the community free reign to develop breakthrough ideas and product concepts. Philips, the Dutch multinational company, adopted open innovation in the late 1990s. It developed a unique community ecosystem, the High Tech Campus Eindhoven, where researchers, start-up founders, and product developers collaborate on new ventures. The endeavor primarily focuses on developing solutions for big societal problems such as climate change and healthcare that can have a real-world impact. Lego created Lego Ideas, which allows users of the community to submit ideas for Lego products. The ideas are voted on by the community and must receive support from 10,000 different users before Lego reviews the proposed idea; if selected, the original designer receives 1% of royalties. To date, more than 40 sets have been launched (e.g., female scientists, working typewriter, treehouse, grand piano), and the community continues to grow in size and engagement.

Recently, hackathons have emerged as a new form of a collaborative community and have gained popularity, particularly in the software industry. A hackathon is typically coordinated by a firm and focuses on large short-term gatherings of individuals to collaborate on a project. Often, hackathons involve intensive sessions lasting 24 hours (or a weekend) and structured as a competition, where participants focus on speed and develop a prototype of a software application or a mobile app. While these events bestow speed, they are also more suited for software-based products and might find limited application in other domains.

The advantage of using a collaborative community is to leverage a passionate user base with the ability and opportunity to generate breakthrough ideas. However, it is essential to recognize that a company must give up control to the community, allowing it to select ideas and designate product design.

15.4.2 Corporate Co-creation Platform

A corporate co-creation platform is a company-driven, technology-based social platform that allows the company to engage with external partners. It is important to note that it is more than an online suggestion box. Instead, value occurs when designed to foster interaction among external users and between users and the company.

There are both advantages and disadvantages to a corporate co-creation platform. One primary benefit is the high control over the type of information sought, how users are involved, and what remains confidential. For example, General Mills' Worldwide Innovation Network (GWIN) requires registration and login that allows the company to determine where and how individuals, suppliers, or other external entities become involved with the company. With this flexibility and control, a company can design the platform to allow continuous engagement. For example, Starbucks launched its platform, My Starbucks Idea, in 2008 as an open platform to collect ideas from anyone. After five years, it received more than 150,000 ideas (including free Wi-Fi, splash sticks, and cake pops) with more than two million votes from its community.

However, an open platform where individuals can discuss different ideas can tip off competitors. In 2017, Starbucks altered its platform to eliminate community interaction and focus on closed idea submissions.[1] Although a corporate co-creation platform enables a company to engage with its consumers and reinforce the brand, it is important to note that the company will need to drive individuals to the platform. It also requires dedicated resources and an internal structure that reinforces engagement with users, or it can drive consumers away.

15.4.3 Crowdfunding

While harnessing resources from the crowd (i.e., crowdsourcing) is beneficial to innovation management (e.g., Mallapragada et al. 2012; Ransbotham et al.

2012), crowdfunding is a particular type of crowdsourcing wherein the resources are monetary contributions. Crowdfunding can enable entrepreneurs and companies to test the viability of their ideas and product concepts by seeking to raise funds from online communities. The premise is that such communities can collectively determine which ideas hold potential as they choose to fund such ideas over others. By funding ideas that appeal to them, the overall community predicts which ideas have the most commercial appeal. Oculus, Pebble Watch, Coolest Cooler, and Exploding Kittens are some examples of crowdfunded projects that raised millions of dollars and were successfully commercialized.

More recently, crowdfunded blockchain projects on the Ethereum platform have raised hundreds of millions in crowdfunding campaigns, setting a new age of collaborative crowdfunding.[2] Crowdfunding may, at one level, seem peripheral to innovation because of its focus on monetary resources. However, it has enabled ideators to raise funds for prototype development and commercialization efforts that are crucial to the overall innovation process (Ordanini et al. 2011).

15.4.4 External Innovation Networking

Establishing an external innovation network enables a company to share knowledge and pursue various cutting-edge innovation challenges with other organizations (e.g., companies, universities, and government agencies). Pfizer developed Centers for Therapeutic Innovation (CTI) Partnerships with leading academic medical institutions to bridge academic research with Pfizer's development expertise for specific diagnostic areas of strategic importance. Some companies develop a network based on specific challenges, where the company has a selective or competitive approach in determining which external partners become part of the network. For example, General Electric launched Ecomagination in 2005 as a groundbreaking strategy to solve environmental sustainability challenges to create a network to inspire companies to work together and co-develop solutions.[3] Additionally, Electrolux announced the creation of the Electrolux Innovation Factory in 2018 as a collaborative space to accelerate innovation with external partners, including start-ups, university spin-offs, SMEs, and others. External innovation networks increase the diversity of ideas and elevate the degree of expertise to evaluate ideas. These networks also enable greater efficiency, increasing innovation development and implementation speed. However, one of the challenges these networks pose involves a form of governance. Given the arm's length nature of the relationships, non-existent a priori contracts specifying the nature of engagement can lead to ex-ante opportunism by one of the parties.

15.4.5 Innovation Contests

Innovation contests are open to individuals (inside or outside the company) to present innovative ideas or designs, with prizes for winning submissions. For these ideation or design contests, the company or the crowd can select the winner, and multiple winners can be chosen. Many companies have pursued different innovation contests, applicable across various industries. Even for space exploration, NASA has an annual BIG Idea Challenge, targeted toward the academic community to help solve a wide range of specific issues. The value of a contest depends on the number of participants and the quality of the ideas or designs submitted. For example, General Electric launched an innovation contest in 2010 in line with its GE Ecomagination strategy as an annual $200 million open innovation contest to uncover breakthrough ideas toward environmental sustainability. In 2021, the challenge generated nearly 4,000 ideas from entrepreneurs across more than 150 countries. An independent panel selected five winners, who each received $100,000 to develop their ideas.[4]

Some companies use both ideation and design contests. For example, Electrolux's DesignLab was established in 2003 as an annual global design competition open to undergraduate and graduate industrial design students, centered on a specific annual theme around sustainable appliances for the future. A panel of employees and external high-level designers select the winner, who receives 5,000 Euros and a six-month paid internship. In 2016, Electrolux created Ideas Lab, which focused on an ideation contest open to all individuals 18 years or older. In 2016, Motorola launched a contest (in collaboration with Indiegogo, Lenovo, and Verizon) for the next generation of Moto Mods apps. It received more than 700 submissions from 55 countries and 30 states and awarded ten grand prize winners.[5]

One of the critical advantages of innovation contests is access to many externally generated ideas, and based on the reward for winning, the ideas can be of higher quality. Innovation contests can also be beneficial in building brand awareness and excitement. For instance, GE's efforts have helped elevate its reputation in environmental sustainability. In a mature consumer packaged goods category where many new products are based on new flavors, textures, or package sizes, Lay's used a contest to reach out to consumers for their ideas for new Lay's potato chip flavors. While the company and a few outside experts helped narrow the ideas, the company developed the prototypes, and the crowd selected the winning flavors. Lays' engagement with consumers helped spark new energy into the brand, generated high social media buzz, and increased sales by 8% in the three months following the competition.[6] Innovation contests work well with a passionate consumer base yet must be publicized to attract a wider audience and more diverse submissions. Moreover, it is essential that the rules for the contest, and the conditions of the winning solutions, are explicitly identified.

15.4.6 Innovation Intermediaries

Innovation intermediaries act as knowledge brokers in the innovation process, connecting companies to external specialists who provide ideas and design solutions.[7] When using an innovation intermediary, a company must have a specific innovation problem it needs to solve. A key benefit is that the company can remain anonymous or reveal its identity. Using an online technological platform (e.g., InnoCentive, NineSigma, yet2.com, HYVE AG), the company will need to specify the type of challenge, the reward, and the period for possible solutions from the expert-based crowd on the intermediary's platform. For example, Innocentive posts a range of challenges that individuals can attempt to solve, including ideation challenges, reduction-to-practice challenges, and grand challenges.

One of the main reasons companies use innovation intermediaries is to reach a wide array of experts with knowledge of analogous domains or industries that can benefit the company's innovation process. In addition to access to experts worldwide, benefits include anonymity, speed, and lower cost. However, the primary disadvantage is that the solutions offered may not be helpful because they are not practical or do not align with the company's resources, capabilities, or strategies. Therefore, there is a risk that the solutions will not be feasible.

15.4.7 Open-Sourcing

While early ideas and implementations of open innovation took shape in open-source software (von Hippel and von Krogh 2003; von Krogh and von Hippel 2006), more recent developments have broadened its scope beyond software, and open-sourcing continues to remain a highly relevant and important context. First, software code, by its very nature, is distributed and can be shared widely with multiple constituents who can edit and then make these changes available to others at little or no cost. Thus, creating open innovation practices at scale is relatively efficient. Second, open-source software licensing allows the development of future products without the threat of IP violations. Early open-source tools such as GNU, python, Linux, and MapReduce have enabled the development of subsequent ideas such as P2P technology, distributed computing, and extensive data analysis, to name a few. Emerging Web 3.0 technologies such as blockchain, NFTs (non-fungible tokens), and DAOs (decentralized autonomous organizations) owe their existence to prior open-source innovations.[8] As more technology-driven business models emerge, open-source innovation remains the defacto platform. While open-sourcing can unleash the power of users in developing code, companies must recognize that all derivative work must remain open-sourced. Some companies, such as Tesla, have embraced this idea,[9] whereas others, such as Apple, have thrived on building closed proprietary systems to safeguard their IP.

15.4.8 Prediction Markets

Prediction markets are a form of crowdsourced forecasting, where participants predict outcomes for various topics. Prediction markets can be managed publicly, where anyone can participate in predicting an outcome through open platforms, such as PredictIt (predictit.org), PredictWise (predictwise.com), and HSX (HSX.com). Another option is to rely on a consulting firm that can create a prediction market solely for the company's employees (i.e., an internal prediction market) or one that also allows for external participation. Many companies have used prediction markets (e.g., Best Buy, Corning, Frito Lay, Google, Intel, Microsoft, Motorola, Pfizer[10]) for various innovation-related decisions. For example, Ford Motor Company used prediction markets for future product planning and restructured their trim packages based on results from the prediction market.

Prediction markets are helpful when there are many independent traders, when traders have dispersed private information about events that they are not otherwise willing to share, and when events are measurable, verifiable, and in the not-too-distant future. Prediction markets generate accurate estimates of future events when designed according to five principles: independence, incentive, indicator, improvement, and a crowd (i.e., I4C, see Ho and Chen 2007). For example, Intel's prediction markets for product demand are 20% more accurate than their other means of forecasting. Some companies have been hesitant to use prediction markets because they remain a mathematical mystery, and managers fear a loss of authority. Although skeptics expect the potential for employees in an internal prediction market to have ulterior motives or biases, research shows the inefficiencies tend to be optimistic, with efficiency improvements over time (Cowgill and Zitzewitz 2015).

15.5 A Roadmap for Managers

Open innovation has attracted interest from companies across various industries. It offers the opportunity to improve the innovation process while finding ways to economize on costs. For companies searching for new avenues to bring in fresh perspectives or are new to engaging in open innovation, there are many options to consider. Across the numerous open innovation tools available, it is important to recognize that each approach offers value and will be differentially valuable at different stages within the innovation process. We offer some guidelines for managers as they navigate the myriad of options.

- Regardless of the open innovation tool, explicit criteria, guidelines, and rules are necessary to attenuate concerns over IP ownership and mitigate potential negative attitudes toward the company when specific stakeholders or partners are not selected.
- Companies that are beginning to consider open innovation need to be cognizant that the costs involved in monitoring and coordinating with outside entities can escalate exponentially based on the innovation tool. For example, managing collaborative communities can be challenging, particularly when consensus outcomes on design challenges are desired. One alternative could

be to establish an in-house team that coordinates such ventures. Similarly, open-sourcing can expand the company's knowledge base but lead to a partial loss of control over intellectual property. In contrast, innovation intermediaries offer a less costly approach. However, the quality of submissions may require a company to revisit the innovation problem if the solution does not align with the company's capabilities and strategies. We recommend that firms choose a particular type after considering the pros and cons of the approach (Table 15.1).

- We recommend starting in smaller increments for a company that has not yet begun to use open innovation. Use one approach for one brand, and then measure and assess outcomes before moving to different approaches and brands. This strategy is not only less risky but also easier to attribute any increase in customer satisfaction, sales, or brand buzz to the specific open innovation approach.

- The risk and costs involved in successfully executing open innovation initiatives tend to increase as a project progresses through the stages of the innovation process. Therefore, we recommend that companies start with open innovation initiatives in the early stages of new product development, that is, ideation and design, before venturing into such endeavors in commercialization.

- If a company has a strong base of passionate consumers, this can be leveraged with innovation contests, a corporate co-creation platform, or a collaborative community. However, it is important to ensure that consumers' attention and interests are not taxed, or it has the potential to harm the brand.

- As a particular open innovation tool becomes more prevalent across companies and industries, a company should search for new ways to stand out from competitors.

When determining whom to involve, consider the degree of involvement; a vital dimension determining the viability of open innovation tools for any given company is its place in the industry value chain (Adner and Kapoor 2010). Although open innovation may involve governance mechanisms that are not traditional in the hierarchy, it may involve increased costs associated with coordinating and monitoring stakeholders with whom contractual terms may not tightly bind as traditional. Furthermore, specific stakeholders in the value chain may be more conducive to engaging in open innovation. For example, partners downstream in the value chain much closer to the customer might be excellent sources of ideation because they have better knowledge of customers. Still, they may not be appropriate partners for design and development. Finally, initiatives that garner initial success may not offer the long-term success desired. Paying close attention to the mechanics of how to design open innovation initiatives in alignment with objectives, along with the willingness to make changes to the innovation tools used over time, might be warranted as more models of innovation emerge.

15.6 Conclusion

As more companies turn to open innovation, it is crucial to be cognizant of consumers, competitors, and the marketplace. If the company has a passionate consumer base, it can leverage this with different open innovation approaches that reinforce how much it values its consumers. At the same time, it is important not to overwhelm consumers or take their involvement for granted with a lack of corporate interaction. It is also essential for a company to be aware of the forms of open innovation used in its industry and look for something that allows it to stand out. Although open innovation offers many advantages, internal innovation remains critical to provide the foundational knowledge to ascertain the value of external knowledge, provide the resources and capabilities to enable the company to become a partner of choice, and help provide a path for differentiation.

Notes

1 http://www.starbucksmelody.com/2017/05/31/starbucks-nixes-mystarbucksidea-community-can-still-submit-ideas.
2 https://www.reuters.com/article/us-blockchain-crowdfunding-idUKKCN0Y82LI?edition-redirect=uk.
3 https://www.ge.com/news/reports/ecomagination-ten-years-later-proving-efficiency-economics-go-hand-hand.
4 https://www.reliableplant.com/Read/27481/GE-winners-ecomagination-challenge.
5 https://www.crowdfundinsider.com/2017/02/95919-motorola-announces-indiegogo-collaborated-transform-smartphone-challenge-finalists-selected.
6 https://digital.hbs.edu/platform-digit/submission/lays-increases-sales-by-asking-customers-to-do-us-a-flavor.
7 Although intermediaries could act as venture capitalists and seek ready-made products (Nambisan and Sawhney 2007), our focus is on how intermediaries can scout for raw or more fully developed ideas and designs.
8 https://www.forbes.com/sites/forbestechcouncil/2022/02/22/the-internet-is-evolving-why-you-should-care-/?sh=71468a686bff.
9 https://www.tesla.com/legal/additional-resources#patent-pledge.
10 "The Rise of Social Forecasting" CrowdWorx Working Paper Series. 2012.

References

Adner, R. and Kapoor, R. (2010). Value creation in innovation ecosystems: how the structure of technological interdependence affects firm performance in new technology generations. *Strategic Management Journal* 31 (3): 306–333.

Bucklin, L.P. and Sengupta, S. (1993). Organizing successful co-marketing alliances. *Journal of Marketing* 57 (2): 32–46.

Chang, W. and Taylor, S.A. (2016). The effectiveness of customer participation in new product development: a meta-analysis. *Journal of Marketing* 80 (1): 47–64.

Chen, P.-Y. and Forman, C. (2006). Can vendors influence switching costs and compatibility in an environment with open standards? *MIS Quarterly* 30 (August): 541–562.

Chesbrough, H.W. (2003). *Open Innovation: The New Imperative for Creating and Profiting from Technology.* Harvard Business Press.

Cooper, R.G. and Kleinschmidt, E.J. (1995). Benchmarking the firm's critical success factors in new product development. *Journal of Product Innovation Management* 12 (5): 374–391.

Coviello, N.E. and Joseph, R.M. (2012). Creating major innovations with customers: insights from small and young technology firms. *Journal of Marketing* 76 (6): 87–104.

Cowgill, B. and Zitzewitz, E. (2015). Corporate prediction markets: evidence from google, ford, and firm x. *The Review of Economic Studies* 82 (4): 1309–1341.

Franke, N., Keinz, P., and Steger, C.J. (2009). Testing the value of customization: when do customers really prefer products tailored to their preferences? *Journal of Marketing* 73 (5): 103–121.

Franke, N., Schreier, M., and Kaiser, U. (2010). The "I designed it myself" effect in mass customization. *Management Science* 56 (1): 125–140.

Griffin, A. and Hauser, J.R. (1996). Integrating R&D and marketing: a review and analysis of the literature. *Journal of Product Innovation Management* 13 (3): 191–215.

Ho, T.-H. and Chen, K.-Y. (2007). New product blockbusters: the magic and science of prediction markets. *California Management Review* 50 (1): 144–158.

Hofstetter, R., Aryobsei, S., and Herrmann, A. (2018). Should you really produce what consumers like online? Empirical evidence for reciprocal voting in open innovation contests. *Journal of Product Innovation Management* 35 (2): 209–229.

King, A. and Lakhani, K.R. (2013). Using open innovation to identify the best ideas. *MIT Sloan Management Review* 55 (1): 41–48.

Lerner, J. and Tirole, J. (2004). Efficient patent pools. *American Economic Review* 94 (3): 691–711.

Lilien, G.L., Morrison, P.D., Searls, K. et al. (2002). Performance assessment of the lead user idea-generation process for new product development. *Management Science* 48 (8): 1042–1059.

Luo, X., Rindfleisch, A., and Tse, D.K. (2007). Working with rivals: the impact of competitor alliances on financial performance. *Journal of Marketing Research* 44 (1): 73–83.

Mallapragada, G., Grewal, R., and Lilien, G. (2012). User-generated open source products: founder's social capital and time to product release. *Marketing Science* 31 (3): 474–492.

Mallapragada, G. and Srinivasan, R. (2017). Innovativeness as an unintended outcome of franchising: insights from restaurant chains. *Decision Sciences* 48 (6): 1164–1197.

Moorman, C. and Slotegraaf, R.J. (1999). The contingency value of complementary capabilities in product development. *Journal of Marketing Research* 36 (2): 239–257.

Moreau, C.P. and Herd, K.B. (2010). To each his own? How comparisons with others influence consumers' evaluations of their self-designed products. *Journal of Consumer Research* 36 (5): 806–819.

Nambisan, S. and Sawhney, M. (2007). A buyer's guide to the innovation bazaar. *Harvard Business Review* 85 (6): 109.

Nerkar, A. and Shane, S. (2007). Determinants of invention commercialization: an empirical examination of academically sourced inventions. *Strategic Management Journal* 28 (11): 1155–1166.

Ordanini, A., Miceli, L., Pizzetti, M., and Parasuraman, A. (2011). Crowd-funding: transforming customers into investors through innovative service platforms. *Journal of Service Management* 22 (4): 443–470.

Poetz, M.K. and Schreier, M. (2012). The value of crowdsourcing: can users really compete with professionals in generating new product ideas? *Journal of Product Innovation Management* 29 (2): 245–256.

Ransbotham, S., Kane, G.C., and Lurie, N.H. (2012). Network characteristics and the value of collaborative user-generated content. *Marketing Science* 31 (3): 387–405.

Raymond, E.S. (1997). *The Cathedral and the Bazaar.* O-Reilly Media.

Rindfleisch, A. and Moorman, C. (2001). The acquisition and utilization of information in new product alliances: a strength-of-ties perspective. *Journal of Marketing* 65 (2): 1–18.

Rindfleisch, A. and Moorman, C. (2003). Interfirm cooperation and customer orientation. *Journal of Marketing Research* 40 (4): 421–436.

Slotegraaf, R.J. (2012). Keep the door open: innovating toward a more sustainable future. *Journal of Product Innovation Management* 29 (3): 349–351.

Stallman, R. (2002). *Free Software, Free Society: Selected Essays of Richard M. Stallman.* Lulu.com.

Timoshenko, A. and Hauser, J.R. (2019). Identifying customer needs from user-generated content. *Marketing Science* 38 (1): 1–20.

Urban, G.L. and von Hippel, E. (1988). Lead user analyses for the development of new industrial products. *Management Science* 34 (5): 569–582.

von Hippel, E. and von Krogh, G. (2003). Open source software and the 'private-collective' innovation model: issues for organization science. *Organization Science* 14 (2): 209–223.

von Krogh, G. and von Hippel, E. (2006). The promise of research on open source software. *Management Science* 52 (7): 975–983.

Wu, Q., Luo, X., Slotegraaf, R.J., and Aspara, J. (2015). Sleeping with competitors: the impact of NPD phases on stock market reactions to horizontal collaboration. *Journal of the Academy of Marketing Science* 43 (4): 490–511.

Rebecca J. Slotegraaf is the Associate Dean for Research and Neal Gilliatt Chair and Professor in Marketing at the Kelley School of Business, Indiana University. Rebecca's research focuses on new product introduction and design, brand strength, alliances, environmental sustainability, and the financial value of marketing actions. Her research can be found in leading journals, including *Journal of Marketing Research, Journal of Marketing, Journal of the Academy of Marketing Science,* the *Journal of Product Innovation Management,* and others. She is an Associate Editor for two leading journals and serves on the editorial review boards for five other journals.

Girish Mallapragada is Weimer Faculty Fellow and Associate professor of Marketing at Kelley School of Business, Indiana University. Girish's research interests include open innovation, new product development, and inter-firm relationships. His research has appeared in premier journals in Marketing including *Marketing Science* and *Journal of Marketing,* among others. He is an Associate Editor at the *Journal of Marketing Research* and an editorial review board member *Journal of Marketing* and *Journal of the Academy of Marketing Science.* He teaches the Kelley Direct Marketing Core at the Kelley School of Business.

CHAPTER SIXTEEN

HOW TO LEVERAGE THE RIGHT USERS AT THE RIGHT TIME WITHIN USER-CENTRIC INNOVATION PROCESSES

Andrea Wöhrl, Sophia Korte, Michael Bartl, Volker Bilgram, and Alexander Brem

16.1 Introduction

User-centricity has become a key imperative in innovation, with many companies emphasizing this principle in their mission and vision statements. In order to successfully create user-centric products and services, it is crucial to integrate users into the innovation process. This article provides innovation managers with guidance on the how and when of user integration, focusing on identifying and motivating the right users at the right time within the innovation process. Based on a literature review and four case studies of user-centric innovation projects, we introduce a framework for user integration.

Empirical and practical evidence demonstrates that users can be a valuable source of innovation at all stages of the innovation process. However, not all users are equally suited for each phase of the innovation process. Depending on the task to be tackled in the innovation project, innovation managers need to recruit and activate different user types and attribute different roles to them in the innovation process (e.g. for idea generation or idea testing) (Frow et al. 2015; Fuchs and Schreier 2011). For example, an average consumer may be overtaxed by the task of evaluating highly innovative ideas as they might require advanced technical knowledge or the capability to rethink established product experiences. Consequently, those users might fail to gauge the future potential of these ideas.

The PDMA Handbook of Innovation and New Product Development, Fourth Edition. Edited by Ludwig Bstieler and Charles H. Noble.

Therefore, companies need to carefully decide which type of user they want to include in their innovation endeavors. Otherwise, they might miss out on good opportunities by asking the "wrong" user type for feedback. The same is true for the generation of ideas. Average consumers may be the right users to create incremental new ideas that build on existing solutions and improve certain aspects of the product. To create ideas that entirely redefine the way we understand and cope with problems, however, companies may want to rely on users with extraordinary skills or extreme unfulfilled needs such as so-called Lead Users.

While the involvement of users has been extensively covered in co-creation and open innovation literature as well as in design thinking (e.g. Randhawa et al. 2021), lean startup (e.g. Scheuenstuhl et al. 2021) and agile project management literature (e.g. Yordanova 2021), there is a lack of research that provides comprehensive insights into the matching of user types and innovation tasks. Currently, user-centricity is primarily described as a "one-fits-all" solution and neglects the importance of selecting the right users for a specific task (Nambisan and Nambisan 2008; Piller et al. 2012). For example, past research on Lead Users, which promise more radical and out-of-the-box ideas than more conventional users, does not specify when it is best to include this extreme type of user in the innovation process. Instead, Lead Users are treated as a universal source of innovation suitable for all stages of the innovation process (Brem et al. 2018). To the knowledge of the authors, no study so far has provided a comparative overview of user types and corresponding innovation tasks.

As each stage of the innovation process poses distinct challenges for innovation practitioners, it is relevant to know which type of user can best support each stage of the innovation process. For this reason, we will provide an overview of user types in the following specifying the phases where these users can best be integrated. In addition, we will exemplify key user concepts in four innovation cases highlighting the who and how of user involvement in various phases of the innovation process.

16.2 What We Can Learn from Users

16.2.1 Relevance

The original notion that users can be a source of innovation from von Hippel (1988) has officially come of age. Research and practice alike have recognized that users are a valuable source of innovation, able to outdo professional producers (Pötz and Schreier 2012). Leveraging the potential provided by users can result in products and services that are novel and highly user-oriented making them more attractive to consumers and, thus, increasing commercial success compared to products solely developed by internal company resources (Bilgram et al. 2011; Pötz and Schreier 2012). When a company succeeds in integrating suitable users into their innovation processes, they can possibly overcome several key issues along the innovation process: they can deal with information stickiness, solve their own functional fixedness, reduce innovation uncertainties and improve innovation results (Schuurman et al. 2011). One of the striking advantages of harnessing users, which makes it superior to traditional ways of innovating under certain circumstances, is the diversity of individuals (Boudreau 2012). In contrast to traditional market research, co-creation does not only focus on companies' existing target groups matching predefined socio-demographics (Brem and Bilgram 2015). Instead, individuals in the periphery that distinguish themselves with high intrinsic motivation, special skills, extensive knowledge and capabilities conducive to the company's goals become valuable partners in co-creation (von Hippel 2005).

More and more companies have started to tap into the creative potential of users by applying agile and iterative development methods such as *Design Thinking*, *SCRUM* or the *Lean Startup* method (Paluch et al. 2020). A key principle these approaches have in common is the user-centricity that guides innovation efforts. For example, one of the core pillars of Design Thinking, besides technological feasibility and viability in terms of business strategy, is desirability which emphasizes the relevance of meeting user needs when coming up with new products or services (Brown 2008). By immersing yourself in the consumer world and by actively working with users, companies are more likely to design products and services that match consumers' desires and, thus, guarantee commercial success. Similarly, the Lean Startup approach relies on the "build-measure-learn" cycle, which is designed to provide validated learning by including potential users in the development process (Ries 2011). The SCRUM software development framework applies a similar principle with its sprints. These fixed-length iterations allow project teams to gather fast feedback from tests that can be done with potential customers or users before moving on to the next sprint and, hence, the consecutive stage of the development process (Takeuchi and Nonaka 1986).

However, the sheer plurality of user types regarding their personal features, expectations, motives, personality traits, backgrounds and origins has become a challenge for companies and requires sophisticated recruitment and management of participants as well as a corresponding co-creation design and empowerment strategy (Frow et al. 2015; Füller 2010). Nevertheless, research in this domain has rather neglected the diversity of user types and only focused on single user concepts such as the popular Lead User concept.

Researchers have argued that individuals may play different roles in different phases of the innovation process (Nambisan and Nambisan 2008; Piller et al. 2012). For instance, the ideal profile of individuals invited to collaboratively generate ideas may be different from one tailored to finding specific solutions to a problem or promoting a new product (Stock et al. 2015). While generating ideas may require in-depth experiential knowledge of the product usage and consumer needs, advocating a product rather demands brand fans with a large network capable of reaching and influencing consumers in an authentic way (Hurmelinna-Laukkanen et al. 2021). As companies' challenges and corresponding requirements for co-creation participants vary widely, the "ideal" members of the crowd need to be reached, attracted and activated (Prpic et al. 2015). Prpic et al. (2015) argue, that participants need to be carefully defined and recruited in light of the specific innovation challenge taking into account size and diversity as well as knowledge and experience. For instance, Frey et al. (2011) guide attention to the question of how companies should compose the group of participants from a motivation and knowledge perspective.

There is a plethora of terminologies to describe individuals involved in the innovation process in both literature and practice. The potential collaboration partners might be described as "customers," "consumers," "users," or "experts by experience," to name a few (McLaughlin 2009). By these definitions, the terms customer, consumer, and user are mutually exclusive but collectively describe the role of agents taking part in the innovation process (Rosenzweig et al. 2015). For the sake of this article, we will use the umbrella term *user* to describe different consumer types.

16.2.2 User-centricity in All Stages of the Innovation Process

The rise of the user-centricity imperative can be attributed to the constantly changing environment today, which requires a rigorous focus on the user and agile processes allowing for learning and pivoting. At the same time, new digital technologies can be used to easily connect to and collaborate with users at all stages of the innovation process as briefly outlined in the following.

16.2.2.1 Exploration. The key goal of the exploration phase in the (fuzzy) front-end of innovation is to discover new avenues for innovation. Potential for innovation can best be found by identifying the needs of consumers that are not yet matched by current offerings in the market. By taking a user-centric approach and by actively integrating users at this stage, innovation managers can gather information on what users actually want and what pain points they are currently facing. This is particularly crucial as research has found that most new product failures are attributed to a firm's inability to accurately assess and satisfy user needs (Ogawa and Piller 2006).

16.2.2.2 Ideation. Users cannot only provide need information, but they can also support ideation (von Hippel 2005). They can either actively come up with creative ideas that match their needs or they can support companies with the selection of promising ideas. Research has shown that generating a large number of high-quality ideas is important for the success of the process since the first set of ideas determine the sample from which winning ideas and concepts can arise (Rosenzweig et al. 2015). By including users in this process, the number of ideas can be increased. Moreover, leveraging users for idea generation can result in products and services that are novel and highly customer-oriented (Nishikawa et al. 2013). Also, there is a high potential for combining internal and external lead users (Schweisfurth 2017).

16.2.2.3 Creation. While the ideation phase concentrates on generating initial ideas with rather limited solution information, the creation of prototypes or minimum viable products (MVPs) aims to find a feasible solution to realize the idea and address the user need (Frederiksen and Brem 2017). Surprisingly, in some industries as well as in households, the majority of innovations are invented and prototyped by users as opposed to companies (von Hippel et al. 2012). Research has found that transferring concepts into prototypes through means of co-creation with consumers can lead to productivity gains through increased efficiency and effectiveness as well as to enhanced innovativeness of the product or service and to a better fit with user needs (Hoyer et al. 2010).

16.2.2.4 Testing. Finally, users can be involved in the testing stage of the innovation process, which represents one of the early forms of customer integration known from traditional market research. This stage reoccurs in agile innovation approaches and is crucial to collect feedback on iterated product versions and ensure the ideal fit with consumers. Research has demonstrated that iterative testing helps managers detect ill-conceived features and aspects at an early stage (Hoyer et al. 2010).

16.3 How to Identify the Best User for Your Project

Based on a literature review, we identified the most frequently cited user types in academic papers between 1980 and 2018. Literature on Open Innovation and user involvement shows a wide variety of different "user types." To select the most relevant concepts for further analysis, a database search, which considers the number of publications mentioning the concept either in the title or abstract, was conducted. To ensure a thematic focus on Open Innovation, results were filtered by the innovation subject, to avoid a bias of marketing-related terms (e.g., Opinion Leader or Dissatisfied Customer).

In our article, we want to concentrate on the five "most popular" user concepts identified in the literature review. The user types we investigate in detail comprise Lead Users, Early Adopters, Expert Users, Ordinary Users, and Creative Consumers. Despite the popularity of the term "prosumer" in academia, we exclude this particular user type from our analysis as the term is mostly used to contrast the original consumer concept (i.e. the obsolete concept of consumers who only consume the products producers sell them) with the "new" understanding of consumers who also incur activities originally subject to producers only (e.g. providing feedback, modifying products or creating their own ones). Thus, the concept of prosumers is typically not used to define a potential partner and co-creator in the innovation process but rather as an umbrella term to describe the entirety of users that can be involved for innovation purposes. We commence our case studies with a short review and description of each user type.

16.3.1 Lead User

One of the most dominant user concepts discussed in scientific research and literature is the concept of Lead Users. *Lead Users* are users who face certain needs earlier than other users and who expect to profit from solving these needs (Brem et al. 2018; von Hippel 2005). Lead users can be found in B2B or B2C settings (Herstatt and von Hippel 1992; Hienerth et al. 2014).

Based on von Hippel's research (1986) there are two main Lead User characteristics: *First*, Lead Users have a strong personal need and are ahead of the target market. These needs are not yet met by existing products or services in the market but have the potential to become relevant later on. Therefore, Lead Users are an important source of innovation as they are ahead of trends (Bilgram et al. 2008;

Franke et al. 2006). Companies can use the potential offered by Lead Users to generate new products and services that will cover future mainstream needs of the market (months or years later). *Second,* Lead Users are motivated to create solutions for the problems they are facing as they expect high benefits from a solution that meets their advanced and domain-specific needs (Franke et al. 2006). Studies on industrial product and process innovations have shown, that the greater the benefit an individual expects to obtain from a needed innovation, the greater the investment will be in obtaining a solution (Urban and von Hippel 1988). This characteristic, thus, can serve as an indicator of innovation likelihood. Innovation managers should bear in mind these characteristics when starting new innovation initiatives. Lead users are well prepared to generate innovations that substantially differ from existing market offers as they face needs that are not yet covered by existing products and services. For this reason, it is recommendable to integrate or monitor Lead Users in the very early stages of the innovation process, that is, in the *exploration* phase as they offer important insights into future needs and demands facing the market. In this context, it is also important to differentiate Lead Users, with needs at the leading edge of the market, from *Extreme Users.* Extreme users might also face strong needs, but it remains unsure if their needs will ever become general for the market (Shih and Venkatesh 2004).

16.3.2 Early Adopters

Another type of users that is often discussed in scientific research as well as in practice are *Early Adopters.* This user classification stems from the model of innovation diffusion introduced by Everett Rogers in the early 1960s (Rogers 1962). According to this model, different groups of consumers adopt new technologies differently (Rogers 2003). Early Adopters start using new technologies after Innovators but they purchase new products soon after launch and well before the average client (Early and Late Majority) (Frattini et al. 2014). Studies on characteristics of Early Adopters found that, compared to others in the social system, Early Adopters tend to be better educated, younger, upwardly mobile, and of higher socioeconomic status. As they can afford to take risks associated with new products or services, they are more likely to adopt a new product or service earlier than other groups of consumers (Brancaleone and Gountas 2007). Furthermore, Early Adopters possess a higher understanding of complex technical product attributes and their use implications, which appears to be crucial for the adoption process (Schreier and Prügl 2008).

Similar to Opinion Leaders, which are likely to influence other persons in their immediate environment, by giving advice and verbal directions for the search, purchase, and use of a product (Kratzer and Lettl 2009). Early Adopters are high information seekers relying especially on word-of-mouth communication, which is considered to have an important role in the adoption of disruptive innovations (Zhao et al. 2021). Their involvement and curiosity motivate them to acquire knowledge about innovative offerings (Clark and Goldsmith 2006). Furthermore, they tend to share gathered information with other members of their local reference groups (Brancaleone and Gountas 2007). As Early Adopters are among the first and very keen to try out new products and services, these types of users can best be involved in the innovation process in the *testing* phase. They can deliver important insights into the first type of prototypes that are relevant for the acceptance of the average consumer, creating a "market before the market" (Brem et al. 2019).

16.3.3 Ordinary Users

Potential for innovation opportunities cannot only be found among groups of more advanced consumers (e.g. Lead Users, Early Adopters) but also among the more average consumers. In the following, we will call this group of users the *Ordinary Users*. Ordinary Users are at the center of most markets. They do not have strong or urgent needs that need to be solved or met by new products and services, and they are also not necessarily dissatisfied with the current offering. Ordinary Users, furthermore, have a "normal" understanding of most products and services without any specific technical knowledge and use these products on a regular basis (Gemser and Perks 2015). Ordinary Users might not seem like the first choice when it comes to innovation because of their limited ability to contribute innovative ideas. Their perception is limited to their current use context, and the ability to conceptualize and verbalize their needs is restricted by their lack of expertise with respect to technical feasibility (Kristensson and Magnusson 2010). For this reason, there can be a risk that Ordinary Users' involvement might lead to the creation of ideas that are merely incremental. However, other studies have found that "too much" know-how regarding limitations of the underlying technology can result in ideas that are less innovative than those originating from users who do not possess this technical information and, thus, do not know of possible limitations (Magnusson et al. 2003).

When integrating this kind of user into innovation projects, one needs to be aware of both the benefits and risks that come with this group. Ordinary Users can be a valuable resource for innovation at the two ends of the innovation process – for *exploration* and for *testing*. When exploring new avenues for innovation, Ordinary Users can offer relevant insights into what mainstream consumers actually want and the current challenges and pain points they are facing. These unmatched needs could be a relevant starting point for innovation activities. In the testing stage, integrating Ordinary Users might be a good strategy as it allows the validation of the developed services and products. Ordinary Users are the key target audience of most companies. For this reason, these consumers can offer first insights into whether the developed service or product will reach mass market acceptance. In addition, research has shown that Ordinary Users can provide innovative ideas under the right circumstances. More specifically, this means when they can try it in a seemingly real context while at the same time not being restricted by "too much" technological information and limitations regarding potential feasibility (Kristensson and Magnusson 2010). These environmental requirements are best applicable in the testing stage of the innovation process.

16.3.4 Expert User

Expert Users, as the name already indicates, can be described as non-professional users that have the knowledge and skills to live up to professional standards (Schuurman et al. 2011). It is assumed that knowledgeable people will be able to make more accurate evaluations than less knowledgeable people (Ozer 2009). This is based on the premise that people with high product knowledge will be more able and willing to process necessary information, focus on the most relevant information and, hence, make more informed product assessments than people with less specific knowledge. The widespread use of expert opinions in the industry provides anecdotal support for this expectation (Ozer 2009). For this reason, the integration of Expert Users is particularly useful when creating new concepts or products. Co-creating with Expert Users can help companies, as the sound product-related knowledge of this consumer group allows for the performance of more complex technical tasks and assignments which are prevalent in the creation phase of the innovation process (Schuurman et al. 2011). Expert Users also show a high understanding of the product or service and expect high benefits from a new and innovative solution. Thus, they are also highly motivated to work on new products and services. In comparison to more need-oriented consumer groups mentioned before (e.g. Lead Users, Early Adopters), Expert Users are characterized by having more knowledge about performance attributes, different physical components of products, and any attribute–performance relationships (for the previous paragraph: Schreier and Prügl 2008). Therefore, they are the ideal partners for the co-creation of products and services.

Selecting and using the most suitable experts can help decision makers in making more informed and accurate product innovation decisions. This notion is supported by scientific research, which found that Experts Users are relevant collaboration partners when it comes to screening new product proposals and selecting which innovation projects to choose out of a given innovation pool (Ozer 2009). Furthermore, it was demonstrated that product expertise is positively related to the accuracy of concept and prototype testing (Ozer 2009).

16.3.5 Creative Consumer

Some users create radically new innovations from scratch, while others, known as *Creative Consumers*, mainly focus on existing product offerings. Berthon et al. (2007) define the Creative Consumer as an individual or group who adapts, modifies, or transforms a proprietary offering, such as a product or service. Even though researchers often relate the concept of Creative Consumers to the more popular construct of Lead Users, they are far from being synonymous (for the following list: Berthon et al. 2007):

- Creative Consumers work with all types of offerings, not only novel or enhanced products. In many cases, they work with old, or even with simplest, defeatured products.
- Creative Consumers do not necessarily face needs that will become general. Instead, they often work on personal interests that can remain personal or

expand in use to a subset of users. Moreover, Creative Consumers often innovate inspired by experimentation and creativity and do not intend to solve specific needs.

- Creative Consumers do not need to benefit directly from their innovations, although they benefit indirectly through peer recognition.
- Companies or researchers tend to use a formal and disciplined process to find, screen, and select Lead Users. In contrast, Creative Consumers innovate without the company being aware of it. This lack of control can represent major challenges to firms.

Even though they often might act in the "underground," the Creative Consumer concept is a concept that companies need to be aware of. As high-technological products become more and more digital and interconnected, the potential for consumers to reprogram, adapt, modify, and transform offerings increases (Berthon et al. 2007). Furthermore, Creative Consumers are a rich source of innovation. They possess imaginative ideas that a company might not have the resources or the time to cultivate itself (Berthon et al. 2007). The recognition and use of Creative Consumers for a given offering can be described as a form of outsourcing with potential cost benefits but also significant risks as the research and development are done in public (Leminen et al. 2014). As Creative Consumers are very likely to experiment and alter existing products and services and generally possess a high level of creativity, they are the ideal partner for ideation. This phase of the innovation process requires coming up with creative ideas for new products and services either from scratch or based on existing products. Creative Consumers are suited for both scenarios. However, in the latter case, Creative Consumers are particularly relevant, as they might have already thought of or worked on enhancing existing offers.

Table 16.1 shows the detailed framework combining both the innovation process perspective with the respective tasks at each step as well as the user profile and user type needed to fulfill these tasks. In the following section, we outline four examples from companies that applied the principle of user-centricity by selecting the right users at the right time within the innovation process.

16.4 Best Practices of User-Centricity from Four Companies

16.4.1 Exploration: Beiersdorf

The first case demonstrating how users-centricity can be achieved in the fuzzy-front end of innovation revolves around the brand NIVEA (Bilgram et al. 2011). NIVEA is the best-known brand of the multinational skin care corporation Beiersdorf based in Hamburg, Germany. The company is a worldwide research and development leader in the area of skin care with more than 130 years of experience. Innovation efforts at NIVEA have been predominantly technology-driven for a long time, harnessing superior internal R&D capabilities. For years innovation in the deodorant sector was dominated by the benefit of efficacy leading to absurd promises such as sweat protection for 96 hours that did not really offer an

TABLE 16.1 THE RIGHT USERS AT EACH STAGE OF THE INNOVATION PROCESS.

	Exploration	Ideation	Creation	Testing
Typical tasks in this phase	• Identification of innovation opportunities • Understanding of consumer needs, new technologies, and trends	• Coming up with ideas for new products and services • Screening and evaluating innovative concepts for new offerings	• Turning concepts into prototypes • Ensuring feasibility and desirability during product/service development	• Evaluating or refining concepts or prototypes to improve feasibility and consumer desirability
Required features and skills	• Dissatisfaction • Use experience • Empathy and sensitivity • Market knowledge • Technological know-how • Trend awareness • Transfer knowledge	• Creativity • Intrinsic motivation • Technological knowledge • Product knowledge • Cross-industry knowledge • Trend awareness • Blue-sky-thinking ability	• Innovation related benefits • Technological knowhow • Ability to interpret concepts • Doing-mentality • Expert knowledge • Entrepreneurial mindset • Prototyping skills	• Non-professional product knowledge • Representing target market • Use experience • Advanced consumer knowledge • Technological knowledge • Innovativeness
Generally suited user types*	• Dissatisfied User • **Ordinary Customer** • Emergent Nature Consumer • **Lead User**	• **Creative Consumer** • Emergent Nature Consumer • Non User	• **Expert User** • User Innovator • Prosumer • User Entrepreneur • Early Adopter	• **Ordinary User** • **Early Adopter** • Non User • Emergent Nature Consumer

*bold = featured in case study.

additional benefit to consumers. Therefore, the prototyping team at NIVEA decided to explore new innovation opportunities in other benefit dimensions from a consumer standpoint. As the goal of the innovation project was to gather insights in the exploration phase, the NIVEA innovation team decided to use an approach to dive deeper into the consumer world. At this stage of the innovation process, *Lead Users* and *Ordinary Users* were identified as relevant partners for co-creation activities. For this reason, the NIVEA team employed a holistic co-creation approach with these types of users. The identification and collaboration with Lead Users were crucial for this project, as these types of users face unmatched deodorant needs earlier than other types of users and can thus offer important insights into what the next type of deodorant should look like. In addition, the NIVEA team wanted to assess whether these consumer needs also apply to the mainstream market. Consequently, the integration of the voice of the Ordinary Users was equally relevant for the project as they offer relevant insights into the distribution of certain needs among a wider audience. In this specific case, NIVEA wanted to find out whether there are deodorant needs that apply to a large group of consumers. In order to leverage the benefits of both user types, *Netnography* was used to dive into the consumers' world. The goal was to draw a landscape of needs, wishes, concerns, consumer language, and potential product solutions by users, which are explicitly and implicitly expressed in online communities and social media. Netnography is a qualitative empathic research methodology utilizing adapted ethnographic research techniques to enable researchers to deeply immerse themselves in online consumer conversations (Bartl et al. 2009; Gutstein and Brem 2018; Kozinets 2002).

Due to the emergence of social media and user-generated content in recent years, the internet has become a relevant source of consumers' voices (Bartl et al. 2015). Consumers exchange personal experiences, concerns, and opinions on a variety of different topics, including products and brands in different online discussion forums or communities. Some users even go one step further and not only discuss their experiences online but also think about or build possible solutions either from scratch or by modifying existing solutions and products themselves. Through the passive and unobtrusive observation of online communities, forums, and other social media content, companies are able to gain unbiased consumer insights. Instead of directly asking and thus inevitably biasing the consumer's response, the netnography approach aims to understand the emotional, social, and cultural context of consumers' product experiences in a merely observant fashion. As opposed to more quantitative web monitoring approaches, listening to consumers rather than asking them, understanding rather than measuring consumers' attitudes and behaviors, are core principles of netnography. One of the greatest challenges for researchers conducting netnographies is to identify "diamonds in the rough" – the most relevant and inspiring insights in the abundance of online user statements. For that reason, a systematic process was conceived.

In the first step of the netnography process, NIVEA defined the key goals of the research, which was to discover new innovation opportunities from a consumer standpoint in the area of deodorants. Next, an initial set of keywords and phrases was conceived, which was continuously developed and refined in the course of the subsequent identification and selection of social media sources. A broad search to screen more than 200 social media sites covering all kinds of social media sources including forums, blogs, advice portals, question and answer sites, and social networks covering various topics (e.g. cosmetics, health, beauty, lifestyle, fashion, sports, and do-it-yourself) was employed. Applying qualitative as well as quantitative selection criteria, for example, the size and activity of communities or the quality of the conversations, the most relevant and insightful communities were observed and analyzed in-depth. In the next step, threads of consumer conversations were retrieved from the social media sites and observed. We systematically analyzed and clustered the content into different topics using software for qualitative data analysis. The netnography revealed that apart from the intensely discussed topics such as sweat and odour prevention, there was one pain point that re-appeared over and over: deodorant stains. Many users complained about unattractive stains on their clothes and suspected several factors to be responsible for this (e.g. nicotine consumption, hormonal fluctuations, or bacteria). They not only discussed pain points but also went as far as to think about possible solutions and workarounds to the problem. "The Undershirt Guy," a Lead User, experimented for example with baking soda or Aspirin in order to try to get rid of this problem. The vast amount of information on needs and solutions from both Lead and Ordinary Users as well as the breadth and depth of information from different communities provided enough evidence to the NIVEA team that deodorant stains were indeed a major concern for a broad mass of consumers. For this reason, NIVEA decided to pursue the idea of developing a deodorant that is specifically designed to target deodorant stains in clothes. When the final product, the "Invisible for Black & White" deodorant, was released, it was immediately well received by consumers, resulting in the most successful deodorant product launch in NIVEA's long brand history.

16.4.2 Ideation: Henkel

A good example of user integration in the ideation phase of the innovation process is the innovation contest the company Henkel did for their adhesive portfolio (Bilgram and Rapp 2013). Henkel is a German chemical and consumer goods company with global brands in the areas of Laundry & Home Care, Beauty Care, and Adhesive Technologies. In order to gather fresh, out-of-the-box ideas and designs for its adhesive portfolio, Henkel wanted to gather ideas on the next-generation adhesive packaging from an external crowd of innovative users. They not only wanted to generate rough ideas but rather detailed concepts with a higher degree of maturity. For this reason, Henkel decided to target *Creative Consumers* as they have creative skills and a knack for design. In addition, they like to experiment and are known for adapting, modifying, or transforming existing products.

In order to attract Creative Consumers to collaborate with Henkel, the company decided to crowdsource the task of creating innovative ideas and designs for consumer adhesive products by launching an innovation contest. Innovation contests utilize the knowledge and creativity of users to answer innovation challenges often on a temporary online platform (Boudreau and Lakhani 2013; Leimeister et al. 2009). They make use of four major principles. First, the crowdsourcing principle is applied to outsource a task or question to a large number of users in the periphery of the company. Second, innovation contests are organized as competitions which offer a reward to a few winners who "take it all." Third, innovation contests are community-based innovation tools which utilize community mechanisms, social ties, and a sense of community to provide an enjoyable collaborative experience (Wang et al. 2021). Fourth, the recruiting strategy takes advantage of the self-selection bias in order to address users with the most creative potential. The recruiting process is a huge success driver. As Creative Consumers are very proactive and experiment a lot with existing products, the innovation contest approach is a good tool to attract them to collaborate with companies. Creative Consumers were recruited by broadly promoting the contest within universities focusing on design, blogs within the design sphere as well as through social media by mobilizing design students and hobby designers. A prize of 5,000€ was offered as a reward to the best three designs to also cater to the extrinsic motivation of participants. Besides the recruiting strategy, the design of the platform was also crucial. It needed to optimally support the creative process while being user-friendly and fun at the same time. The platform encompasses community features such as profiles and commenting to encourage discussions and exchange between participants. To foster transparency, Henkel further used the platform to communicate with the participants regarding selection criteria and the selection of ideas made by a jury. In order to provide users with some guidance, Henkel suggested specific idea categories and underlying consumer needs. The contest explicitly asked for ideas in the fields of adhesive dosage and application, two-component adhesives and mixing as well as easy and convenient opening and closing of adhesive packaging. For more guidance and transparency, Henkel also outlined the key criteria for the selection of winners: sustainability, innovativeness, convenience, performance, and aesthetic appeal. Within six weeks, more than 1,000 users registered on the platform and submitted 385 ideas. Most of these users could be categorized as Creative Consumers. A large share of the ideas and designs were of outstanding quality and contained sophisticated technical details as well as design aspects. The best ideas submitted by users were integrated into Henkel's internal innovation process to be further elaborated. In addition to its primary goal of collecting fresh ideas, the contest also had significant effects on the user–brand relationship and the image of the company.

16.4.3 Creation: Pöttinger

Pöttinger is a medium-sized family business from Austria with more than 140 years of experience in the field of agricultural machinery. Today, the family business employs 1,370 people, and annual sales amount to 282€ million. A key challenge Pöttinger's engineers faced at the time of the project was to improve the uniformity of the cut grass within the cutting unit of the in-house haulage vehicle. A well-sorted alignment of the blades of grass facilitates a more consistent cut of the animal feed, and this correlates with an increase in the milk production of the cows. The challenge clearly demonstrated that high product knowledge and expertise were key requirements to solve this intricate problem. While ideation challenges focus on generating ideas which address relevant consumer needs, the creation phase emphasizes the actual solving of a problem. In other words, the creation phase specifies an idea, provides technical information, combines various aspects of ideas, and investigates the feasibility of the idea in greater detail. Consequently, *Expert Users* were chosen as collaboration partners. They possess sound product-related knowledge and are able to process relevant information quickly, allowing them to make informed assessments regarding new cutting technology. Mechanical engineers, designers, farmers, and students in these domains were among the participants that matched Pöttinger's search profile as they possessed the relevant detailed knowledge and problem-solving skills. Pöttinger used the HYVE Crowd, a community of motivated users with profound technical knowledge as a source for recruiting and additionally promoted the contest in relevant online communities and blogs. During the contest live phase, Pöttinger made sure to involve internal R&D and technology experts to moderate the community and provide feedback. The technical challenge and complexity of the problem required close cooperation between internal industry experts and external Expert Users to clarify technical requirements and improve initial ideas by adding domain and industry knowledge. Throughout the contest period, 574 people registered on the platform and submitted a total of 112 ideas. Encouraged by the continuous feedback from experts, the users generated 819 comments and 571 evaluations of the ideas. The jury selected three solution submissions which were awarded with 8,000€ in total. The jury praised the winning idea, "aligning crop by buckets," as being simple to realize while adding value to the current harvesting process. Given the elaboration and professionalism of the submitted ideas, Pöttinger further developed the three winning ideas. An extensive analysis and patent research were conducted for each idea, and the company started working on the prototypes of the ideas.

16.4.4 Testing: Vorwerk

In the testing stage of the innovation process, user integration can help improve the feasibility and consumer desirability, as the following case of the company Vorwerk demonstrates (Kröper et al. 2013). Vorwerk is an international company which is best known for household appliances, in particular the Thermomix. The Thermomix is a multi-functional food processor enabling consumers not only to mix ingredients but also to cook, knead, mix, beat, chop, grind, purée, weigh, and roast them. The "fuel" of the Thermomix usage are the special recipes which are specifically tailored to the Thermomix appliance. These recipes enable users to get the most out of the appliance. Over the years, an enormous variety of recipes has been generated both by the company as well as by users. In the Vorwerk Thermomix brand community, these recipes can be explored, shared, discussed, and downloaded. More than 600,000 users worldwide are registered in the brand's community. The search for recipes represents an important phase in users' cooking journey as they need guidance to select the recipe that fits their taste, the season, the situation, and other decision factors. Vorwerk's goal was to analyze user habits and identify different types of search behavior among users. In addition, Vorwerk wanted to collaboratively develop and collect feedback on a newly developed search function for the community. As the Thermomix brand community hosts different types of consumers, it was relevant to not only focus on integrating one distinct type of user. For the purpose of this project, both *Ordinary Users* and *Early Adopters* were identified as relevant sources for innovation but for different reasons. Early Adopters were chosen as co-creation partners for this project as the key goal was to develop a new search function with the users. Early Adopters were particularly suited for this task as they are not only more likely to adopt new products or services, but they are also better at understanding product and service attributes. Ordinary Users can also be relevant partners for testing as they represent the more average consumer, who is also present in the brand community. They can help validate the acceptance, performance, and usability of the new search functionality. Vorwerk Thermomix leveraged an online research community to involve its customers. Online research communities offer a selected number of consumers a closed online environment in which they interact and co-create with the company not only in "one-off" projects but over an extended period of time. The approach distinguishes itself by the flexibility and diversity of market research methodologies that can be applied. An agenda for each day comprising various questions and tasks helps to guide the collaboration. Accordingly, participants of this collaborative research platform are also rewarded differently. As opposed to the tournament principle in innovation contests, participants of a research community receive a small incentive as a "thank you" and compensation for their efforts. One of the key challenges at the beginning of a co-creation project is to identify and recruit suitable participants to collaborate with. Here, the Vorwerk Thermomix brand community served as a base for recruiting. A pre-screening questionnaire distributed to the members of the brand community was used to identify the right participants. As the goal was to recruit both Ordinary Users and Early Adopters, several selection criteria were used. An important indicator for both user types was the activity level within the brand community. Activities in the brand community comprised the upload and download of recipes, comments, and evaluations of recipes, and exchange of messages with other community members or friends. Usage data allowed the identification of Ordinary Users who engage with the platform in an average way. In addition, the activity information further supported the detection of Early Adopters. They showed considerably higher activity levels in all previously mentioned dimensions. For instance, a typical Early Adopter in the brand community uploaded approximately ten times as many recipes as the average Ordinary User. Due to the participants' strong identification with and passion for the brand, a non-monetary incentive closely related to the Thermomix world was chosen to compensate them for their efforts. Each participant could choose one of the Thermomix cookbooks which are highly coveted by Thermomix fans. The amount of data produced in the online research community over the duration of the research project was very rich with 1,540 posts. This would amount to a total number of 191,346 words which equals approximately 350 written A4 pages. Feedback on the future Vorwerk community was collected to eventually refine features and functions. The research team and software developers worked closely together during the whole research project and were able to directly transfer user feedback into the new community platform.

16.5 Conclusion and Key Takeaways

The cases demonstrate four examples of user-centric approaches by different companies (see Table 16.2 for an overview) clearly showing that there is no one-size-fits-all approach of involving users. All cases highlight the fact that choosing the right user is deeply intertwined with the task at hand and the respective

TABLE 16.2 OVERVIEW OF THE FOUR CASES SHOWCASING USER-CENTRICITY AT DIFFERENT STAGES OF THE INNOVATION PROCESS.

	Exploration	Ideation	Creation	Testing
Company	BEIERSDORF	HENKEL	PÖTTINGER	VORWERK
Topic	Next-generation deodorant	Next-generation adhesive packaging	Agricultural Engineering	Thermomix kitchen appliance
Method	Netnography	Idea Contest	Crowdsourcing Challenge	Online Research Community
User Type	Lead User Ordinary Consumer	Creative Consumer	Expert User	Ordinary User Early Adopter

phase in the innovation process. When trying to get the most out of the collaboration with different user types, innovation managers need to answer several questions. *First,* managers need to ask themselves *what* innovation tasks they intend to concentrate on in the current phase of the innovation process, that is, an exploration, ideation, creation, or testing task. Only if these are clearly specified, can managers answer the following questions. Next, it is important to define which skills and qualities are needed to help them perform these tasks. This step is crucial as it defines *who* is the best user type suited for the innovation project. Even within one phase in the innovation process, managers may be advised to involve different user types. For example, when companies want to learn about the problems of average customers today and at the same time understand what the future might look like, they could involve both Ordinary Consumers for today's perspective and Lead Users for a vision of future needs (see case 1). Once the target group has been defined, managers need to determine *where* to find these users and motivate them to participate. In the case of Ordinary Consumers, of course, the obvious choice for many companies might be to engage their own customers (e.g. via customer databases) or brand fans (e.g. via social media channels). These sources are usually "owned" by the company and, therefore, can be easily accessed and utilized for recruiting. Our cases, however, also show how users with special qualities can be involved in innovation projects. For instance, to solve technical problems that require specific domain knowledge and skills, managers may typically tap into communities of interest or professional networks (see case 3) for recruiting purposes. Moreover, innovation managers need to specify *how* they can best integrate the selected users (i.e. which method to apply) to utilize their skills most effectively. The cases provide insights into how other companies have spurred innovation by putting users at the center of innovation activities. The methods range from observant market research for exploratory purposes (case 1) to interactive research approaches for testing (case 4) and crowdsourcing-based initiatives for idea generation (case 2) or problem-solving (case 3). Table 16.3 below summarizes guiding questions for managers and gives tips and tricks on how to best design user-centric innovation projects that consider the innovation tasks, user

TABLE 16.3 KEY QUESTIONS FOR INNOVATION MANAGERS TO ANSWER PRIOR TO USER-CENTRIC INNOVATION PROJECTS.

	Key Questions	Tips and Tricks
WHAT?	What is the specific innovation task in the respective phase of the innovation process?	Specify the innovation task(s) along the innovation process: • *Exploration:* Understanding the technological feasibility and consumer desirability • *Ideation:* Developing value propositions that combine desirability and viability • *Creation:* Ensuring feasibility and desirability during product/service development • *Testing:* Evaluating and refining concepts or prototypes to improve feasibility and desirability
WHO?	Who has the skills and qualities to help you perform these tasks?	Match your requirements with the skill, knowledge, and motivation profile of the various user types. Go carefully through the description of each user type to decide which user can best solve your innovation tasks
WHERE?	Where can you find these users and how can you approach and motivate them?	Screen possible relevant blogs, online communities, and social media platforms or use customer databases and partnerships with universities or other institutions to look for the right users for your project
HOW?	How can you integrate the selected users to best utilize their skills for your purposes?	Determine prior to the project what kind of information you need from your selected user type. Do you require need information, that is, information on what consumers want or do you want to know how to realize a solution addressing that need, that is, solution information?

types, recruiting strategies, and methods. Finally, it will be interesting to see the future development of Lead User applications in the context of artificial intelligence. Apparently, with these emerging technologies, corporate management of innovation in general, but also specifically Lead User research and practice might be revolutionized by extending, complementing, or even substituting capabilities of human beings with AI-based platforms (Battisti et al. 2022; Brem et al. 2021; Füller et al. 2022; Kakatkar et al. 2020).

References

Bartl, M., Hück, S., and Ruppert, S. (2009). Netnography research: community insights in the cosmetics industry. *Conference Proceedings ESOMAR Consumer Insights 2009*, Dubai.

Bartl, M., Kumar, V., and Stockinger, H. (2015). A review and analysis of literature on netnography research. *International Journal of Technology Marketing* 11 (2), 2016.

Battisti, S., Agarwal, N., and Brem, A. (2022). Creating new tech entrepreneurs with digital platforms: meta-organizations for shared value in data-driven retail ecosystems. *Technological Forecasting and Social Change* 175: 121392.

Berthon, P.R., Pitt, L.F., McCarthy, I., and Kates, S.M. (2007). When customers get clever: managerial approaches to dealing with creative consumers. *Business Horizons* 50 (1): 39–47.

Bilgram, V., Bartl, M., and Biel, S. (2011). Getting closer to the consumer: how Nivea co-creates new products. *Marketing Review St. Gallen* 1: 34–40.

Bilgram, V., Brem, A., and Voigt, K.I. (2008). User-centric innovations in new product development: systematic identification of lead users harnessing interactive and collaborative online-tools. *International Journal of Innovation Management* 12 (3): 419–458.

Bilgram, V. and Rapp, M. (2013). Phänomen Crowdsourcing: was privater und öffentlicher Sektor voneinander lernen können. *Business + Innovation*, March 2013.

Boudreau, K.J. (2012). Let a thousand flowers bloom? An early look at large numbers of software app developers and patterns of innovation. *Organization Science* 23 (5): 1409–1427.

Boudreau, K.J. and Lakhani, K.R. (2013). Using the crowd as an innovation partner. *Harvard Business Review* 91 (4): 61–69.

Brancaleone, V. and Gountas, J. (2007). Personality characteristics of market mavens. *ACR North American Advances*, NA-34. http://acrwebsite.org/volumes/12859/volumes/v34/NA-34.

Brem, A. and Bilgram, V. (2015). The search for innovative partners in co-creation: identifying lead users in social media through netnography and crowdsourcing. *Journal of Engineering and Technology Management* 37: 40–51.

Brem, A., Bilgram, V., and Gutstein, A. (2018). Involving lead users in innovation: a structured summary of research on the lead user method. *International Journal of Innovation and Technology Management* 15 (3): 1850022.

Brem, A., Bilgram, V., and Marchuk, A. (2019). How crowdfunding platforms change the nature of user innovation–from problem solving to entrepreneurship. *Technological Forecasting and Social Change* 144: 348–360.

Brem, A., Giones, F., and Werle, M. (2021). The AI digital revolution in innovation: a conceptual framework of artificial intelligence technologies for the management of innovation. *IEEE Transactions on Engineering Management.* doi: 10.1109/TEM.2021.3109983.

Brown, T. (2008 June 1). Design thinking. Retrieved July 9, 2018, from https://hbr.org/2008/06/design-thinking.

Clark, R.A. and Goldsmith, R.E. (2006). Interpersonal influence and consumer innovativeness. *International Journal of Consumer Studies* 30 (1): 34–43.

Franke, N., von Hippel, E., and Schreier, M. (2006). Finding commercially attractive user innovations: a test of lead-user theory*. *Journal of Product Innovation Management* 23 (4): 301–315.

Frattini, F., Bianchi, M., Massis, A.D., and Sikimic, U. (2014). The role of early adopters in the diffusion of new products: differences between platform and nonplatform innovations. *Journal of Product Innovation Management* 31 (3): 466–488.

Frederiksen, D.L. and Brem, A. (2017). How do entrepreneurs think they create value? A scientific reflection of Eric Ries' Lean Startup approach. *International Entrepreneurship and Management Journal* 13 (1): 169–189.

Frey, K., Lüthje, C., and Haag, S. (2011). Whom should firms attract to open innovation platforms? The role of knowledge diversity and motivation. *Long Range Planning* 44 (5–6): 397–420.

Frow, P., Nenonen, S., Payne, A., and Storbacka, K. (2015). Managing co-creation design: a strategic approach to innovation. *British Journal of Management* 26 (3): 463–483.

Fuchs, C. and Schreier, M. (2011). Customer empowerment in new product development. *Journal of Product Innovation Management* 28 (1): 17–32.

Füller, J. (2010). Refining virtual co-creation from a consumer perspective. *California Management Review* 52 (2): 98–122.

Füller, J., Hutter, K., Wahl, J. et al. (2022). How AI revolutionizes innovation management–Perceptions and implementation preferences of AI-based innovators. *Technological Forecasting and Social Change* 178: 121598.

Gemser, G. and Perks, H. (2015). Co-creation with customers: an evolving innovation research field. *Journal of Product Innovation Management* 32 (5): 660–665.

Gutstein, A. and Brem, A. (2018). Lead user projects in practice—results from an analysis of an open innovation accelerator. *International Journal of Innovation and Technology Management* 15 (02): 1850015.

Herstatt, C. and von Hippel, E. (1992). From experience: developing new product concepts via the lead user method. *Journal of Product Innovation Management* 9 (3): 213–221.

Hienerth, C., von Hippel, E., and Berg Jensen, M. (2014). User community vs. producer innovation development efficiency: a first empirical study. *Research Policy* 43 (1): 190–201.

Hoyer, W.D., Chandy, R., Dorotic, M. et al. (2010). Consumer cocreation in new product development. *Journal of Service Research* 13 (3): 283–296.

Hurmelinna-Laukkanen, P., Nätti, S., and Pikkarainen, M. (2021). Orchestrating for lead user involvement in innovation networks. *Technovation* 108: 102326.

Kakatkar, C., Bilgram, V., and Füller, J. (2020). Innovation analytics: leveraging artificial intelligence in the innovation process. *Business Horizons* 63 (2): 171–181.

Kozinets, R.V. (2002). The field behind the screen: using netnography for marketing research in online communities. *Journal of Marketing Research* 39 (1): 61–72.

Kratzer, J. and Lettl, C. (2009). Distinctive roles of lead users and opinion leaders in the social networks of schoolchildren. *Journal of Consumer Research* 36 (4): 646–659.

Kristensson, P. and Magnusson, P.R. (2010). Tuning users' innovativeness during ideation. *Creativity and Innovation Management* 19 (2): 147–159.

Kröper, M., Bilgram, V., and Wehlig, R. (2013). Empowering members of a brand community to gain consumer insights and create new products: the case of the Vorwerk Thermomix Research Community. In: *Strategy and Communication for Innovation* (ed. N. Pfeffermann, T. Minshall, and L. Mortara), 415–426. Berlin, Heidelberg: Springer Berlin Heidelberg.

Leimeister, J.M., Huber, M., Bretschneider, U., and Krcmar, H. (2009). Leveraging crowdsourcing: activation-supporting components for IT-based ideas competition. *Journal of Management Information Systems* 26 (1): 197–224.

Leminen, S., Westerlund, M., and Nyström, A.G. (2014). On becoming creative consumers – user roles in living labs networks. *International Journal of Technology Marketing* 9 (1): 33.

Magnusson, P.R., Matthing, J., and Kristensson, P. (2003). Managing user involvement in service innovation. *Journal of Service Research: JSR* 6 (2): 111–124.

McLaughlin, H. (2009). What's in a name: 'client', 'patient', 'customer', 'consumer', 'expert by experience', 'service user'—what's next? *The British Journal of Social Work* 39 (6): 1101–1117.

Nambisan, S. and Nambisan, P. (2008). How to profit from a better 'virtual customer environment'. *MIT Sloan Management Review* 49 (3): 53–61.

Nishikawa, H., Schreier, M., and Ogawa, S. (2013). User-generated versus designer-generated products: a performance assessment at Muji. *International Journal of Research in Marketing* 30 (2): 160–167.

Ogawa, S. and Piller, F.T. (2006). Reducing the risks of new product development. *MIT Sloan Management Review* 47 (2): 65–71.

Ozer, M. (2009). The roles of product lead-users and product experts in new product evaluation. *Research Policy* 38 (8): 1340–1349.

Paluch, S., Antons, D., Brettel, M. et al. (2020). Stage-gate and agile development in the digital age: promises, perils, and boundary conditions. *Journal of Business Research* 110 (3): al495–501.

Piller, F.T., Vossen, A., and Ihl, C. (2012). From social media to social product development: the impact of social media on co-creation of innovation. *Die Unternehmung* 65 (1): 7–27.

Pötz, M.K. and Schreier, M. (2012). The value of crowdsourcing: can users really compete with professionals in generating new product ideas? *Journal of Product Innovation Management* 29 (2): 245–256.

Prpic, J., Shukla, P.P., Kietzmann, J.H., and McCarthy, I.P. (2015). How to work a crowd: developing crowd capital through crowdsourcing. *Business Horizon* 58 (1): 77–85.

Randhawa, K., Nikolova, N., Ahuja, S., and Schweitzer, J. (2021). Design thinking implementation for innovation: an organization's journey to ambidexterity. *Journal of Product Innovation Management* 38 (6): 668–700.

Ries, E. (2011). *The Lean Startup: How Today's Entrepreneurs Use Continuous Innovation to Create Radically Successful Businesses*. New York: Crown Business.

Rogers, E.M. (1962). *Diffusion of Innovations*. New York: Free Press of Glencoe.

Rogers, E.M. (2003). *Diffusion of Innovations*, 5e. New York: Free Press. Retrieved from Sample text http://catdir.loc.gov/catdir/enhancements/fy0641/2003049022-s.html.

Rosenzweig, S., Tellis, G.J., and Mazursky, D. (2015). *Where Does Innovation Start: With Customers, Users, or Inventors?* (Report summary No. 15–108). 2015 Marketing Science Institute.

Scheuenstuhl, F., Bican, P.M., and Brem, A. (2021). How can the lean startup approach improve the innovation process of established companies? An experimental approach. *International Journal of Innovation Management* 25 (03): 2150029.

Schreier, M. and Prügl, R. (2008). Extending lead-user theory: antecedents and consequences of consumers' lead userness*. *Journal of Product Innovation Management* 25 (4): 331–346.

Schuurman, D., Mahr, D., and De Marez, L. (2011). User characteristics for customer involvement in innovation processes: deconstructing the Lead User-concept. *ISPIM XXII, Proceedings*. International Society for Professional Innovation Management (ISPIM). http://hdl.handle.net/1854/LU-1887184.

Schweisfurth, T.G. (2017). Comparing internal and external lead users as sources of innovation. *Research Policy* 46 (1): 238–248.

Shih, C.-F. and Venkatesh, A. (2004). Beyond adoption: development and application of a use-diffusion model. *Journal of Marketing* 68 (1): 59–72.

Stock, R.M., Oliveira, P., and von Hippel, E. (2015). Impacts of hedonic and utilitarian user motives on the innovativeness of user-developed solutions. *Journal of Product Innovation Management* 32 (3): 389–403.

Takeuchi, H. and Nonaka, I. (1986 January 1). The new new product development game. Retrieved 11 July 2018, from https://hbr.org/1986/01/the-new-new-product-development-game.

Urban, G.I. and von Hippel, E. (1988). Lead user analyses for the development of new industrial products. *Management Science* 34 (5): 569–582.

von Hippel, E. (1986). Lead users: a source of novel product concepts. *Management Science* 32 (7): 791–805.

von Hippel, E. (1988). *The Sources of Innovation*. Oxford University Press.

von Hippel, E. (2005). *Democratizing Innovation*. Cambridge, MA: MIT Press.

von Hippel, E., de Jong, J.P., and Flowers, S. (2012). Comparing business and household sector innovation in consumer products: findings from a representative study in the United Kingdom. *Management Science* 58 (9): 1669–1681.

Wang, N., Tiberius, V., Chen, X. et al. (2021). Idea selection and adoption by users – a process model in an online innovation community. *Technology Analysis & Strategic Management* 33 (9): 1036–1051.

Yordanova, Z. (2021 February). Agile application for innovation projects in science organizations-knowledge gap and state of art. *International Conference on Information Technology & Systems*, 108–117. Springer, Cham.

Zhao, Z., Haikel-Elsabeh, M., Baudier, P. et al. (2021). Need for uniqueness and word of mouth in disruptive innovation adoption: the context of self-quantification. *IEEE Transactions on Engineering Management*. doi: 10.1109/TEM.2021.3067639.

Andrea Wöhrl holds a B.A in communication science and economics from Ludwig Maximilian University of Munich and an M.Sc. in Consumer Affairs from the Technical University of Munich, specializing in user research, user-centric innovation, co-creation, as well as leadership for innovation. She currently works in the public sector as cross innovation manager for the city of Regensburg (Germany). In a previous role, she was an Innovation Researcher at HYVE.

Sophia Korte is currently working as Strategic Design Researcher for INGKA Digital in Amsterdam with the focus of bringing Human-Centred-Design into Ikea Retail. Prior to her position at IKEA she worked as a Design Researcher and Innovation Consultant at HYVE in Munich where she also researched the topic of integrating different external competencies for open innovation approaches. She completed her Masters Degree in International Business Development Studies M.Sc. at Reutlingen Business School.

Michael Bartl is CEO of HYVE, an innovation company creating new products and services, and Managing Director of the Emotion AI startup TAWNY. Prior starting HYVE, Michael worked for the carmaker AUDI in the R&D division. He started his academic career with a Bachelor of Science from the University of Westminster London, a Dipl.-Kfm. from the University of Munich and a PhD from the Otto Beisheim Graduate School in Vallendar. In the research field of innovation management, he published in international journals such as Journal of Product Innovation Management, Electronic Commerce Research Journal and Harvard Business Manager.

Volker Bilgram is Professor of Global Innovation Management at Nuremberg Institute of Technology (THN) and Partner at HYVE, an innovation and design company helping clients create new products and services. He holds a diploma degree in International Business Law from Friedrich Alexander University of Erlangen-Nuremberg and a PhD from RWTH Aachen University. Volker's research focus is on AI-based innovation, user-centricity and service design and his work has been published in journals such as the International Journal of Innovation Management, Journal of Management Information Systems, Technological Forecasting and Social Change, and Harvard Business Manager.

Alexander Brem is an Endowed Chaired Professor and Institute Head with the University of Stuttgart, Stuttgart, Germany. In addition, he is an Honorary Professor with the University of Southern Denmark, Sønderborg, Denmark. Alexander Brem studied business administration with a major in entrepreneurship at Friedrich-Alexander-University Erlangen-Nürnberg (FAU), where he also received his PhD in 2007.

CHAPTER SEVENTEEN

HARNESSING ORDINARY USERS' IDEAS FOR INNOVATION

Peter R. Magnusson

Ordinary people just don't comprehend.

— VAN MORRISON (SONG ORDINARY PEOPLE, 1998)

Many product/service developers consider ordinary people unable to contribute much when it comes to innovation. This chapter challenges that assumption – it is about how ordinary people, or more specifically ordinary users, can contribute to the front-end of innovation (FEI).

Many of us have heard about lead users and how they can contribute to great innovations. Lead user innovation was discovered and introduced by Eric von Hippel in the '80s (von Hippel 1986). Lead users are either B2B users or consumers who have, and are aware of, needs that cannot be fulfilled by existing offerings. Most importantly, they are also able to come up with a solution to fulfill these needs and solve their problems. The Melitta coffee brewing filter, the Camelback, Liquid Paper, and the Windsurfing board are all examples of successful lead-user innovations.

However, there are also several difficulties linked to lead-user innovation, for instance, how to identify potential lead users. Whether a person is a lead user can be established first in retrospect. There is also a challenge with intellectual property rights, that is, obtaining ownership of lead users' innovations. Many lead-user innovations have initiated startup companies with the innovator as an entrepreneur. Furthermore, only a small fraction of all users are expected to be potential lead users. Linking to Rogers' (1995) famous model for diffusion of innovation, only the innovators and perhaps some of the early adopters

The PDMA Handbook of Innovation and New Product Development, Fourth Edition. Edited by Ludwig Bstieler and Charles H. Noble.

FIGURE 17.1 LEAD USERS VS ORDINARY USERS.

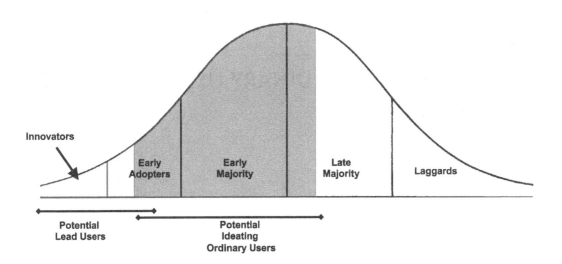

Source: Adapted from Rogers (1995).

might be considered potential lead users. Much of the attention on involving users in innovation has been focused exclusively on lead users.

What can the rest of the users, non-lead users, contribute with regarding innovation? This chapter addresses the potential benefits of involving non-lead users, that is, *ordinary users*, in the early stages of product and service development.

The term "ordinary users" refers to users, or potential ones, who are not considered lead users. They are familiar with the context of use but are not expected to possess special technical expertise in order to come up with technically feasible ideas/solutions (Magnusson 2009). Usually, ordinary users are expected to be found among the majority of individuals in Roger's (1995) diffusion model. However, they could also be expected to occur among the "early adopters," having a restricted understanding of the underlying technology. Moreover, "lead users" may also be found among the early adopters (see Figure 17.1).

More specifically, this chapter discusses how ordinary users can contribute to a company's idea management activities. As we will see, there are at least three benefits to involving ordinary users, namely suggesting unique ideas, gaining valuable use knowledge, and promoting goodwill.

17.1 What Is Ordinary User Innovation?

The origin of all innovations is creative ideas. Tidd and Bessant (2014, p. 3) describe innovation as the "the process of creating value from ideas." To understand what users can contribute with, let us take a deeper look into what constitutes an idea for innovation. Simply put, an idea for a new innovation (product or service) consists of a *solution* to a relevant *problem*. The relevance of the problem is determined by the potential users of the intended innovation. There is a strong relationship between knowledge and innovation. It is widely agreed that at least two different domains of knowledge are needed to accomplish a good innovation, namely *use knowledge* and *technology knowledge*.

Use knowledge is linked to the problem or the use side. It is the knowledge that renders a detailed understanding of the problem/need that should be solved/satisfied; thus, an understanding of why the product/service creates value for its users is gained. This includes a contextual understanding of the use side, that is, the environment where the product or service should operate.

Technology knowledge is linked to an understanding of how to solve the given problem, often called the supply side, and includes an understanding of technology, but also how to organize the internal resources to reach a solution. For services, this is rather how to design the service processes including employee resources. The relation between problem/solution and knowledge distribution is illustrated in Figure 17.2.

To summarize, problems relate to the use side, and solutions to the supply side of an innovation. The problem reflects a need, and the solution a potential approach to satisfy this need. Solving the problem is expected to generate *value* for the intended user.

Use and technology knowledge are both needed and must be in resonance to accomplish successful products/services. As Steve Jobs once put it "*I think*

FIGURE 17.2 IDEA COMPONENTS AND RELATED KNOWLEDGE, AND ITS DISTRIBUTION.

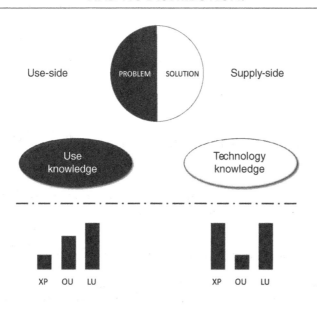

(XP=Professional experts; OU=Ordinary Users; LU=Lead Users)

really great products come from melding two points of view – the technology point of view and the customer point of view. You need both." (Burlingham and Gendron 1989)

From a knowledge perspective, there is an imbalance between company experts and ordinary users. Figure 17.2 illustrates the approximate relative distribution of use and technology knowledge among the three groups: (1) in-house professional experts (XP), (2) ordinary users (OU), and (3) lead users (LU).

Lead users stand out, being in possession of both high technology and use knowledge. Ordinary users and company experts are, on the other hand, each other's opposites regarding knowledge distribution. Experts possess more technology knowledge than do ordinary users. The inverted relationship applies to use knowledge, where professional experts normally have a rather poor understanding of the problem side, or at least the details of the problem (Magnusson 2009). Understanding the problem side is also the crux of the matter for most marketeers. As we will see, ordinary users' ideas, can in fact unveil some of the users' problems and needs.

Involving ordinary users in the front-end of innovation is essentially about acquiring more use knowledge, in other words, a better understanding of the users' problems that need solving.

Ordinary users do, by definition, have low technology knowledge. Nevertheless, they can use the product/service without understanding what is happening behind the curtain. As an example, many of us drive a car daily and can do this without understanding the details about how the car actually works on a technical engineering level. Without having the technical competence, we can still provide many suggestions on how to improve certain details of the car to better fit our needs. The lack of technology knowledge can, as will be seen, even be an asset when it comes to generating new creative ideas.

On the other hand, a deep understanding of the technology does not mean that one understands what different ordinary drivers intend to use the car for, nor what problems they encounter when they use the car to get their hundreds of daily "small jobs done."

17.2 Merits and Challenges of Involving Ordinary Users in Ideation?

There are several reasons for involving ordinary users in ideation, but there are also some hazards. Let us look into these.

17.2.1 Ignorance Is Bliss – Ordinary Users Think Outside the Box

Research has shown that ordinary users' ideas are perceived as more original than those of professional experts (e.g., Magnusson 2009). One reason for the high originality is that these users do not have the "burden" of knowing what is currently technically feasible. We have all heard the saying, "be creative – think outside the box!" What box? One way of seeing this is that the box is your knowledge of what currently works and what not, that is, your technology knowledge. Ordinary users have very little technology knowledge; for them virtually nothing is technically impossible. This is also why they can suggest very original, but perhaps not so feasible ideas. Ignorance is bliss.

This has, of course, a flipside as it often renders ordinary users' ideas difficult, if not impossible, to be readily implemented. In contrast, lead users' ideas are more elaborated and feasible because of their high technology knowledge. Ordinary user ideas – at least the suggested solutions – should thus not generally be regarded as direct seeds for new innovations. Nevertheless, their ideas should not be rejected because of their eventual lack of technical feasibility. Users' "crazy" ideas might be a valuable source of inspiration for a company's developers and even make them think outside the box, just as science fiction can be a means to think in new, creative ways (Michaud and Appio 2022).

An example of being inspired by a "crazy idea" is the newspaper boy idea that occurred in one of our studies, together with Telia Mobile, where we had gathered numerous user ideas for new mobile services. A group of eight professional experts were presented with a sample of ideas. One idea was deemed the most original, but simultaneously the least feasible. The creator described this idea as follows:

> *I had received the wrong Sunday paper for the third consecutive week. It suddenly occurred to me that the phone ought to have a function that would enable me to enter my newspaper subscription number and send an electric shock of 440 V into the body of the newspaper boy so that he would learn to give me the right newspaper.*

The intention was to show how "crazy" some of the users' ideas could be. After presenting the idea, the professional experts' first reaction was laughter; the idea had to be a joke, and it hopefully was. Nevertheless, after a few seconds, one of the experts said loudly and clearly, "that could be an interesting service." It went quiet in the room. "Oh sorry, I was only thinking aloud," he continued. The silence became even more embarrassing. Then, he understood that the others had taken him for a sadist. "Oh, I have to explain. I was thinking about how the idea could be developed into something interesting." He explained how the proposed lethal service idea could be redesigned into an innocuous one that would solve the core problem. Instead of sending an electrical shock, it could automatically identify the subscriber and the newspaper boy and send a text message notification that the wrong paper had been delivered. The rest of the experts were relieved and collectively started to refine the idea even further.

From being something very original but completely unfeasible, the group had, within the space of a few minutes, come up with a new, realizable design solving the problem. The idea had thus *triggered a creative design process*, which led to a new idea that was feasible yet novel.

17.2.2 Call for a New Idea Management

The episode just described also illustrates that idea management must be rather different to utilize the potential of ordinary user ideas. Traditionally, ideas are assessed in a filtering process toward several different criteria, such as, user value, originality, and feasibility. However, owing to the ordinary users' lack of technology knowledge, their ideas cannot be expected to be as elaborated as lead users' or professional experts' ideas. Feasibility is thus a poor criterion to apply on these ideas. Instead, ordinary user ideas should be handled as potential *rough diamonds*.

The value of a diamond cannot be determined until it has been cut and polished. Accordingly, it would be ignorant to think that the potential of users' ideas could be fairly assessed in their initial form. Managers should thus not expect the ideas of ordinary users to be immediately suitable for introduction into the NPD process; rather, their ideas should be used as inspiration for further elaboration. The ideas need to be refined. In most cases, an idea's solution is not as absurd as in the example above. More common is that the user-suggested solution is not the best, and there are new technologies that can be used for a smarter and more cost-competitive solution. In these instances, an expert looking at the proposed user idea will most likely see what can be done to refine the solution into something better that still solves the problem. Idea assessment thus becomes a generative process.

17.2.3 User Ideas as a Probe to Understand the Ordinary User

Even if the solution side is often rather immature and can sometimes even be regarded as ridiculous, user-created ideas reflect *real relevant user problems* that exist, at least for the individual who proposed the idea. Accordingly, the strength of ordinary-user ideas is normally the problem side. We can learn more about the users' situations, problems, and needs.

User ideas can thus be a *probe* for understanding ordinary users' needs and problems. As can be understood from Figure 17.2, each user idea's problem part is embedded with use knowledge that can be captured and assimilated. By studying a user idea, and specifically reflecting on the problem side, one can gain new use knowledge, just as in the previous example of the newspaper boy, where the professional expert gained new insights into a real user problem. However, the originally proposed solution was quite impossible for many reasons. Still, the problem side became a catalyst to come up with a new solution. Professional experts, especially engineers, are trained and adept at solving problems. They are much less proficient in understanding true customer problems.

It has been suggested that company experts should "be a user" to envision what problems and needs should be addressed. This is, however, rather difficult. It often seems that professionals are not a good proxy for ordinary users. Correspondingly, Veryzer (1998, p. 149) cited a development team manager who expressed the following opinion:

> *Engineers are NOT real people! Don't rely on an engineering test sample—they know too much and often think in a way that differs from the people that will be using the product— you need to test the naive user.*

I have had the same experience myself. For a decade in the '90s I worked in research and development of mobile services at Telia Mobile, Sweden's largest mobile telephone operator. The service development in the company was by that time driven by technological opportunities. The services, needs, and problems that needed to be addressed were decided by the developers. They were all demanding users of mobile telephony and did of course understand how new services should be designed. A wake-up call was a comparative study where the company experts' ideas for new mobile services for a targeted group were compared with ideas created by the users in the target group. The expert group's ideas were on average best when it came to feasibility, but the users outscored the experts both in perceived user value and idea originality; for more details see Magnusson (2009). Telia Mobile realized it was time for a change and to start taking the users' ideas seriously.

The key to successful new products/services is to come up with cost-competitive solutions to customers' problems (Griffin 2013). The traditional approaches to gain customer need/problem information are to be involved with, observe, and interview users. This information is then transformed into product/service concepts on which the customers are asked what they think. As these concepts do not reflect the respondents' personal experiences (needs and problems), their answers will likely have low credibility (Griffin 2013).

An alternative route is to encourage the users to come up with new ideas of new product/services that would be useful for themselves. This will provide user-originated ideas where real user problems are defined and use knowledge is gained. There are also positive tradeoffs to engaging users in the development process. Users have been found to be willing to pay more for the products if they have been engaged as co-creators or co-designers (e.g., Franke and Piller 2004).

17.2.4 Involving Users in Ideation as a Marketing Technique

Involving users and asking for their ideas is also a way of showing interest, empathy, and care for the users. It can thus create goodwill. Numerous crowdsourcing platforms exist, and have existed, that enable users to submit their ideas for new products/services or express feedback for improving existing products/services; DELL, LEGO, and Proctor & Gamble, just to mention a few.

However, goodwill can also turn to the opposite, especially on open platforms. If most of the users express their approval for an idea that the members of the hosting company dislike, a delicate situation will arise. There have been quite a few idea contests where the users have voted for suggestions that are more or less a joke (Blohm et al. 2013). Giving the crowd control of the final decision is seldom a smart strategy. To minimize the risk of ending up in these situations, I will present a method to collect user ideas that reduces these risks.

17.3 OUI – A Method of Involving Users in Ideation

OUI (Ordinary User Ideation) aims at grasping "natural" problems (and their solutions) that occur among ordinary users. It is a qualitative user ideation method that has been developed over two decades of research on user involvement in the front-end of innovation. The research has comprised idea creation, refinement, and evaluation in cooperation with different sets of users. Still, it can indeed in this modified version be used as a practical method for gaining new ideas for innovation, gaining use knowledge about specific user segments, and building good relations. It has, for instance, been used in the contexts of wireless services and public transportation services, and for improving primary healthcare. The projects have been in close cooperation with external companies and public organizations. The OUI that is presented here is a synthesis and adaption of some of the variants that we have used over the years for research purposes.

17.3.1 Preparation and Recruitment

OUI aims at acquiring detailed information about problems and potential solutions for a pre-defined target group in a specific context. The goal is to grasp real relevant problems when they occur (*in situ* and *in context*). It is thus important that the ideation takes place in the participants' own context. The number of participants must be adapted accordingly so that it can be handled. In our projects we have handled around 40–85 participants in each study, and the ideation periods have been about two weeks.

What should managers look for when sampling ordinary users? First of all, there is no harm in eventually including some lead users in the sample; it is actually quite the opposite. However, this chapter focuses on how to handle ordinary users. Use knowledge is probably the most important factor, that is, *experience of context* for which the products/services are intended. Ordinary users can have different degrees of context experience which have both pros and cons. In brief one can say that experience has a priming effect. With experience comes increased knowledge and understanding, both concerning use and technology. An experienced user might become "institutionalized" and might not react to things that a greenhorn user perceives as surprising. Different degrees of experience, as well as variety in demographic background, contribute to a more holistic coverage of ideas. For a more detailed discussion see Edvardsson et al. (2012).

When the participants were invited, they were informed that this is an opportunity to contribute with ideas that can lead to improvements that they could benefit from; for example, to improve the healthcare center they belong to. Normally, we use some kind of reimbursement to attract participating users. It is though important that the reimbursement itself is not the main reason for participation. Participants should have an intrinsic motivation to really offer suggestions for improvement that will solve their real and relevant problems.

17.3.2 Start-up Meeting – Initiation

The start-up meeting is a physical meeting aiming to inform the participants about the assignment and provide them with potential equipment for their ideation. It is recommended to gather no more than 20 people at a time to form manageable groups. For example, a project with 50 participants would be divided into three start-up meetings.

Even if informed at recruitment, the participants are reminded of which context they should focus their ideation upon. To urge the participants to primarily come up with real and relevant problems, they are instructed to be "egoistical," that is, to propose ideas that they would consider valuable for themselves as users. Within the given context, the assignment is to come up with ideas that produce an added value for the participants.

The participants are provided with a pocket-sized notebook (A6 format) where the challenge is repeated. They are instructed to bring the notebook with them during the study and write down in it every idea impulse and thought that comes to their mind. It is emphasized that they should *not reject any ideas*. It is also asked that the participants make notes in the diary of *how their ideas emerged*. The purpose of the narratives is to enable an even better understanding of the ideas and why they would be valuable.

17.3.3 Ideation Period

During the ideation period (approximately two weeks) the participants are on their own. They are instructed to not sit down and try to envision new ideas. As pointed out, it is important that the participants have time to "stumble" over problems and ideas in their natural environment. It seems that many participants unconsciously activate a background process in the brain ready to alert when a new problem occurs or unfulfilled needs are discovered following problem solving.

It is also recommended to give the participants a reminder after a week with the following message: "hope your ideation is going fine, and please do not hesitate to contact us if you have any questions."

17.3.4 Idea Gathering

The participants are asked to transfer their idea notes to a pre-defined template to ease further processing. We have used the following fields: *name of idea creator, idea name, target group, problem, solution,* and *added value.* The template thus forces the participants to separate the problems and solutions descriptions. As previously mentioned, the problems are often more interesting than the proposed solutions that users come up with. That said, not all solutions are uninteresting. Remember that the users' lack of technology knowledge can also be a source of inspiration and a trigger to really think outside the box.

In the template one can also ask the users to quantify the added value. It could, for instance, be time saving in minutes or cost saving in dollars. Value can also be hedonic, that is, the idea would provide the users with a great experience hard to specify in monetary terms.

Besides collecting the ideas, also ensure to collect the participants' notebooks, which can give richer descriptions to the background of how the idea emerged.

17.3.5 Processing and Refinement of Ideas

A good advice would be to first take the gathered ideas and screen for the unacceptable ones to be excluded from further processing. For example, ideas that are out of scope for the assignment or those that can directly be deemed as not sufficiently useful. It is also helpful to try and categorize them to accomplish different clusters of related ideas which makes the processing easier.

If a large number of ideas are present, it might also be necessary to prioritize the most interesting ideas. Since users mainly contribute with knowledge regarding the problem-side, my advice is to prioritize focusing on the problem relevancy, that is, how important it would be to solve the problem described in the idea. To ease this process, as users have been proven a good proxy for assessing ideas (Magnusson et al. 2016), you could ask the participating users to rate or rank the ideas.

So far, the ideas are intact but, as previously discussed, those coming from ordinary users must be managed differently than those of professional experts or lead users. Ordinary user ideas cannot be expected to be technically feasible and must first be considered for refinement.

In our research, we have found that refinement is preferably a collective activity where each idea is considered regarding potential improvements. We have used groups of 6–8 people in one-to-two-hour meetings. A person facilitates the meeting toward the ideas' refinement. Usually, it is mainly the solution part of the idea that need to be scrutinized. This focuses mainly on how appropriate the proposed solution is and how it can be changed. Sometimes the problem needs to be further clarified to thoroughly understand its potential. The facilitator also takes notes during the meeting regarding eventual improvements to update the idea description before assessment.

Refinement can be done either in a closed or open manner. Closed means that only in-house experts perform the refinement. In some studies, we have invited the participating users to take part in the refinement process. There are pros and cons associated with this procedure. The main advantage is that the organizer will gain an even deeper understanding of the users' problems and needs, and the refinement will be a true co-creation between users and experts. A positive effect of this is that the experts acquire deeper use knowledge from these sessions. In the same way, the users learn more about feasibility issues. Thus, it is a mutual learning experience. This is beneficial if you choose to enable more than one ideation period, as the users will then be able to adapt their ideas to be more feasible. It is also a goodwill opportunity to strengthen the bonds with the participating users. The disadvantage is that it requires more resources.

17.3.6 Assessment of Ideas

After the final gathering of the ideas, they are assessed using the organization's preferred way of assessment. Those who pass will move on to implementation. At least three assessment dimensions have been used in all our studies, relating to the idea's *user value, feasibility,* and *originality. User value* is linked to the relevance of the problem, that is, how valuable it would be to have it solved. *Feasibility* is connected to the solution part, valuing the merits of the solution. *Originality* is a holistic measure of the innovativeness of the idea.

17.3.7 OUI–Some Reflections and Practical Advice

The OUI method provides higher integrity than using crowdsourcing where ideas are openly displayed for anyone to take part of. The risk of unintentional leaking of information is thus reduced. The assessment is not left to the users, as in many crowdsourcing projects, which also removes the drawback of evil voting and gives control to the organizer to choose which ideas to accept.

It is essential to give participants *feedback* on their suggestions. If user refinement groups are used, this will be completed during those sessions. Not giving feedback on proposals will likely result in frustrated participants.

It should be made clear early in the process that there is no guarantee that the users' suggestions will be implemented. In some of our studies, some participants expected their ideas to also be implemented by the organizing company. There should be transparency regarding the IPR (intellectual property rights) policy. Will there, for instance, be any reimbursement if an idea is further developed into a product/service? This will be especially delicate if user refinement groups are used. The refined idea will be the result of a collective effort. A recommendation is to make it clear that the organizer will have the right to exploit the submitted ideas.

It should also be noted that involving ordinary users in contributing ideas implies new routines which are summarized in Figure 17.3. Idea refinement is the most essential new activity needed compared to working with expert, or lead user ideas. As described in the chapter, this is mainly due to the fact that you cannot expect ordinary user ideas to be technically feasible. If the elaboration is done together with users, it will also be an opportunity for mutual learning – experts can learn about the users' needs/problems and the users can understand more about what is possible to implement.

FIGURE 17.3 OUI – A METHOD OF INVOLVING USERS IN IDEATION.

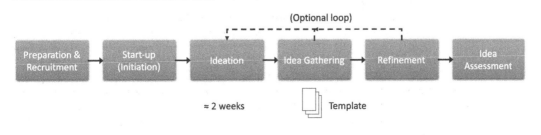

17.4 Summary

Lead users have long been acknowledged in the innovation literature. This chapter has dealt with the involvement of ordinary users in ideation. Ordinary users are not a substitute to lead users, but rather play a complementary or alternative role. The contributions of ordinary users are different to those of their better-known counterparts, the lead users, and so is the management of involving them.

Lead users come up with ready or almost complete innovations. The nature of their ideas is often radical. They have the necessary knowledge and skills to do this without interacting with any companies. This is both a strength and a problem for companies wanting to obtain access to lead user innovations, as IPR becomes an intriguing issue.

Ordinary users do not provide fully elaborated innovations, that is, prototypes. Their ideas, at least the solutions, might not be fully elaborated. Nevertheless, they have proven to contribute with valuable input. With their in-depth contextual knowledge regarding the use side, they can provide ideas that address real and relevant problems and needs among the target group. Most of the ideas are incremental in nature.

Ordinary users' ideas often lack full feasibility due to their reduced technology knowledge. However, by using a refinement procedure, concentrating on the solution, this can be overcome. Even if the submitted ideas are unfeasible, they are embedded with valuable use knowledge, reflecting true user problems and needs. The lack of technology knowledge can also give birth to really original ideas that can inspire professional developers. The chapter has provided a research-based method that can be used to (1) obtain ordinary user ideas (2) gain use knowledge, and (3) potentially promote goodwill. Good luck in your endeavors of involving ordinary users in your innovation process!

References

Blohm, I., Leimeister, J.M., and Krcmar, H. (2013). Crowdsourcing: how to benefit from (too) many great ideas. *MIS Quarterly Executive* 12 (4): 199–211.

Burlingham, B. and Gendron, G. (1989). The entrepreneur of the decade – an interview with Steve Jobs, Inc.'s entrepreneur of the decade. https://www.inc.com/magazine/19890401/5602.html. (Accessed on 2nd April 2022.)

Edvardsson, B., Kristensson, P., Magnusson, P., and Sundström, E. (2012). Customer integration within service development—a review of methods and an analysis of insitu and exsitu contributions. *Technovation* 32 (7–8): 419–429.

Franke, N. and Piller, F. (2004). Value creation by toolkits for user innovation and design: the case of the watch market. *Journal of Product Innovation Management* 21 (6): 401–415.

Griffin, A. (2013). Obtaining customer needs for product development. In: *PDMA Handbook of New Product Development*, 3e (ed. K.B. Kahn), 213–230. Hoboken, NJ: Wiley & Sons.

Magnusson, P.R. (2009). Exploring the contributions of involving ordinary users in ideation of technology-based services. *Journal of Product Innovation Management* 26 (5): 578–593.

Magnusson, P.R., Wästlund, E., and Netz, J. (2016). Exploring users' appropriateness as a proxy for experts when screening new product/service ideas. *Journal of Product Innovation Management* 33 (1): 4–18.

Michaud, T. and Appio, F.P. (2022). Envisioning innovation opportunities through science fiction. *Journal of Product Innovation Management* 39 (2): 121–131.

Rogers, E.M. (1995). *Diffusion of Innovations*, 4e. New York: Free Press.

Tidd, J. and Bessant, J.R. (2014). *Strategic Innovation Management.* Wiley.

Veryzer, R.W.J. (1998). Key factors affecting customer evaluation of discontinuous new products. *Journal of Product Innovation Management* 15 (2): 136–150.

von Hippel, E. (1986). Lead users: a source of novel product concepts. *Management Science* 32 (7): 791–805.

Prof. Dr. Peter R. Magnusson is a professor at Karlstad Business School, Karlstad University, Sweden. He is also affiliated to the CTF Service Research Center at Karlstad University. He holds an M.Sc. in Electrical engineering from Chalmers University, and a PhD from the Stockholm School of Economics. Dr. Magnusson has 20 years' practical experience in R&D in the computing and telecommunications industries. His research focuses on innovation management and servitization. Dr. Magnusson has received several nominations and rewards for his research. His research has published in leading refereed journals, including Journal of Product Innovation Management, Journal of Service Research, and Journal of the Academy of Marketing Science.

CHAPTER EIGHTEEN

NEW PRODUCT CO-CREATION: KEY INSIGHTS AND SUCCESS FACTORS

Gregory J. Fisher and Aric Rindfleisch

18.1 Introduction

New product development (NPD) is essentially a process of creation. According to Jane Fulton Suri (a partner at IDEO), humans are innately creative and commonly engage in a wide variety of innovative activities (Suri 2005). For instance, over 12 million US consumers participate in some type of product-related innovation and their expenditures on these activities equal one third of the combined R&D budgets of all US consumer product manufacturers (von Hippel et al. 2011). Thus, it is not surprising that a large number of organizations are seeking to harness the power of individual creativity to enhance their NPD activities, ranging from early-stage idea generation to downstream development and product launch (Athaide and Zhang 2011; Fisher and Fang 2018; Roberts and Candi 2014).

Since its advent in the early 2000s, new product co-creation has been widely heralded as a useful tool for enhancing organizational innovation (Prahalad and Ramaswamy 2002, 2004). Moreover, a sizable body of academic research suggests that co-creation can enhance a firm's ability to generate novel ideas, evaluate new product concepts, and commercialize new offerings (Füller et al. 2009; Gemser and Perks 2015; Mahr et al. 2014). For example, the logistics services giant DHL has successfully employed co-creation to develop new technology initiatives that have improved their on-time delivery performance and decreased their rate of customer churn (Crandell 2016).

Despite new product co-creation's considerable promise, many firms have struggled (and even failed) to achieve gains from engaging outsiders in their

The PDMA Handbook of Innovation and New Product Development, Fourth Edition. Edited by Ludwig Bstieler and Charles H. Noble.

NPD processes (Mahr et al. 2015). Several early exemplars have either shut down their co-creation platforms or are no longer in business (e.g., Dell IdeaStorm, MyStarbucksIdea, Local Motors). The fact that these stars of the co-creation constellation are no longer shining casts a shadow upon the viability of new product co-creation and suggests that this approach may be more challenging that commonly believed. Indeed, prior research in the NPD domain suggests that converting customer insights into new product outcomes is a difficult task (Bayus 2013). As evidence of this difficulty, a recent survey by McKinsey reveals that over 40% of firms do not incorporate customers in their NPD activities at all (Loughlin et al. 2020). Consequently, the implementation of new product co-creation, like most innovation activities, appears to be both risky and prone to failure.

In this chapter, our objective is to help firms identify best practices associated with new product co-creation in order to improve their chance of success. Although co-creation has attracted considerable inquiry over the past two decades, research in this domain tends to focus on providing descriptive frameworks of co-creation (Prahalad and Ramaswamy 2004), demonstrating the potential benefits of co-creation to firms (Gemser and Perks 2015) and consumers (Franke and Piller 2004), or understanding participants' motivations to co-create with firms (Mandolfo et al. 2020; Nambisan and Baron 2009). In contrast, this chapter focuses on offering guidelines for enhancing new product co-creation success.

We begin by defining the phenomenon of new product co-creation and identifying its two key dimensions (i.e., contribution and selection). We also illustrate the difficulties of putting co-creation into practice by highlighting three notable co-creation failures (e.g., Dell IdeaStorm, MyStarbucksIdea, Local Motors). We then identify six key co-creation challenges (three for each dimension) and suggest a set of best practices for overcoming these challenges. Finally, we conclude by offering a Co-Creation Readiness Scorecard to help firms assess the degree to which they are prepared to engage in this intriguing approach to NPD.

18.2 Overview of New Product Co-Creation

18.2.1 Definition and Dimensions of New Product Co-Creation

As noted earlier, new product co-creation has been heralded as a useful tool for enhancing NPD by soliciting and incorporating external ideas. According to O'Hern and Rindfleisch (2010), new product co-creation is "a collaborative NPD activity in which customers actively contribute and/or select the content of a new product offering" (p. 86). As seen by this definition, new product co-creation involves two fundamental steps: contribution and selection. In the first step (i.e., contribution), a firm must motivate external participants to devote time and energy to submit contributions. Although firms often turn to customers for new product ideas, co-creation contributors can also include users, researchers, suppliers, and professionals working in related industries. In the second step (i.e., selection), a firm employs a process for selecting the few (if any) valuable contributions offered by external participants. In most co-creation initiatives, participants digitally engage in contribution and selection via a web-based platform that is either created by a firm (e.g., ideas.lego.com) or managed by an intermediary (e.g., innocentive.com).

In essence, a firm's ability to execute these two key steps (i.e., contribution and selection) is critical to new product co-creation success. Both steps can be quite challenging for different reasons. Contributions are hard to solicit because external participants are often quite busy and have little incentive to devote their valuable time and energy to help a company enhance its NPD efforts. Consequently, many co-creation efforts fail because firms receive too few contributions. Conversely, selection is challenging because most contributions are not very useful. External ideas are often too idiosyncratic, too expensive, or too impractical to be viable for development and commercialization. Hence, firms face the difficult task of having to reject many (if not most) submissions, which risks creating ill will with customers and other external stakeholders. In sum, to succeed at new product co-creation, a firm must be able to motivate external participants to submit valuable ideas and be able to reject most of these ideas without alienating its customer base or other collaborators.

18.2.2 New Product Co-Creation Failure

In theory, these two steps may seem straightforward and easy to implement. However, they are exceedingly difficult to put into practice. As a testament to this difficulty, some widely celebrated examples of successful new product co-creation have fallen on hard times. For illustrative purposes, we briefly detail three notable examples of co-creation failure: Dell IdeaStorm, MyStarbucksIdea, and Local Motors. Although all three of these initiatives were initially proclaimed as co-creation success stories, their success appears to have been short-lived.

Dell IdeaStorm: Dell began this co-creation initiative in 2007 as a forum for soliciting customer suggestions. This initiative quickly attracted a large following as well as a plethora of submissions. Dell IdeaStorm was revamped in 2012 (IdeaStorm 2.0) with great fanfare and offered a sophisticated platform that provided various levels of contributor recognition (https://www.dell.com/en-us/blog/ideastorm-2-0-launched-today). According to many observers, this platform was the epitome of co-creation success. For example, Bayus (2013) proclaimed that, "Dell's IdeaStorm represents the gold standard for new product idea crowdsourcing applications" (p. 241). By 2013, this platform had attracted approximately 2,000 users and 15,000 submitted ideas (Westerski et al. 2013). To its credit, this initiative resulted in a number of successful NPD-related ideas such as new laptop covers and wireless headphones. However, these successes appear to be few and far between. According to Bayus (2013), less than 4% of all of the ideas submitted to IdeaStorm were actually implemented by Dell and most of these implemented ideas seem to be incremental in nature (e.g., a personal computer aimed at children). Despite its considerable attention and large following, Dell quietly closed the doors on IdeaStorm circa 2019 and its website now directs traffic to Dell's homepage.

MyStarbucksIdea: Starbucks launched this co-creation platform in 2008 to solicit new product (and process) ideas from its customers. According to Starbucks, this initiative provided "an online community for people to share, vote, discuss and put into action ideas on how to enhance the Starbucks experience" (Starbucks 2013). Given Starbucks' considerable visibility, this initiative was widely viewed as verification of co-creation's arrival and was frequently cited as an exemplar of this new NPD approach. For example, Lee and Suh (2016) proclaimed this initiative to be a "success" and noted that, "Starbucks has been gathering valuable ideas directly from its customers through the community … and these ideas have contributed to innovating Starbucks" (p. 170). By 2013, this platform received over 150,000 submissions and led to a number of new offerings, such as "cake pops" and coffee "splash sticks" (Starbucks 2013). However, some commentators suggested that the success of MyStarbucksIdea was overblown and that many of the ideas that Starbucks implemented were actually internally derived (Moore 2012). In 2017, Starbucks suddenly removed the communal aspect from this platform and converted MyStarbucksIdea into a simple submission portal in which individuals could still submit new product-related ideas, but these ideas could only be viewed or commented upon by Starbucks management. In essence, Starbucks converted this platform into an electronic suggestion box. In hindsight, this conversion to a closed platform appears to have been a harbinger of things to come, as MyStarbucksIdea.com was permanently shut down in 2021.

Local Motors: This innovative new vehicle manufacturer was established in 2007 with the ambitious goal of offering "the world's first co-created car, the Rally Fighter; the world's first 3D-printed car, Strati; and the world's first co-created, self-driving, cognitive, electric shuttle, Olli" (Peels 2019). Thus, co-creation was an essential part of Local Motors' DNA. To its credit, Local Motors successfully managed a co-creation platform that resulted in the design, development and manufacture of the Rally Fighter within its first two years of operation. This early success captured widespread attention. For example, Jay Rogers, the CEO of Local Motors was featured on the cover of *Wired Magazine* in 2010, and Harvard Business School crafted a case study about the development of the Rally Fighter (Norton and Dann 2011). However, Local Motors was only able to sell a handful of Rally Fighters and its plans to manufacture a portfolio of other vehicles never materialized. In 2012, Local Motors opened its co-creation platform and network of contributors to other firms. This revamped initiative (rebranded as "The Forge") attracted a number of clients, including Dominos, Reebok, and the US Army, and led to the design and development of a new pizza delivery vehicle manufactured by General Motors and currently used by Dominos. Unfortunately, much like Dell IdeaStorm and MyStarbucksIdea, this co-creation initiative failed the test of time and Local Motors shut down not just its co-creation platform but its entire company in 2022 (Tarantola 2022).

In sum, although new product co-creation is an intriguing tool for enhancing a firm's NPD activities, putting its promise into practice is a difficult task as witnessed by the cases of Dell IdeaStorm, MyStarbucksIdea and Local Motors. Thus, our next section identifies a set of key challenges and success factors to help managers navigate past these common pitfalls.

18.3 Challenges and Success Factors

As noted earlier, new product co-creation consists of two key steps: contribution and selection. In this section, we identify common challenges that firms face when trying to implement each of these steps and then identify success factors to help NPD managers navigate these challenges. Exhibit 18.1 provides a summary overview of the challenges, success factors, examples, and best practices.

18.3.1 Contribution Challenges and Success Factors

18.3.1.1 Attracting and Motivating External Contributors (Nonmonetary Patches and Badges).
In today's information-rich environment, people are busier than ever and face an array of digital distractions. Thus, attracting and motivating external contributors is a considerable challenge for most new product co-creation initiatives. Obtaining contributions is difficult for many co-creation platforms, while the majority of contributions are offered by only a small portion of its users. For example, nearly 80% of Wikipedia's content is contributed by about 1% of its users.

We all like to be rewarded for our efforts. Participants who engage in new product co-creation are no exception. Thus, incentivizing contributors is an essential element of successful co-creation initiatives. A healthy body of research suggests that contributors are largely motivated by two types of rewards: extrinsic and intrinsic (Benkler 2008). Typically, extrinsic rewards entail some type of financial incentive, while intrinsic rewards focus more on providing social recognition. Most successful new product co-creation initiatives provide both types of rewards. For example, contributors who submit winning designs to Threadless (a co-creation-based apparel firm) receive a percentage of the sales of any

EXHIBIT 18.1 NEW PRODUCT CO-CREATION SUCCESS FACTORS.

Phases	Challenges	Success Factors	Examples	Best Practices
Contribution	Attracting and motivating external contributors	Patches and badges	NASA	Offer nonmonetary rewards
	Incremental ideas	Look beyond your customers	InnoCentive	Cast a wide net
	The "Rule of One"	Attract new contributors	General Mills	View churn as good
Selection	Protecting contributors' egos	Avoid being the corporate villain	LEGO	Let participants do the dirty work
	Harvesting co-created value	Control the process	Threadless	Own the key NPD elements that enable sufficient value capture
	Maintaining the peace	Share the wealth	General Electric	Create a winners' circle

products that employ their design and also gain the intrinsic benefit of having their name and design featured as a winner on Threadless' website. These types of public accolades are fairly inexpensive, offer considerable prestige and help motivate new contributors. Usually, these rewards are directed toward individuals whose contributions have been selected as winners. However (as detailed later), it is good practice to offer rewards to a more expansive set of contributors, if possible.

Although both financial and social rewards may be employed to attract potential contributors, the most talented external contributors are more likely to be motivated by social recognition rather than by financial rewards. Research on co-creation suggests that contributors are often intrinsically motivated by social rewards, so the publicity received from winning a co-creation contest is likely to be an effective way to incentivize contributors (Hoyer et al. 2010). In addition to satisfying the need to be recognized for their accomplishments, intrinsic rewards (such as having their name associated with a new product) also provide contributors with a variety of other benefits, such as social standing and tangible evidence of their creativity skills, which can be used to build their resume and enhance their future employment opportunities.

Beyond simply acknowledging the names of successful contributors, firms should also consider providing meaningful intrinsic rewards such as patches and badges. These types of rewards can be either digital and/or physical in nature and can help incentivize contribution by providing a form of recognition that is more meaningful and tangible. For example, the US space agency NASA has historically rewarded successful contributors by giving them a special patch that contains a space-related logo that they can proudly display on their clothing. In fact, NASA has found that this visible symbol of recognition is a more effective incentive than a cash prize (Davis et al. 2015). Thus, firms seeking to attract contributors to their new product co-creation initiatives should consider issuing nonmonetary rewards such as patches and badges. In addition to providing a powerful incentive, this approach should help reduce the costs of implementing co-creation.

18.3.1.2 Incremental Ideas (Look beyond Your Customers). Many firms' new product co-creation initiatives start by seeking ideas from their existing customers. This is a natural tendency, as the notion of leveraging customer insights to generate new ideas about unmet customer needs is often recommended by NPD gurus. Unfortunately, customers are often constrained by functional fixedness and only a relatively small number of customers possess the motivation and ability to generate ideas that will lead to novel new products (Mahr et al. 2014). Consequently, co-creation initiatives that target a firm's existing customer base often fail to generate ideas for radical innovations and are typically flooded with contributions that focus on making minor modifications to existing products, such as a new color or flavor (Menguc et al. 2014).

To avoid this dilemma, firms should cast a wide net that goes beyond their existing customers and solicits ideas from a broader and more diverse pool of contributors such as users, suppliers, retailers, and professionals from adjacent industries. In some cases, the most valuable contributions emerge from individuals who have little or no prior connection with the co-creating firm. For example, one of the most successful examples of new product co-creation is InnoCentive (innocentive.com), which provides a platform in which firms facing difficult innovation challenges can anonymously seek contributions from a wide array of technicians, scientists, and professionals who are often far outside of the sponsoring firm's domain. In most cases, these contributors are not customers and typically do not even know the name of the firm (since many firms choose to remain anonymous). Due to this lack of familiarity, these contributors are less likely to be constrained by functional fixedness and, instead, offer the benefit of a fresh perspective.

While a lack of familiarity may provide new insights from contributors, co-creating firms that look beyond their customers may face a new set of challenges. In particular, firms that cast a wide net may find it difficult to convey their NPD-related needs to individuals who are largely unfamiliar with their technologies and products. To help overcome this obstacle, firms seeking to expand their contributor base should offer templates and tools to help plant the seeds of innovation. In order to garner relevant and actionable submissions, external idea solicitations should include starter kits, guides, and basic information that makes the innovation problem easy to understand. For example, crowdsourcing platforms such as IdeaScale and InnoCentive provide structured, well-defined innovation problems that are translated to accessible language and, thus, avoid industry-specific terminology that might deter potential contributors who lack experience in a particular industry yet hold innovation expertise that is helpful to solving the challenge.

Firms can also cast a wide net by broadcasting their co-creation initiatives through social media platforms or promoting their innovation challenge to expert communities who may hold relevant knowledge and can approach the problem with a new perspective. For example, Reddit hosts numerous communities centered on specific knowledge interests such as pharmaceuticals, energy, and artificial intelligence. Another important social tool of co-creating firms is to provide a forum for contributors to exchange ideas with each other, which should ultimately improve the overall quality of ideas generated. Some co-creation platforms allow contributors to form communities within the brand's website (e.g., Threadless), which may help generate a market eager to try the eventual co-created product. As a result of casting a wide net, a firm not only benefits by organically developing a test market for a co-created new product but also by freeing up valuable, limited internal resources that can be redeployed to other NPD activities.

18.3.1.3 The Rule of One (Attract New Contributors). Some new product co-creation initiatives are short-term in nature and may focus on obtaining ideas for a particular NPD-related problem. However, co-creation can also be a longer-term strategy for soliciting NPD ideas on a recurring basis. These enduring co-creation platforms are employed by a number of firms such as Threadless and General Mills. In these ongoing co-creation efforts, obtaining a steady supply of contributions is a critical and challenging task. This task is further complicated by what we refer to as "The Rule of One." As noted earlier, within most co-creation initiatives, only a small percentage (often less than 2%) of contributions are selected for eventual product development. Thus, the number of "winners" within any given co-creation challenge is typically quite small. To make matters worse, contributors who win a particular co-creation challenge are unlikely to win future challenges (Bayus 2013). Thus, it is rare to find repeat innovators that can contribute a steady stream of viable new product ideas. In contrast with the appealing notion of the "wisdom of crowds," the vast majority of crowdsourced ideas are not wise. In essence, most external contributors seem to have one good idea (if any). Hence, "The Rule of One."

The Rule of One presents a vexing challenge for sustaining ongoing co-creation initiatives. If only a small percentage of contributed ideas are selected for inclusion in an NPD project, individuals whose ideas are not selected may be unlikely to continue to contribute to future initiatives. Moreover, since the winners of a prior challenge are unlikely to provide fruitful ideas for future challenges, their value to a co-creation platform is greatly diminished. These twin factors put co-creating firms in a rather precarious situation. In essence, most contributors are unlikely to offer viable ideas, while the few that do offer a viable idea are unlikely to have viable ideas in the future.

In order to deal with this dilemma, firms engaged in ongoing co-creation efforts need to have a steady inflow of new contributors (who bring in new ideas). This inflow is necessary to replace the outflow of prior contributors that leave the co-creation platform due to frustration from their ideas repeatedly not being selected. In brief, in order to make an ongoing co-creation initiative sustainable, a firm needs a considerable amount of contributor churn. While most firms are likely to view customer churn as something that is very negative, contributor churn should be viewed in a more positive light in the context of new product co-creation. Firms seeking to employ co-creation as an ongoing strategic initiative should anticipate contributor churn and should regularly promote their co-creation efforts in order to attract new participants.

Firms can also avoid the dangers of The Rule of One by implementing a more focused approach to co-creation. For example, rather than hosting an ongoing co-creation initiative seeking to attract random new product ideas, General Mills divides its new product-related needs into a series of specific challenges. This more targeted approach allows General Mills to attract different sets of contributors to different challenges. In addition to broadening its base of contributors, this tactic helps General Mills to avoid the hazards of repeated rejections since contributors are likely to be more discerning in responding to challenges that more closely match their interests and expertise.

According to Griffin et al. (2014), "serial innovators" comprise a small fraction of the total pool of potential innovators. Thus, firms engaged in ongoing co-creation initiatives should be on the lookout for these rare and talented contributors. Firms that have been engaged in co-creation for a lengthy period of time can identify these individuals by searching their database to identify contributors who have won multiple co-creation challenges and then offer an appropriate incentive to secure their services for future co-creation efforts. For example, Threadless offers individuals that have contributed multiple winning designs the opportunity to fast-track their subsequent design submissions and bypass the typical community-driven idea selection process.

18.3.2 Selection Challenges and Success Factors

18.3.2.1 Protecting Contributors' Egos (Avoid Being the Corporate Villain). As noted earlier, the vast majority of ideas submitted to any new product co-creation initiative are unlikely to be selected. Thus, most contributors will encounter rejection. At some point, having one's ideas repeatedly rejected is likely to damage a contributor's ego. At best, a contributor with a bruised ego is unwilling to keep contributing to a firm's co-creation efforts. At worst, a contributor with a bruised ego may spread negative word-of-mouth (WOM) about the co-creating firm, or perhaps even seek some form of retaliation (Kähr et al. 2016). Thus, protecting the ego of contributors is an important aspect of the co-creation selection process.

One possible solution to this dilemma is to provide contributors with objective and timely feedback about their ideas (regardless of their selection status). If contributors do not receive this type of feedback or are not provided with information about how winning ideas are selected, they will likely become frustrated and may leave the platform. While some degree of churn is good, a firm should still seek to retain and motivate contributors because some may learn from the feedback provided and later suggest useful ideas in subsequent submissions.

Although this feedback principle sounds good in theory, it may be difficult to put into practice, especially if a firm is inundated with thousands of contributions. Indeed, most co-creation initiatives provide little, if any, direct feedback to contributors whose ideas are not selected. Another way that firms can avoid negative backlash from rejection is to post their co-creation challenges on third-party intermediary platforms such as InnoCentive. By using these types of intermediaries, a firm can remain largely anonymous. However, this anonymity comes at a cost, as an intermediary will charge a sizable fee for playing this role. Although providing feedback or transacting through an intermediary are costly endeavors, the benefits may be worth the cost if a firm can avoid being the corporate villain.

Another way that firms can reduce this risk is to delegate (at least in part) the selection process to their broader co-creation communities. For example, LEGO relies upon members of its co-creation platform to help select contributions. Specifically, LEGO posts external contributions on its webpage (ideas.

lego.com) and invites its co-creation community to vote and comment upon all submissions. In order to reach the next stage of its NPD process (i.e., internal review by LEGO managers), a submission must receive 10,000 votes (each community member has one vote) within six months. This is a democratic selection process that avoids the negative stigma sometimes associated with firms that select ideas via an internal closed process. In addition, this approach relieves LEGO from being the bearer of bad news to contributors whose ideas are not selected. As a result, contributors whose ideas are rejected are less likely to pin the blame on LEGO. In addition to letting its co-creation community do the dirty work, this approach also allows LEGO to obtain a sense of the potential market demand for any particular contribution. This tactic can be replicated by other firms and should avoid some of the sting of rejection, and thus, help firms avoid being the corporate villain.

18.3.2.2 Harvesting Co-Created Value (Control the Process). Capturing value from new product co-creation is a key concern for many firms (Almirall and Casadesus-Masanell 2010). A co-created idea gains value as its progresses from contribution to selection. If a firm leverages its community as part of the selection process (e.g., voting), the value of an idea may become apparent to its contributor. This awareness may result in conflict regarding the ownership of an idea and who should reap the harvest of any NPD-related outcomes that it inspires. This potential conflict creates a challenging situation. While an idea may have originated from an external contributor, it typically requires considerable investment from a firm's internal resources to develop, test, produce, and launch any new products that emerge from this idea.

This potential conflict may raise concerns regarding property rights, especially for firms that are accustomed to developing products completely in-house. In addition, firms may feel pressure about the responsibility for ensuring that a co-created product is not only well-developed but also delivered on time. Thus, firms need to establish clear policies and procedures for capturing and transferring intellectual property for ideas selected for further development (Alexy et al. 2009). In essence, firms face the challenge of how to harvest an appropriate degree of value from ideas garnered from a co-creation platform.

One way that firms can harvest the value of new product co-creation is to employ procedures that enable them to maintain control over the most critical elements of an NPD process (which may vary from one firm to another). For example, Threadless allows its co-creation community to play a large role in selecting ideas via commenting and voting. However, once an idea is selected, Threadless takes complete control over the later stages of the NPD process, including production, sale, and delivery. As a result, Threadless maintains not only operational control but also control over the costs, payment mechanism, and revenue associated with its co-creation efforts, while sharing a small portion of each sale with the contributor who submitted the selected idea. On the other hand, like nearly all firms that engage in co-creation, Threadless has little control over the early stages (i.e., fuzzy front end) of the co-creation process. While this lack of control may be a bit unsettling, ceding some control may provide benefits such as a plethora of new ideas as well as a greater sense of perceived ownership among contributors. As a result, customers engaged in a co-creation community may be more willing to accept some responsibility for offerings that may not fully meet their expectations and may be less likely to generate negative WOM toward the firm. Thus, firms should consider controlling only the activities that enable them to harvest sufficient value from their co-creation initiatives.

18.3.2.3 *Maintaining the Peace (Share the Wealth).*

In addition to the need for systems and processes that extract value from new product co-creation, firms also need systems and processes that fairly allocate value to co-creation participants. If co-creation contributors form the impression that a firm is taking advantage of their submissions, they will likely leave and never return. Moreover, negative fallout could spill over to a firm's broader customer base, which could damage its brand image and customer relationships. In essence, firms engaged in new product co-creation need to maintain the peace.

One solution to this challenge is to share the wealth by distributing some of the value of a selected co-creation initiative to multiple contributors and/or providing contributors with widespread public recognition (Hofstetter et al. 2018). For example, the LEGO Ideas program sends winning contributors a variety of LEGO prizes, recognizes them on its website, and also treats them as stars at LEGO autograph signing events. In addition to awarding an overall winner with a unique grand prize package, LEGO also recognizes runner-ups on its website and sends them a smaller prize package. Moreover, LEGO regularly gives public recognition to several high-quality contributions, even if they are not selected as winners. Similarly, GE's open innovation competitions routinely award multiple contributors with substantial monetary awards as well as public recognition. In essence, the selection mechanisms employed by GE and LEGO facilitate a winners' (not a winner's) circle.

By creating a wide winners' circle, a firm's co-creation platform is able to not only maintain the peace among current participants but also make its platform more attractive to future co-creation participants (Hofstetter et al. 2018). By doling out a healthy dose of recognition during selection and broadening the notion of winning, LEGO has been able to build a considerable degree of trust with its contributors, who feel assured that LEGO will treat them fairly. As noted earlier, most co-creation contributors are intrinsically motivated (O'Hern and Rindfleisch 2010). However, this motivation can quickly dissipate if contributors believe that a selection process is opaque or unfair (Mahr et al. 2015). Thus, in addition to establishing a winners' circle, a firm must also clearly communicate how winners were selected. The selection process should enhance a contributor's overall experience to such a degree that they may recommend or even recruit friends to a firm's co-creation platform via positive WOM. This positive WOM, in turn, should help a firm obtain the necessary churn for renewing its participant community and generating a self-sustaining fountain of ideas.

18.4 The New Product Co-Creation Readiness Scorecard

In order to help NPD managers put these key success factors into operation, we developed a Co-Creation Readiness Scorecard to provide managers with a means to assess their organization's aptitude to engage in contribution and selection activities. Specifically, this scorecard (which is depicted in Exhibit 18.2) is comprised of a set of 12 Likert-scale (i.e., 1 = strongly disagree, 5 = strongly agree) items. Half of these items assess contribution readiness and half assess selection readiness. Moreover, effective utilization of new product co-creation depends upon both an organization's internal resources and processes as well as the nature of its external environment (Johnson et al. 2019). Internal resources and processes provide organizations with the capability to solicit a wide array of contributions and successfully select the most viable and promising ideas for potential product development. The external environment plays an important role in terms of generating a ready supply of potential contributions and may provide assistance in terms of selecting the most desirable contributions. Consequently, the scorecard serves as a diagnostic tool to help managers identify both internal and external barriers to new product co-creation.

For each of the key success factors, we provide items that assess these internal and external factors. This structure allows managers to examine their co-creation readiness from a number of different perspectives (i.e., overall readiness, contribution readiness, selection readiness, internal readiness, external readiness). Thus, although our scoring system focuses on overall co-creation readiness, managers can self-score their organization's readiness across these various components (if they so desire). We also provide a scoring guide (i.e., grades) to help managers calibrate their organization's degree of co-creation readiness and use this information to decide if and how best to move forward with new product co-creation. In order to obtain a more accurate and well-rounded assessment, we strongly encourage managers to invite both internal (e.g., NPD managers, R&D managers, marketing managers) and external (e.g., key customers, brand ambassadors, key suppliers) constituents to complete this scorecard and then examine both the mean and standard deviation of each item. Moreover, firms that have a wide product portfolio and/or a diverse customer base may wish to complete separate assessments for various business units, market segments, products, and/or customers.

18.5 Conclusion

New product co-creation is employed by many firms as a popular tool for enhancing their innovation activities. Unfortunately, achieving co-creation's promise is a challenging task that is fraught with potential failure. Our chapter seeks to enhance co-creation success by clearly defining co-creation, identifying its two key stages (i.e., contribution and selection), outlining a set of six key challenges, and suggesting strategies for tackling these challenges. In addition, we offer a scorecard to help managers gauge their organization's readiness for

EXHIBIT 18.2 THE NEW PRODUCT CO-CREATION READINESS SCORECARD.

Contribution Questions

1 *Internal*: My organization has established a set of strong brand associations (Use patches and badges)
2 *Internal*: My organization's products have a high degree of public visibility (Look beyond your customers)
3 *Internal*: My organization is connected to a large population of external contributors who can provide high quality new product ideas (The rule of one)

Internal Contribution Score (higher score indicates lower internal barriers to implementing the contribution phase of co-creation)

4 *External*: Our customers highly value being associated with our company (Use patches and badges)
5 *External*: There are a large number of non-customers who are knowledgeable about my firm and its products (Look beyond your customers)
6 *External*: Anyone can provide valuable product-related ideas to our organization even if they lack specialized knowledge (The rule of one)

External Contribution Score (higher score indicates lower external barriers to implementing the contribution phase of co-creation)

Total Contribution Score (higher score indicates a higher degree of contribution readiness)

Selection Questions

7 *Internal*: My organization has a network of external collaborators to help screen externally generated ideas (Let participants do the dirty work)
8 *Internal*: My organization has processes for capturing value from innovations co-created with external contributions (Harvest co-created value)
9 *Internal*: My organization has processes to recognize the contributions of external co-creators (Share the wealth)

Internal Selection Score (higher score indicates lowers internal barriers to implementing the selection phase of co-creation)

10 *External*: My organization is connected to external contributors that are highly knowledgeable about our products (Let participants do the dirty work)
11 *External*: My organization has a positive reputation among external contributors (Harvest co-created value)
12 *External*: My organization's external contributors are strongly motivated by non-financial rewards (Share the wealth)

External Selection Score (higher score indicates lower external barriers to implementing the selection phase of co-creation)

Total Selection Score (higher score indicates a higher degree of selection readiness)

Combined Contribution & Selection Score (higher score indicates a higher degree of co-creation readiness)

Score Readiness Level
55–60   Good to Go
50–54   Looks Promising
45–49   Exercise Caution
35–44   Danger Zone
0–34   Abandon Mission

Please rate how strongly you agree with each statement on a scale of 1 to 5, where 1 indicates strongly disagree, 3 indicates neither agree nor disagree, and 5 indicates strongly agree.

moving forward with a co-creation initiative. By clearly understanding co-creation's nature, challenges, and solutions, managers should be more likely to utilize this intriguing tool to maximize NPD success.

References

Alexy, O., Criscuolo, P., and Salter, A. (2009). Does IP strategy have to cripple open innovation? *MIT Sloan Management Review* 51 (1): 71–77.

Almirall, E. and Casadesus-Masanell, R. (2010). Open versus closed innovation: a model of discovery and divergence. *Academy of Management Review* 35 (1): 27–47.

Athaide, G.A. and Zhang, J.Q. (2011). The determinants of seller-buyer interactions during new product development in technology-based industrial markets. *Journal of Product Innovation Management* 28 (S1): 146–158.

Bayus, B.L. (2013). Crowdsourcing new product ideas over time: an analysis of the Dell IdeaStorm community. *Management Science* 59 (1): 226–244.

Benkler, Y. (2008). *The Wealth of Networks*. Yale University press.

Crandell, C. (2016). Customer co-creation is the secret sauce to success. *Forbes*, June 10.

Davis, J.R., Richard, E.E., and Keeton, K.E. (2015). Open innovation at NASA: a new business model for advancing human health and performance innovations. *Research-Technology Management* 58 (3): 52–58.

Fisher, G.J. and Fang, E.E. (2018). Customer-driven innovation: a conceptual typology, review of theoretical perspectives, and future research directions. In: *Handbook of Research on New Product Development* (ed. P.N. Golder and D. Mitra), 60–80. Cheltenham, UK: Edward Elgar Publishing.

Franke, N. and Piller, F. (2004). Value creation by toolkits for user innovation and design: the case of the watch market. *Journal of Product Innovation Management* 21 (6): 401–415.

Füller, J., Mühlbacher, H., Matzler, K., and Jawecki, G. (2009). Consumer empowerment through internet-based co-creation. *Journal of Management Information Systems* 26 (3): 71–102.

Gemser, G. and Perks, H. (2015). Co-Creation with customers: an evolving innovation research field: virtual issue editorial. *Journal of Product Innovation Management* 32 (5): 660–665.

Griffin, A., Price, R.L., Vojak, B.A., and Hoffman, N. (2014). Serial Innovators' processes: how they overcome barriers to creating radical innovations. *Industrial Marketing Management* 43 (8): 1362–1371.

Hofstetter, R., Zhang, J.Z., and Herrmann, A. (2018). Successive open innovation contests and incentives: winner-take-all or multiple prizes? *Journal of Product Innovation Management* 35 (4): 492–517.

Hoyer, W.D., Chandy, R., Dorotic, M. et al. (2010). Consumer cocreation in new product development. *Journal of Service Research* 13 (3): 283–296.

Johnson, J.S., Fisher, G.J., and Friend, S.B. (2019). Crowdsourcing service innovation creativity: environmental influences and contingencies. *Journal of Marketing Theory and Practice* 27 (3): 251–268.

Kähr, A., Nyffenegger, B., Krohmer, H., and Hoyer, W.D. (2016). When hostile consumers wreak havoc on your brand: the phenomenon of consumer brand sabotage. *Journal of Marketing* 80 (3): 25–41.

Lee, H. and Suh, Y. (2016). Who creates value in a user innovation community? A case study of MyStarbucksIdea.com. *Online Information Review* 40 (2): 170–186.

Loughlin, R., Salazar, J., Srivastava, S., and Woodruff, S. (2020). Modern CPG product development calls for a new kind of product manager. McKinsey & Company Design Insights. October 22. https://www.mckinsey.com/business-functions/mckinsey-design/our-insights/modern-cpg-product-development-calls-for-a-new-kind-of-product-manager.

Mahr, D., Lievens, A., and Blazevic, V. (2014). The value of customer cocreated knowledge during the innovation process. *Journal of Product Innovation Management* 31 (3): 599–615.

Mahr, D., Rindfleisch, A., and Slotegraaf, R.J. (2015). Enhancing crowdsourcing success: the role of creative and deliberate problem-solving styles. *Customer Needs and Solutions* 2 (3): 209–221.

Mandolfo, M., Chen, S., and Noci, G. (2020). Co-creation in new product development: which drivers of consumer participation? *International Journal of Engineering Business Management* 12: 1–14.

Menguc, B., Auh, S., and Yannopoulos, P. (2014). Customer and supplier involvement in design: the moderating role of incremental and radical innovation capability. *Journal of Product Innovation Management* 31 (2): 313–328.

Moore, J. (2012). Revisiting My Starbucks Idea. https://brandautopsy.com/2012/01/revisiting-my-starbucks-idea.html.

Nambisan, S. and Baron, R. (2009). Virtual customer environments: testing a model of voluntary participation in value co-creation activities. *Journal of Product Innovation Management* 26 (4): 388–406.

Norton, M.I. and Dann, J. (2011). Local Motors: designed by the crowd, built by the customer. *Harvard Business School Marketing Unit Case* (510-062).

O'Hern, M.S. and Rindfleisch, A. (2010). Customer co-creation: a typology and research agenda. *Review of Marketing Research* 6 (1): 84–106.

Peels, J. (2019). Interview with Greg Haye of Local Motors about co-creation and 3D printing. 3DPrint.com: The Voice of 3D Printing/Additive Manufacturing. https://3dprint.com/235395/interview-with-greg-haye-of-local-motors-about-co-creation-and-3d-printing.

Prahalad, C.K. and Ramaswamy, V. (2002). The co-creation connection. *Strategy and Business*, 50–61.

Prahalad, C.K. and Ramaswamy, V. (2004). Co-creation experiences: the next practice in value creation. *Journal of Interactive Marketing* 18 (3): 5–14.

Roberts, D.L. and Candi, M. (2014). Leveraging social network sites in new product development: opportunity or hype? *Journal of Product Innovation Management* 31 (S1): 105–117.

Starbucks. (2013). Starbucks celebrates five-year anniversary of MyStarbucksIdea. https://stories.starbucks.com/stories/2013/starbucks-celebrates-five-year-anniversary-of-my-starbucks-idea/.

Suri, J.F. (2005). *Thoughtless Acts? Observations on Intuitive Design.* Chronicle Books.

Tarantola, A. (2022). After 15 years, Local Motors will reportedly cease operations this Friday. Yahoo! Finance. January 13. https://finance.yahoo.com/news/after-15-years-local-motors-will-reportedly-cease-operations-this-friday-224540158.html?fr=sychp_catchall.

von Hippel, E., Ogawa, S., and De Jong, J. (2011). The age of the consumer-innovator. *MIT Sloan Management Review* 53 (1): 1–16.

Westerski, A., Dalamagas, T., and Iglesias, C.A. (2013). Classifying and comparing community innovation in Idea Management Systems. *Decision Support Systems* 54 (3): 1316–1326.

Gregory J. Fisher is Associate Professor of Marketing at Miami University's Farmer School of Business. Greg earned a PhD in business administration (marketing, strategic management) at the University of Illinois. His research has been published in outlets such as the *Journal of Product Innovation Management, Journal of the Academy of Marketing Science, Journal of Business Research, Industrial Marketing Management, International Business Review,* among others. His research interests include marketing strategy, new product development, open innovation, co-creation, 3D printing, user innovation, marketing alliances, and interorganizational learning.

Aric Rindfleisch is the John M. Jones Professor of Marketing and Executive Director of the Illinois MakerLab at the University of Illinois. Aric's research focuses on understanding our new digital age and has been published in several leading academic journals including the *Journal of Product Innovation Management, Journal of Marketing, Journal of Marketing Research, Journal of Consumer Research, Journal of Operations Management, Strategic Management Journal,* among others. Aric is also an award-winning teacher and teaches three popular Coursera classes (Marketing in a Digital World, The 3D Printing Revolution & The Digital Marketing Revolution) that have collectively enrolled over 600,000 learners.

CHAPTER NINETEEN

CROWDSOURCING AND CROWDFUNDING: EMERGING APPROACHES FOR NEW PRODUCT CONCEPT GENERATION AND MARKET TESTING

Mohammad Hossein Tajvarpour and Devashish Pujari

19.1 Introduction

Crowdsourcing and crowdfunding have become incredibly popular ways of inbound open innovation that allowed firms to take advantage of external innovation sources (Chesbrough 2003). These two approaches of inbound innovation have gained significant traction in the past decade (Pollok et al. 2019; Zhu et al. 2017). One popular platform alone, has collected more than $6 billion dollars from more than 20 million different individuals to actualize more than 219,000 innovative ideas (Kickstarter 2022).

 Both crowdfunding and crowdsourcing are tapping into the wisdom of crowds to generate, test, and refine ideas and new products (Mollick and Nanda 2015). While crowdsourcing collects and refines the ideas from the publics outside the firm, crowdfunding is considered a form of market-testing for the new products and ideas from far and wide. Internet has eased spatial frictions (Agrawal et al., 2015) and made it possible for firms and innovators to collect funds and ideas from very distant individuals. In crowdfunding average distance between supporters and innovators was estimated to be around 2500 km and famous projects such as Oculus Rift, had an average distance of 5000 km from their supporters (Tajvarpour and Pujari 2022).

The PDMA Handbook of Innovation and New Product Development, Fourth Edition. Edited by Ludwig Bstieler and Charles H. Noble.
© 2023 John Wiley & Sons, Inc. Published 2023 by John Wiley & Sons, Inc.

Evidence from corporate world shows that big brands have started incorporating crowdsourcing and crowdfunding into their new product development process. For example, P&G launched its latest Gillette razors after testing the waters in a crowdfunding campaign on Indiegogo, while Lego have been using the same crowdfunding platform to test and validate its new concepts and ideas (Vizard 2019). The same is happening with crowdsourcing, many Fortune 500 businesses are using crowdsourcing to generate creative ideas. Lego Ideas[1] is a crowdsourcing website ran by Lego that runs design contests with the aim of collecting creative ideas from the crowd of customers (Chen et al. 2020). Even NASA[2] is tapping into the wisdom of the crowds for problem solving (Nevo and Kotlarsky 2020).

While these two modern-day approaches have significantly influenced how companies ideate, refine, market-test, and finance their new product and service concepts (Stanko and Henard 2017), there is not yet a comprehensive review of crowdfunding and crowdsourcing. This chapter is intended to address the need for an integrated, state-of-the-art review article in the field of innovation that covers the attributes and applications of crowdfunding and crowdsourcing as two main open-source methods of innovation. It is of utmost importance for practitioners to embrace and incorporate these two crucial phenomena into their innovation strategy as well as for researchers to further investigate these phenomena in their scholarly endeavors.

19.2 Crowdfunding Definition and Typology

19.2.1 Crowdfunding Definition

Crowdfunding, which is also referred to as crowd-sourced funding is an open model of securing funds from a crowd of supporters. One of the very first definitions for crowdfunding was suggested by Schwienbacher and Larralde (2010). According to their definition crowdfunding is an "open call, essentially through internet for the provision of financial resources either in form of donation or in exchange for some form of reward and/or voting rights in order to support initiatives for specific purposes."

A more recent and well-adopted definition is the one suggested by Mollick (2014) who describes crowdfunding as "the efforts by entrepreneurial individuals and groups – cultural, social, and for profit- to fund their ventures by drawing on relatively small contributions from a relatively large number of individuals using the internet, without standard financial intermediaries."

It is clear from both definitions that crowdfunding is based on collective contributions of countless backers. What is important in typology of crowdfunding platforms is the type of reward or return that the supporters expect from the campaign owners. The purpose of crowdfunding varies from non-profit donations (Greenberg and Mollick 2017) to for-profit investments (Ahlers et al. 2015). Those who donate money through crowdfunding have more altruistic motivations, while those who invest in crowdfunding expect financial returns. Accordingly, motivations of contributors are different among different crowdfunding types.

19.2.2 Crowdfunding Typology

Crowdfunding can be classified into four categories of donation-based, reward-based, lending-based, and equity-based. A summary of these categories is provided in Table 19.1.

19.2.2.1 Donation-based Crowdfunding is a crowdfunding model, in which support happens without the expectation of any financial returns, and it is mostly used for charitable or cause related initiatives. Backer motivation in this type of crowdfunding is mainly philanthropic. The backer is supporting a project, a cause, or even another person in-need and does not expect any form of financial return. Examples of such platforms are GoFundMe[3] and DonorsChoose[4] (Xiao and Yue 2021).

19.2.2.2 Reward-based Crowdfunding is a pre-payment crowdfunding model in which although backers do not have expectation of financial return, they expect to receive some form of rewards in return for their contributions (Tajvarpour and Pujari 2022). The reward can be an early-bird version of the product, a discount on the product, or even a thank you letter from the founders. In this crowdfunding model the backers support a product, service or idea that is yet to be developed. It is similar to donation-based crowdfunding, but the goal is not to support a cause, it is to support innovation. Backers know that the risk of failure is high, but they risk their money to support innovation. In this method backers are mainly early adopters who inherently enjoy and value innovation. Examples of such reward-based crowdfunding platforms are Kickstarter[5] and Indiegogo.[6]

19.2.2.3 Lending-based Crowdfunding (Peer-to-Peer Lending) is a model that considers some form of financial return for the lenders that may be higher or equal to the money they lent. The motivation of the contributors can be both philanthropic (prosocial lending) and pecuniary (for-profit lending).

In philanthropic mode, the numerous supporters lend money to a low credit borrower. Although the lenders are expecting to receive their money back plus an interest, they are philanthropically risking their money by

TABLE 19.1 CROWDFUNDING TYPOLOGIES BASED ON THE BACKER MOTIVATIONS.

Crowdfunding Type	Example	Backer Motivation		
		Pecuniary Motivation	Philanthropic Motivation	Backer Innate Innovativeness
Donation-based	Gofundme.com	Low	High	Low
Reward-based	Kickstarter.com	Low	Low	High
Lending-based	*Prosocial lending:* Kiva.org	Low	High	Low
	For-profit lending: Prosper.com	High	Low	Low
Equity-based	Seedinvest.com	High	Low	Low

supporting a low credit borrower, or an entrepreneur in a developing country. One of the best examples of such prosocial lending-based crowdfunding platforms is Kiva[7] that has enabled loans to more than one million people (Berns et al. 2020; Burtch et al. 2014).

The other group of P2P crowdfunding platforms focus on pecuniary motive of collecting an interest rate higher than what the banks pay. For example, lenders risk their money by lending to a person who wants to pay off their credit card at a lower interest rate. The higher the risk of default, the higher the expected interest rate. One of the best-known examples of such platforms is the famous P2P lending website Prosper[8] that has enabled $14 billion dollars of unsecured loans (Netzer et al. 2019).

19.2.2.4 Equity-based Crowdfunding is a model that mimics the angel investment or even a stock market. In this model, the supporters are investors who contribute money to a start-up and in return receive a certain amount of equity or bond-like share of that start-up (Ahlers et al. 2015). Although equity-based crowdfunding has coexisted along with other types of crowdfunding in the UK; and Europe, in the US Equity-based crowdfunding had not been legalized until 2016. Under the title III of JOBS act (Jumpstart Our Business Start-ups), equity-based crowdfunding became available in the US since January 29, 2016. In this model, the main motivation for the backers is to invest in a campaign that brings them substantial financial return in future. Since the main motivation in this model is pecuniary returns, those who support equity-based crowdfunding projects are investors rather than backers. A good example of equity crowdfunding is SeedInvest.[9]

It is clear that the motivations to contribute to crowdfunding campaigns changes based on two important dimensions of innate innovativeness and pecuniary/philanthropic motivations.

Innate innovativeness is an inherent personal trait that affects the rate of innovation adoption among individuals (Im et al. 2003). More innovative consumers are more likely to be early adopters who embrace new products and concepts. These are people who have a more innovative character to embrace innovation and are willing to take risks of adopting new products.

Donation-based crowdfunding is mainly a philanthropic approach to crowdfunding. The backers make their donations to support other human beings, or different cause-related projects. Lending-based crowding is a method that falls in between philanthropic and pecuniary spectrum. In this method lenders can have pecuniary and philanthropic motivations. In prosocial lending, philanthropic supporters gain financial return while helping those in need. In for-profit lending, investors with pecuniary motives receive high interest rates by lending money to their peers. Equity-based crowdfunding is the most similar to regular investments. The main motivation in equity-based crowdfunding is pecuniary. Reward-based crowdfunding is the most interesting type of crowdfunding in which crowdfunding backers commit financial support to an unfinished and even unproved product that is yet to be made (Davis et al. 2017; Tajvarpour and Pujari 2022). In this method, the backers support projects without pecuniary or philanthropic motivations. In the reward-based model creative projects are supported by innovative backers (Davis et al. 2017; Oo et al. 2019).

19.3 The Role of Crowdfunding in Innovation and New Product Marketing

Crowdfunding has become a very effective method of test-marketing new products and ideas. Crowdfunding can help new product development process in four ways: first, it can help with refinement of the product concepts based on backers' comments. Second, it works as a channel for preselling and testing the potential market for the new products. Third, it helps the innovators find the best pricing strategy for their new products. Lastly, it helps with the promotion of the new product.

19.3.1 Crowdfunding Affecting Product Innovation

Since the backers actively engage in conversations that help refinement of innovation and even affect entrepreneurs' future innovation activity, crowdfunding can be considered as an open search for external ideas (Stanko and Henard 2017). Crowdfunding backers are not just prototype testers, but they are user innovators who engage in the co-creation and refinement of a new product or service (Brem et al. 2019). Crowdfunding enables user innovation by creating a community of backers who provide feedback and ideas that help improve the initial product.

19.3.2 Crowdfunding As a Channel for Sales and Test-marketing

With respect to testing potential market, crowdfunding acts as a crowd pre-selling that helps the innovator to not only collect the funds to make their new product, but also gives them a good idea of the market potential (Crosetto and Regner 2018; Stanko and Henard 2017). Crowdfunding is a test market and backers can act as need identifiers. Interestingly, studies show that it is not the amount of funds collected through crowdfunding that predict its future success, but it is the total number of backers that supported them (Stanko and Henard 2017).

19.3.3 Crowdfunding Affecting Pricing Strategy

Crowdfunding is a pre-selling method that can help the innovators with pricing their new products (Crosetto and Regner 2018). Unlike traditional systems in which an innovator makes a product and then sells it for a price, in crowdfunding the innovators are selling a new product at a certain price first and then begin making the new product (Hu et al. 2015). This change in the sequence of pricing and product creation process changes the pricing dynamics for crowdfunded products (Hu et al. 2015). The decisions regarding new product pricing are affected both during the campaign (Guan et al. 2020) and in retail pricing after the campaign (Blaseg et al. 2020).

19.3.4 Crowdfunding Affecting Promotions

Crowdfunding also affects the promotion of the new products (Guan et al. 2020). Crowdfunding backers are early adopters who are highly important for the diffusion of new products. Crowdfunding is a great method of bonding with customers and connecting with them (Bitterl and Schreier 2018). Backers that support a crowdfunding campaign are later likely to become ambassadors that generate positive word of mouth in support of that new product (Stanko and Henard 2017). Crowdfunding backers are more prone to identify and connect with the venture that they funded and this increases their future purchases and their active participation in generating positive word of mouth and marketing buzz around a new product (Bitterl and Schreier 2018; Sahaym et al. 2021).

19.4 Crowdfunding and Gender and Ethnic Equity

It is argued that women are less likely to get support from the traditional entrepreneurial systems dominated by men who control money and power (Gamba and Kleiner 2001). Research shows that ethnic minorities have limited access to credit and have to rely on their personal savings to start their ventures (Freeland and Keister 2016). Women and minority groups are underrepresented in gatekeeping financial and entrepreneurial systems such as venture capitals (Younkin and Kashkooli 2016). Crowdfunding is expected to be a method for democratizing capital (Mollick and Robb 2016) as it removes the traditional gatekeepers (Greenberg and Mollick 2017). Since open-source systems are not controlled by any specific group, it is expected that racial and gender biases should be minimal in crowdfunding. Crowdfunding has proved itself as a medium that reduces barriers against women innovators (Greenberg and Mollick 2017). Crowdfunding's role in reducing ethnic biases is still ambiguous. Studies show that there are still ethnic biases in crowdfunding systems and certain ethnicities have lower chances of success (Younkin and Kuppuswamy 2018). Interestingly, the racial biases in crowdfunding are not taste-based and can be effectively mitigated or even reversed if the minority innovator provides relevant information and reliable quality signals (Younkin and Kuppuswamy 2018, 2019).

19.5 Crowdsourcing Definition and Typology

19.5.1 Crowdsourcing Definition

Crowdsourcing is a term that was first coined by Howe (2006). In this definition, crowdsourcing is referred to as "the act of a company or institution taking a function once performed by employees and outsourcing it to an undefined (and generally large) network of people in the form of an open call. This can take the form of peer-production (when the job is performed collaboratively) but is also often undertaken by sole individuals. The crucial prerequisite is the use of the open call format and the large network of potential laborers." (Howe 2006)

In crowdsourcing, we have two sides, the seeker, and the solvers (similar to creator and backers in crowdfunding). The seeker has a task and is looking for a solution from the crowd of solvers. The seeker will make an open call mainly enabled by the internet (Ghezzi et al. 2018) to a crowd of solvers who perform the task. Compared to the traditional ways of solving the tasks within the organization, crowdsourcing is a very efficient method since it is much faster and much less costly.

19.5.2 Crowdsourcing Typology

The main goal among all crowdsourcing platforms is to use the crowd to perform a task and address a challenge. Although the main idea behind all crowdsourcing platforms is the same, there are different tasks and challenges that they address. There are four different categories of crowdsourcing platforms (Boudreau and Lakhani 2013):

19.5.2.1 Contest Crowdsourcing. It is a process in which the seeker is asking for an innovative task. In this method the solvers need specific skills to be able to participate in the crowdsourcing activity. A good example of such contest platforms is InnoCentive,[10] an online platform that posts challenges and seeks innovative solutions to solve problems from different areas of engineering, health sciences, and even math.

19.5.2.2 Collaborative Communities. In this method the collective and collaborative activities of solvers create a new whole. The system is an aggregate of all contributions and collaborations which mainly benefits from the diversity of skills among solvers (Boudreau and Lakhani 2013). The best examples of collaborative communities in which participants collectively solve a problem and integrate their contributions (Kolbjørnsrud 2017) are Wikipedia and Linux.

19.5.2.3 Complementors. In this method of crowdsourcing, solvers provide innovative user generated solutions to product challenges (Nevo and Kotlarsky 2020). In this model, solvers are not addressing one problem, but they are providing solutions to numerous product problems (Boudreau and Lakhani 2013). The more complementors in a platform the more useful and demanded the product becomes (Boudreau and Lakhani 2013). The best example is the Apple's App Store. The developers use the platform created by the company to create complementary solutions that address a variety of problems from entertainment to education, and even health care. The users' experience with Apple products to a large degree is the result of creativity and innovation of applications that are made by innovators outside of Apple company walls (Kornberger 2017).

19.5.2.4 Service Markets. It is a matching platform that pairs the seeker with relevant solvers. These markets can be categorized into two types of labor-intensive and knowledge-intensive markets (Du and Mao 2018). In the labor-intensive markets, the goal is not innovation, but performing a routine task. On the other hand, in knowledge-intensive service markets the goal is to request innovative tasks from highly skilled solvers.

The best example for labor-intensive markets is Amazon Mechanical Turk (MTurk) that is an open-source platform of matching seekers with workers that perform mundane tasks. A good example of knowledge-intensive market is Upwork,[11] which is a platform that matches seekers with freelancer skilled workers in different areas such as engineering, design, and IT.

19.6 The Role of Crowdsourcing in Innovation and New Product Marketing

19.6.1 Crowdsourcing and New Product/Service Idea Generation

One of the most important roles of crowdsourcing is idea generation. Crowdsourcing has become a very current method of generating, perfecting, and even voting on innovative ideas (Cheng et al. 2020). Many companies are consistently using crowdsourcing communities as a tool for collecting ideas about new products and services from the crowds (Bayus 2013). The crowd that helps with the ideation could be internal or external to the organization. In external crowdsourcing, the company seeks innovative ideas from the general public and mainly its own community of customers. For example, P&G uses its Connect + Develop platform to collect ideas from external participants, mainly its customers (Brokaw 2014). In internal crowdsourcing, the company uses its own internal publics as a source of idea generation and idea refinement (Zuchowski et al. 2016). The rationale for internal crowdsourcing is to leverage the diverse skills and talents of the firm's employees (Simula and Ahola 2014). In the internal crowdsourcing, innovation is not limited to the R&D department and is open to all employees (Simula and Ahola 2014).

In crowdsourcing, the company can instantaneously communicate with the crowd of customers (Devece et al. 2017; Whitla 2009). Crowdsourcing reduces the total amount of time for new product development, reduces the financial costs of this process, and eventually increases the perceived newness of the product idea (Devece et al. 2017).

19.6.2 Crowdsourcing and Promotion and Customer Engagement

Interestingly, crowdsourcing has gone beyond idea generation and has been used in multiple promotion related marketing activities. Companies are widely using crowdsourcing methods to promote their products online (Kim et al. 2018). To be more specific, crowdsourcing has been used for advertisement creativity and brand management. We explain, how companies can benefit from crowdsourcing in their promotional activities.

19.6.2.1 Crowdsourcing for Brand Management in Online Communication Channels. The concept of user generated branding (UGB) underscores the important role of crowdsourcing in promotion and brand management in the Web 2.0 era (Burmann 2010). Brands are using their online social networks (OSN) and Fan Pages to crowdsource many of their brand management tasks (Banerjee and Chua 2019; Hassan Zadeh and Sharda 2014). Many companies are assigning the promotional task of delivering brand messages in online channels to crowdsourced workers (Kim et al. 2018). Brands use campaigns, challenges, and contests that motivate consumer generated branding by asking fans to create and deliver creative brand messages. This not only generates the brand related content but also encourages wide dissemination of that content.

19.6.2.2 Crowdsourcing Advertisement Content.

Consumer generated advertising (CGA) is a method that is being widely used by firms to outsource their advertisement content (Martínez-Navarro and Bigné 2021). Many small and large firms have started outsourcing their advertisement content creation to crowdsourcing platforms (Whitla 2009). These advertising tasks that are being crowdsourced vary from photography and copywriting, to creation of creative visual content (Chen et al. 2020; Whitla 2009). For example, Designhill[12] is a crowdsourcing website that enables small businesses to run logo design contests. Another crowdsourcing platform Zooppa,[13] lets seekers to request creative video and print advertising content from a variety of solvers (Chen et al. 2020). The benefit to the firms is twofold: 1. lower production costs, and 2. more creative content. The advertising content that is generated through crowdsourcing is generally much less expensive than the same content created by ad agencies. At the same time, the diversity in the crowds increases the creativity and newness of the ideas used in the crowdsourced advertising content.

19.7 Barriers to Crowdsourcing and Crowdfunding

Although both crowdfunding and crowdsourcing are very well-adopted by companies and innovators all around the globe, there are still many barriers that impede the ability of companies to take advantage of these two open innovation models. We categorize these barriers into regulatory (legal) barriers, trust issues, and skills required to benefit from open innovation.

19.7.1 Regulatory Barriers

This is specifically affecting crowdfunding method. There are numerous concerns about fraud in crowdfunding that make regulators suspicious of this method. Campaign owners can misrepresent the risks, and gamble on a very risky project with backers' money (Schwienbacher 2018). In the US, the federal government tried to regulate the enterprise crowdfunding under the CROWDFUNDING Act which is part of the Title III of JOBS Act (Hornuf and Schwienbacher 2017). It is important to note that this regulation is mainly toward enterprise crowdfunding (allowing startups to sell stocks using crowdfunding) and does not include reward or donation crowdfunding platforms such as Kickstarter (Connor 2014). The regulations in the US currently forbid regular

TABLE 19.2 RESTRICTIONS TO EQUITY CROWDFUNDING.

Region/ State	Total Collection in a 12-month Period	Regulation	Example
US	$1,000,000 USD	Title III of JOBS Act	seedinvest.com
Canada	$500,000 CAN	Start-up crowdfunding exemptions (limited to six provinces)	vested.ca
UK	Unlimited	Financial Services and Markets Act 2000	seedrs.com
Australia	$5,000,000 AUD ($10,000 per retail investor)	The Corporations Amendment (Crowd-sourced Funding for Proprietary Companies) Act 2018	birchal.com
France	• €100,000 • €1,000,000 if it doesn't exceed 50% of the current capital	Prospectus Directive 2010/73/EU	wiseed.com
Belgium	• €100,000 • €300,000 (limited to less than €1000 per retail investor)	Prospectus Directive 2010/73/EU	spreds.com
Germany	• €2,500,000 (subordinated profit-participating loans) • €100,000 (other crowdfunding types)	Kleinanlegerschutzgesetz	katrim.de
Italy	€5,000,000	Decreto Legge n. 179/2012	backtowork24.com

investors (salary/net worth of less than $100,000) from investing more than 5% or their annual income (or net worth) in equity-based crowdfunding and the annual investment amount should not exceed $2,000 (Hornuf and Schwienbacher 2017). The total amount collected by issuers cannot surpass $1,000,000 (otherwise it needs to be registered with SEC) and the maximum amount of equity sold through crowdfunding to an investor must not surpass $100,000 threshold regardless of their salary or net worth (Hornuf and Schwienbacher 2017). Although equity-based crowdfunding has been legalized in the US and many other countries, the total amount collected and the total investment by retail clients are restricted. Table 19.2 summarizes equity-crowdfunding limitations in eight different countries in North America, Europe, and Oceania.[14]

19.7.2 Intellectual Property and Trust Issues

Protecting intellectual property is a major concern in both crowdfunding and crowdsourcing. The managers need to come up with governance methods that help protect their IP when using external and even internal crowdsourcing (Zuchowski et al. 2016). The disclosure requirements under enterprise crowdfunding regulations in the US create a risk of IP infringement for startups who want to sell securities using crowdfunding (Connor 2014). Public disclosure of IP has very high risks of copycat companies commercializing the idea even before the end of the campaign (Wells 2013).

The issue of intellectual property rights can even be more conflicting when it comes to using backers' and solvers' ideas in innovation. In crowdfunding most of the time backers can provide comments for each campaign, and some of those comments will later be incorporated into the product development by the campaign owners (Wells 2013). A campaign owner that uses publicly posted ideas may be at the risk of future litigations (Wells 2013). Many platforms such as Indiegogo have terms that gives the IP ownership of backer posted ideas to the campaign owners, but many other platforms do not have such provisions (Wells 2013).

IP ownership problem exists in crowdsourcing method as well, and it has been a source of conflict between seekers and solvers (Mazzola et al. 2018). Crowdsourcing platforms need to regulate the method through which seekers can acquire ownership of a solution that was developed by crowds or by winning solvers (Mazzola et al. 2018). The seekers need to come up with arrangements that can help them protect the crowdsourced idea for the purpose of value capture. These IP ownership arrangement methods should not be so strict to demotivate or even scare the solvers away from sharing their solutions (Mazzola et al. 2018).

19.7.3 Skills, Knowledge, and Competencies

Both crowdfunding and crowdsourcing require certain skills and competencies. Each online platform has its own rules and procedures that the entrepreneurs need to educate themselves about. For example, Kickstarter is an AON (All-Or-Nothing) platform which means the entrepreneurs will receive the funds only and only if they reach the amount that they set as their goal in the beginning of the campaign. GoFundMe on the other hand has a Keep-It-All policy, which means the crowdfunder will receive all the money that backers contributed even if the campaign fails to reach its goal. Entrepreneurs need to be clear about the crowdfunding process in each platform before deciding about the platform they want to use.

The other important skill that the entrepreneurs and innovators need is content creation. Successful crowdfunding campaigns need a good narrative and a high-quality video pitch at a minimum. Making such content requires time, budget, and skills. Practitioners who want to use crowdfunding and crowdsourcing also need to decide about the goal/prize and the duration that maximizes backers/solvers contribution (Bernardino and Santos 2020).

Another very important factor is the technological skills for collecting ideas and suggestions from the crowds. Crowdfunding and crowdsourcing campaigns can collect hundreds and even thousands of ideas and comments from the crowd. The collected ideas are not useful unless the company has an efficient way of organizing and analyzing big unstructured data.

19.8 Conclusions and Managerial Recommendations

Crowdfunding and crowdsourcing have been widely used in the business world and their use is growing at a very fast pace. Although both of these methods fall under the umbrella of open innovation (Stanko et al. 2017), and sometimes offer similar benefits, their use and their role in innovation and new product development varies vastly. Table 19.3 provides a comparison of these two open innovation models. There are three required components for any project on creating a new product. The first is ideation. No new product can be created without a new idea. The second is financial/funding support. Without financial/funding support, firms/entrepreneurs cannot turn an idea into a product. The third component is managerial services to execute the project. Firms/entrepreneurs need to have managerial capabilities to execute the new product development tasks. In the process of creating a new product, firms/entrepreneurs can rely on the crowd for part of these three components.

Below we explain and differentiate the role that crowd can play through crowdfunding and crowdsourcing with regards to these three components (ideation, financial support, and managerial services).

19.8.1 The Role of Crowdfunding and Crowdsourcing in Ideation

Entrepreneurs and firms who wish to open their innovation process need to consider that crowdsourcing is a method that is beneficial for the ideation phase of new product development, while crowdfunding is a method that should be used at the later stages of new product development. Crowdfunders can still improve their idea based on comments and feedback from their backers, but the main ideation should have happened in the firm before crowdfunding takes place.

19.8.2 The Role of Crowdfunding and Crowdsourcing in Providing Financial Support

Crowdfunding's main goal is to provide financial support for innovation, but crowdsourcing has no role in that matter. All typologies of crowdfunding are built around collecting financial support/funding while no crowdsourcing method includes collecting financial support from the crowd. In crowdsourcing, capital is provided by the entrepreneur/firm.

TABLE 19.3 CROWDFUNDING AND CROWDSOURCING COMPARISON.

	Crowdsourcing	Crowdfunding
Ideation	Crowd	Firm (Entrepreneur)
Managerial services	Firm (Entrepreneur)	Firm (Entrepreneur)
Financial support	Firm (Entrepreneur)	Crowd

19.8.3 The Role of Crowdfunding and Crowdsourcing in Providing Managerial Support

Both of these methods require skilled managerial and administrative support. Although in crowdsourcing some basic administrative tasks can be outsourced to the crowds, still it is the firm's managerial capabilities that make a crowdsourced product successful. Crowdsourcing is most beneficial as an ideation method used at the very early stages of new product development. It is crucial for firms/entrepreneurs who want to benefit from crowdfunding and crowdsourcing to have managerial skills of directing the crowd and using their ideas and finances to actualize a new product.

19.8.4 Managerial Recommendations

Crowdfunding and crowdsourcing are two modern mechanisms for open innovation. Based on what we discussed so far, each method has its own advantages and limitations. Mangers who want to benefit from crowdsourcing and crowdfunding should be aware and familiar with the advantages and purpose of each method. In Table 19.4 we have summarized the recommendations that can help managers get the most from each of these two open innovation methods.

TABLE 19.4 MANAGERIAL RECOMMENDATIONS.

Managerial Recommendations	Crowdsourcing	Crowdfunding
Ideation/Idea testing	• Crowdsourcing can help in ideation and idea improvement/refinement. Firm/entrepreneur as a solution seeker can ask the crowdsourcing solvers to propose ideas and solutions for very simple to very complex tasks.	• Crowdfunding can be used as a test market to predict potential demand for a new product/idea. • Firms/entrepreneurs can refine their new product ideas based of backers' feedback/comments, but crowdfunding is not an effective method of idea generation.
Intellectual Property Rights (IPR)	• Ownership of the ideas and solutions that are provided by the crowd must be clarified. • Managers need to make sure that they respect the intellectual property rights of the solvers who contribute to performing crowdsourced tasks. Otherwise, the crowd will not be motivated to participate.	• The publicly crowdfunded idea can be duplicated by copycats around the world. Non-disclosure agreement (NDA) can be used to protect your idea. • It is important to have clauses that clarify the ownership of ideas that are provided by the crowd through comments, messages, etc.

(Continued)

TABLE 19.4 (CONTINUED)

Managerial Recommendations	Crowdsourcing	Crowdfunding
Financial caps	• N/A	Each country/region has certain limits for the total funds that can be collected through crowdfunding. These regulations get updated frequently. Make sure to abide the most up-to-date regulations.
Platform policy	• Platforms have certain mechanism for running crowdsourcing contests and selection of the winning solution. These mechanisms vary from one platform to another. • Platforms may have certain policies regarding the ownership of ideas that are generated by the crowd. Make sure to check these policies before posting a crowdsourcing task.	• Each crowdfunding platform has certain policies regarding the transfer of funds from backers to the campaign owners. The fees that the platform charges, and the timing of the fund transfer can vary from one platform to another. • Platforms may have or have not set policies regarding the ownership of ideas that are provided by the crowd through comments, messages, etc.
Software	• Thousands of comments and solutions that are suggested by the crowd are useless unless the company uses advanced software to organize and use those comments and solutions.	• Backers not only make financial contributions, but also post comments and suggestions. Software capabilities are required to collect and use the crowd comments.
Financial support	• N/A	Managers need to pick the most relevant type of crowdfunding for their purpose: • Start-ups can use equity-based crowdfunding. • Innovative and risky product ideas can benefit from reward-based crowdfunding. • Small businesses looking for loans can use lending-based crowdfunding. • Charity, and cause related activities can use donation-based crowdfunding.

Notes

1   https://ideas.lego.com.
2   https://www.nasa.gov/solve/index.html.
3   https://www.gofundme.com.
4   https://www.donorschoose.org.
5   https://www.kickstarter.com.
6   https://www.indiegogo.com.

7   https://kiva.org.
8   https://prosper.com.
9   https://seedinvest.com.
10   https://www.innocentive.com.
11   https://www.upwork.com.
12   https://designhill.com.
13   https://zooppa.com.
14   Based on Hornuf and Schwienbacher (2017) and local government websites.

References

Agrawal, A., Catalini, C., and Goldfarb, A. (2015). Crowdfunding: geography, social networks, and the timing of investment decisions. *Journal of Economics & Management Strategy* 24 (2): 253–274. doi:10.1111/jems.12093.

Ahlers, G.K.C., Cumming, D., Günther, C., and Schweizer, D. (2015). Signaling in equity crowdfunding. *Entrepreneurship Theory and Practice* 39 (4): 955–980. doi:10.1111/etap.12157.

Banerjee, S. and Chua, A.Y.K. (2019). Identifying the antecedents of posts' popularity on Facebook Fan Pages. *Journal of Brand Management* 26 (6): 621–633. doi:10.1057/s41262-019-00157-7.

Bayus, B.L. (2013). Crowdsourcing new product ideas over time: an analysis of the Dell IdeaStorm community. *Management Science* 59 (1): 226–244. doi:10.1287/mnsc.1120.1599.

Bernardino, S. and Santos, J.F. (2020). Crowdfunding: an exploratory study on knowledge, benefits and barriers perceived by young potential entrepreneurs. *Journal of Risk and Financial Management* 13 (4): 81. doi:10.3390/jrfm13040081.

Berns, J.P., Figueroa-Armijos, M., da Motta Veiga, S.P., and Dunne, T.C. (2020). Dynamics of lending-based prosocial crowdfunding: using a social responsibility lens. *Journal of Business Ethics* 161 (1): 169–185. doi:10.1007/s10551-018-3932-0.

Bitterl, S. and Schreier, M. (2018). When consumers become project backers: the psychological consequences of participation in crowdfunding. *International Journal of Research in Marketing* 35 (4): 673–685. doi:10.1016/j.ijresmar.2018.07.001.

Blaseg, D., Schulze, C., and Skiera, B. (2020). Consumer protection on Kickstarter. *Marketing Science* 39 (1): 211–233. doi:10.1287/mksc.2019.1203.

Boudreau, K.J. and Lakhani, K.R. (2013). Using the crowd as an innovation partner. *Harvard Business Review*, 61–69.

Brem, A., Bilgram, V., and Marchuk, A. (2019). How crowdfunding platforms change the nature of user innovation – from problem solving to entrepreneurship. *Technological Forecasting & Social Change* 144 (April 2016): 348–360. doi:10.1016/j.techfore.2017.11.020.

Brokaw, L. (2014). *How procter & gamble uses external ideas for internal innovation*. MIT Sloan Management Review. https://sloanreview.mit.edu/article/how-procter-gamble-uses-external-ideas-for-internal-innovation (accessed June 14, 2022).

Burmann, C. (2010). A call for 'User-Generated Branding'. *Journal of Brand Management* 18 (1): 1–4. doi:10.1057/bm.2010.30.

Burtch, G., Ghose, A., and Wattal, S. (2014). Cultural differences and geography as determinants of online prosocial lending. *MIS Quarterly* 38 (3): 773–794. doi:10.25300/MISQ/2014/38.3.07.

Chen, L., Xu, P., and Liu, D. (2020). Effect of crowd voting on participation in crowdsourcing contests. *Journal of Management Information Systems* 37 (2): 510–535. doi:10.1080/07421222.2020.1759342.

Cheng, X., Fu, S., de Vreede, T. et al. (2020). Idea convergence quality in open innovation crowdsourcing: a cognitive load perspective. *Journal of Management Information Systems* 37 (2): 349–376. doi:10.1080/07421222.2020.1759344.

Chesbrough, H. (2003). The era of open innovation. *MIT Sloan Management Review* 44 (3): 35–41.

Connor, S.M.O. (2014). Crowdfunding's impact on start-up IP strategy. *George Mason Law Review* 21 (4): 895–918.

Crosetto, P. and Regner, T. (2018). It's never too late: funding dynamics and self pledges in reward-based crowdfunding. *Research Policy* 47 (8): 1463–1477. doi:10.1016/j.respol.2018. 04.020.

Davis, B.C., Hmieleski, K.M., Webb, J.W., and Coombs, J.E. (2017). Funders' positive affective reactions to entrepreneurs' crowdfunding pitches: the influence of perceived product creativity and entrepreneurial passion. *Journal of Business Venturing* 32 (1): 90–106. doi:10.1016/j.jbusvent.2016.10.006.

Devece, C., Llopis-Albert, C., and Palacios-Marqués, D. (2017). Market orientation, organizational performance, and the mediating role of crowdsourcing in knowledge-based firms. *Psychology and Marketing* 34 (12): 1127–1134. doi:10.1002/mar.21053.

Du, W.D. and Mao, J.-Y. (2018). Developing and maintaining clients' trust through institutional mechanisms in online service markets for digital entrepreneurs: a process model. *The Journal of Strategic Information Systems* 27 (4): 296–310. doi:10.1016/j. jsis.2018.07.001.

Freeland, R.E. and Keister, L.A. (2016). How does race and ethnicity affect persistence in immature ventures? *Journal of Small Business Management* 54 (1): 210–228. doi:10.1111/ jsbm.12138.

Gamba, M. and Kleiner, B.H. (2001). The old boys' network today. *International Journal of Sociology and Social Policy*. doi:10.1108/01443330110789853.

Ghezzi, A., Gabelloni, D., Martini, A., and Natalicchio, A. (2018). Crowdsourcing: a review and suggestions for future research. *International Journal of Management Reviews* 20 (2): 343–363. doi:10.1111/ijmr.12135.

Greenberg, J. and Mollick, E. (2017). Activist choice homophily and the crowdfunding of female founders. *Administrative Science Quarterly* 62 (2): 341–374. doi:10.1177/ 0001839216678847.

Guan, X., Deng, W.-J., Jiang, -Z.-Z., and Huang, M. (2020). Pricing and advertising for reward-based crowdfunding products in E-commerce. *Decision Support Systems* 131 (February 2019): 113231. doi:10.1016/j.dss.2019.113231.

Hassan Zadeh, A. and Sharda, R. (2014). Modeling brand post popularity dynamics in online social networks. *Decision Support Systems* 65 (C): 59–68. doi:10.1016/j.dss.2014.05.003.

Hornuf, L. and Schwienbacher, A. (2017). Should securities regulation promote equity crowdfunding? *Small Business Economics* 49 (3): 579–593. doi:10.1007/s11187-017-9839-9.

Howe, J. (2006). *Crowdsourcing: A definition*. https://crowdsourcing.typepad.com/cs/2006/06/ crowdsourcing_a.html.

Hu, M., Li, X., and Shi, M. (2015). Product and pricing decisions in crowdfunding. *Marketing Science* 34 (3): 331–345. doi:10.1287/mksc.2014.0900.

Im, S., Bayus, B.L., and Mason, C.H. (2003). An empirical study of innate consumer innovativeness, personal characteristics, and new-product adoption behavior. *Journal of the Academy of Marketing Science* 31 (1): 61–73. doi:10.1177/0092070302238602.

Kickstarter (2022). *Kickstarter Stats*. https://www.kickstarter.com/help/stats.

Kim, H.-J., Lee, J., Chae, D.-K., and Kim, S.-W. (2018). Crowdsourced promotions in doubt: analyzing effective crowdsourced promotions. *Information Sciences* 432: 185–198. doi:10.1016/j.ins.2017.12.004.

Kolbjørnsrud, V. (2017). Agency problems and governance mechanisms in collaborative communities. *Strategic Organization* 15 (2): 141–173. doi:10.1177/1476127016653727.

Kornberger, M. (2017). The visible hand and the crowd: analyzing organization design in distributed innovation systems. *Strategic Organization* 15 (2): 174–193. doi:10.1177/ 1476127016648499.

Martínez-Navarro, J. and Bigné, E. (2021). Sponsored consumer-generated advertising in the digital era: what prompts individuals to generate video ads, and what creative strategies do they adopt? *International Journal of Advertising* 1–32. https://doi.org/10.1080/026504 87.2021.1972586.

Mazzola, E., Acur, N., Piazza, M., and Perrone, G. (2018). "To own or not to own?" A study on the determinants and consequences of alternative intellectual property rights arrangements in crowdsourcing for innovation contests. *Journal of Product Innovation Management* 35 (6): 908–929. doi:10.1111/jpim.12467.

Mollick, E. (2014). The dynamics of crowdfunding: an exploratory study. *Journal of Business Venturing* 29 (1): 1–16. doi:10.1016/j.jbusvent.2013.06.005.

Mollick, E. and Nanda, R. (2015). Wisdom or madness? Comparing crowds with expert evaluation in funding the arts. *Management Science* 62 (6): 1533–1553. doi:10.1287/mnsc.2015.2207.

Mollick, E. and Robb, A. (2016). Democratizing innovation and capital access: the role of crowdfunding. *California Management Review* 58 (2): 72–87. doi:10.1525/cmr.2016.58.2.72.

Netzer, O., Lemaire, A., and Herzenstein, M. (2019). When words sweat: identifying signals for loan default in the text of loan applications. *Journal of Marketing Research* 56 (6): 002224371985295. doi:10.1177/0022243719852959.

Nevo, D. and Kotlarsky, J. (2020). Crowdsourcing as a strategic IS sourcing phenomenon: critical review and insights for future research. *The Journal of Strategic Information Systems* 29 (4): 101593. doi:10.1016/j.jsis.2020.101593.

Oo, P.P., Allison, T.H., Sahaym, A., and Juasrikul, S. (2019). User entrepreneurs' multiple identities and crowdfunding performance: effects through product innovativeness, perceived passion, and need similarity. *Journal of Business Venturing* 34 (5): 105895. doi:10.1016/j.jbusvent.2018.08.005.

Pollok, P., Lüttgens, D., and Piller, F.T. (2019). How firms develop capabilities for crowdsourcing to increase open innovation performance: the interplay between organizational roles and knowledge processes. *Journal of Product Innovation Management* 36 (4): 512–441.

Sahaym, A., (Avi) Datta, A., and Brooks, S. (2021). Crowdfunding success through social media: going beyond entrepreneurial orientation in the context of small and medium-sized enterprises. *Journal of Business Research* 125 (September 2019): 483–494. doi:10.1016/j.jbusres.2019.09.026.

Schwienbacher, A. (2018). Entrepreneurial risk-taking in crowdfunding campaigns. *Small Business Economics* 51 (4): 843–859. doi:10.1007/s11187-017-9965-4.

Schwienbacher, A. and Larralde, B. (2010). Crowdfunding of small entrepreneurial ventures. *Handbook of Entrepreneurial Finance* 2010: 1–23. doi:10.2139/ssrn.1699183.

Simula, H. and Ahola, T. (2014). A network perspective on idea and innovation crowdsourcing in industrial firms. *Industrial Marketing Management* 43 (3): 400–408. doi:10.1016/j.indmarman.2013.12.008.

Stanko, M.A., Fisher, G.J., and Bogers, M. (2017). Under the wide umbrella of open innovation. *Journal of Product Innovation Management* 34 (4): 543–558. doi:10.1111/jpim.12392.

Stanko, M.A. and Henard, D.H. (2017). Toward a better understanding of crowdfunding, openness and the consequences for innovation. *Research Policy* 46 (4): 784–798. doi:10.1016/j.respol.2017.02.003.

Tajvarpour, M.H. and Pujari, D. (2022). Bigger from a distance: the moderating role of spatial distance on the importance of traditional and rhetorical quality signals for transactions in crowdfunding. *Decision Support Systems* 156: 113742. doi:10.1016/j.dss.2022.113742.

Vizard, S. (2019). *How Coca-Cola, Lego and Gillette tapped into the wisdom of crowds.* Marketing Week. https://www.marketingweek.com/coca-cola-lego-gillette-crowdfunding.

Wells, N. (2013). The risks of crowdfunding: most have the best intentions when it comes to crowdfunding an ambitious project, but intellectual property issues, ownership rights and perk obligations present potential hurdles to making a dream become reality. *Risk Managemen* 60 (2): 26–30.

Whitla, P. (2009). Crowdsourcing and its application in marketing activities. *Contemporary Management Research* 5 (1): 15–28. doi:10.7903/cmr.1145.

Xiao, S. and Yue, Q. (2021). The role you play, the life you have: donor retention in online charitable crowdfunding platform. *Decision Support Systems* 140 (October 2020): 113427. doi:10.1016/j.dss.2020.113427.

Younkin, P. and Kashkooli, K. (2016). What problems does crowdfunding solve? *California Management Review* 58 (2): 20–43. doi:10.1525/cmr.2016.58.2.20.

Younkin, P. and Kuppuswamy, V. (2018). The colorblind crowd? Founder race and performance in crowdfunding. *Management Science* 64 (7): 3269–3287. doi:10.1287/mnsc.2017.2774.

Younkin, P. and Kuppuswamy, V. (2019). Discounted: the effect of founder race on the price of new products. *Journal of Business Venturing* 34 (2): 389–412. doi:10.1016/j.jbusvent.2018.02.004.

Zhu, J.J., Li, S.Y., and Andrews, M. (2017). Ideator expertise and cocreator inputs in crowdsourcing-based new product development. *Journal of Product Innovation Management* 34 (5): 598–616.

Zuchowski, O., Posegga, O., Schlagwein, D., and Fischbach, K. (2016). Internal crowdsourcing: conceptual framework, structured review, and research agenda. *Journal of Information Technology* 31 (2): 166–184. doi:10.1057/jit.2016.14.

Mohammad Hossein Tajvarpour, PhD, is an assistant professor at the School of Business, at the State University of New York at Oswego. His research is focused on Data Analytics, Innovation, Crowdfunding, and Entrepreneurial Marketing. His crowdfunding research has been published in academic journals such as *Journal of Business Research*, and *Decision Support Systems*.

Devashish Pujari, PhD, is a professor at the DeGroote School of Business, McMaster University, Canada. He has published in journals among others *Journal of Academy of Marketing Science, Journal of Product Innovation Management, Journal of Business Research, Decision Support Systems* and *Journal of Business Ethics*.

TRANSFORMATIVE FORCES OF NEW PRODUCT DEVELOPMENT AND INNOVATION

CHAPTER TWENTY

DIGITAL TRANSFORMATION IN THE MAKING: LESSONS FROM A LARGE ENERGY COMPANY

Luigi M. De Luca, Andrea Rossi, Zahir Sumar, and Gabriele Troilo

20.1 Summary

The objective of this chapter is to describe the journey of Centrica Plc from a traditional energy retailer toward a digital service company. The shift follows the threat of commoditization of the core business but is also driven by the huge innovation opportunities afforded by digital technologies, big data and analytics that are part of the assets available to the organization. The journey is characterized by both successes and failures. Far from having been fully realized, this transformation is still in the making. The insights reported in this chapter are backed by a longitudinal case study of the company's digital transformation, which includes empirical evidence from participant observation, project reviews, and other sources of primary data from the company.[1] The overarching question discussed in this chapter is: *How does a large incumbent organization transform itself through a digital transformation?* In describing the key phases of Centrica's digital transformation journey over the last ten years, we focus on a specific innovation team (the *Accelerator Hub*) and describe how they have successfully promoted new product/service development approaches such as Design Thinking, agile, and other techniques to support the organizational transition from a traditional to a digital innovation environment. The chapter will also discuss challenges, problems, and limitations met by the team and the organization, thus offering

The PDMA Handbook of Innovation and New Product Development, Fourth Edition. Edited by Ludwig Bstieler and Charles H. Noble.

a realistic account that resonates with the experience of practitioners tasked with managing digital transformation in other companies.

20.2　Digital Transformation in the Energy Industry

Over the last ten years, the traditional services, business models and organizational systems in the energy sector have been increasingly impacted by digitalization and the big data revolution (De Luca et al. 2021). The advent of smart meters and domestic IoT (i.e., Google Nest) gave energy companies a leading role within an ecosystem of digital technologies revolving around energy, which generate huge volumes of customer data in real time. Such an increased rate of digital innovation in the industry has spurred market entry by a new generation of pure digital players targeting tech-savvy and price-sensitive customers with innovations such as smart offers and flexible tariffs (Zhou et al. 2016). In addition, growing sales of private hybrid and fully electric vehicles, and the increasing electrification of commercial fleet and public transport networks are driving a seamless integration between energy and mobility, adding a new source of big data from geo-localization for energy companies. Finally, enduring concerns and mounting awareness of the dangers related to climate change have been driving a shift to private and public investments in renewable technologies as well as innovations in energy production that push toward household self-sufficiency and peer-to-peer energy trading powered by blockchain. All these factors have converged on incumbent players in the energy industry, creating a perfect storm for the disruption of traditional services, business models, and organizational structures (McKinsey 2020).

Centrica is part of the so-called big six energy players in the United Kingdom. The company's strategy is heavily focused on excellent customer service as well as driving for higher returns on investment through greater efficiency. As of March 2018, *British Gas* (Centrica's main brand in the UK) served 5.5 million electricity customers and 6.9 million gas customers in the UK. As far back as the early 2000s, Centrica had started investing in their internet-based customer services, introducing electronic processing of meter readings, bills and payments. More recently, Centrica responded proactively to the UK Government mandate[2] to install smart meters and was one of the first big industry players to embrace digital transformation and launch new digital services in the energy market.

20.3　Methodology

The insights presented in this chapter emerge from a longitudinal and multi-method data collection process that spanned at least five years. In 2017, the research team conducted four semi-structured interviews with Centrica's management and data science team as part of a project on data-driven service innovation. These interviews were fully transcribed and manually coded. As a follow-up to that first interaction, Centrica entered into an industry–academia knowledge transfer partnership (KTP) with the UK-based research team, funded by a government grant, between 2018 and 2021. A requirement of receiving the grant was that the team should use rigorous methodological approaches to create new knowledge to be embedded in the organization. To this end, the KTP research associate conducted semi-structured interviews, participant and non-participant observations over three years, while working on a daily basis as part of the company. Interviews were fully transcribed and coded to identify success factors for (and barriers to) innovation within the data science team. Participant and non-participant observation generated a large quantity of field notes and memos. Finally, the project database included different types of company documents (i.e., archival data, reports, organizational charts, emails, and other written communication), which were triangulated with primary data. All the evidence collected was presented quarterly to the project committee, which included the academic team, Centrica's management, an independent project adviser to represent the funder, and a project secretary who took full minutes of each discussion. Together, the 11 project committee meetings (from October 2018 to December 2021) provide a full trail of evidence, which includes project reviews, reports of emerging evidence, benefit/risk analyses, and action plans. Thus, the insights presented in this chapter are supported by a longitudinal and multi-method action research process. Our protracted engagement with the case study offered the chance to move beyond short-term observation toward a more in-depth investigation of the digital transformation process (Marion and Fixson 2021). The next section outlines how Centrica's digital transformation began.

20.4 Early Innovation Successes

At the start of 2013, Centrica created a team of young and talented data scientists. This team was very entrepreneurial and started experimenting with a small-scale Hadoop cluster running on Raspberry Pis, championed by the Director of Strategic Systems and the CIO. This system initially ran in parallel with the traditional data warehouses, but quickly ended up replacing the old technology stack. This successful experiment resonated within the organization, and soon the nascent data science team became involved with several innovation projects centered around digital technologies and data (see Table 20.1).

Digital innovation in this phase focused on few but highly relevant projects that propelled the company toward a future as a data company. In fact, despite being an incumbent, Centrica found itself at the forefront of digital innovation in the energy industry, where only a few players were active at the time. As stated earlier, the engine behind these early successes was a strong *IT-enabled entrepreneurial spirit* among the data science team. Interestingly, to mark their entrepreneurial identity as different from the prevailing culture within the organization

TABLE 20.1 EARLY-STAGE DIGITAL INNOVATION SUCCESSES FOR CENTRICA.

New Service	Description	Concept
Hive	Ecosystem of Smart Devices that can be controlled through a single app.	Hive's mission is to make customers' life easier when it comes to home management. They offer a range of IoT devices, including heating (smart thermostats), security (smart cameras), lighting (smart lights), sensors (doors and windows), leak monitoring and, more recently, smart electric vehicle charging solutions.
		Hive was spun off from Centrica in 2012 and is now a strong player in the IoT/Smart-home spaces. All the devices can be connected to each other and remotely controlled through a single app, and can seamlessly integrate and interact with Amazon Alexa, Google Assistant, and Siri. More recently, Hive has been brought back into the core business, under the British Gas Services & Solution Space.
Local Heroes	Digital platform to connect local tradespeople and customers in need for a service and repair job.	Local Heroes is a local tradespeople platform backed by British Gas. The initial idea was to build an "Uber experience" for heating maintenance, and gas/electrical repair. Instead of providing these services via their own engineers' fleet, British Gas created a platform to connect customers and local tradespeople. By vetting the traders, underwriting the work, and managing secured payment, Local Heroes enables a fully digital experience to its customers. By entering the type of job needed (e.g., "boiler service," "blocked sink") and the postcode, the platform returns a series of options for local traders (plumbers, electrician, boiler engineers) available to do the job and allows customers to book them online in the space of minutes. Since it was spun off from Centrica in 2016, Local heroes has developed a network of 7,000 local companies across the UK and helped almost 150,000 customers in their homes. Local Heroes has now been brought back into the core business, under the British Gas Services & Solution Space.
Centrica Business Solutions	Distributed energy trial to empower local communities to sell and buy energy through a range of advanced technologies.	Local Energy Market (LEM) was a three-year trial to pilot future energy solutions using blockchain and AI technologies. Centrica installed a range of advanced technologies such as combined heat and power (CHP) systems, smart batteries, solar panels, and monitoring equipment in 100 homes and 81 businesses across Cornwall, England. Powered by solar panels, the energy in the batteries was aggregated and controlled remotely to provide a single block of power and flexibility to the local grid and national system, making it one of the most advanced examples of Local Power Plants in the UK. Customers could log onto the LEM platform, interact directly with the network operators to vary their power use in a way that would help them to manage the grid and receive monetary compensation. This resulted in about 900 tons of CO_2/year saved by residential customers and about 8,500 tons of CO_2/year for business participants. This venture is now part of the "Distributed Energy & Power" business unit, under the Centrica Business Solutions umbrella.

(Continued)

TABLE 20.1 (CONTINUED)

New Service	Description	Concept
Io-Tahoe	Automated Data Lake solution to discover insights and connections from unstructured data.	Io-Tahoe is an automated Data Lake solution enabling unstructured data discovery, categorization, and mapping. This solution enables commercial customers (particularly large organizations) to automate their data management and realize strategic and operational benefits. The platform was initially developed by Centrica to respond to the increasing demand for data storage, data analytics, and data governance within the company. Once the solution was developed and rolled out internally by the data science team, they realized the potential market value of selling the service to external organizations (including competitors) facing similar challenges. Unlike Hive and Local Heroes, Io-Tahoe was sold externally in 2016 and is no longer part of the Centrica ecosystem.

(which tended to be slow-moving and risk adverse), the data science team operated as a profit center within their department: they did not have a fixed budget, but instead they went out to the business lines hunting for projects and "billed" the internal customers for their time. In addition, the data science team also started to pursue external consulting opportunities, with the idea of commercializing their expertise outside Centrica. Later, the data science director reflected that while this "transactional" approach was meant to stimulate the entrepreneurial mindset within the team, it also limited internal collaboration and the emergence of a coherent and long-term oriented innovation strategy within the team.

Despite their significance for Centrica, the innovations that emerged during this stage resulted from an ad-hoc, slow, and expensive innovation process, characterized by difficulties and mistakes, and not supported by systematic innovation capabilities within the firm. In hindsight, delivering innovation was not a pleasant experience for those involved. Eventually, several new businesses had to be spun off or sold, after costly attempts to grow and develop them internally. Through this experience, Centrica started to realize that its culture, structures, and processes were not enabling innovation, and this realization marked the next stage of the company's innovation journey, which is described in the following section.

20.5 The Attention Turns to Organizational Change

The struggles experienced with the first generation of digital innovation projects triggered a wider effort toward a series of organizational changes, both structural and cultural, seen as necessary to fully unlock Centrica's innovation potential in future projects. This phase was characterized by a slower pace of innovation, during which no significant new product was introduced, and a shift of emphasis from the external environment (entrepreneurial opportunities) to the internal environment (organizational barriers).

Business and innovation literature has widely documented the importance of the people and other organizational enablers of digital innovation in established organizations (O'Connor et al. 2018). These include various dimensions of organizational design, such as: leadership and top-management support (Lanzolla et al. 2021), cross-functional collaboration between IT and other departments (Troilo et al. 2017), data-oriented culture (Davenport and Bean 2018), agile working methods (Augustine 2005), and data science talent management (Davenport and Patil 2012). In particular, a recent stream described and highlighted the advantages of integrating traditional stage-gate methodologies with agile, to form a new generation of hybrid agile stage-gate approaches to new product development that can effectively serve the digital transformation of incumbent companies (Cooper and Sommer 2016). Within Centrica, the main organizational barriers that were experienced during the initial innovation projects were related to:

- *Siloed organizational structures and lack of cross-functional collaboration*: the coordination between the data science team and the other departments involved in the launch of the new services was complex and often characterized by politics and internal resistance. There was a lack of shared goals, and information was not disseminated effectively across projects and between organizational units.
- *Predominance of a waterfall approach*: the prevalence of a linear and sequential approach to innovation projects slowed down the pace of prototyping, testing and commercialization, reduced experimentation, and increased product costs.
- *Risk aversion*: Small and incremental innovations (or "low-hanging fruits") were prioritized over more radical and risky ones; there was a strong emphasis on avoiding failure and short-term losses (i.e., prevention focus), and a very tight scrutiny of the supporting business case for every new idea.

To address the above problems, Centrica resorted to creating new teams, units, initiatives, and schemes to support its innovation ability. These included an internal venture capital unit, a bottom-up process to collect innovative ideas and suggestions from employees, a consumer lab and a team tasked to promote the generation of new IP within the firm, among others (see Table 20.2).

These organizational interventions were both instigated and backed by the data science team. Interestingly, members of the data science team increasingly perceived their role as "evangelists" of a new way of working, primarily based on scrum and agile approaches. In particular, the data science team promoted the adoption of agile squads, guilds, and tribes to resemble the agile organization structure popularized by Spotify (Kniberg and Ivarsson 2012) and others. An "Innovation Tribe" was set up in 2018 as "collaborative space" where diverse teams could share their progress and identify opportunities for cross-collaboration (see Table 20.2). The Innovation Tribe had monthly meetings, but these were

TABLE 20.2 ORGANIZATIONAL CHANGES TO SUPPORT INNOVATION.

Name (Business Unit)	Initial Expectations	History
Centrica Innovations (Group Function)	Venture Capital to invest in emerging start-ups and transfer knowledge internally.	Centrica Innovations was a group function set up in 2016 as an internal venture capital unit to monitor the external environment and invest in start-ups. By 2018, the VC units started funding internal projects as well, incubating internal ideas for new products and services across the organization (although the main focus remained external start-ups investments).
Accelerator Hub (Digital Technology Services)	Accelerate internal innovation through the generation of IP from data science initiatives.	The Accelerator Hub was set up in 2017 as a spin-off of the data science team. With the initial goal to accelerate the innovation process internally and get more ideas to market, the Hub was initially conceived as an "IP machine" for the data science team. However, the team went through a series of iterations, and is now the only team left of the four discussed in this table.
MAGIC (British Gas – Energy)	Collect innovative ideas and suggestion for improvement from employees.	MAGIC was an initiative set up in 2016 to collect ideas within Centrica, mainly from Front Line employees (Customer Service Advisors and Engineers) to improve customer journeys and satisfaction. This team was using an internal social network (Yammer) to collect ideas and thoughts on how to potentially improve customer journeys. This team disappeared as part of the 2019 reorganization, while as of today, idea-gathering challenges are conducted by individual business units.
Consumer Lab (British Gas – Services)	Ideate, prototype and test new service and solution concepts.	Organization unit set up in 2016 within the British Gas Services & Solutions space to promote customer journey innovations for both new and existing markets. The main goal for this team was to evaluate new service concepts for British Gas through smoke tests and innovation sprints, adopting a "Test & Learn" approach. The Consumer Lab was discontinued as part of the 2019 reorganization.

often characterized by loose agendas and no clear leadership, and over time became just a routine for checking-in with others, but not to promote any significant collaboration.[3] About one year after its launch, the Innovation Tribe was discontinued, as the employee who started it left the company.

As these new structures and processes were created, the innovative sparks and big ideas that characterized the early innovation successes started fading away, and organizational attention turned to many, often unrelated, projects with a limited degree of coherence and a high rate of failure. In the meantime, the newly created initiatives were increasingly perceived by employees as "empty boxes" as they lacked a solid pipeline of projects to work with. Lacking coordination, the new teams and organizational units which should have supported innovation started to step on each other's toes, which ultimately reduced collaboration. Meanwhile, the number of competitors embracing digital innovation in the industry had grown, and as a result Centrica had lost the leadership gained earlier on in terms of innovative services and business models. By the end of 2019, Centrica abandoned all these organizational initiatives, and closed the data science team. The only exception was represented by the Accelerator Hub, which survived a series of significant internal reorganizations, and emerged with a new role and a new vision, as we describe in the following section.

20.6 Evolution of the Accelerator Hub: From Filing Patents to Empowering People

The Accelerator Hub (from here on "the Hub") is a 4-people team embedded in Centrica's Digital Technology Services (DTS) group function. Despite the name, the Hub is not a traditional corporate accelerator, designed for the incubation and growth of external ventures (Kohler 2016). Instead, the Hub's aim is to accelerate Centrica's path to better products, services, and teamwork, by allowing innovative ideas and solutions to be explored, structured, and quickly tested while always keeping *people* at the forefront of every activity.

Despite the Hub being now recognized inside the organization as the "go-to place" for exploring and testing new ideas, it went on a five-year journey through diverse ways of working before a clear vision and direction of travel for the team was established.

The Hub was launched in 2017 as a "squad" of the data science team, with the objective of becoming the "IP machine" of Centrica's digital innovation. IP was relatively unfamiliar territory for Centrica, so the Hub's initial goal was to systematically generate patents from data science and blockchain applications developed by the data science team; the initial ambition was to generate as many as 100 such patents per year. The data science leadership team assumed that creating new IP would generate a hard metric for the innovative ability of their team, hence driving more strategic investments in data-enabled innovation. However, this iteration of the Hub was not successful for various reasons[4]: first, the company did not have a coherent IP strategy or the need for such a high rate of patent submissions; second, team members were assigned to projects in a top-down fashion, and experienced low levels of fun and intrinsic motivation on the job, both factors data scientists considered very important; finally, the Hub's work was very technical and obscure for the rest of the organization, which led to the team becoming marginalized, despite their name.

In early 2018 the role and function of the Hub changed dramatically, as the data science team increasingly realized that they did not have an effective process in place to manage the front end of their innovation process. At the same time, the data science team started to perceive a lack of integration with the rest of the company, due to the highly technical nature of their work. As a result, the data science team innovation pipeline started to run out of new projects. Then, the Hub was "repurposed" to become the "front door" to the data science team and moved to a new physical space in the headquarters, where it played host to surgeries and ideation sessions focused on data-driven innovation with other teams from across the organization.

Concurrently, the Hub and data science team also entered a knowledge transfer partnership with academia to "identify, implement, and embed an accelerated innovation process" (i.e., from idea to market) for the data science team. This marked a shift from a technical to a more *relational* focus, both within the organization and with external partners.

By engaging with distinct parts of the organization and external stake-holders, the Hub team started to understand the need to create a more inclusive innovation environment and culture within Centrica that was not narrowly focused on data-led innovations. This factor, coupled with a major reorganiza-tion that "killed" the other initiatives described in Table 20.2, created a gap in the organizational innovation structure, culture, and competences that the Hub team was ready to fill by adopting a completely novel approach. While retaining their link to data-driven solutions, the Hub broadened their scope to cover a much more diverse range of innovation opportunities, engaging with a wider audience of actors within the company.

In 2019, the Hub team created their own framework for innovation based on five stages: Feel, Learn, Identify, Test, Evangelize (FLITE Framework, Figure 20.1), which incorporated and adapted tools and frameworks from Design Thinking.

Alongside the FLITE framework, the Hub developed a tool to operational-ize the approach: the FLITE Plan (Figure 20.2). The FLITE Plan can be applied to any new idea or solution at any stage of delivery. When used at the start of a project, it provides structure around the concept and a template for exploring ideas and de-risking their execution; in more advanced stages of a project, the FLITE Plan acts as a point of reference to make sure the team delivers on the intended objectives, keeps seeking answers to the right ques-tions, and ultimately achieves the intended outcomes. Both the FLITE

FIGURE 20.1 FLITE FRAMEWORK.

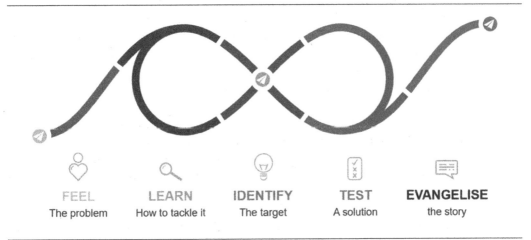

FEEL	LEARN	IDENTIFY	TEST	**EVANGELISE**
The problem	How to tackle it	The target	A solution	the story

FIGURE 20.2 THE FLITE PLAN TOOL.

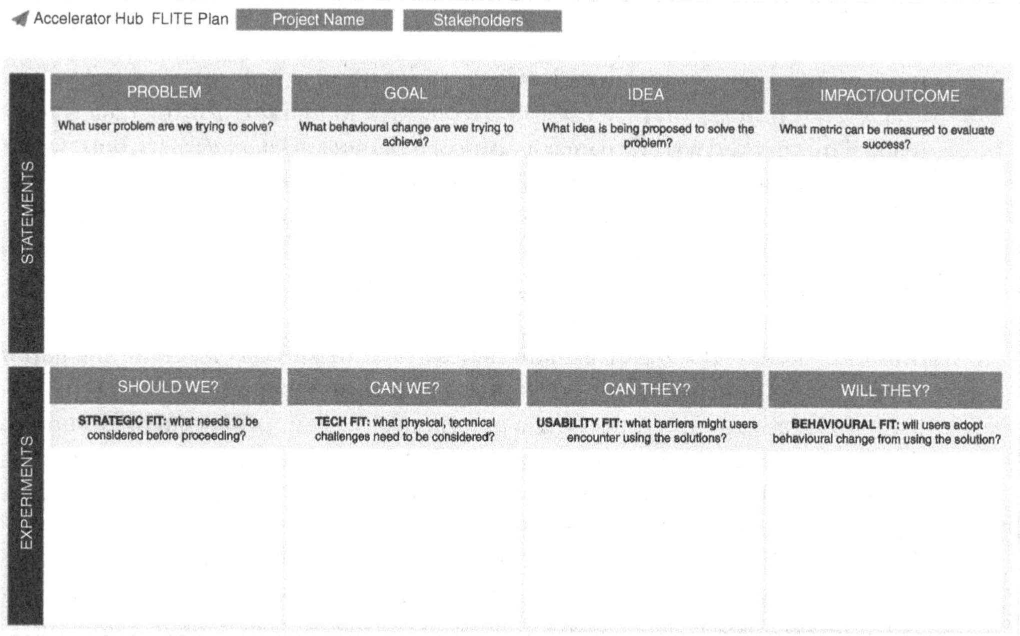

framework and FLITE Plan build on a Design Thinking approach and established project management techniques, so they are not "new to the world." However, they are significant as they are "new to the firm" and represented the first structured attempt by the company to adopt a shared process to support the generation and development of new ideas. Another novel aspect of FLITE was that this resource was developed internally by the Accelerator Hub, and hence was more contextually relevant than previous frameworks Centrica had adopted from consultancy firms.

In 2020, the Hub started to raise awareness of their role of "Design Thinking consultancy" within Centrica, collaborating with various parts of the organization to test ideas quickly and to accelerate the development process of new products and services. The Hub team built their reputation as *innovation facilitators*, working alongside internal business lines and connecting them with external partners to identify and shape opportunities for new products and services.

As they became more popular internally, in 2021 the Hub team started to receive a sustained high level of requests from various parts of the organization to share and transfer their knowledge of innovation tools and methodologies. These requests were often driven by positive word-of-mouth among employees and were boosted even more after the Hub was featured as a best practice example in the CEO's monthly newsletter. The requests received by the Hub tended to be of three types: (i) to consult with other teams on how to improve

FIGURE 20.3 THE ACCELERATOR HUB ACTIVITIES (CONSULT, COLLABORATE, COACH) AND MAIN STAKEHOLDERS.

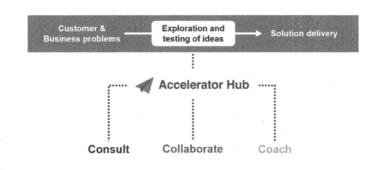

and optimize their creativity and innovation; (ii) to help and guide innovation collaborations with other departments or external partners; (iii) to upskill and train individuals and teams on new product and service development methods and tools. By listening and reflecting on these signals, the Hub structured their activities in three primary areas: Consulting, Collaborating and Coaching (Figure 20.3).

20.6.1 Consult

When consulting, the Hub maintains a high-level, holistic view of the ideas and problems under consideration, rather than entering into any specific technical or operational aspect. In this type of engagement, one or more members of the Hub work in partnership with internal stakeholders to help them structure their challenges, and reach a shared understanding of pain points, opportunities, priorities, and actions. Consulting projects are typically short-term (2–12 weeks) and lead to well-defined outcomes and outputs. Once a consulting project is completed, the Hub hands it over to the stakeholder and exits the project. One example of the Hub's consulting activity is a workshop run in 2021 to help Board Gáis Energy (Centrica's brand in Ireland) to reassess the alignment between their business and technology strategies. First, the Hub helped to identify and recruit the relevant participants for the workshop, to make sure the right people were "in the room." Then, the Hub consulted on different exercises through which the workshop participants could achieve their objectives, and finally facilitated the session on the day. After the workshop, the Hub team created a report summarizing the main findings and actions agreed, allowing the Bord Gáis Energy team to track progress and results. The feedback was particularly positive. One of the Bord Gáis Energy executives in charge of the workshop commented: "*sincere thanks to you for facilitating our IT/Business workshop ... and for all the helpful prep and advice pre-workshop. Your assistance and guidance were invaluable. What took [major consulting company] 6 weeks to deliver, the Accelerator Hub was able to achieve in 90 minutes!*".[5]

20.6.2 Collaborate

Collaborations typically follow on from consultations whereby one or more members of the Hub actively participate in the identification and testing of a solution for a specific customer/business problem. These pieces of work are more granular, usually looking at a specific challenge rather than the high-level approach described earlier. Collaborations are particularly suited to new products/services development where the Hub embeds itself among the stakeholder and their teams on an ongoing basis. These projects tend to be open-ended and typically last between 12 and 24 weeks. One example of such work is a Design Sprint run with British Gas Energy, aimed at improving their current customer onboarding journey. More specifically, the goal was to "remove moments of doubts from the customer experience while keeping them informed proactively." By engaging in a series of collaborative activities, a cross-functional team explored and tested different solutions in the space of five days, which resulted in tangible changes to the current journey within British Gas and a significant increase in the Net Promoter Score. A key stakeholder of this project (Head of Customer Complaints) commented "*This is the best thing I've seen at Centrica.*"[6]

20.6.3 Coach

With a vision to "*empower colleagues to deliver for our customers,*" the Hub set up the FLITE School, a means to coach people about product development techniques, facilitation, team building, and leadership. The Hub is currently offering three different pathways as part of FLITE School:

- *Facilitation*: A series of sessions and classes that help employees to design, facilitate, and lead innovation workshops in their team.
- *Design Thinking*: A course covering the fundamentals of Design Thinking, its impact in real-world case studies, and practical applications and key learnings gained through the team's experience within Centrica.
- *Effective Team Leadership*: A course looking at what makes a truly effective team and some of the building blocks needed to sustain it, and what it takes to be an effective leader of any team, big or small.

Very simply, the idea behind this activity is to *empower individuals* and teams to function autonomously and independently as effective innovators.

The FLITE School started in 2021 as an experiment, with a limited range of content and an initial goal of running one session per month. The Hub is now offering six different courses (two for each pathway) and running six sessions every month. Within a year of its launch, the FLITE School delivered training to over 250 employees and received an average rating of 4.8/5. In line with the "evangelize" element of the FLITE framework, the Hub team publish all their projects (including successes, failures, and lessons learned), reports and stakeholder feedback in an internal repository which can be accessed by every employee in Centrica. A key feature of the FLITE School is the aspiration to bring the stakeholders fully onboard with the analysis, solutions, and decisions, regardless of their technical competencies or innovation expertise; if there are knowledge gaps, these are addressed via the FLITE School. The Hub team tries to avoid technical jargon and terminology that may create communication and engagement barriers. Finally, the Hub team made an effort to make FLITE School sessions open and informal, providing psychological safety for both the trainers and participants. Despite this, the team faced several challenges, which required frequent adaptations in terms of the content, number of participants, duration and activities of the FLITE School before they could achieve the desired level of engagement and impact for the participants. Since FLITE School sessions started during the pandemic, the biggest challenge was to help participants actively engage and work together during and after the workshops, with no chance of interacting face to face. These initiatives included the creation of development pathways (linking individual sessions and giving the participants a sense of direction) and a FLITE community (connecting participants and the Hub team). As evidence of these efforts, in 2021, the Hub won Centrica's Digital Transformation Services Award for "Diversity and Inclusion."

20.7 Implications

This chapter described the digital transformation journey of Centrica Plc, a global energy company, over the last ten years. We focused on the contribution made by an innovative organizational unit (the Accelerator Hub), to the development of a more dynamic and innovative culture and way of working within the firm. Although the chapter builds on a single case company, there are several implications and take-aways that are relevant for a wider audience of new product/service development managers, as they can be applied to other large organizations facing a similar process of transformation. Also, as the work of the Accelerator Hub is live and constantly evolving, our implications are meant to provide new product/service developers with a range of opportunities rather than set-in-stone conclusions or decisions.

First, the experience of the Accelerator Hub highlights the power of creating "in-house" frameworks and tools that work well in the context of the focal organization, to foster collaboration and enhance innovation culture. This approach may complement or be a substitute for the use of external consultancies or expertise, which may be more expensive and not always able to fully capture the soft dimensions of the context in which they operate. Our case highlights how the main strength of the Hub is to build *trust* by focusing on people (rather than technologies) and acting as a catalyst for collaborative innovation by "walking the talk" daily (Bstieler 2006). In contrast, previous attempts by data scientists to revolutionize the company by "pushing" scrum and agile working methods were perceived as overly technical by the business functions and did not take off. While an entrepreneurial approach that mimics a start-up mentality was appealing for the data science team, other departments did not identify with it (or even perceived it as a threat), and this situation eventually emerged as a bigger barrier. In sum, our first implication for incumbent firms is that *putting people at the center of digital transformation – instead of technologies – will generate a stronger momentum for innovation which will benefit the organization beyond the specific domain of digital innovation.* This insight resonates with growing evidence from research and practice supporting the idea that focusing on people, re-skilling and upskilling the workforce, and providing an incentive to experiment, fail fast or succeed slowly (Frankiewicz and Chamorro-Premuzic 2020) stimulates employees and teams to develop new ways to coexist and coevolve with the new digital technologies that transform their working environment.

Second, the evidence provided by Centrica's digital transformation caution other companies against the risk linked to the attempts by leaders to reorganize, reframe, recast, reform, and restructure the organizational structures and innovation processes without an overall innovation strategy. Blank (2019) describes these as organizational, innovation, or process "*theaters*". These efforts, though supported by good intentions, are often ineffective and short-lived; in fact, the results of these reorganizations are quickly reset when the next reorganization starts, like sandcastles on a beach. Many large and incumbent organizations may fall into the same situation experienced by Centrica, whereby an overabundance of initiatives created to support digital transformation resulted in costly but ineffective experiments, which were mostly abandoned. According to our observations, the Accelerator Hub survived several reorganizations (unlike other units, including the data science team) because they were able to keep a clear focus on product, solutions, and deliverables, albeit on a small scale. So, our second implication for incumbent firms is to *avoid organizational and innovation theaters whereby process takes precedence over product, and to ensure that all organizational interventions aimed at transforming the culture and systems are subordinate to a clear innovation strategy.* In this respect, the Accelerator Hub is still a "work in progress" as they are still mostly active in small-scale projects; to address this issue, the Hub is currently changing its strategy to grow its role and contribution within big and strategic innovation partnerships, for example in the electric vehicles' ecosystem.

Finally, the learnings from Centrica's case study inform the general understanding of the process of digital transformation in incumbent companies. Leveraging existing definitions, and the evidence presented in this chapter, we suggest that *digital transformation in established organizations is akin to a process of corporate entrepreneurship afforded by and realized through digital innovation.* Understanding the components of digital transformations and their dynamics provides a series of insights on the success of such initiatives. First, digital innovation can be an input, a process and/or an output of digital transformation in incumbent firms (Nambisan et al. 2017). In addition, the prevailing corporate entrepreneurship frameworks pinpoint four forms of corporate entrepreneurship that are reflected in Centrica's experience: (i) a stream of new product/service innovations; (ii) the rejuvenation of organizational structures, culture, and processes; (iii) a striving for strategic renewal; and (iv) the redefinition of the competitive domain that shifts the sources of competitive advantage away from the traditional territories (e.g., Covin and Miles 1999). These forms may be interpreted as a linear sequence, or as multiple routes which can be pursued in parallel, but there are very few normative guidelines for firms. Centrica's experience indeed points to a non-linear pattern that started earlier on with an attempt to redefine the competitive arena with disruptive digital innovations aimed at cannibalizing energy retailing. In doing so, Centrica has leapfrogged the opportunity to connect the existing core business with the new digital opportunities. The implication is that *incumbent organizations may be blinded by the illusion of disrupting their industries, which if unrealized can backfire on their ability to leverage and exploit new digital technologies in their current markets.* In short, digital transformation may lead to overplaying disruption and over-exploration. Companies may realize this mistake when it is already too late to fix it, as both existing and new competitors might have capitalized on digital innovation within traditional domains. A counterexample of this approach is offered by Bosch, which has been able to create a very strong synergy between its traditional product–market combinations and the new opportunities offered by digital technologies (particularly IoT), before venturing into new domains. For example, by embedding sensors and smart capabilities in appliances, mechanical components, and systems, Bosch built over time the knowledge base to support a new ecosystem of digital business models, thereby transforming and renewing the firm's competitive advantage in the digital era (e.g., Bosch IoT Lab). This points once again to the importance of a strong and coordinated effort of strategic renewal that binds the different forms of digital transformation together.

The Accelerator Hub over the last few years has supported Centrica's efforts to re-establish a clear innovation strategy for the organization centered around the customer. For example, the Hub is currently supporting the company's strategy to re-integrate the new businesses of smart-home technologies, digital services, and electric mobility with the mainstream energy retailing market, to create customer value where the company's competitive advantage and sources of profitability are still strongly rooted. Hence, our final implication is that *maintaining a strong customer-centricity is a necessary condition for the success of digital transformation initiatives in incumbent organizations: customer-centricity is a core organizational enabler that connects technologies, innovation, and strategy.*

20.8 Conclusions and Key Learnings

To conclude, many companies know the destination of their digital transformation journey, but struggle with the route and get lost along the way. This chapter provides the experience lived by Centrica through their digital transformation and focuses on the role of the Accelerator Hub. Far from being perfect or complete, this account aims to stimulate reflection and offer actionable implications for practitioners that may help when contextualized in other organizations facing similar scenarios.

While caution is needed alongside any effort to generalize our insights beyond the focal organization, some of the implications from Centrica's case study may resonate with other large and established service organizations competing in industries with medium-to-high degrees of digitalization (i.e. insurance and financial services, retail, transportation, and travel). Our insights may also be valuable for organizations with different characteristics, in particular smaller companies. In fact, small and medium companies may not have the same issues related to siloed organizational structures and lack of interaction/cooperation between IT/data experts and other departments. Also, in small companies, any attempts to replicate the FLITE School activities may be redundant. However, the example of the Accelerator Hub may inspire the leaders of small companies to identify, promote and support a "go-to place" to accelerate the development of innovative ideas within the organization. In this respect, the key features of the Accelerator Hub approach are very applicable to small companies for several reasons. First, the Hub team is very small, so their activities and capabilities are applicable to a small-scale context; second, the Hub operates very efficiently, without big financial investments; finally, the Hub's people-centric approach could work even better in a small working environment with flatter structures and tighter interpersonal networks.

Our insights may also be adapted to other types of contexts, for example to companies offering tangible products (or a mix of products and services). These companies may have a more structured new product development process and are also probably in a better position to integrate the physical elements of digital transformation (i.e. IoT and sensors) into their product/service innovation. The challenge in this context could be to establish and maintain a continuous and efficient connection between customers and technology, which could be led/facilitated by the tools and approaches promoted by the Hub team. In this context, the benefits related to collaboration and skill development could be extended to customers, suppliers, and partners in the supply chain.

Finally, we believe our results may provide some insights also to organizations competing in service industries with lower degrees of digitalization (e.g. hospitality, education, culture).

An Accelerator Hub-like team tasked with roles of internal consulting, promotion of inter-unit collaboration, and coaching may support the initial transition of such organizations to a higher degree of digitalization. In fact, the establishment of a team like the Accelerator Hub may demonstrate to organizations still unfamiliar with digital innovation that this does not necessarily require strong consultancy support, huge technological investments, and a large dedicated team. On the contrary, with a focus on product innovation, backed by a people-centered and customer-oriented approach, a small but competent and passionate team may ignite internal organizational energies and help direct them to digital innovation projects with the aim of increasing operational efficiency and enhancing customer experience (De Luca et al. 2021).

Beyond the managerial implications highlighted earlier, the chapter presents new research-led insights that chime with the emerging research interest around the evolution of NPD/NSD capabilities in data-rich environments (Bharadwaj and Noble 2017).

Notes

1　The case study is based on the Knowledge Transfer Partnership (KTP) grant n. 11248 between Centrica Plc and Cardiff University Business School, funded by Innovate UK. The project lasted between November 2018 and November 2021. The first three authors were Academic Lead, Research Associate, and Company Supervisor in the project. Additional information on the KTP project is available from the authors.
2　The smart meter adoption in the UK energy sector, otherwise known as "the Smart Metering Implementation Programme," is led by the Department for Business, Energy, and Industrial Strategy (BEIS), regulated by the Office of Gas and Electricity Markets (Ofgem), and delivered by energy suppliers.
3　This evidence was gathered from participant observation and meeting documentation.
4　*Source*: Interviews with the data science team.
5　*Source*: Post-project feedback form.
6　*Source*: Post-project feedback form.

References

Augustine, S. (2005). *Managing Agile Projects*. Prentice Hall.

Bharadwaj, N. and Noble, C. (2017). Finding innovation in data rich environments. *Journal of Product Innovation Management* 34 (5): 560–564.

Blank, S. (2019). Why companies do 'Innovation Theater' instead of actual innovation. Harvard Business Review, 7.

Bstieler, L. (2006). Trust formation in collaborative new product development. *Journal of Product Innovation Management* 23 (1): 56–72.

Cooper, R.G. and Sommer, A.F. (2016). The agile–stage-gate hybrid model: a promising new approach and a new research opportunity. *Journal of Product Innovation Management* 33 (5): 513–526.

Covin, J.G. and Miles, M.P. (1999). Corporate entrepreneurship and the pursuit of competitive advantage. *Entrepreneurship Theory and Practice* 23 (3): 47–63.

Davenport, T.H. and Bean, R. (2018). Big companies are embracing analytics, but most still don't have a data-driven culture. *Harvard Business Review* 6: 1–4.

Davenport, T.H. and Patil, D.J. (2012). Data scientist. *Harvard Business Review* 90 (5): 70–76.

De Luca, L.M., Herhausen, D., Troilo, G., and Rossi, A. (2021). How and when do big data investments pay off? The role of marketing affordances and service innovation. *Journal of the Academy of Marketing Science* 49 (4): 790–810.

Frankiewicz, B. and Chamorro-Premuzic, T. (2020). Digital transformation is about talent, not technology. *Harvard Business Review* 6: 3.

Kniberg, H. and Ivarsson, A. (2012). Scaling Agile @ Spotify with Tribes, Squads, Chapters & Guilds. https://blog.crisp.se/wp-content/uploads/2012/11/SpotifyScaling.pdf.

Kohler, T. (2016). Corporate accelerators: building bridges between corporations and startups. *Business Horizons* 59 (3): 347–357.

Lanzolla, G., Pesce, D., and Tucci, C.L. (2021). The digital transformation of search and recombination in the innovation function: tensions and an integrative framework. *Journal of Product Innovation Management* 38 (1): 90–113.

Marion, T.J. and Fixson, S.K. (2021). The transformation of the innovation process: how digital tools are changing work, collaboration, and organizations in new product development. *Journal of Product Innovation Management* 38 (1): 192–215.

McKinsey (2020). Digital transformation in energy: achieving escape velocity. https://www.mckinsey.com/industries/oil-and-gas/our-insights/digital-transformation-in-energy-achieving-escape-velocity.

Nambisan, S., Lyytinen, K., Majchrzak, A., and Song, M. (2017). Digital Innovation Management: reinventing innovation management research in a digital world. *MIS Quarterly* 41 (1).

O'Connor, G.C., Corbett, A.C., and Peters, L.S. (2018). *Beyond the Champion: Institutionalizing Innovation through People.* Stanford University Press.

Troilo, G., De Luca, L.M., and Guenzi, P. (2017). Linking data-rich environments with service innovation in incumbent firms: a conceptual framework and research propositions. *Journal of Product Innovation Management* 34 (5): 617–639.

Zhou, K., Fu, C., and Yang, S. (2016). Big data driven smart energy management: from big data to big insights. *Renewable and Sustainable Energy Reviews* 56: 215–225.

Luigi M. De Luca (PhD, Bocconi University) is Professor of Marketing and Innovation at Cardiff Business School, where he is also Pro-Dean for Doctoral Studies. His research interests are digital transformation, big Data, marketing strategy, and innovation. His work has been published in the *Journal of Marketing, Journal of the Academy of Marketing Science, Journal of Product Innovation Management, Research Policy, British Journal of Management, Journal of Service Research, Industrial Marketing Management,* and *Journal of Business Research* among others. He received the 2014 Thomas Hustad Best Paper Award for *Journal of Product Innovation Management.* In 2021 he has been appointed to the PDMA Board of Directors.

Andrea Rossi has an MSc in Management Engineering from the University of Pisa (Italy), with a strong interest in Product Development, Innovation, and Data Science. Before joining the Accelerator Hub, Andrea worked as product manager for several SMEs in different countries. He is passionate about delivering human-centered innovative solutions to improve customer lives and loves coaching corporate teams on their path to developing great products. From a research perspective, he is particularly interested in how to combine business model Innovation and data science to exploit cutting-edge technologies, and how digital transformation is re-shaping traditional organizations.

Zahir Sumar is the Head of the Accelerator Hub at Centrica – leading a team that bridges the gap between ideas and problems, and delivering solutions that meet the needs of customers.

He does this using his background in mechanical engineering and data science which gives him a technical grounding, coupled with a fierce passion for human-centered design and delivering products and services that truly add value to people's lives.

Zahir is a firm believer that anyone can be an innovator in their own right and wants to grow a community of problem solvers and design thinkers from the grassroots through to the boardroom.

Gabriele Troilo (PhD, Bocconi University) is Associate Professor of Marketing at Bocconi University, Milan. He is also Associate Dean of SDA Bocconi School of Management. His main research areas are marketing organization, marketing cross-functional collaboration, and the relations between marketing, creativity, and innovation. He has published in several marketing and management journals on these topics, including *Journal of Product Innovation Management, Journal of the Academy of Marketing Science, Industrial Marketing Management, Journal of Business Research, Research Policy,* and *Psychology and Marketing.* He received the 2014 Thomas Hustad Best Paper Award for Journal of Product Innovation Management, for his article co-authored with Luigi M. De Luca and Kwaku Atuahene-Gima.

CHAPTER TWENTY-ONE

HYBRID INTELLIGENCE FOR INNOVATION: AUGMENTING NPD TEAMS WITH ARTIFICIAL INTELLIGENCE AND MACHINE LEARNING

Frank T. Piller, Sebastian G. Bouschery, and Vera Blazevic

21.1 Innovation in the Age of Artificial Intelligence

Artificial intelligence (AI) and machine learning (ML) are perhaps the technologies with the most impact on industries and societies. But Cockburn et al. (2019) argue that AI's greatest economic impact is still to come: its potential as a new method of invention. New methods of invention that can reshape the nature of the innovation process are relatively rare, and AI could be one of these rare cases. Two opening case examples may serve as an illustration of this change. **Choosy**, a New York-based fashion brand, delivers algorithmically informed fashion items in as little as two weeks (Eldor 2020). Founded by Jessie Zeng in 2018, the company's core assets are a group of algorithms that basically do most of its NPD work. First, a predictive algorithm using natural language processing spots top-trending fashion on Instagram by creating a database of all posts from a large group of influencers and visually tracking not just their posts, but also all comments received. This allows Choosy to rank the popularity of specific items and their underlying design features. Once the team (and algorithm) is sure that they discovered a hot fashion trend not covered by mainstream fashion brands yet, they use a generative algorithm to actually design new fashion items incorporating the identified trends. At this stage, human

The PDMA Handbook of Innovation and New Product Development, Fourth Edition. Edited by Ludwig Bstieler and Charles H. Noble.

fashion designers and algorithms work hand in hand. Next, while the firm creates a small batch of the new styles in-house, a third set of algorithms finds trendsetting customers who might be interested in buying items from the new collection. If an item proves popular in this initial group of customers, it is either offered by Choosy (manufacturing then is outsourced to larger suppliers) or licensed to a large fashion retailer as an "approved trendsetting collection without fashion risk."

AI is not just revolutionizing soft fabrics, but also the hard sciences. The mission here is to address an R&D productivity crisis. Discovering inventions with major impact has become more and more resource-intensive. Over the last two decades, firms' R&D productivity declined in many fields by 65 percent (Cummings and Knott 2018; see Bloom et al. 2019, for a review of R&D productivity in different industries). Take **material sciences**, a technology field with large impact for many applications. Breakthroughs in the discovery of new materials have become more difficult and costly, as the field became more complex and saturated with data. But the complexity that has slowed progress is where AI (and especially deep-learning algorithms) excels: Searching through a multidimensional space to come up with valuable predictions is one of AI's core competences (Rotman 2019). AI-augmented discovery of new materials is potentially much faster and cheaper by orders of magnitude (Gómez-Bombarelli et al. 2018). In an experiment published in *Nature*, Tshitoyan et al. (2019) demonstrate this performance. They first trained a language processing algorithm to screen large sets of research publications to identify open opportunities for new compounds with specific desired characteristics in thermoelectric compositions, an important field in material sciences. From the literature, the algorithm extracted technical terms and concepts from thermoelectrics, but also learned to correctly interpret general concepts such as the periodic table, crystal and band structures, and other material properties and to place them in context. Another set of algorithms, so called Generative Adversarial Network (GAN), then used this input to suggest ("invent") promising new material compositions. To test the quality of this procedure, the authors provided the algorithm in their experiment only publications from 1922 to 2008, and compared the new material compositions suggested by the algorithm with the discoveries by the human science community in the years 2008–2018. Using only publications before 2008, the system actually suggested novel thermoelectric substances that were unknown in the cut-off year. Five of these compounds have today been identified as very promising candidates, including copper-gallium telluride, a critical material for more efficient solar cells. This compound was only described by materials researchers in 2012. Hence, an unsupervised AI preceded human discoveries in an important scientific field by an astonishing five years, just using latent knowledge embedded in past publications that human scientists did not discover before.

These two examples are not exceptions. While not in the toolbox of most firms yet, AI and ML are entering the innovation lab rapidly, with the promise to address some of the largest challenges of innovation. Faster product lifecycles, high demand uncertainty, supply chain disruptions, intense competitive pressure from converging industries, and new sustainability requirements are increasing the complexity and cost of NPD and its underlying activities today. But as our opening case examples demonstrated, AI and ML bear new opportunities. The objective of this chapter is to explore different ways how AI and ML can augment the innovation process. We start by briefly reviewing the basic abilities of AI and ML to improve NPD. We then introduce the concept of hybrid intelligence as the most likely scenario how AI and ML will augment human innovation teams. In the main part of our chapter, we explore exemplary use cases of AI and ML along the stages of a NPD project (from insight generation via ideation and development toward launch) and discuss how hybrid intelligence can mitigate current constraints challenging NPD performance. We conclude by discussing limitations and organizational challenges of implementing of AI and ML in NPD.

21.2 From Machine Intelligence to Hybrid Intelligence

We have been talking about AI for decades, but notice significant improvements only recently. Consider speech recognition software, invented in the 1950s in the Bell Labs (its "Audrey" system was able to recognize the numbers 1 to 9), but until a few years ago, not really functioning well (van der Aalst 2021). Fast forward to today, when many of us cannot imagine their life without Amazon's Alexa, Apple's Siri, Google's Assistant, etc. A similar development can be seen in image recognition and other tasks that before could only be done by humans. A core technology behind these achievements are deep learning algorithms with artificial neural networks, systems with multiple layers able to progressively extract higher-level insights from a raw input. Hence, when we speak about AI today, we often refer to ML based on artificial neural networks (van der Aalst 2021). This is why we use the term *machine intelligence*, that is, mixtures of AI and ML, in the following.

In the corporate context, the use of machine intelligence attempts to improve the efficiency of business processes. As a result, applications like automated translation, image recognition in quality control, speech recognition in service environments, robotic process automation of administrative tasks, autonomous robots in manufacturing, or predictions in customer relationship management have blurred the classical divide between human and machine tasks. However, while machine intelligence can deal amazingly well with unstructured data for such clearly defined tasks (as long as there are enough training data), it is not foreseeable that it will become capable of fully mapping complex business problems in organizational contexts (Dellermann et al. 2019), solving multiple tasks simultaneously (Raj and Seamans 2019), or autonomously creating novel outputs where no clear solution criteria exist – all situations typical for NPD. Although current machine intelligence technologies outperform humans in many areas, human intelligence is still considered superior in performing tasks requiring common sense, contextual knowledge, creativity, adaptively, or empathy (Amabile 2020). Machine intelligence, on the contrary, is about data and algorithms and can be characterized by terms such as fast, efficient, cheap, scalable, and consistent (Piller et al. 2022), enabled by its core abilities of pattern recognition, predictions, and generative abilities, as outlined in Table 21.1. Our opening examples already illustrated how these abilities can also contribute to an R&D or NPD process. *Choosy* combines pattern recognition abilities to identify trends in social media and generative abilities of machine intelligence to create new fashion designs. It also deploys prediction algorithms to identify potential early adopters of a new collection. The material science example builds largely on the usage of language processing algorithms to detect open opportunities beyond the state-of-the-art, and then uses generative algorithms to develop new material compositions.

But in both examples, humans remained in the loop. There was no autonomous "invention machine" that developed new products or materials without any human input or intervention. On the contrary, machine intelligence will augment human innovation teams – combining the best of human intelligence and machine intelligence in the understanding of a *hybrid intelligence* (Piller et al. 2022). Hybrid intelligence can be defined as "the ability to achieve complex goals by combining human and artificial intelligence, thereby reaching superior results to those each of them could have accomplished separately, and continuously improve by learning from each other" (Dellermann et al. 2019, p. 638). As things stand today, it is the most likely deployment scenario of machine intelligence in companies over the next few decades (Shrestha et al. 2019). From an information systems perspective, hybrid intelligence is a "human-in-the-loop" system (Zheng et al. 2017). Humans first influence the outcome of a machine

TABLE 21.1 CORE ABILITIES OF MACHINE INTELLIGENCE AND POSSIBLE APPLICATIONS IN NPD.

Machine Intelligence Ability	
Pattern recognition abilities	An algorithm's ability to learn relevant relationships in data and spot emerging patterns, e.g. to recognize certain objects in a picture after being trained on large amounts of labeled data, to find emerging topics in large repositories of texts, to analyze relationships in data about social networks or technologies (like patent data).
Predictive abilities	An algorithm's ability to dynamically predict trends or future occurrences by techniques like support vector machines, random forest algorithms, or regression trees. Prediction tasks are perhaps the most fundamental ability of machine intelligence and have been subject of some of the most significant recent advances in ML.
Generative ("creative") abilities	An algorithm's ability to produce not just mere numerical predictions, but output considered novel and understandable by humans. Generative adversarial networks (GANs) or (transformer) language models allow machines to generate outcomes indistinguishable from human-generated output, for example artificial pictures instead of photographs, virtual objects instead of craft or design work, and computer-generated text instead of authored articles.

intelligence ability in such a way that they provide further judgment if a low confident result is given by the algorithm. But the collaboration goes further. The idea is to "realize a close coupling between the analysis-response advanced cognitive mechanisms in fuzzy and uncertain problems and the intelligent systems of a machine" (Zheng et al. 2017). Hence, human and machine intelligence adapt to and collaborate with each other, forming a two-way information exchange and control.

An insightful illustration of this collaboration provides AlphaGo, a Go-playing computer developed by Google. Commonly seen as a breakthrough in AI, AlphaGo learned the game by just playing against itself. It was able to defeat the best-ranked Go player Ke Jie and any other human player in late 2017. However, humans are still in the loop. The interplay between human and machine intelligence actually led to new insights. AlphaGo showed human players new strategies. As one of the world's leading players Shi Yue acknowledged, "AlphaGo's game transformed the industry of Go and its players. The way AlphaGo showed its level was far above our expectations and brought many new elements to the game" (as recorded in Baker and Hui 2017). At the same time, the new strategies explored by human players informed future iterations of the algorithm. Humans can learn from machines, and machines from humans: "We look forward with great excitement to AlphaGo and human professionals striving together to discover the true nature of Go," Baker and Hui (2017) conclude a review of the innovations to the gameplay by the AlphaGo machine.

Hybrid intelligence aims to combine the best of both worlds, as illustrated in Figure 21.1. The recent developments in machine intelligence extended the reach of software and hardware automation. Once an algorithm is able to

FIGURE 21.1 HYBRID INTELLIGENCE COMBINING THE BEST OF HUMAN AND MACHINE INTELLIGENCE.

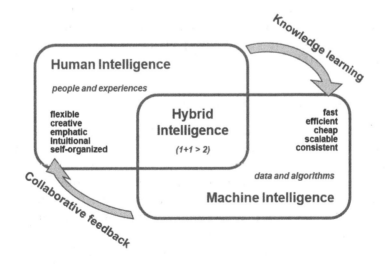

Source: Piller et al., 2022.

perform a repetitive task at a similar level of quality, it is often also more reliable and cost-effective. However, humans still have unique capabilities. For example, we have the ability to transfer experiences from one problem domain to another. As van der Aalst et al. (2021) argue, machines cannot deal with disruptions. The Corona pandemic or events of severe weather have shown that when there is a sudden dramatic change, predictive models fail, no matter how much data was there before. Especially at the beginning of the Corona pandemic, the established algorithms predicting demand in supply chains failed because of the unforeseen demand for certain products (e.g., pasta and toilet paper) combined with simultaneous restrictions for travel, work, and business. In such a situation, machine intelligence needs to be complemented by human intelligence.

But also in non-catastrophic events, humans need to remain in the loop. The idea of hybrid intelligence is not just to use humans when machine intelligence fails due to disruptions. The allocation of machine intelligence in decision-making processes often leads to more efficient, but sometimes also to unreflective or non-transparent solutions with unintended biases. Consider situations that need empathy, creativity, or ethics. Decisions in these situations cannot entirely be executed by a machine (Piller et al. 2022). Machine intelligence and human intelligence will complement each other. Understanding these factors as well as the mechanisms of interaction between humans and machine intelligence is a domain that opens a wide demand for further research. We will explore these dynamics at the example of NPD processes in larger detail in the following section.

21.3 Increasing NPD Performance by Hybrid Intelligence

Building on the unique abilities of state-of-the-art machine intelligence and the idea of hybrid intelligence as the most likely deployment scenario, we now present a number of concrete use cases along the stages of a typical NPD project. Table 21.2 provides an overview, outlining first some typical challenges present in a stage and illustrating then exemplary applications of machine intelligence to address these challenges in a novel way.

21.3.1 Insight and Opportunity Generation

NPD starts with exploring the problem space, that is, identifying open customer needs, market opportunities, and screening new technologies. Possessing and successfully incorporating such often latent information can have a significant impact on NPD success. However, it is notoriously difficult to access such information, as it often is "sticky" (von Hippel 1994), that is, not easily transferable and therefore hard to acquire. Traditionally, gathering such relevant information built on extensive market research, trend scouting, or technology forecasting activities, which could be very time consuming and cost intensive. Today, increased digitization creates a far greater availability of large volumes of

TABLE 21.2 OPPORTUNITIES OF AI AND ML TO OVERCOME TYPICAL CONSTRAINS AND CHALLENGES ALONG THE NPD PROCESS.

	Insight Generation	Ideation	Development	Launch
Challenges in NPD	• Stickiness of information about latent customer needs • Amount and complexity of potentially relevant information about technological, market, and regulatory requirements • Limited cognitive abilities and human biases in interpreting and prioritizing this knowledge	• Declining creativity scores of human ideation tasks • Limited search and association space due to lack of domain-related expertise • Myopic search and local search bias in divergent thinking	• Growing resource constraints; stronger time- and cost-to-market requirements • Increasing complexity of multi-dimensional requirement spaces, and growing technological uncertainty at the same time • Not-invented-here syndrome in utilizing external inputs	• Uncertainty about market acceptance and diffusion partners • Limited ability in finding relevant opinion leaders (influencers) • Predictions of market demand • Increasing complexity of regulatory and, documentation requirements
Exemplary AI-enabled Opportunities	**Pattern recognition:** • Finding market insights from large sets of social media streams or customer feedbacks • Identifying technical opportunities from large sets of technical documents like patent data, project reports, scientific literature **Prediction ability:** • Predicting paths of technology development (technology forecasting and road-mapping) **Generative ability:** • Generating solution proposals based on the identification of blind spots in state-of-the art	**Pattern recognition:** • Finding patterns in idea pools • Identifying experts/users with specific characteristics like lead userness **Prediction ability:** • Prioritizing ideas based on given criteria **Generative ability:** • Brainstorming ideas or concepts (tuning the expected degree of creativity) • Generating visual representations or stimuli	**Pattern recognition:** • Increasing the solution space by identifying relevant external technical knowledge • Understanding usage patterns in data streams generated by connected (smart) products **Prediction ability:** • Simulating reactions or performance outputs in technical problem solving **Generative ability:** • Engineering and design of components with specific characteristics, e.g. lightweight structures or material optimization • Autonomous generation of software code	**Pattern recognition:** • Identifying relevant influencers and early adopters in social media **Prediction ability:** • Predicting market adoption and user acceptance • Predicting the (licensing) value of a patent application **Generative ability:** • Targeting diffusion through advanced recommender systems • Generating service manuals and product documentation

relevant information that is available for analysis. In our opening case of using machine intelligence for the discovery of new materials, the starting point were algorithms that were able to navigate the vast amounts of existing scientific literature in various fields unmanageable by human capacities any longer (Bloom et al. 2019; Cummings and Knott 2018).

Capitalizing and spanning existing knowledge are core practices to access relevant opportunities (De Silva et al. 2018). Machine intelligence can augment innovation teams in knowledge extraction. At their core, these tasks are classification problems that both supervised and unsupervised machine learning models are well equipped to handle. Especially data and text mining techniques have great potential to be useful for such practices. In AI, this domain is called natural language processing (NLP), that is, a range of computational techniques for analyzing and representing naturally occurring texts at one or more levels of linguistic analysis (Liddy 2018). In general, NLP is not new to innovation management. Previous research has explored such techniques in the area of text mining (Antons et al. 2020). Whereas these models have typically been very task-specific, newer NLP technologies like transformer-based language models are a promising technology for such knowledge extraction practices. These models are context-aware, which allows them to better understand connections within a given text, extract relevant information, and summarize it in human-understandable form. Commercial applications like Google's BERT or OpenAI's line of GPT models have shown remarkable skills to extract relevant insights from large repositories of text-based data (Bouschery et al. 2023).

Regarding customer insights, billions of customers create data every day either by explicitly expressing opinions on social media or leaving customer reviews (Roberts et al. 2016). While it would not be possible for humans to analyze this plethora of data manually, due to their limited cognitive abilities, machine intelligence provides a highly scalable solution to extract insights from available data. Machines can autonomously analyze data from online communities to identify trending topics, identify novel ideas, and analyze the sentiment of the content at the same time (Christensen et al. 2017). From customer reviews (e.g., on Amazon or Yelp), algorithms can extract features existing or lacking in existing products and driving customer (dis)satisfaction, proving important insights for NPD (Hu and Liu 2004; Zhang et al. 2019). Other algorithms can identify customers with lead user characteristics from statements in online forums (Kaminski et al. 2017). In addition, observing how users interact with products creates valuable information that innovating firms can leverage to improve their product development activities (Blazevic and Lievens 2008). As Verganti et al. (2020) point out, AI enables so-called problem-solving loops where forms of unsupervised, supervised, and reinforcement learning techniques extract insights as customers are using a product. Companies like Netflix and Airbnb, for example, employ such AI to change certain aspects of their products automatically to improve the user experience.

But humans will stay in the loop. Consider the example of *SpringerNature*, one of the largest publishers of scientific information. In 2019, Springer published its first monograph entirely written by an algorithm (Writer 2019), also listing the algorithm (Beta Writer) as the author in bibliographic databases. Still, for its second machine-generated book (Visconti 2021), a human editor was on the cover. Springer learned that the collaboration of the capacities of a human expert with the scalability and processing capacity of the algorithm when interpreting the initial output of an algorithm, refining the questions to prioritize the findings, and also copy-editing the final text delivered a much better output – a typical illustration of hybrid intelligence and probably also the role model how NPD teams will work with the new algorithmic capabilities accessible to them.

21.3.2 Ideation and Concept Generation

In the next stage of a typical NPD project, the innovation team would create ideas and aggregate them into concepts to capitalize the identified opportunities. The likelihood of finding good concepts increases with the quantity and quality of the initial idea generation (Marion and Fixson 2019). For this, innovation teams employ their creative potential, a trait considered inherently human. However, new forms of machine intelligence for creative tasks changed this perception (Amabile 2020). Especially unsupervised learning algorithms can emulate or augment divergent thinking capabilities typical for idea generation (Brown et al. 2020; Griebel et al. 2020). For example, transformer-based language models can autonomously generate ideas in brainstorming sessions (after being promoted with a task description in natural language), also taking human-generated input into consideration to build on (see Bouschery et al. 2023, for a discussion of examples of this ability). Other generative algorithms can provide visual output. One example is DALL-E, a commercial transformer-based language model that can automatically generate pictures based on textual descriptions. As an example, we prompted the algorithm to provide ideas for the cover of a book on innovation. Selected results are given in Figure 21.2. Interestingly, these outputs may not just help team members to express themselves creatively ("why not use Picasso-like picture for our book cover?") and inspire others ("I like the glass window style. Let's build on this."), but actually also provide visual insights how advanced machine intelligence sees and understands our world, a central contribution for a critical perspective on the opportunities and limits of AI and ML.

Using such technologies can help innovation teams to quickly generate and communicate initial ideas and concepts – and then take this input as a starting point for further iterations between the members of the innovation team, but also the algorithm. Koch et al. (2019) provide an example of such a hybrid intelligence. They use cooperative contextual bandits, a type of interactive ML model that works together with humans, providing suggestions of relevant content based on content previously selected by the team and optimizing its suggestions by learning from feedback. In Koch et al.'s example, the AI suggested images to add to a mood board that human designers developed during an ideation

FIGURE 21.2 USING A GENERATIVE ALGORITHM TO PRODUCE GENUINE VISUAL REPRESENTATIONS OF VERBAL INPUT.

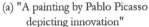

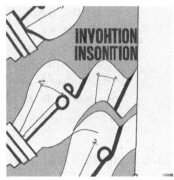

(a) "A painting by Pablo Picasso depicting innovation"

(b) "A stained glass window depicting an innovator"

(c) "Picture for a cover of a book about innovation"

Source: Images generated by the authors using the DALL-E 2 algorithm with the prompts listed below the pictures.

session. This adaptability highlights one of the important strengths of machine intelligence. Not only can such AI find patterns in data that match human preferences, it can also then make adequate predictions of what input would most benefit the designer in achieving her goal and improve its predictive abilities during the interaction. Interestingly, it also is possible to manipulate whether the AI suggests semantically similar or distant ideas. While sometimes a deep exploration of a train of thought might be fruitful, other times diverging ideas might lead to more desirable results (Stroebe et al. 2010). For some algorithms, innovation teams can set the desired level of creativity of the AI during a hybrid ideation session (see Bouschery et al. 2023, for a demonstration of this ability).

Conceptually, the promise of utilizing machine intelligence for ideation can be placed in the discourse of different search and solution finding mechanisms (Afuah and Tucci 2012). In this literature, a *local search* refers to an effort that primarily looks for solutions to an innovative task in the vicinity of previous solutions, that is, it only considers prior knowledge that the firm is already familiar with. Such a local search can efficiently yield answers to problems similar to previous ones, such as improvements to existing products (Rosenkopf and Almeida 2003). However, it rarely leads to radical advances. In contrast, a *distant search* refers to a systematic effort to transcend boundaries, avoiding the pitfall of industry blindness (Lopez-Vega et al. 2016). Prior literature has shown that firms need to exploit ideas and knowledge from outside their knowledge sphere to offset the negative biases of a local search (Jeppesen and Lakhani 2010; Salter et al. 2015). The conventional strategy to achieve a more distant search is to either use creative-thinking techniques to enlarge one's association ability (Burroughs et al. 2011) or to engage in open innovation practices like crowdsourcing to incorporate the input of others

(Pollok et al. 2019). Machine intelligence offers an alternative. The sheer amount of data used for training purposes of many AI models provides the basis for a very broad exploration of relevant ideas from various domains, inherently mitigating a bias for local search.

21.3.3 Development, Engineering, and Design

Using machine intelligence in the frontend of NPD is not a common practice in many firms yet. Technical development and design, however, are a domain where digital tools and early forms of machine intelligence have been used to a much larger extent for many years. The objective of their usage are performance requirements in terms of reducing cost- and time-to-market, increasing the efficiency of the development stage. Consider the pharmaceutical industry. Pharma companies nowadays rely on AI's predictive abilities to find promising compounds for treating and curing various diseases (Paul et al. 2021). The development of new drugs is highly complex, as research has to find the right compounds out of billions of potential molecule combinations that show the desired effects when taken by patients without side effects. Hence, this field has been an early adopter of AI and ML technologies, as also illustrated recently in the race to discover the COVID-19 vaccine (von Krogh et al. 2020).

Already earlier, companies like Autodesk, PTC, or Siemens have made generative design algorithms part of their engineering software packages. One type of AI model particularly relevant in this context are GANs, which have driven a new generation of technical problem solving and design approaches. GANs are consisting of two competing neural networks: one is generating ideas and the other is discriminating to determine if the idea meets a predefined performance function (Shu et al. 2020; Wolfe et al. 2019). Engineers can use such types of AI for example, for topology optimization of car or airplane parts that have to meet high safety standards while simultaneously minimizing factors like weight and cost. After providing certain design goals, like relevant forces, material parameters or other constraints, the AI automatically optimizes designs to reach those goals. These algorithms have dramatically decreased the time to design a new component with novel functionality. But beyond being much faster and cheaper than human engineers, the AI-generated designs also often have a different design quality with regard to functionality performance (and hence, sometimes do not look intuitive to humans at first glance). But human designers again can take the machine's output as an inspiration, iterating it with their experience and intuition into an even higher quality solution.

21.3.4 Launch and Nurture

Finally, machine intelligence opens up new opportunities for improving NPD performance during the launch stage. Due to a growing heterogeneity of demand and more diverse consumer demographics, the uncertainty about market acceptance of new products is continuously increasing, challenging traditional models of predicting market demand. In addition, in many consumer markets, identifying the right opinion leaders (especially social media influencers) has become a central mandate (Hughes et al. 2019) – a task that machine intelligence is well equipped to perform, as also our opening case example of Choosy has shown. Related to this, using targeted advertisements based on sophisticated personalization mechanisms can support the diffusion of a new product (Alalwan 2018). This can also improve the utilization of marketing resources and allow a NPD team to focus on relevant sales activities. Also, AI enables a far more active customer engagement. Similar to how it can be used to identify customer insights at the frontend of innovation, firms could use the same technologies to, for example, identify unhappy early adopters that seem to have a problem with a newly launched product by analyzing social media and flagging relevant posts for customer service representatives, who then can actively engage with those customers to prevent bad word-of-mouth and use this early feedback for rapid product improvements. In the domain of digital products, companies like Netflix use AI not only to improve their marketing, but directly incorporate AI into their (digital) products, allowing them to adapt to user behavior in real time (Verganti et al. 2020). This highlights a paradigm shift in how innovation is approached, as products are no longer static, but can change and adapt to customer needs even after market launch.

Obviously, machine intelligence also offers creative abilities for launch marketing, using similar models we discussed in the ideation stage. Companies like CopyAI offer solutions for marketers that augment human copywriting using language models. These models can learn to write different types of text, like blog posts, sales copies, or social media posts, etc. based on short descriptions of the desired content. Providing different versions of automatically generated content can speed up the communication of a new product to different audiences. These kind of generative abilities of machine intelligence also provide large support for meeting documentation or regulatory requirements. One of the core drivers of complexity and cost in NPD in many industries is the increasing extent of regulatory, compliance, and sustainability standards, which always come with extensive demands for documentation. AI can be used to facilitate the creation of documentation by predictive approaches to identify relevant standards and legal requirements that need to be matched, by using deep-learning-based generative approaches to automatize documentation creation, but also by prompting human developers with AI-generated nudges to write documentation (Wang et al. 2022). Especially in the medical field, these approaches are already used and validated to a larger extent (Luh et al. 2019; Quiroz et al. 2019).

21.4 Limitations, Conclusions, and Implications for the NPD Profession

As our exploration of different use cases of machine intelligence in NPD has shown, the unique ability of machine intelligence to process massive amounts of data in very short time creates important speed advantages and allows NPD teams to incorporate far larger and broader amounts of information and knowledge into their innovation and decision-making processes. Machine intelligence solutions are also relatively easy to scale, as compared to adding more human capacity when working on a problem. Whereas knowledge that is contained within an AI can be shared easily and even be duplicated, simply by instantiating another instance of the model on another computer, human knowledge cannot be shared in the same way. Likewise, knowledge that a model has acquired stays within a company indefinitely, whereas human knowledge is lost if an employee leaves the company. Overall, machine intelligence allows firms to tackle many highly complex problems that would be overwhelming for our limited cognitive abilities.

But as exciting as the potentials of machine intelligence are to augment NPD, one has to be aware of several limitations. While discussing these limitations would deserve an own chapter, for the sake of keeping the focus of this chapter on the opportunities that AI enables for NPD we only briefly address some of the most pressing limitations. First, it is important to understand that data quality plays an important role for the value of the resulting AI-generated output. Not only does this entail the precision of algorithmic predictions, but also potential biases that might be present in the training data. Because ML algorithms attempt to best represent the data they were trained on, they will also learn biases that are present in the data (Srinivasan and Chander 2021). Therefore, users of such algorithms will benefit from critically evaluating AI-generated outputs for such potential flaws (Akter et al. 2021).

Second, it is important to consider what data was actually included in the database used for training the applied models. The fact that innovation generally happens at the frontier of knowledge and oftentimes requires firm-specific knowledge that is not encoded in text corpora used for training of general language models might exacerbate this problem. Also, training data has a natural cut-off point after which new knowledge is no longer contained in the training set used for unsupervised learning. This means that critical information might not be included in the model's knowledge base. For example, OpenAI released its GPT-3 algorithm in early 2020 and trained the model only on data up to 2019 (Bouschery et al. 2023). Consequently, the original model contains no knowledge of important events like the COVID-19 pandemic, consumer trends, or technological breakthroughs made after 2019. Users have to be conscious of this aspect when interacting with a model. Generally, this natural data cut-off point calls for a continued re-training of models in use. Even though this aspect might not be a vulnerability for autonomous music composition or writing novels, it might provide a problem for NPD innovators to have the latest knowledge available, especially in research fields where a lot of new knowledge is created rapidly. While the need for continued retraining of language models is one limitation, it also means that these models can very easily acquire new knowledge and bring it into innovation processes, as long as the information is available in a machine-readable form.

Noteworthy, too, is that many AI models are black boxes to the user, that is, it is not necessarily clear how the AI reached a certain conclusion and how reliable the output is (Rai 2020). Modern language models, for example, are currently able to generate very plausible text passages that sound as though they could be true, but might not necessarily be. Therefore, the fact that transformer-based language models do not just retrieve answers to a question from a set of pre-defined responses, but generate an original response from scratch can also become a weakness. For example, a language model will not reveal to a user if it does not know the answer for a given task (as a human hopefully would do) – it will merely generate the most probable answer. This is important because it makes auditing output more difficult and could hinder acceptance of users – a kind of new "not invented here" syndrome in the age of machine intelligence. Such problems arise specifically in areas where rather limited amounts of relevant knowledge on a subject were included in the initial training data. NPD teams using such technology to innovate need to be aware that this limitation exists and have to develop expertise to screen machine generated output for such problems. All these limitations emphasize the need to see machine intelligence for NPD through the lens of hybrid intelligence, keeping humans in the loop.

The rise of hybrid intelligence in NPD asks us to reconsider one of the most fundamental of all NPD management tasks: organizing the division of labor and task allocation in the NPD function. While the development of machine intelligence is a field of computer science (decision routines and data structures) and research on corresponding technical applications of AI is primarily located in the engineering sciences, the implementation of hybrid intelligence is an (innovation) management phenomenon (Bailey and Barley 2020; von Krogh 2018). It asks the question how to efficiently design the NPD organization of the future. Past research has developed a comprehensive understanding of the conditions under which NPD teams are likely to be efficient and effective (Groysberg et al. 2011; Phillips et al. 2013). Especially interdisciplinary and cross-functional, also including external stakeholders, have been shown to be a central success factor of NPD performance. Members of diverse teams will more likely approach things from a number of different angles, think "outside the box," challenge accepted approaches, and follow new pathways of thought (Aggarwal and Woolley 2019). The rise of machine intelligence and its integration in the NPD process asks us to extend this established understanding. Machine intelligence can be more than a tool, but could become a work partner or even a supervisor, as suggested in the debate of algorithmic management (Lee et al. 2015). In the future NPD system, human and machine intelligence will complement each other as members of one team.

For managers, an important task will be to understand the respective strengths and weaknesses of human and machine intelligence and orchestrate them accordingly. Acknowledging that human and machine intelligence both have unique skillsets that can complement each other is an important step in successfully integrating machine intelligence into NPD. Ultimately, companies that successfully orchestrate hybrid intelligence teams will have a competitive advantage over those that do not use machine intelligence at all, or simply see it as an automation technology to replace human labor.

References

Afuah, A. and Tucci, C.L. (2012). Crowdsourcing as a solution to distant search. *Academy of Management Review* 37 (3): 355–375.

Aggarwal, I. and Woolley, A.W. (2019). Team creativity, cognition, and cognitive style diversity. *Management Science* 65 (4): 1586–1599.

Akter, S., McCarthy, G., Sajib, S. et al. (2021). Algorithmic bias in data-driven innovation in the age of AI. *International Journal of Information Management* 60: 102387.

Alalwan, A. (2018). Investigating the impact of social media advertising features on customer purchase intention. *International Journal of Information Management* 42 (1): 65–77.

Amabile, T.M. (2020). Creativity, artificial intelligence, and a world of surprises. *Academy of Management Discoveries* 6 (3): 351–354.

Antons, D., Grünwald, E., Cichy, P., and Salge, T.O. (2020). The application of text mining methods in innovation research. *R&D Management* 50 (3): 329–351.

Bailey, D.E. and Barley, S.R. (2020). Beyond design and use: how scholars should study intelligent technologies. *Information and Organization* 30 (2): 100286.

Baker, L. and Hui, F. (2017). Innovations of AlphaGo. Research blog by Deepmind, Inc., April 2017. https://deepmind.com/blog/article/innovations-alphago.

Blazevic, V. and Lievens, A. (2008). Managing innovation through customer coproduced knowledge in electronic services. *Journal of the Academy of Marketing Science* 36 (1): 138–151.

Bloom, N., Jones, C.I., Van Reenen, J., and Webb, M. (2019). Are ideas getting harder to find? Version 4.0 of the National Bureau of Economic Research Working Paper w23782.

Bouschery, S., Blazevic, V., and Piller, F. (2023). Augmenting human innovation teams with artificial intelligence: Exploring transformer-based language models. *Journal of Product Innovation Management* 40 (2): DOI 10.1111/jpim.12656.

Brown, T.B., Mann, B., Ryder, N. et al. (2020). Language models are few-shot learners. ArXiv 2005.14165.

Burroughs, J.E., Dahl, D.W., Moreau, C.P. et al. (2011). Facilitating and rewarding creativity during new product development. *Journal of Marketing* 75 (4): 53–67.

Christensen, K., Nørskov, S., Frederiksen, L., and Scholderer, J. (2017). In search of new product ideas: identifying ideas in online communities by machine learning and text mining. *Creativity and Innovation Management* 26 (1): 17–30.

Cockburn, I.M., Henderson, R., and Stern, S. (2019). The impact of artificial intelligence on innovation: an exploratory analysis. In: *The Economics of Artificial Intelligence* (ed. A. Goldfarb, J. Gans, and A. Agrawal), 115–148. University of Chicago Press.

Cummings, T. and Knott, A.M. (2018). Outside CEOs and innovation. *Strategic Management Journal* 39 (8): 2095–2119.

De Silva, M., Howells, J., and Meyer, M. (2018). Innovation intermediaries and collaboration: knowledge-based practices and internal value creation. *Research Policy* 47 (1): 70–87.

Dellermann, D., Ebel, P., Söllner, M., and Leimeister, J.M. (2019). Hybrid intelligence. *Business & Information Systems Engineering* 61 (5): 637–643.

Eldor, K. (2020). Why the founder of startup Choosy believes reinvention is critical. Forbes Magazin, April 24, 2020.

Gómez-Bombarelli, R., Wei, J.N., Duvenaud, D. et al. (2018). Automatic chemical design using a data-driven continuous representation of molecules. *ACS Central Science* 4 (2): 268–276.

Griebel, M., Flath, C., and Friesike, S. (2020). Augmented creativity: leveraging artificial intelligence for idea generation in the creative sphere. *Proceedings of the ECIS 2020 Conference.*

Groysberg, B., Polzer, J.T., and Elfenbein, H.A. (2011). Too many cooks spoil the broth: how high-status individuals decrease group effectiveness. *Organization Science* 22 (3): 722–737.

Hu, M. and Liu, B. (2004 August). Mining and summarizing customer reviews. *Proceedings of the 10th ACM International Conference on Knowledge Discovery & Data Mining*, 168–177.

Hughes, C., Swaminathan, V., and Brooks, G. (2019). Driving brand engagement through online social influencers. *Journal of Marketing* 83 (5): 78–96.

Jeppesen, L. and Lakhani, K. (2010). Marginality and problem-solving effectiveness in broadcast search. *Organization Science* 21 (5): 1016–1033.

Kaminski, J., Jiang, Y., Piller, F., and Hopp, C. (2017). Do user entrepreneurs speak different? Applying natural language processing to crowdfunding videos. *Proceedings of the 2017 CHI Conference on Human Factors in Computing Systems*.

Koch, J., Lucero, A., Hegemann, L., and Oulasvirta, A. (2019). May AI? Design ideation with cooperative contextual bandits. *Proceedings of the 2019 CHI Conference on Human Factors in Computing Systems*.

Lee, M.K., Kusbit, D., Metsky, E., and Dabbish, L. (2015). Working with machines: the impact of algorithmic and data-driven management on human workers. *Proceedings of the 33rd ACM Conference on Human Factors in Computing Systems*, 1603–1612.

Liddy, E.D. (2018). Natural language processing for information retrieval. In: *Encyclopedia of Library and Information Sciences*, 4e, 5 (ed. J.D. McDonald and M. Levine-Clark), 3346–3355. Boca Raton: CRC Press.

Lopez-Vega, H., Tell, F., and Vanhaverbeke, W. (2016). Where and how to search? Search paths in open innovation. *Research Policy* 45 (1): 125–136.

Luh, J.Y., Thompson, R.F., and Lin, S. (2019). Clinical documentation and patient care using artificial intelligence in radiation oncology. *Journal of the American College of Radiology* 16 (9): 1343–1346.

Marion, T. and Fixson, S. (2019). *The Innovation Navigator: Transforming Your Organization in the Era of Digital Design and Collaborative Culture*. Toronto: University of Toronto Press.

Paul, D., Sanap, G., Shenoy, S. et al. (2021). Artificial intelligence in drug discovery and development. *Drug Discovery Today* 26 (1): 80.

Phillips, K.W., Duguid, M.M., Thomas-Hunt, M., and Uparna, J. (2013). Diversity as knowledge exchange. In: *The Oxford Handbook of Diversity and Work* (ed. Q.M. Roberson), 157–178. Oxford: Oxford University Press.

Piller, F., Nitsch, V., and van der Aalst, W. (2022). Hybrid intelligence in next generation manufacturing: an outlook on new forms of collaboration between human and algorithmic decision-makers in the factory of the future. In: *Forecasting Next Generation Manufacturing. Contributions to Management Science* (ed. F.T. Piller, V. Nitsch, D. Lüttgens et al.), 139–158. Cham: Springer. https://doi.org/10.1007/978-3-031-07734-0_10.

Pollok, P., Lüttgens, D., and Piller, F. (2019). Attracting solutions in crowdsourcing contests: the role of knowledge distance, identity disclosure, and seeker status. *Research Policy* 48 (1): 98–114.

Quiroz, J.C., Laranjo, L., Kocaballi, A.B. et al. (2019). Challenges of developing a digital scribe to reduce clinical documentation burden. *NPJ Digital Medicine* 2 (1): 1–6.

Rai, A. (2020). Explainable AI: from black box to glass box. *Journal of the Academy of Marketing Science* 48 (1): 137–141.

Raj, M. and Seamans, R. (2019). Primer on artificial intelligence and robotics. *Journal of Organization Design* 8 (1): 1–14.

Roberts, D.L., Piller, F.T., and Lüttgens, D. (2016). Mapping the impact of social media for innovation. *Journal of Product Innovation Management* 33 (S1): 117–135.

Rosenkopf, L. and Almeida, P. (2003). Overcoming local search through alliances and mobility. *Management Science* 49 (6): 751–766.

Rotman, D. (2019). AI is reinventing the way we invent. MIT Technology Review (15 February 2019). https://www.technologyreview.com/s/612898/ai-is-reinventing-the-way-we-invent.

Salter, A., Ter Wal, A.L.J., Criscuolo, P., and Alexy, O. (2015). Open for ideation: individual-level openness and idea generation in R&D. *Journal of Product Innovation Management* 32 (4): 488–504.

Shrestha, Y.R., Ben-Menahem, S., and von Krogh, G. (2019). Organizational decision-making structures in the age of artificial intelligence. *California Management Review* 61 (4): 66–83.

Shu, D., Cunningham, J., Stump, G. et al. (2020). 3D design using generative adversarial networks and physics-based validation. *Journal of Mechanical Design* 142 (7): 071701.

Srinivasan, R. and Chander, A. (2021). Biases in AI systems. *Communications of the ACM* 64 (8): 44–49.

Stroebe, W., Nijstad, B.A., and Rietzschel, E.F. (2010). Beyond productivity loss in brainstorming groups: the evolution of a question. *Advances in Experimental Social Psychology* 43: 157–203.

Tshitoyan, V., Dagdelen, J., Weston, L. et al. (2019). Unsupervised word embeddings capture latent knowledge from materials science literature. *Nature* 571 (7763): 95–98.

van der Aalst, W.M. (2021). Hybrid Intelligence: to automate or not to automate, that is the question. *International Journal of Information Systems and Project Management* 9 (2): 5–20.

van der Aalst, W.M., Hinz, O., and Weinhardt, C. (2021). Resilient digital twins. *Business & Information Systems Engineering* 63 (6): 615–619.

Verganti, R., Vendraminelli, L., and Iansiti, M. (2020). Innovation and design in the age of artificial intelligence. *Journal of Product Innovation Management* 37 (3): 212–227.

Visconti, G. (2021). *Climate, Planetary and Evolutionary Sciences: A Machine-Generated Literature Overview.* Cham: Springer Nature.

von Hippel, E. (1994). "Sticky information" and the locus of problem solving: implications for innovation. *Management Science* 40 (4): 429–439.

von Krogh, G. (2018). Artificial intelligence in organizations: new opportunities for phenomenon-based theorizing. *Academy of Management Discoveries* 4 (4): 404–409.

von Krogh, G., Kucukkeles, B., and Ben-Menahem, S.M. (2020). Lessons in rapid innovation from the COVID-19 pandemic. *MIT Sloan Management Review* 61 (4): 8–10.

Wang, A.Y., Wang, D., Drozdal, J. et al. (2022). Documentation matters: human-centered AI system to assist data science code documentation in computational notebooks. *ACM Transactions on Computer-Human Interaction* 29 (2): 1–33.

Wolfe, C.R., Tutum, C.C., and Miikkulainen, R. (2019). Functional generative design of mechanisms with recurrent neural networks and novelty search. *Proceedings of the Genetic and Evolutionary Computation Conference*, 1373–1380.

Writer, B. (2019). *Lithium-Ion Batteries: A Machine-Generated Summary of Current Research.* Cham: Springer.

Zhang, L., Chu, X., and Xue, D. (2019). Identification of the to-be-improved product features based on online reviews for product redesign. *International Journal of Production Research* 57 (8): 2464–2479.

Zheng, N.N., Liu, Z.Y., Ren, P.J. et al. (2017). Hybrid-augmented intelligence: collaboration and cognition. *Frontiers of Information Technology & Electronic Engineering* 18 (2): 153–179.

Prof. Dr. Frank T. Piller is a professor of management at RWTH Aachen University, where he co-directs the Institute for Technology and Innovation Management. His research interests include open and user innovation, mass customization, managing disruptive innovation, and implications of new information technologies for new product development. Frank's research has been published in Journal of Product Innovation Management, R&D Management, Research Policy, Academy of Management Perspectives, Journal of Operations Management, MIT Sloan Management Review, amongst others. Frank obtained a PhD in Operations Management from University of Würzburg and previously worked at TU Munich, HKUST, and MIT.

Sebastian Bouschery is a doctoral student at the Institute for Technology and Innovation Management at RWTH Aachen University. He received a Bachelor's and Master's degree in Business Administration and Mechanical Engineering from RWTH Aachen University. His research focuses on the collaboration of humans and artificial intelligence in a form of hybrid intelligence. Particularly, he investigates how artificial intelligence can augment human innovation teams in creative innovation processes and how this affects innovation management.

Dr. Vera Blažević is associate professor of marketing at Radboud University Nijmegen and visiting professor at RWTH Aachen University at the Institute for Technology and Innovation Management. Her research interests include stakeholder co-creation, social processes in digital innovation management and responsible innovation. Her prior work has been published in Journal of Product Innovation Management, Journal of Marketing, Journal of Service Research, Journal of the Academy of Marketing Science, Industrial Marketing Management and Journal of Interactive Marketing, amongst others. She obtained her PhD from Maastricht University.

CHAPTER TWENTY-TWO

AI FOR USER-CENTERED NEW PRODUCT DEVELOPMENT—FROM LARGE-SCALE NEED ELICITATION TO GENERATIVE DESIGN

Tucker J. Marion, Mohsen Moghaddam, Paolo Ciuccarelli, and Lu Wang

22.1 Introduction

The motivation to utilize Artificial Intelligence (AI) in New Product Development (NPD) stems from the growing abundance of user-generated data, such as online reviews and technological advances in computational methods for drawing useful design knowledge and insights from that data. As NPD is a knowledge and data-intensive effort, AI offers the ability to augment historically human-centered activities, just as AI is beginning to augment other human activities like highway driving (e.g., Tesla's Autopilot driver-assist system). Current advances make AI very capable of prediction and comparison of attributes within large data sets.

The potential applications for AI and NPD are broad, spanning from user needs identification to aiding engineers in generating new designs. For example, AI can improve the process of analyzing customer needs by the adoption of Natural Language Processing (NLP). Such services are already being applied in organizations throughout the industrial spectrum. In design, AI is being experimented with to generate new designs and improved solutions. For example, NASA has used generative design computer-aided software to make the structural elements of Mars rovers more robust yet lighter.[1] As we progress through the third decade of the 21st century, new applications of AI within and in

The PDMA Handbook of Innovation and New Product Development, Fourth Edition. Edited by Ludwig Bstieler and Charles H. Noble.
© 2023 John Wiley & Sons, Inc. Published 2023 by John Wiley & Sons, Inc.

support of NPD will be increasingly important, for both the process and resulting outputs.

This chapter illustrates how AI can augment the performance and creativity of designers and NPD practitioners in innovating user-centered products. The chapter gives an overview of the potential of AI in NPD, illustrates several state-of-the-art examples, and discusses the potential process and managerial implications on integration. The chapter concludes with a discussion of the implications on the opportunities and challenges of integrating AI within NPD from both a design process and managerial perspective.

22.2 AI in New Product Development Processes

NPD has a history of integrating new technologies in the process of developing new products. From Computer-Aided-Design (CAD) to marketing analytics, the foundations of NPD as an expression of knowledge generation tend to leverage new methods and tools as they become available. Over the last twenty years, advances in computing power and software have opened new avenues for improvements in AI. Today, AI has been integrated into services and digital platforms such as smart assistants (e.g., Siri, Alexa), predictive text (e.g., Grammarly, Gmail), and self-driving cars (e.g., Tesla Autopilot). Computational support systems have been used by designers since the 1960s to aid in the creation process (Gero 2000). Similarly, from its beginning in 1957 until now, CAD has been a mainstay in physical product design. New features and capabilities have been continuously added to the platforms and current research is exploring how AI can be integrated to better support design and engineering efforts (Marion and Fixson 2021). As an example, engineering sketches – a methodology of drawing for conceptualizing designs – aims to preserve the strengths of paper-pencil sketches (Company et al. 2009). Taking this a step further with AI, researchers are creating a computational support system that proposes innovative opportunities to assist designers in creating better solutions (Bertãoa and Joo 2021). Current approaches employ generative algorithms to support the design process (Chen, Wang et al. 2019), or apply algorithms to automate design tasks that in the past required intense manual design manipulation (Cautela et al. 2019). An example would be automatically adding repetitive features to a CAD model or performing analysis on the strength of a part in real-time as it is being designed.

As innovations in AI continue to progress, designers are at the front line questioning the role of AI in the design process and the tasks in which it will be involved (Verganti et al. 2020; Zhang et al. 2021). The collaboration between AI and design can take on many forms, from proposing new design concepts to obtaining knowledge to expand the scope of the design field (Wu and Gary Wang 2020). Designers will have to play multiple, and new, higher-level roles. It will be less about shaping the final solution and more about defining the goals and constraints for AI, curating, guiding, and auditing its evolution, increasing transparency in the AI's developmental process, and identifying responsible incentives (Brundage et al. 2020). Tasks will be less about design, and more about "meta-design," defined as *the process of orchestrating the elements of a system, structuring norms and requirements, and shaping a generative strategy* (Mendini 1969; Van Onck 1965). Table 22.1 shows a summary of some of the key terms surrounding AI, with a brief description.

In terms of the NPD process, AI has the potential to be integrated into nearly all facets of the process. In Figure 22.1 is a classic "double diamond" process flow, split into the need-finding and design stage. In the figure, we highlight four areas where current AI research is being focused. These include Problem Articulation, Problem Selection, Concept Generation, and Concept Selection (Marion and Fixson 2018). In this chapter, we detail state-of-the-art research and examples in each of these four areas.

In the New Finding Space, NLP can be used to elicit customer needs from large datasets. For example, NLP can be used to determine the most relevant

TABLE 22.1 KEY AI TERMS DISCUSSED IN THIS CHAPTER ALONG WITH THEIR DESCRIPTIONS.

Term	Description
Artificial Intelligence (AI)	AI refers to the capability of a computer to mimic the cognitive processes of humans such as problem solving, decision-making, and learning.
Machine Learning (ML)	A subfield of AI, ML refers to the process of enabling computers to learn from data using mathematical models with minimal inputs from humans.
Feature Learning	A subfield of ML, feature learning refers to the process of learning compact numerical features/representations for different signals (e.g., text, image, audio).
Neural Networks (NNs)	A subfield of ML, NNs are computational models composed of networks of artificial "neurons" which loosely model how biological brains process complex relationships between inputs (e.g., stimuli) and outputs (e.g., actions).
Deep Learning (DL)	A subfield of ML, DL utilizes deep NNs (i.e., NNs with multiple layers) along with feature learning to "learn from examples," a lot of them!
Natural Language Processing (NLP)	NLP refers to the capability of a computer to process and represent natural human language using ML/DL (e.g., classify or summarize text, caption images/videos, translate, answer questions).
Computer Vision (CV)	CV refers to the capability of a computer to process and represent images, videos, or other visual inputs using ML/DL (for, e.g., facial recognition, self-driving cars, augmented reality, robotics).
Generative Modeling	Generative modeling refers to the capability of a computer to automatically learn the patterns in input data and generative samples that could have been plausibly drawn from the original input dataset.
Generative Adversarial Networks (GANs)	GANs are a generative modeling approach that utilize DL. GANs are capable of generating realistic images by pairing two NNs, a "generator" which learns to generate new images from scratch, and a "discriminator" which learns to differentiate real samples from fake samples. The generator's goal is to fool the discriminator, and the discriminator's goal is to avoid being fooled.

FIGURE 22.1 AI INTEGRATION INTO THE DESIGN PROCESS.[2]

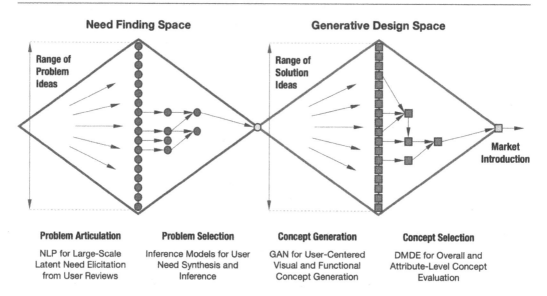

features within a given product using data from a social media application. Within Problem Selection, inference models using ML can be used to synthesize results from NLP to determine, with high probability, what most important attributes should be included in a new conceptual design. Next, GANs and generative design can be used to develop new to the world concepts within the Generative Design Space, which ultimately can be evaluated using results and data from Problem Articulation phase. In this example, multiple forms of AI are integrated into different parts of the NPD process (ML, GANs, NLP). In the next section, we show in greater detail how AI can impact and influence a better understanding of user needs within the Need Finding Space.

22.3 AI in the Need Finding Space

Understanding user needs has been a hallmark of successful NPD efforts for decades. During the front-end of innovation, best practice dictates a deep understanding of the user (Griffin and Hauser 1993). Given the amount of user data available and generated, from social media to digital platforms, there is a substantial opportunity to utilize machine learning and AI to better inform the front-end of new product and service development, where user sentiment, emotion, and the identification of latent needs play a central role in developing successful innovations. This is an area of intense research within academia and industry.

Data-driven empathetic design that augments the ability of designers to uncover and address the critical yet latent needs of users at scale is one of the most active research areas within AI-augmented design research. Current need-finding approaches are based primarily on qualitative analysis of former designs, interviews, surveys and focus-group studies, or web-based configurators, which are inherently biased due to targeting a small fraction of users and product instances with structured and direct inquiries. The inability of users to clearly articulate their needs further exacerbates this gap. Motivated by this knowledge gap, state-of-the-art in NLP for scalable and computationally efficient elicitation of explicit and latent user needs from myriad reviews on e-commerce and social media platforms is at the leading edge of academic and industry research. Within the Need Finding Space, the overall and attribute-level analysis of user sentiment polarity and intensity, opinion summarization, extraction of product usage contexts and respective emotions, and inference of extreme use-cases to facilitate latent need elicitation can be used to inform the designer (Figure 22.2).

Recent work focuses on detecting fine-grained sentiment toward specific product aspects or characteristics (Li et al. 2019; Wei 2021). Moreover, text spans corresponding to the product or its aspects, such as "shoes," "the fit," "the laces," are extracted along with the aspect category and sentiment polarity (Yan et al. 2021). Opinion expressions that support the prediction of sentiment, such as "too tight," or "wore out within a month" are also detected, since these rationale expressions are valuable within the needs-finding product design step (Siddharth et al. 2021; Zhang et al. 2021). Among aspect-based sentiment analysis work, indirect subjective language usage has only been recently studied, though it has excellent potential in need finding problems (Ganganwar and Rajalakshmi 2019). For instance, Cai et al. (2021) aims to extract implicit aspect

FIGURE 22.2 A GENERIC CONCEPTUAL FRAMEWORK FOR GAN-BASED GENERATIVE DESIGN.

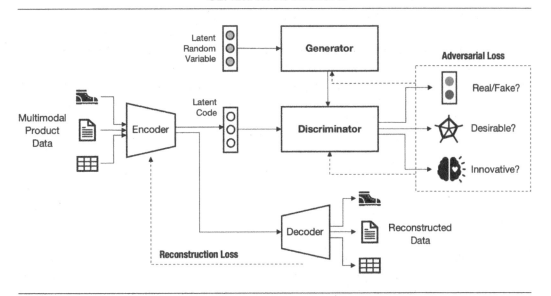

and opinion expressions, namely, examples where one expresses sentiment regarding a particular aspect category without explicitly using sentiment words.

Many researchers have investigated how and where to elicit user value using several theoretical models to understand where this knowledge resides (Martin and Hanington 2012), and develop methods that aim to look at user emotional attachment to objects as a way to enlighten the search (Schifferstein and Zwartkruis-Pelgrim 2008; Turkle 2007). The state-of-the-art in design concept evaluation predominantly relies on the judgment and expertise of the design team – either through subjective concept rating and selection or using the aforementioned rule-based quantitative methods (e.g., fuzzy sets, AHP). Yet, user needs and opinions are proven to play a critical role in successful concept evaluation and selection. The growing popularity of online reviews on e-commerce platforms as a medium for users to express their sentiments and feedback about their experience with previous products provides an unprecedented opportunity to rethink design concept evaluation and actively engage users in the creative design process.

Traditionally, customer sentiments have been integrated into the NPD process in a sequential fashion before the concept development process begins or after concepts have been developed. Recent research at Northeastern University focuses on analysis of customer sentiment that can be integrated more fluidly into the creation process. Researchers have developed a Deep Multimodal Design Evaluation (DMDE) model (Yuan et al. 2022). DMDE is a data-driven design concept evaluation tool based on user sentiments. DMDE represents a deep regression model that predicts the overall and attribute-level user sentiments associated with a new product based on its images and technical descriptions.

To inform the design process based on users' sentiments and feedback, DMDE builds on advanced computational methods to translate large-scale, unstructured natural language data into valuable design knowledge and insights. Hence, the first step of DMDE is to process individual reviews to extract aspect-level sentiment intensity of users (i.e., the positivity or negativity of their emotions) associated with different aspects of the product (Han and Moghaddam 2021a, 2021b). To do this, the system is trained on user reviews from common product evaluation sites and blogs. User reviews and product descriptions are input into the system, and values on customer sentiment are predicted. These values are then aggregated for each product and used as labels for further training DMDE's algorithms. DMDE is proven effective in predicting the overall and aspect-level sentiments of users for new products with over 98% accuracy

22.4 AI in the Design Space

AI offers the potential to augment the NPD design profession by developing and evaluating new concepts without human intervention. State-of-the-art data-driven generative design architectures and algorithms for automated creation of new design concepts – both aesthetic and functional attributes – conditioned on the user needs elicited from online reviews are being developed now. Generative Adversarial Networks (GANs) are an emerging technology that has shown tremendous recent success in a variety of generative design tasks, from topology optimization to material design and shape parametrization. In line with Osborn's rules for brainstorming (Osborn 1957), these generative models have proven effective in increasing the quantity of options at the designer's disposal to inspire her ideas exploration and avoid investing too heavily in few of them.

22.4.1 AI Design Concept Generation

Recent advances in AI offer remarkable capabilities to automatically generate design ideas that are innovative, realistic, and potentially desirable. AI can serve as an inspiration tool in the creative process and act as a generative tool to assist designers for design concept generation. AI-powered generative design tools can potentially augment designers' ability to create concepts faster and more quantitatively due to the increased speed and efficiency they offer. The power of AI lies in the speed in which it can analyze vast amounts of data and suggest design adjustments. A designer can then choose and approve adjustments based on that data. The most effective designs to test can be created expediently, and multiple prototype versions can be tested with users. Deep generative models have been recently adopted for design automation (Oh et al. 2019; Shu et al. 2020) to improve designers' performance through *co-creation with AI*. Specifically, GANs (Goodfellow et al. 2014) have shown tremendous success in a variety of generative design tasks, from topology optimization (Oh et al. 2019) to material design (Yang et al. 2018) and shape parameterization (Shu et al. 2020). Extant approaches for assessing the quality of GAN-generated samples typically apply manual assessment and the use of various convergence criteria and distance metrics for comparing real and generated images in the feature space. Some recent studies have proposed using physics-based simulators for performance assessment of generative design with respect to form and function (Shu et al. 2020).

In a standard GAN architecture, a *generator* neural network is trained to generate samples that resemble real samples in a way that would be indistinguishable, while a *discriminator* neural network learns to differentiate between the real and fake samples. The main distinction between discriminative and generative models is that the former only has to capture the conditional probability $P(X \mid Y)$ while the latter must capture $P(X)$, where X and Y denote data and labels, respectively. In other words, the generator's job is to fool the discriminator by generating samples that look real, and the discriminator's job is to try not to get fooled by the generator. The generator's job is therefore much harder than the discriminator–it must capture correlations like "nose appears below and in between eyes" or "eyelets probably appear near shoelaces." The discriminator, on the other hand, only has to decide whether a sample (e.g., a human face or a sneaker) is real or fake (i.e., made by the generator).

GANs have made significant progress in synthesizing and generating *realistic* images as their central objective. Various successful GAN architectures have been proposed in recent years mostly for facial image synthesis and generation. Examples include CycleGAN (Teng et al. 2020), StyleGAN (Karras et al. 2019), PixelRNN (Chen, Maddox et al. 2019), Text2Image (Reed et al. 2016), and DiscoGAN (Kim et al. 2017). These are powerful image synthesis models that can generate a large amount of high-quality and high-resolution images that are often hard to distinguish from real images without close inspection. Yet, the question remains on how to leverage these models in early-stage product design to generate realistic but also innovative and user-centered concepts.

Figure 22.3 depicts a conceptual architecture for GAN-based, multimodal generative design, where a discriminator along with an "evaluator" guides the

FIGURE 22.3 EXAMPLES OF FOOTWEAR GENERATED BY STYLEGAN2.

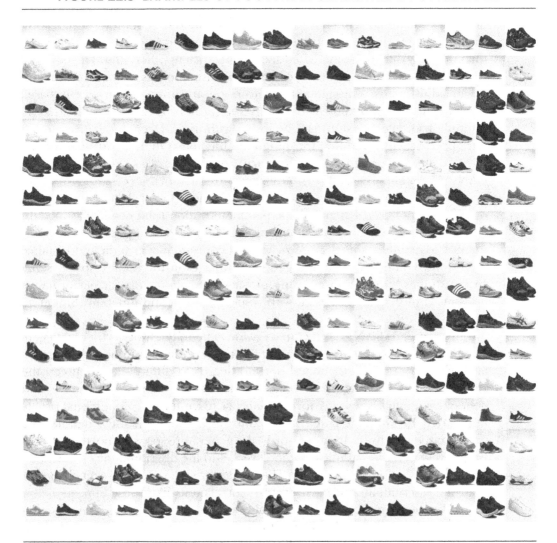

generative process based not only on realism but also on the desirability and novelty of the generated samples. The generative design process can output various modes of data such as 2D images, 3D renderings, descriptions, numerical values, or combinations thereof. That is, given a large set of real product data that contains one mode (e.g., CAD models of aircraft bodies) or multiple modes (e.g., images and textual descriptions of shoes) of data, a GAN can be trained to learn and replicate those complex distributions and generate similar samples from scratch (i.e., random inputs). The key difference between the proposed conceptual architecture in Figure 22.3 and existing GAN architectures is that the generator must generate samples that are not only realistic (as in standard GANs) but also *desirable* and *innovative*. In this context, a generated sample is considered (a) desirable if it satisfies the needs of users and (b) innovative if it is significantly different from the training dataset (i.e., real product data). The remainder of this section and the following section respectively address the issues of innovativeness and desirability in GAN-based generative design.

Although the GAN-based generative design architecture depicted in Figure 22.3 has yet to be developed, various extensions of GANs have been adopted in recent years for generative design. In the standard GANs model (Goodfellow et al. 2014), there is no control over the modes of the data being generated. GANs are notoriously difficult to train and may often be unstable due to *mode collapse*, a situation where the generator can only produce a single type or a small set of outputs. Hence, it is often recommended to use more advanced GAN architectures for generating realistic designs especially considering the significant developments of GANs, which established a new state-of-the-art in generated images with high-quality and high-resolution.

An example of a cutting-edge GAN architecture for artificial image generation is StyleGAN2 (Karras et al. 2020), StyleGAN, created by NVIDIA, produces facial images in high resolution with unprecedented quality, and is capable of synthesizing and mixing non-existent photorealistic images; that is, combining the latent codes of two generated samples to create a mix of them. *Style mixing* is an operation where a given percentage of images are generated using two random latent codes instead of one during training. Figure 22.4 several examples of footwear generated by a modified StyleGAN2 model trained on a large dataset of existing footwear images. Although all samples are 100% AI-generated, many resemble existing shoes, which points to the limitation of state-of-the-art GANs in terms of innovativeness. Further, a mechanism is needed to evaluate the quality of each sample and guide the generative design process toward producing more desirable and user-centered concepts, as described next.

22.4.2 Design Concept Evaluation

Although generating a large number of novel concepts is necessary for successful innovative design in NPD processes, it is not sufficient without rigorous evaluation against a set of performance metrics that reflect users' needs (i.e., desirability). To make the evaluation decisions more effective and to avoid the

FIGURE 22.4 THE DEEP MULTIMODAL DESIGN EVALUATION (DMDE) MODEL.

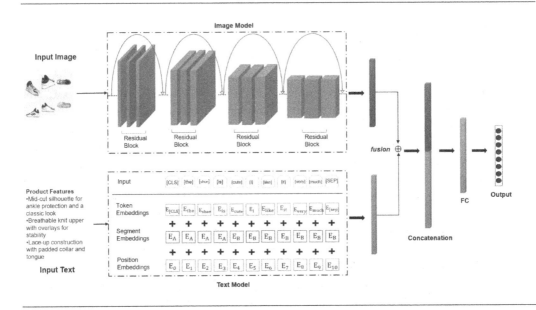

Source: Yuan et al., 2022/American Society of Mechanical Engineers.

vagueness and uncertainty of experts' subjective judgments in conventional ways, various fuzzy set-based decision-making methods and algorithms have been proposed in the literature. In the early 2000s, research (Lee and Kim 2000) was conducted on the Analytical Network Process (ANP), to address the problem of accommodating the dependencies between higher and lower-level elements. Fuzzy logic has also been proposed in conjunction with ANP to evaluate a set of conceptual design alternatives. The Fuzzy-Weighted Average (FWA) (Vanegas and Labib 2001) method was developed to calculate desirability levels in engineering design evaluation, which suggests a new method of measuring design candidates by computing an aggregate fuzzy set (Vanegas and Labib 2005). A systematic decision process via the fuzzy set method (Malekly et al. 2010) as also proposed in the literature to identify and choose the best design concept based upon expert knowledge and experience combined with optimization-based methodologies. Fuzzy analysis-based multi-criteria group decision-making methods (Thurston and Carnahan 1992; Zhang and Chu 2009) have also been employed for evaluating the performance of design alternatives, where all design alternatives are ranked and then selected according to the multiplied evaluation scores of concepts along with their weights.

To improve the evaluation process based on the fuzzy set method, interval arithmetic, rough sets, ranking design alternatives, and new methods were developed and integrated with other methods. An interval-based method (Chin et al. 2009) was proposed to effectively address uncertain and incomplete data and information in various instances of product design evaluation. Owing to the strength of rough sets in handling vagueness, a gray relation analysis integrated multi-criteria decision-making method (Zhai et al. 2009) was proposed to evaluate design concepts to improve the effectiveness and objectivity of the design concept evaluation process. Other rough sets-based methods (Li et al. 2009) were also developed to reduce evaluation bias in the pairwise comparison process in criteria weighting or rule mining. Integrated fuzzy sets (Huang et al. 2006) with genetic algorithms and neural networks were developed to identify the optimal concepts from a group of satisfactory concepts. Many experts apply methods of evaluating design concepts by ranking design alternatives in a qualitative fashion, such as multi-attribute utility theory (Ashour and Kremer 2013; Jimenez et al. 2013), preference ranking organization method for enrichment evaluations (Kilic et al. 2015; Vetschera and De Almeida 2012), and technique for order preference by similarity to an ideal solution (Ayağ 2016; Song et al. 2013).

The Deep Multimodal Design Evaluation (DMDE) model introduced in Section 2 includes elements (Yuan et al. 2022) for data-driven design concept evaluation based on user sentiments culled as part of the Need Finding Space (Figure 22.1). The DMDE model is composed of four major elements:

- *Attribute-level sentiment analysis.* The very first element of the DMDE model is to compute the attribute-level sentiment intensities of users associated with each existing product.
- *Image processing.* In DMDE, the image features must be processed and extracted to estimate the expected user-centered desirability of a concept based on its visual features.
- *Natural language processing.* Online product catalogs typically comprise brief textual descriptions of the product attributes. For example, almost every single item sold on Amazon, or similar e-commerce platforms, includes a brief textual description of the product. To identify the relationship between the technical descriptions of the products and the sentiment intensity of the users, DMDE utilizes a pretrained language model, BERT (Bidirectional Encoder Representation from Transformers) (Devlin et al. 2018).
- *Multimodal data fusion.* The product images/renderings and textual descriptions are used as inputs to the DMDE model (Figure 22.5). The text features and image features are extracted simultaneously by the fine-tuned ResNet-50 model and BERT model, respectively. Once the two modality features are identified, they are integrated using the multimodal fusion layer of the DMDE architecture (see (Yuan et al. 2022) for details).

Once the DMDE model is trained, with as few as four to six rendered images of a new product along with a short, itemized description of its different aspects (see Figure 22.5 for example), design teams can confidently predict the prospects of their new design and proactively refine their decisions during concept generation which are known to determine up to 70% of the total cost over the entire product life cycle (Pahl and Beitz 1996). Figure 22.5 shows three examples of DMDE predictions compared with ground truth customer sentiments

FIGURE 22.5 EXAMPLES OF OVERALL AND ASPECT-CATEGORY-LEVEL SENTIMENT PREDICTIONS USING DMDE.

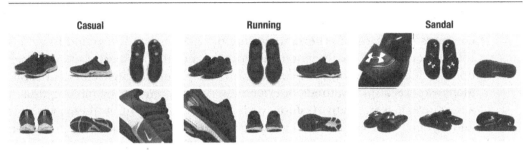

Attributes	User Sentiments	
	Predicted	Actual
Appearance	0.2012	0.1659
Comfort	0.2973	0.3518
Performance	0.1849	0.1657
Quality	0.3546	0.3877
Overall	4.615	4.7

Attributes	User Sentiments	
	Predicted	Actual
Appearance	0.1968	0.2083
Comfort	0.3017	0.2422
Performance	0.2254	0.2232
Quality	0.1663	0.1649
Overall	4.341	4.2

Attributes	User Sentiments	
	Predicted	Actual
Appearance	0.102	0
Comfort	0.2651	0.3476
Performance	0.109	0
Quality	0.268	0.387
Overall	4.804	4.8

(product descriptions are omitted for brevity). DMDE is proven to be capable of predicting the overall and attribute-level user sentiment values for a new concept with over 95% accuracy which is a potentially game-changing capability for design and NPD teams (see Yuan et al. (2022) for details).

In addition to evaluating human-generated concepts, DMDE can be incorporated in GAN architectures (e.g., see Figure 22.5) to generate not only realistic but also desirable samples. Current approaches for assessing the quality of GAN-generated samples are limited to manual assessment and the use of various convergence criteria and distance metrics for comparing real and generated images in the feature space. Some recent studies have proposed using physics-based simulators for performance assessment of generative design with respect to form and function (Shu et al. 2020); however, those mechanisms are domain-specific and applicable to a limited set of functional attributes (e.g., aerodynamic performance). The DMDE model can potentially bridge this knowledge gap by serving as a disruptive tool for accurate, data-driven evaluation of GAN-generated design concepts. This effectively closes the loop between user need elicitation and concept generation and evaluation.

22.5 Discussion and Implications

For a new product to succeed, it must resonate with the needs and desires of large and diverse populations of users (Cooper 2008). Thus, there is a need for devising new methods that capture user sentiments and feedback on a large scale and leverage that information to project the success of new designs from the perspectives of potential users. AI and the use of sentiment analysis and NLP offer the ability to accomplish this during need-finding, while generative design and GANs offer the ability to develop need designs without human intervention. This, in turn, would augment the ability of designers to make informed judgments and decisions during concept development and evaluation processes. This process has the potential to increase the quality, quantity, and diversity of user feedback versus traditional methods such as interviews, focus groups, and surveys, which are often used to inform traditional NPD methods such as Quality Function Deployment (QFD) (Cristiano et al. 2001). It also allows the product development professional to move from the process of design creation to design selection. These ramifications for individual skills and training, organizational design and management, and how the development process is arranged. Ultimately, this could reshape organizational structure and process. This may reduce the size of R&D departments concomitantly with increasing design output. Over the coming decade, expect many tools and platforms to propagate AI through the various phases of NPD. We are just at the beginning of this transformation. There are parallels with the adoption of new communication and collaboration tools that took place beginning in the early 2000s (Marion and Fixson 2021). We can expect a deluge of AI-centered new ventures, applications, and tools to be available for the NPD professional over the next decade.

However, there are serious challenges that remain in integrating AI into NPD. For example, while there have been recent advances in GAN-based generative design, there are three major knowledge gaps that still limit their ability to assist designers in early-stage product development processes. These are: (1) a focus of extant literature on the generative design of form, disregarding other non-visual aspects associated with its function (e.g., architecture, materials, performance), (2) a lack of formal methods for conditioning the generative design process on large-scale user needs, and (3) a lack of a standardized method of assessing the performance of the generated design concepts. Models such as the DMDE model described chapter is a first step in solving these challenges.

Currently, AI systems require a substantial amount of set-up and generally work well with defined constraints. To be generalizable, much work will be needed to improve ease-of-use and allow systems to more flexible for their users. Another challenge is understanding and defining the new role of the designer in an AI-augmented process. There still remains the challenge in supporting designers' roles and interactions from a systems perspective (Bernal et al. 2015). Liao et al. (2020) have recently developed a framework in which AI's role in ideation is related to creating representation, triggering empathy, and promoting engagement. Developments like these will see the designers' role change as Girling (2017) observed that – designers will become curators and auditors and not necessarily be creators in future AI contexts. With this in mind, we need to understand where computational support intersects in the course of design if we hope to assess and understand potential outcomes with an ethical lens (Pangaro et al. 2021). Few adequate AI methods can support designers by communicating their goals. In this sense, the collaboration between AI and designers becomes a platform for design communication, conversations, and conversational alignment (Dubberly and Pangaro 2019). An AI application should appreciate the explorative and evolving character of their thinking, just like designers, who generate solutions not only to solve a problem, but also to learn about it, including its objective and constraints – enhancing designers' problem-solving ability and their creative process (De Peuter et al. 2021).

To maintain the human-to-human aspect of design value, NPD professionals and designers must become an auditor for the training of AI models, with intervention incentive actions working toward the expected utility, so that they do not unconsciously stray too far by relying too heavily on user generative models. Value resides between users and artifacts and is catalyzed by the interactions we have. De Peuter et al. (2021) propose incorporating AI into the design process by defining the design decision process as proposed designs being states, designer changes as actions, and the goal being a utility function that is to be maximized. The aim of their AI integration procedure is that of AI as an assistant for the designer and what they desire. To successfully cooperate, two capabilities must be met: the AI must be able to estimate the best outcome of expected utility for the designer, and, given that utility, predict how the designer will behave in future situations after receiving the state (Dafoe et al. 2021). A further proposition that De Peuter et al. (2021) make for the collaborative effort is that the AI has a generative user model, modeling how human reasoning translates an internal utility function into behavior. This allows the AI "assistant" to evaluate possible actions by forward simulating their effect on the user.

The possibilities for using AI in the design process seem to be boundless. As such, the question of how AI can assist designers to better understand the enigma of users' values, and their best use in the design process, seems especially fertile for designers' creative endeavors, as the AI community also puts much emphasis on the need for aligning AI models with human values (Butlin 2021). And the human values consideration is another challenge. Considerations for human values exist not only on the design side, but also on the user or customer side. Much work has been focused on building learning algorithms that can deal with the uncertainty of a user using different forms of probabilistic reasoning (Evans et al. 2016). The focus on the learning algorithm in achieving user value estimation with precision comes with an abundance of failure modes, notwithstanding safety, security, privacy, and fairness. It is that last point, fairness, which brings us to the last challenge – fostering non-bias and inclusion within AI systems. For AI to realize its potential, efforts must be made to make sure algorithms and their resulting output have reduced bias, negative influence, and increased inclusivity. For all NPD professional going forward, this should be an important consideration when evaluating AI systems for inclusion in the process.

22.6 Conclusions and Thoughts on Future Research

Integrating AI into NPD has enormous potential and will most likely become one of the most important features in the NPD toolbox in the near future. By collaborating with an AI system and presenting a wider design space, designers will benefit with more accurate and desirable creations. However, AI-augmented design is limited in its current form as it is most relevant to incremental innovations or improvements to existing products in the market. Current models have a limited ability to predict more radical innovation in which the customer or user has limited or no knowledge. Radical innovation is more related to technology-push innovation, in which a market is not fully or developed at all. To address this limitation, researchers and industry should plan to investigate concept development and evaluation approaches using more latent needs and behavioral-based approaches, which are not tied explicitly to attributes or features common among users.

Acknowledgments

This material is based upon work supported by the National Science Foundation under the Engineering Design and System Engineering (EDSE) Grant No. 2050052. Any opinions, findings, and conclusions or recommendations expressed in this material are those of the author(s) and do not necessarily reflect the views of the National Science Foundation. We would like to thank graduate students Estefania Ciliotta, Chenxi Yuan, Yi Han, Joseph Peper, and Ryan Bruggeman for their assistance in completing this research.

Notes

1   https://www.autodesk.com/campaigns/generative-design/lander#:~:text=NASA's%20
    JPL%20used%20generative%20design,design%20better%20for%20space%20
    exploration%3F, accessed 3/30/22.
2   Double diamond process graphic is based on figures noted in *The Innovation Navigator*,
    Marion and Fixson (2018), Rotman-University of Toronto Press.

References

Ashour, O.M. and Kremer, G.E.O. (2013). A simulation analysis of the impact of FAHP–MAUT triage algorithm on the Emergency Department performance measures. *Expert Systems with Applications* 40 (1): 177–187.

Ayağ, Z. (2016). An integrated approach to concept evaluation in a new product development. *Journal of Intelligent Manufacturing* 27 (5): 991–1005.

Bernal, M., Haymaker, J.R., and Eastman, C. (2015). On the role of computational support for designers in action. *Design Studies* 41: 163–182.

Bertão, R.A. and Joo, J. (2021). Artificial intelligence in UX/UI design: a survey on current adoption and [future] practices. *Safe Harbors for Design Research*, 14th EAD Conference, 1–10.

Brundage, M., Avin, S., Wang, J. et al. (2020). Toward trustworthy AI development: mechanisms for supporting verifiable claims. *ArXiv:2004.07213 [Cs]*.

Butlin, P. (2021). AI alignment and human reward. *Proceedings of the 2021 AAAI/ACM Conference on AI, Ethics, and Society*, 437–445.

Cai, H., Xia, R., and Yu, J. (2021 August). Aspect-category-opinion-sentiment quadruple extraction with implicit aspects and opinions. *Proceedings of the 59th Annual Meeting of the*

Association for Computational Linguistics and the 11th International Joint Conference on Natural Language Processing (Volume 1: Long Papers), 340–350.

Cautela, C., Mortati, M., Dell'Era, C., and Gastaldi, L. (2019). The impact of artificial intelligence on design thinking practice: insights from the ecosystem of startups. *Strategic Design Research Journal* 12 (1): 114–134.

Chen, L., Maddox, R.K., Duan, Z., and Xu, C. (2019). Hierarchical cross-modal talking face generation with dynamic pixel-wise loss. *Proceedings of the IEEE Conference on Computer Vision and Pattern Recognition*, 7832–7841.

Chen, L., Wang, P., Dong, H. et al. (2019). An artificial intelligence based data-driven approach for design ideation. *Journal of Visual Communication and Image Representation* 61: 10–22.

Chin, K.S., Yang, J.B., Guo, M., and Lam, J.P.K. (2009). An evidential-reasoning-interval-based method for new product design assessment. *IEEE Transactions on Engineering Management* 56 (1): 142–156.

Company, P., Contero, M., Varley, P. et al. (2009). Computer-aided sketching as a tool to promote innovation in the new product development process. *Computers in Industry* 60 (8): 592–603.

Cooper, R.G. (2008). Perspective: the stage-gate® idea-to-launch process—update, what's new, and nexgen systems. *Journal of Product Innovation Management* 25 (3): 213–232.

Cristiano, J.J., Liker, J.K., and White, C.C., III. (2001). Key factors in the successful application of quality function deployment (QFD). *IEEE Transactions on Engineering Management* 48 (1): 81–95.

Dafoe, A., Bachrach, A., Hadfield, G. et al. (2021). Cooperative AI: machines must learn to find common ground. *Nature* 593: 33–36.

De Peuter, S., Oulasvirta, A., and Kaski, S. (2021). Toward AI assistants that let designers design. arXiv preprint arXiv:2107.13074.

Devlin, J., Chang, M.-W., Lee, K., and Toutanova, K. (2018). Bert: pre-training of deep bidirectional transformers for language understanding. arXiv preprint arXiv:1810.04805.

Dubberly, H. and Pangaro, P. (2019). Cybernetics and design: conversations for action. In: *Design Cybernetics*, 85–99. Springer, Cham.

Evans, O., Stuhlmüller, A., and Goodman, N. (2016 February). Learning the preferences of ignorant, inconsistent agents. In: *Thirtieth AAAI Conference on Artificial Intelligence*.

Ganganwar, V. and Rajalakshmi, R. (2019). Implicit aspect extraction for sentiment analysis: a survey of recent approaches. *Procedia Computer Science* 165: 485–491.

Gero, J.S. (2000). Computational models of innovative and creative design processes. *Technological Forecasting and Social Change* 64 (2): 183–196.

Girling, R. (2017). AI and the future of design: what will the designer of 2025 look like. Recuperado de https://www.artefactgroup.com/ideas/ai_design_2025 (consulado el 18 de Febrero de 2020).

Goodfellow, I.J., Pouget-Abadie, J., Mirza, M. et al. (2014). Generative adversarial nets. *Proceedings of the 27th International Conference on Neural Information Processing Systems*, Montreal, Canada, December, 8–13.

Griffin, A. and Hauser, J.R. (1993). The voice of the customer. *Marketing Science* 12 (1): 1–27.

Han, Y. and Moghaddam, M. (2021a). Eliciting attribute-level user needs from online reviews with deep language models and information extraction. *Journal of Mechanical Design* 143 (6): 061403.

Han, Y. and Moghaddam, M. (2021b). Analysis of sentiment expressions for user-centered design. *Expert Systems with Applications* 171: 114604.

Huang, H.Z., Bo, R., and Chen, W. (2006). An integrated computational intelligence approach to product concept generation and evaluation. *Mechanism and Machine Theory* 41 (5): 567–583.

Jimenez, A., Mateos, A., and Sabio, P. (2013). Dominance intensity measure within fuzzy weight oriented MAUT: an application. *Omega* 41 (2): 397–405.

Karras, T., Laine, S., and Aila, T. (2019). A style-based generator architecture for generative adversarial networks. *Proceedings of the IEEE/CVF Conference on Computer Vision and Pattern Recognition*, 4401–4410.

Karras, T., Aittala, M., Hellsten, J., et al. (2020). Training generative adversarial networks with limited data. *Advances in Neural Information Processing Systems* 33: 12104–12114.

Kilic, H.S., Zaim, S., and Delen, D. (2015). Selecting "The Best" ERP system for SMEs using a combination of ANP and PROMETHEE methods. *Expert Systems with Applications* 42 (5): 2343–2352.

Kim, T., Cha, M., Kim, H. et al. (2017). Learning to discover cross-domain relations with generative adversarial networks. *International Conference on Machine Learning*, PMLR, pp. 1857–1865.

Lee, J.W. and Kim, S.H. (2000). Using analytic network process and goal programming for interdependent information system project selection. *Computers and Operations Research* 27: 367–382.

Li, X., Bing, L., Zhang, W., and Lam, W. (2019 November). Exploiting BERT for end-to-end aspect-based sentiment analysis. *Proceedings of the 5th Workshop on Noisy User-generated Text (W-NUT 2019)*, 34–41.

Li, Y., Tang, J., Luo, X., and Xu, J. (2009). An integrated method of rough set, Kano's model and AHP for rating customer requirements' final importance. *Expert Systems with Applications* 36 (3): 7045–7053.

Liao, J., Hansen, P., and Chai, C. (2020). A framework of artificial intelligence augmented design support. *Human-Computer Interaction* 35 (5–6): 511–544.

Malekly, H., Mousavi, S.M., and Hashemi, H. (2010). A fuzzy integrated methodology for evaluating conceptual bridge design. *Expert Systems with Applications* 37 (7): 4910–4920.

Marion, T.J. and Fixson, S. (2018). *The Innovation Navigator: Transforming Your Organization in the Era of Digital Design and Collaborative Culture*. University of Toronto Press.

Marion, T.J. and Fixson, S.K. (2021). The transformation of the innovation process: how digital tools are changing work, collaboration, and organizations in new product development. *Journal of Product Innovation Management* 38 (1): 192–215.

Martin, B. and Hanington, B. (2012). *Universal methods of design: 100 ways to research complex problems, develop innovative ideas, and design effective solutions*. Rockport Publishers. http:// choicereviews.org/review/10.5860/CHOICE.49-5403.

Mendini, A. (1969). Metaprogetto, si e no. *Editorial. Casabella* (333): 4–15.

Oh, S., Jung, Y., Kim, S. et al. (2019). Deep generative design: integration of topology optimization and generative models. *ASME Journal of Mechanical Design* 141 (11): 111405.

Osborn, A.F. (1957). *Applied Imagination*, 1e. New York: Scribner.

Pahl, G. and Beitz, W. (1996). *Engineering Design: A Systematic Approach*. London, UK: Springer-Verlag.

Pangaro, P.P., Russman, T., and Barron, T. (2021). What do you want from your AI?: What does it want from you?. *Journal of the International Society for the Systems Sciences* 65 (1).

Reed, S., Akata, Z., Yan, X. et al. (2016). Generative adversarial text to image synthesis.

Schifferstein, H.N.J. and Zwartkruis-Pelgrim, E.P.H. (2008). Consumer-product attachment: measurement and design implications. *International Journal of Design* 2 (3): 1–13.

Shu, D., Cunningham, J., Stump, G. et al. (2020). 3d design using generative adversarial networks and physics-based validation. *ASME Journal of Mechanical Design* 142 (7): 071701.

Siddharth, L., Blessing, L., and Luo, J. (2021). Natural language processing in-and-for design research. *arXiv preprint arXiv:2111.13827*.

Song, W., Ming, X., and Wu, Z. (2013). An integrated rough number-based approach to design concept evaluation under subjective environments. *Journal of Engineering Design* 24 (5): 320–341.

Teng, L., Fu, Z., and Yao, Y. (2020). Interactive translation in echocardiography training system with enhanced cycle-gan. *IEEE Access* 8: 106147–106156.

Thurston, D.L. and Carnahan, J.V. (1992). Fuzzy ratings and utility analysis in preliminary design evaluation of multiple attributes.

Turkle, S. (ed.) (2007). *Evocative Objects: Things We Think With.* MIT Press.

Van Onck, A. (1965). Metadesign. *Produto e linguagem* 1 (2): 27–31.

Vanegas, L.V. and Labib, A.W. (2001). Application of new fuzzy-weighted average (NFWA) method to engineering design evaluation. *International Journal of Production Research* 39 (6): 1147–1162.

Vanegas, L.V. and Labib, A.W. (2005). Fuzzy approaches to evaluation in engineering design. *Journal of Mechanical Design* 127 (1): 24–33.

Verganti, R., Vendraminelli, L., and Iansiti, M. (2020). Innovation and design in the age of artificial intelligence. *Journal of Product Innovation Management* 37 (3): 212–227.

Vetschera, R. and De Almeida, A.T. (2012). A PROMETHEE-based approach to portfolio selection problems. *Computers & Operations Research* 39 (5): 1010–1020.

Wei, Y. (2021 January). A survey of sentiment analysis based on product review. *2021 2nd International Conference on Computing and Data Science (CDS)*, 57–63. IEEE.

Wu, D. and Gary Wang, G. (2020). Knowledge-assisted optimization for large-scale design problems: a review and proposition. *Journal of Mechanical Design* 142 (1): 010801.

Yan, H., Dai, J., Qiu, X., and Zhang, Z. (2021). A unified generative framework for aspect-based sentiment analysis. *arXiv preprint arXiv:2106.04300.*

Yang, Z., Li, X., Catherine Brinson, L. et al. (2018). Microstructural materials design via deep adversarial learning methodology. *ASME Journal of Mechanical Design* 140 (11): 111416.

Yuan, C., Marion, T., and Moghaddam, M. (2022). Leveraging end-user data for enhanced design concept evaluation: a multimodal deep regression model. *Journal of Mechanical Design* 144 (2): 021403.

Zhai, L.Y., Khoo, L.P., and Zhong, Z.W. (2009). Design concept evaluation in product development using rough sets and grey relation analysis. *Expert Systems with Applications* 36 (3): 7072–7079.

Zhang, G., Raina, A., Cagan, J., and McComb, C. (2021). A cautionary tale about the impact of AI on human design teams. *Design Studies* 72: 100990.

Zhang, W., Deng, Y., Li, X. et al. (2021). Aspect sentiment quad prediction as paraphrase generation. *arXiv preprint arXiv:2110.00796.*

Zhang, Z. and Chu, X. (2009). A new integrated decision-making approach for design alternative selection for supporting complex product development. *International Journal of Computer Integrated Manufacturing* 22 (3): 179–198.

Architect and Communication Designer, Paolo Ciuccarelli is Professor of Design at Northeastern after 20 years at Politecnico di Milano in Italy. At Politecnico he coordinated the Communication Design program (BSc and MSc), has been member of the board at the PhD in Design, and founded the DensityDesign Research Lab, an award winning laboratory for data visualization and information design. Paolo's research focuses on the design transformations that help making sense of data and information to improve decision making processes, especially with non-expert stakeholders and for controversial complex social issues where he's also experimenting on the role of rhetoric and visual poetry for a deeper engagement. He also works in developing tools and methods to understanding the evolution of the design discipline in the frame of a meta-design.

Mohsen Moghaddam is an Assistant Professor of Mechanical and Industrial Engineering at Northeastern University, Boston. He received his PhD from Purdue University in 2016 and served the GE-Purdue Partnership in Research and Innovation in Advanced Manufacturing as a Postdoctoral Associate prior to joining Northeastern. His areas of research interest include human-technology teaming through extended reality, user-centered design, and use-inspired AI.

Tucker Marion's interdisciplinary research is centered on the new product development (NPD) process, for both startups and corporate ventures. Specifically, he researches critical attributes that can influence improvements in efficiency and efficacy from a process and organizational standpoint. These include lean processes used for innovation, Information Technology (IT) and design tools used in NPD, and sourcing and cost engineering. His work has appeared in many publications, including the *Journal of Business Venturing, Journal of Product Innovation Management, R&D Management, Harvard Business Review, MIT Sloan Management Review, Research-Technology Management, Design Studies, International Journal of Production Research, Journal of Concurrent Engineering,* and others.

Lu Wang is an Assistant Professor of Computer Science and Engineering at University of Michigan. Previously until 2020, she was at Khoury College of Computer Sciences, Northeastern University. She completed her PhD in the Department of Computer Science at Cornell University, under supervision of Professor Claire Cardie in 2015. Lu's research is focused on natural language processing, computational social science, and machine learning. More specifically, Lu works on algorithms for text summarization, language generation, reasoning, argument mining, information extraction, and discourse analysis, as well as novel applications that apply such techniques to understand media bias and polarization and other interdisciplinary subjects. Her work won outstanding short paper award at ACL 2017, and best paper nomination award at SIGDIAL 2012.

CHAPTER TWENTY-THREE

RE-THINKING DESIGN THINKING: THE TRANSFORMATIVE ROLE OF DESIGN THINKING IN NEW PRODUCT DEVELOPMENT

Marina Candi, Claudio Dell'Era, Stefano Magistretti, K. Scott Swan, and Roberto Verganti

23.1 What Is Design Thinking?

Scholars and practitioners acknowledge the central role of design as a driver of innovation and change (Brown 2008; Candi 2016; Liedtka 2015; Magistretti et al. 2022; Martin 2009). The importance of design as a source of value creation has been explored for decades (Hirschman 1986; Peterson et al. 1986). Early on, most investigations viewed design as the aesthetic and symbolic dimension of products, that is, design as form, identity, and emotions, attributing it only a marginal role in innovation studies (Capaldo 2007; Dell'Era and Verganti 2009). What has driven the recent and considerable increase in attention to design in the business community is a change in perspective: design not only as an aesthetic component of innovation but as a comprehensive innovation management practice fostering new processes, mindsets, and capabilities within organizations (Dell'Era and Verganti 2010; Verganti 2009, 2017). Design Thinking encompasses many principles and tools to address innovative problem-solving from a design perspective and has gained substantial interest in the business world, particularly in the last decade. Design Thinking is now practiced not only by designers but by anyone seeking to innovate (Dell'Era et al. 2020).

The PDMA Handbook of Innovation and New Product Development, Fourth Edition. Edited by Ludwig Bstieler and Charles H. Noble.
© 2023 John Wiley & Sons, Inc. Published 2023 by John Wiley & Sons, Inc.

The creativity and substance of the Design Thinking paradigm have transformed it into one of the preferred methodologies to address seemingly intractable problems for which other business tools and processes are less suited. The increasing attention of practitioners to Design Thinking is evident when examining the recent actions of strategy and innovation consultancies. The acquisition of Lunar by McKinsey and Fjord by Accenture are just two examples of the broader phenomenon. Accenture, Deloitte, IBM, KPMG, and PwC rank among the most aggressive players in acquiring design agencies to renew their innovation service offerings. Design Thinking is booming, especially in industries where digital transformation requires new competencies and capabilities to develop compelling customer experiences. Even software developers and integrators, such as Adobe, Microsoft, and Oracle, extensively adopt Design Thinking practices.

This rapid and remarkable success of Design Thinking has spurred enthusiasm but also confusion about what Design Thinking is, how it differs from design in the broad sense, and whether it can be used as a general approach to complex innovation beyond redesigning digital experiences. First to note is the profound difference between design and Design Thinking, analogous to the significant difference between management and Total Quality Management (TQM). Design, like management, is a practice. It is whatever humans (and increasingly machines) do to devise "courses of action aimed at changing existing situations into preferred ones" (Simon 1969). In other words, design is the decision-making side of problem-solving: How we frame a problem, create possible solutions, and test them. From a more constructivist perspective, design is the practice of "making sense of things" (Krippendorff 1989). In particular, as design is concerned with artifacts, its sense-making activity focuses on things rather than organizations (which are the focus of management practice). In both cases, these definitions do not indicate a tool, a method, or a process whose effectiveness needs to be demonstrated. Design simply "happens," similarly to medicine (the practice of caring for patients) or management (the practice of administering resources and organizations).

As any practice can be performed in different ways (for example, in medicine, back pain can be treated with drugs, manipulation, or surgery), design can also be practiced variously. One of these options is Design Thinking. Therein lies an important message: Design Thinking is one of the many possible ways to practice design. In other words, it is a paradigm, or as Jeanne Liedtka (2020) puts it, "a social technology." There are other ways to practice design, such as engineering design, with its strong focus on technology as the driver of change. So, what makes the Design Thinking paradigm unique in the design space and particularly germane in driving innovation? Regardless of the specific techniques or tools falling under the Design Thinking umbrella, three specific principles can be distinguished (Liedtka 2015; Micheli et al. 2019; Seidel and Fixson 2013; Verganti et al. 2020): user-centeredness, imagination, and iterative prototyping.

User-centeredness implies that design decisions are driven by the goal of *maximizing meaningfulness for users.* This can be compared to business viability, the driver of the business design paradigm, and technical excellence, the driver of the engineering design paradigm. This does not mean that Design Thinking does not advocate viability or feasibility, but priority is given to meaningfulness when facing a trade-off.

Imagination implies that solutions are not found deductively by applying science to problems or inductively by taking inspiration from other cases but abductively through novel hypotheses about how things could be. It also means that subject-level expertise does not always matter, that is, good solutions may well emerge by approaching a problem from unusual perspectives.

Iterative Prototyping implies two things: (a) design can be practiced as a trial-and-error learning process, often engaging users. Early design mistakes are simply ways to iterate toward better solutions; (b) learning iterations are based on visual and material representations of solutions rather than abstract design models and representations.

Many other paradigms support the practice of design. Design Thinking shares some tools and principles with other paradigms (for example, the practice of iterations is commonly used in contemporary design practice) but also proposes a contrasting way of practicing design. Put differently, the "system engineering" design practice departs from a user-centered focus to embrace the needs of complex systems of stakeholders and "actants," including devices and natural elements (Latour 1987). It also values field expertise and abstract modeling. The "reflective practice," typical of architecture or policy design, departs from the creative and ideative view of design to embrace a process based on inquiry and critical reflection where exploration quality and depth matter more than the number of ideas (Schön and Rein 1994).

Of particular interest for this handbook is the potential of Design Thinking not only as an approach to design but also (and probably even more so) as an alternative way to practice NPD. In other words, as an innovation management paradigm. Some even proposed Design Thinking as an alternative for managing organizational and cultural transformation, sharing some principles with novel innovation paradigms, for example, the iterative view of agile development (MacCormack et al. 2001) or the generative approach of dialogic organizational transformation (Bushe and Marshak 2014). Design Thinking can support each of the phases of the NPD (see Figure 23.1) as outlined next.

The first phase of the NPD is **Opportunity Identification** defined as the moment in which options within a set of strategies are detected and shared, and Design Thinking has been proven to assist in envisioning new opportunities (Verganti et al. 2021).

Idea Generation is the purview of creativity and the phase in which options and alternatives are crafted, bold ideas are conceived, and outside-the-box thinking takes place. Indeed, as Design Thinking is rooted in creativity, it is expedient in this NPD phase (Carlgren et al. 2016).

FIGURE 23.1 NEW PRODUCT DEVELOPMENT (NPD) PROCESS AND PHASES.

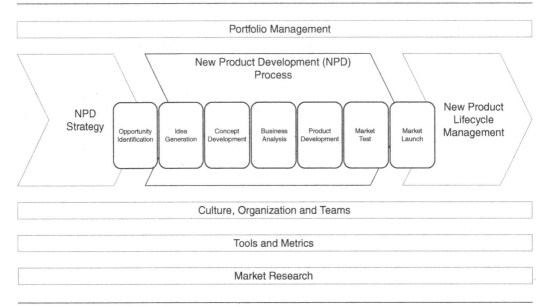

Source: Adapted from PDMA website at https://www.pdma.org.

Concept Development is when ideas become concrete and the phase in which solution-building occurs. By leveraging prototyping techniques and mindsets (Micheli et al. 2019), Design Thinking can support concept development.

Business Analysis is the phase in which analytical thinking takes place, economic evaluations are performed, and options are verified in terms of viability. Design Thinking can play a significant role here, thanks to its ability to sustain complex conversations through visualization and action (Liedtka 2020).

Product Development is the phase in which concepts are created, production lines are defined, and the first embodiments of solutions emerge. In this phase, Design Thinking assists in crafting and anticipating user needs and finding creative solutions to specific development problems (Magistretti et al. 2021b), ultimately making products more successful.

Market Test is the phase in which a product is introduced to the market on a trial basis, almost ready to be launched, to verify the market's desire and fine-tune details. Testing is crucial and, as part of the Design Thinking framework, can assist in this NPD phase (Dell'Era et al. 2020).

Market Launch is when a new product is introduced to the market, and end-users can start to use and adapt it. Design Thinking is valuable as a strategic element supporting marketing activities (Beverland et al. 2015) in these initial steps of a product entering the market.

This introduction has highlighted that Design Thinking is a paradigm that can support the NPD process across its phases, as will be further detailed in the next section.

23.2 Do I Need Design Thinking?

Design Thinking can assume different forms and interpretations. As recently highlighted in Dell'Era et al. (2020), four kinds of Design Thinking respond to different challenges: *Creative Problem Solving* (Brown 2009; Martin 2009); *Sprint Execution* (Knapp et al. 2016; Magistretti et al. 2021a); *Creative Confidence* (Kelley and Kelley 2013); and *Innovation of Meaning* (Artusi and Bellini 2022; Verganti 2009, 2017). While these four kinds of Design Thinking have features in common, they also embody nuances in interpreting and adopting specific practices, demonstrating the kaleidoscopic nature of Design Thinking.

The main aim of **Creative Problem Solving** is to leverage creativity in solving problems, and the underlying assumption is that users have a need, problem, and desire for which a solution can be found. This approach implies that organizations innovate by profoundly understanding user needs and desires and then creating ideas to solve these problems better. The increasing complexity and dynamism of user behaviors, on the one hand, and the growing demand for more sophisticated and personalized solutions, on the other hand, have propelled the rapid diffusion of the Creative Problem Solving approach. In this form of Design Thinking, human-centered design is based on a deep understanding of user needs and desires. In other words, this is an outside-in approach that starts by going out and observing how users use existing products, then interpreting these observations to inspire original solutions, seeing "with a fresh set of eyes" through empathy. Solving problems is the main aim of this form of Design Thinking, and creative ideating is the core practice of framing and reframing a problem iteratively. Ideating is about sharing insights within a team, making sense of vast data, and identifying opportunities to conceive new solutions. Creative Problem Solving is not about generating the "right" idea but rather a wide range of ideas. Creative ideation means exploring broad landscapes in terms of concepts and opportunities and crossing the bridge between identifying the problem and creating a solution through a combination of understanding the problem and the team's imagination to generate solutions.

Sprint Execution aims to accelerate the development process and reduce market uncertainty to quickly and effectively launch new solutions in the market. This form of Design Thinking can be viewed as a hybridization of Creative Problem Solving and the Lean/Agile movement. In particular, creativity is crucial to boosting innovation in the Sprint Execution approach. Like the Lean Start-up approach and the Agile-Stage-Gate hybrid model, Sprint Execution accentuates the crucial role of time constraints and iterations in terms of the effectiveness of the process. The Sprint Execution approach aims not merely to design a product concept or an innovative idea but to identify business questions and prepare products for the market. The product is the primary vehicle to capture value and learn from the market's reactions. Sprint Execution responds to the acceleration required by digital transformation through "making" and building minimum viable products (MVP) is one of its core practices. Defined as products with just enough features to satisfy early customers, the aim of MVPs is not only to create physical embodiments of ideas but to serve as means to learn from feedback. While prototypes are non-committal and created to explore problems and concepts, understand the relevance of functionalities, and get stakeholders on board, MVPs are used to gain insights from early adopters and explore market viability. Prototypes help to understand the feasibility of an idea, whereas MVPs are more about validated learning. MVPs are not final products but serve as means to learn the reactions of potential users early in the development process.

Leveraging the core features of Creative Problem Solving and fine-tuning the complementary traits, the **Creative Confidence** approach aims to nurture innovation mindsets based on a set of approaches, practices, and methodologies that foster innovation and change. Human-centeredness is the key principle of the Creative Confidence approach to stimulate co-creation practices and co-design a shared sense of purpose that inspires action. The main aim of this approach is to engender confidence in facing organizational changes and activate transformation processes. Organizations face new and significant challenges in engaging and keeping stakeholders motivated, and Design Thinking is increasingly adopted to reshape organizational culture and enable digital transformation.

On the one hand, the digital revolution has led to incredible entrepreneurial opportunities for individuals and small businesses. Indeed, technological developments in the last few decades have undeniably reshaped our economy, and the last ten years have seen several young start-ups develop into billion-dollar businesses. In this era of entrepreneurship, such businesses will no longer be the exception. On the other hand, people ascribe more and more importance to work-life balance or the opportunity to discover a personal and intimate purpose in their work.

The **Innovation of Meaning** approach mainly aims to create innovative visions supporting new strategic directions. While the Creative Problem Solving approach favors the development of better ideas able to solve problems, the Innovation of Meaning approach identifies a novel purpose that redefines the problems worth addressing. In other words, a creative solution may provide incremental or even radical improvements, but usually in an existing direction. Instead, an innovative meaning is about a novel purpose, not only a new "how," but especially a new "why," a new interpretation of what is meaningful to people. The Innovation of Meaning approach is based on a particular interpretation of human-centered design in contrast to the Creative Problem Solving approach: users are a valuable source of inspiration for new solutions, but their contribution is less effective in supporting the development of new meanings. New meanings are not usually called for by the market but are instead gifted by organizations responsible for interpreting what is good and what is not. People will never love a product that is not loved by its designers and developers. Indeed, if they do not love it, the market recognizes the weak relationship. According to the Innovation of Meaning approach, organizations envision scenarios to support the search for new meaning and make people fall in love.

The different practices characterizing the four kinds of Design Thinking contribute differently to the NPD process. Below, we illustrate inspirational cases of Design Thinking projects to unpack why managers dealing with NPD need Design Thinking.

Starting from **Innovation of Meaning** and the case of Alfa Romeo 4C,[1] we illustrate the relevance of this form of Design Thinking in the first two phases of the NPD process: Opportunity Identification and Idea Generation. Alfa Romeo, the world-famous Italian carmaker, among the first to win a Formula One race, and producing a series of venerated cars (e.g., Alfa Giulia, Alfa Spider), launched a project to find a way to compete with German car manufacturers in the premium sector. In particular, the project's aim was not just NPD but also positioning. Alfa brought together a team of 20 individuals and asked them to discover new opportunities. Instead of starting from the market, they began by questioning why people buy cars. In so doing, the team came up with a new direction for Alfa Romeo and its new supercar targeted at skilled and passionate drivers, a completely different direction to German car manufacturers focused on electronics and pursuing perfection. The Alfa team envisioned a new meaning for the industry: "a car that would allow people to express a passion for driving, rather than their wealth, and would be agile and responsive, rather than superfast and super powerful" (Verganti 2016). The new car was the Alfa Romeo 4C, a supercar born from this new meaning and direction that enabled the company to compete in the market. Thus, designing a new direction and envisioning a new meaning for Alfa Romeo exemplifies the adoption of Innovation of Meaning for opportunity identification and idea generation.

For **Creative Problem Solving**, we highlight the case of the multinational company PepsiCo and the SodaStream[2] brand. In 2018, PepsiCo acquired SodaStream, an acquisition that boosted the brand founded in 1903 with the mission of letting people obtain fresh sparkling water using water from their taps. The company attained huge market success by continually adapting to market needs, boosting creativity and idea generation. Supported by PepsiCo Design, a unit that evolved from one designer in 2012 to more than 300 in 2022, SodaStream started looking at market needs in 2020 and noted the increasing presence of water fountains in offices. By empathizing with users, observing them, and studying their daily drinking habits, they framed the hydration problem from an entirely new perspective. When they started developing the concept, they did not begin by proposing another water fountain as many competitors were already doing but envisioned a new fully digital and physical ("phygital") experience that led to the SodaStream Professional solution being launched in 2020. SodaStream Professional is an NFC-enabled fountain that recognizes how much each person drinks and what they drink and, via an app, suggests improved drinking habits. Thus, Design Thinking in the form of Creative Problem Solving helped PepsiCo start idea generation not from scratch but through user analysis and framing the problem as the neglected hydration of employees rather than a way to sell more cans, thereby proposing a new and innovative phygital solution.

The case of HO,[3] a telecommunications operator owned and managed by Vodafone, offers insights into the use of **Sprint Execution** Design Thinking in NPD. In 2018, Vodafone saw the entry of Iliad in the Italian market, a low-cost company in the telecommunications industry already operating in France. By asking questions, such as, "Should we introduce a new Vodafone offer and tariff to compete with Iliad?", "How can we make this sustainable, and what will the impact be on customer loyalty and our brand image?" Vodafone started developing the idea of proposing a second brand or even a new company with a completely different positioning and scope to Vodafone. Hence, they asked relevant business questions, such as "Is it truly possible to build a new brand in just six months and launch a new company?" Their answer was "yes." So, they started a series of sprints to build a minimum viable product (MVP) of an app and new solutions. HO was envisioned as a fully digital telco provider, without stores, no call center, and with everything managed in an app. HO in Italian means "have," and the slogan and motto were "I have everything under control." The app was the door to the company and its offer. However, time was ticking. Iliad was to enter the Italian market within six months, and Vodafone needed to react quickly. Creating focus groups to learn from feedback as fast as possible, they tested the brand positioning, the app, the back-end, and the front-end, then validated it to launch the solution. Vodafone introduced the HO brand within six months, beating all other companies. It proposed a new product to the market that relied on Vodafone's expertise but reduced call center costs and simplified management, activating a SIM card through online onboarding and introducing innovation in a reluctant field. Thus, Design Thinking in the form of Sprint Execution helped Vodafone identify the business questions to ask and introduce a new second brand as an MVP that, after testing and launching on the market, grew in market share year after year.

Finally, to illustrate the use of the **Creative Confidence** form of Design Thinking in the NPD process, we present the Steelcase Learning Innovation Center (LINC). Steelcase is an American multinational furniture company that embraced the Design Thinking paradigm to develop a space that would drive innovation and NPD with an entirely new approach. Indeed, established in 2017 in Munich as the European headquarters, LINC is a place that nurtures different mindsets. The idea was not traditional headquarters where offices are fixed, and spaces are defined but a flexible space that enables employees to experiment with new solutions and products, sharing new uses with clients and colleagues. Under the constant observation of R&D experts, a series of sensors monitor the use of the different spaces and stimulate co-creation. Indeed, people use the newly conceived products according to their needs, and by observing these uses, R&D obtains sometimes unexpected insights for alternative creative solutions. Employees constantly engage in experimentation, and LINC is an always-changing environment that engages stakeholders from different perspectives. Through direct interviews or indirect sensor monitoring, constant feedback is provided on the value and relevance of the proposed solutions. By its very nature, this space supports continuous product development that enables and activates transformations. As the people at Steelcase routinely mention, if visiting LINC, you must visit again and again because it will be different each time. Thus, Design Thinking in the form of Creative Confidence helps Steelcase spot opportunities and develop new concepts addressing new business

FIGURE 23.2 THE CONTRIBUTIONS OF FOUR TYPES OF DESIGN THINKING TO THE NEW PRODUCT DEVELOPMENT (NPD) PROCESS.

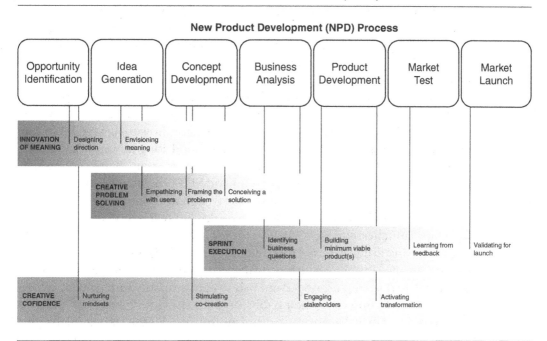

challenges by focusing on humans and valuing the role of endless iterative prototyping.

In summary, the Alfa Romeo 4C, SodaStream Professional, HO, and LINC cases exemplify why Design Thinking is needed in NPD, illustrating that the adoption of different forms of Design Thinking can help companies pivot the NPD process and unveil different and, at times counterintuitive solutions, as shown in Figure 23.2.

23.3 How Can I Apply Design Thinking?

Following from the value and reasons for embracing Design Thinking in NPD processes highlighted above, this section presents the practices companies adopt in each NPD phase. We refer to a set of cases to illustrate the application of Design Thinking in a number of different situations and industries.

MSC Cruises adopted the Innovation of Meaning type of Design Thinking to innovate in the cruise industry. Starting with a new opportunity brought about by digital transformation, the company launched the MSC Smart Ship program in 2016 in some vessels to set a new direction for developing cruise services. The company started by envisioning a new direction with the support of Deloitte Digital[4] through a series of workshops to identify a relevant new direction for the cruise market and provide new meaning for passengers based on the "pampering" principle. Proposing technological solutions onboard, they designed an entirely new experience called MSC for ME, a digital platform that supports passengers during the journey. MSC leveraged Design Thinking principles to envision a new direction by exploiting internal knowledge, in line with the Innovation of Meaning framework (Verganti 2008). MSC proposed a solution that digitalizes the concierge concept and offers passengers an entirely new reason to be on board, enabled by focusing intensely on users as humans and identifying new opportunities. Indeed, although MSC knew that most of their passengers were repeat customers, they did not know how to engage them over time. The MSC for ME strategy helped them build long-lasting relationships with their passengers supported by the Smart Ship program. Thus, every single ship has a new purpose and meaning embedded. For instance, Meraviglia has Cirque du Soleil onboard, and Seaside was designed to let people experience more the see through a series of rounded decks and balconies, just some of the smart ships launched through NPD.

Another inspiring case is Uber Eats. Uber[5] used Design Thinking in the form of Creative Problem Solving to delve deeper into their drivers' needs. They performed a deep user-need analysis to gain empathy with drivers, categorizing their needs, observing their habits, and searching for inspiration. In 2014, they found that Uber drivers wanted to fruitfully use all the time available and not just wait in their cars like traditional taxi drivers. Thus, Uber started thinking about exploiting the waiting time and reframing the problem. By leveraging the framing and reframing techniques (Micheli et al. 2019), they observed the problems of waiting from a different perspective (Dell'Era et al. 2008), from annoying and money-wasting, to time to invest in different purposes. Searching for different solutions, they observed that the hours around dinner and lunch time were those with fewer Uber rides requested. Thus, adopting Integrative Thinking (Martin 2009), they proposed introducing Uber Eats. This idea was conceived to support drivers in making food deliveries in times of lower demand, a solution that has seen tremendous success.

Moving to the viability and feasibility NPD phases, the Savioke case is inspirational, illustrating how the Sprint Execution form of Design Thinking can be enacted. In 2016 this robotics company started to evaluate the business opportunity of introducing robots in the service industry. This business potential arose from the assumption that there are situations in which robots can either substitute or help humans. However, Savioke's problem was uncertainty about the market's reaction; they were unsure whether people were ready for robots in hotels and similar environments. Therefore, they started building an MVP and tested it in a simple use case leveraging fast prototyping techniques, typical of Design Thinking (Magistretti et al. 2021b), envisioning a situation through a storyboard of a person calling for room service. This exercise helped them craft and design the features of the robot.[6] They initially sketched it on paper, built a rough version, tested it in a live context to observe the reaction of users, and clustered the findings into positive, negative, and neutral aspects. They then fine-tuned the solutions. Savioke now has many robots operating in more than

70 hotels around the world as well as in hospitals for medical exchanges in different parts of hospitals so that nurses are not required to leave their areas. This has been enabled by implementing iterative prototyping and continuously learning from the market (Carlgren et al. 2016).

Finally, the PepsiCo Design unit's growth exemplifies the value of Design Thinking in the form of Creative Confidence. PepsiCo started shifting from a marketing-oriented to a design-based company in 2012, beginning its transformation journey by changing its organizational mindset. In particular, PepsiCo introduced a Chief Design Officer, Mauro Porcini, dedicated to pushing the company to adopt a design culture. Porcini started as the first designer in the company and then created design centers where PepsiCo employees could exchange ideas and be inspired by one another. The idea was to break down silos and allow business units to exchange views in an informal space. The centers were created in strategic cities like New York and London. This physical initiative was coupled with the idea of diffusing the value of design and the creation of an organization within the organization. With the support of PepsiCo's President (initially Nooyi and then Laguarta), Porcini hired more designers to bring the number to over 300. Instilling the practice of sharing, he defined a motto and a manifesto, together with a dedicated website for the initiative.[7] This highlights the tangible aspects that allowed PepsiCo to transform into a company now rooted in design and fully embracing the Creative Confidence perspective. This was enabled by focusing on employees, nurturing their growth, and rewarding them for their transformation.

To summarize, the MSC for ME, Uber Eats, Savioke, and PepsiCo Design cases illustrate the adoption of Design Thinking in NPD and how these companies enacted user-centeredness, ideation, and iterative prototyping in different NPD moments and with different practices, from observing users and employee needs to fast prototyping and continuous learning. Indeed, the cases show that Design Thinking overlaps with and supports the NPD phases.

23.4 What Is the Future of Design Thinking?

The popularity of Design Thinking has grown with digitalization, leading to increasingly complex technologies and products, hence the urgent need to understand and serve users. Digital technologies are pervasive and encroach on all territories of business and personal life, from products and services to organizational processes. The challenge is making an ever-morphing overabundance of technologies (e.g., artificial intelligence, metaverse, data, virtual and augmented reality) accessible and usable. In response to this challenge, Design Thinking offers a bridge between humans and technology.

Innovation has left the boundaries of the R&D laboratory to diffuse to all stages of the value chain. Consequently, everyone is concerned with innovating and designing, hence the need for a design paradigm that anyone can use. Enter Design Thinking, which provides design methodologies that can be readily learned and practiced by anyone. In fact, Design Thinking can be thought of as a democratizer of design, just like open innovation has been cast as a democratizer of innovation.

Design Thinking, or at least the user-centered mentality that underpins Design Thinking, is already widely accepted, and employed in NPD practice. With the widespread diffusion of Design Thinking, the boundaries between Design Thinking and NPD practice have become blurred. In fact, it is possible to envision a future in which the NPD process becomes a series of creativity and innovation double diamonds intensely focused on users. In such a future, Design Thinking may even disappear as a separate paradigm, while nevertheless leaving its indelible fingerprint on NPD practice. Thus, NPD managers are encouraged to absorb Design Thinking capabilities and practices to boost their NPD processes.

Notes

1 https://hbr.org/2016/01/the-innovative-power-of-criticism.
2 https://design.pepsico.com/case-studies/brands/sodastream.
3 https://tangity.design/en/case-study/ho-mobile.
4 https://www2.deloitte.com/it/it/pages/technology/articles/msc-for-me—deloitte-italy--deloitte-digital.html.
5 https://voltagecontrol.com/blog/8-great-design-thinking-examples; https://medium.com/uber-design/how-we-design-on-the-ubereats-team-ff7c41fffb76.
6 https://www.fastcompany.com/3057075/how-savioke-labs-built-a-robot-personality-in-5-days.
7 https://design.pepsico.com/.

References

Artusi, F. and Bellini, E. (2022). From vision to innovation: new service development through front-line employee engagement. *Innovation* 24 (3): 433–458.

Beverland, M.B., Wilner, S.J.S., and Micheli, P. (2015). Reconciling the tension between consistency and relevance: design thinking as a mechanism for brand ambidexterity. *Journal of the Academy of Marketing Science* 43 (5): 589–609.

Brown, T. (2008). Design thinking. *Harvard Business Review* 86 (6): 84–93.

Brown, T. (2009). *Change by Design: How Design Thinking Transforms Organizations and Inspires Innovation.* HarperCollins.

Bushe, G.R. and Marshak, R.J. (2014). The dialogic mindset in organization development. *Research in Organizational Change and Development* 22: 55–97.

Candi, M. (2016). Contributions of design emphasis, design resources and design excellence to market performance in technology-based service innovation. *Technovation* 55: 33–41.

Capaldo, A. (2007). Network structure and innovation: the leveraging of a dual network as a distinctive relational capability. *Strategic Management Journal* 28 (6): 585–608.

Carlgren, L., Rauth, I., and Elmquist, M. (2016). Framing design thinking: the concept in idea and enactment. *Creativity and Innovation Management* 25 (1): 38–57.

Dell'Era, C., Magistretti, S., Cautela, C. et al. (2020). Four kinds of design thinking: from ideating to making, engaging, and criticizing. *Creativity and Innovation Management* 29 (2): 324–344.

Dell'Era, C., Marchesi, A., and Verganti, R. (2008). Linguistic Network Configurations: management of innovation in design-intensive firms. *International Journal of Innovation Management* 12 (01): 1–19.

Dell'Era, C. and Verganti, R. (2009). The impact of international designers on firm innovation capability and consumer interest. *International Journal of Operations & Production Management* 29 (9): 870–893. https://doi.org/10.1108/01443570910986201

Dell'Era, C. and Verganti, R. (2010). Collaborative strategies in design-intensive industries: knowledge diversity and innovation. *Long Range Planning* 43 (1): 123–141.

Hirschman, E. (1986). The creation of product symbolism. *Advances in Consumer Research* 13: 327–331.

Kelley, T. and Kelley, D. (2013). *Creative Confidence: Unleashing the Creative Potential within Us All.* New York, NY: Currency.

Knapp, J., Zeratsky, J., and Kowitz, B. (2016). *Sprint: How to Solve Big Problems and Test New Ideas in Just Five Days.* New York, NY: Simon and Schuster.

Krippendorff, K. (1989). On the essential contexts of artifacts or on the proposition that 'design is making sense (of things)'. *Design Issues* 5 (2): 9–38.

Latour, B. (1987). *Science in Action: How to Follow Scientists and Engineers through Society.* Cambridge, MA: Harvard University Press.

Liedtka, J. (2015). Perspective: linking design thinking with innovation outcomes through cognitive bias reduction. *Journal of Product Innovation Management* 32 (6): 925–938.

Liedtka, J. (2020). Putting technology in its place: design thinking's social technology at work. *California Management Review* 62 (2): 53–83.

MacCormack, A., Verganti, R., and Iansiti, M. (2001). Developing products on 'internet time': the anatomy of a flexible development process. *Management Science* 47 (1): 133–150.

Magistretti, S., Allo, L., Verganti, R. et al. (2021a). The microfoundations of design sprint: how Johnson & Johnson cultivates innovation in a highly regulated market. *Journal of Knowledge Management* 25 (11): 88–104.

Magistretti, S., Ardito, L., and Petruzzelli, A.M. (2021b). Framing the microfoundations of design thinking as a dynamic capability for innovation: reconciling theory and practice. *Journal of Product Innovation Management* 38 (6): 645–667.

Magistretti, S., Bianchi, M., Calabretta, G. et al. (2022). Framing the multifaceted nature of design thinking in addressing different innovation purposes. *Long Range Planning* 55 (5): 102163. https://doi.org/10.1016/j.lrp.2021.102163.

Martin, R. (2009). *The Design of Business: Why Design Thinking Is the Next Competitive Advantage.* Harvard Business Press.

Micheli, P., Wilner, S.J.S., Bhatti, S.H. et al. (2019). Doing design thinking: conceptual review, synthesis and research agenda. *Journal of Product Innovation Management* 36 (2): 124–148.

Peterson, R.A., Hoyer, W.D., and Wilson, W.R. (1986). *The Role of Affect in Consumer Behaviour: Emerging Theories and Applications.* Lexington, MA: Lexington Books.

Schön, D.A. and Rein, M. (1994). *Frame Reflection: Towards the Resolution of Intractable Policy Controversies.* New York: Basic Books.

Seidel, V.P. and Fixson, S.K. (2013). Adopting design thinking in novice multidisciplinary teams: the application and limits of design methods and reflexive practices. *Journal of Product Innovation Management* 30 (S1): 19–33.

Simon, H.A. (1969). *The Sciences of the Artificial.* Cambridge, MA: The MIT Press.

Verganti, R. (2008). Design, meanings, and radical innovation: a metamodel and a research agenda. *Journal of Product Innovation Management* 25 (5): 436–456.

Verganti, R. (2009). *Design-driven Innovation – Changing the Rules of Competition by Radically Innovating What Things Mean.* Boston, MA: Harvard Business Press.

Verganti, R. (2016). The innovative power of criticism. *Harvard Business Review* 94 (1): 88–95.

Verganti, R. (2017). *Overcrowded – Designing Meaningful Products in a World Awash with Ideas.* Boston, MA: MIT Press.

Verganti, R., Dell'Era, C., and Scott Swan, K. (2021). Design thinking: critical analysis and future evolution. *Journal of Product Innovation Management* 38 (6): 603–622.

Verganti, R., Vendraminelli, L., and Iansiti, M. (2020). Innovation and design in the age of artificial intelligence. *Journal of Product Innovation Management* 37 (3): 212–227.

Marina Candi, PhD, is a professor of Innovation Management at Reykjavik University's Department of Business Administration and director of the Reykjavik University Center for Research on Innovation and Entrepreneurship. Her research interests include innovation management, disruptive innovation, design-driven innovation, experience-based innovation, and Industry 4.0. She is a fractional Professor at the University of Edinburgh Business School.

Claudio Dell'Era, PhD, is Professor in Design Thinking for Business at the School of Management – Politecnico di Milano, where he serves also as Co-Founder of LEADIN'Lab, the Laboratory of LEAdership, Design and INnovation. He is also Director of the Observatory "Design Thinking for Business" of the School of Management –Politecnico di Milano. Research activities developed by Claudio Dell'Era are concentrated in the areas of Design Thinking and Design Strategy. He has published more than 100 chapters in edited books, papers published in conference proceedings and leading international journals such as Entrepreneurship Theory and Practice, Journal of Product Innovation Management.

Stefano Magistretti, PhD, is Assistant Professor in Agile Innovation at the School of Management, Politecnico di Milano, and a senior researcher in the LEADIN'Lab, the Laboratory of LEAdership, Design, and INnovation. Within the School of Management, he also serves as Director for the Observatory Design Thinking for Business. He has published conference articles and a chapter in an edited book, as well as articles in journals such as Journal of Product Innovation Management, Industrial Marketing Management, Technological Forecasting and Social Change, Industry and Innovation, Business Horizons, Creativity and Innovation Management, Management Decision, Journal of Knowledge Management, Research Technology Management.

K. Scott Swan, PhD, is the David L. Peebles Professor of Business and Head of the Marketing and Innovation Area in the Raymond A. Mason School of Business at William and Mary. He serves on the Advisory Board for the Alan B. Miller Entrepreneurship Center (EC). He was awarded a Senior Fulbright Chair: the 2015–2016 Hall Chair for Entrepreneurship in Central Europe at WU (Vienna, Austria) and The University of Bratislava, Slovakia. Prof. Swan publishes widely and serves on the editorial board of The Design Journal and the Journal of Product Innovation Management (JPIM) along with authoring books.

Roberto Verganti is Professor of Leadership and Innovation at the *Stockholm School of Economics – House of Innovation*, where he is Director of *The Garden – Center for Design and Leadership*. He is also in the Faculty of the *Harvard Business School*, and is a co-founder of Leadin'Lab, the laboratory on LEAdership, Design and Innovation at the School of Management of *Politecnico di Milano*. Roberto serves on the Advisory Board of the European Innovation Council, at the *European Commission*. Roberto is the author of "*Overcrowded*," published by MIT Press, and "*Design-Driven Innovation*," Harvard Business Press. Roberto has issued more than 150 scientific articles.

SECTION FIVE

SERVICE INNOVATION

CHAPTER TWENTY-FOUR

INNOVATION WHEN ALL PRODUCTS ARE SERVICES

Anders Gustafsson, Per Kristensson, Gary R. Schirr, and Lars Witell

All products are services. This message from "Service Dominant Logic of Marketing," (Vargo and Lusch 2004) nearly two decades ago, has had a major impact on marketing theory and practice. A shift to a service dominant logic paradigm affects new product development ("NPD") theory and practice going forward. As we shall establish, highlighted trends in the use and delivery of goods have reinforced the need to consider service innovation ideas in NPD.

Before Vargo and Lusch (2004), the study of service innovation had often been framed by the factors said to distinguish services from goods: intangibility, heterogeneity, inseparability of production and consumption, and perishability (Zeithaml et al. 1985). The inseparability of production and consumption meant that service innovation needed to focus on both (Johne and Storey 1998). In contrast, NPD often focused on goods, traditionally treating product innovation and process innovation as separate processes (Page and Schirr 2008). Changes in product features could mandate changes in production processes, but otherwise, product and process were distinct.

A recurring theme of this chapter will be that products are increasingly services as offerings are "servitized." Responding to the inseparability of production and consumption, purchasers weigh buying decisions not only on price and features but also on production factors such as their carbon footprint, sustainability, country of origin, fair wages and labor conditions, technology employed, and transparency. From a customer's point of view, and therefore from a marketing view, a good and its production are no longer separable.

The PDMA Handbook of Innovation and New Product Development, Fourth Edition. Edited by Ludwig Bstieler and Charles H. Noble.

Goods producers are accordingly under pressure to make their production processes transparent and attractive to modern customers. An example from the experience of a heavy truck manufacturer may serve as an illustration:

Example: Transparency in goods production

One of the authors oversaw several semester-long class consulting projects with a factory manufacturing large trucks. He was impressed by the assembly process. There was ample spacing between stations, and a few stations would rotate or even flip the trucks so that assembly personnel could comfortably perform multiple tasks at their station. Everything was amazingly clean.

There were picnic tables on the line for coffee and lunch breaks; office workers might join line associates in using the tables when it was raining outside. There was a shared work-space where associates from assembly could collaborate with engineers and designers on process or product improvement ideas during the workday.

The company management noticed how impressed visitors were with the facility and built a multi-million-dollar visitor center where sales prospects could observe the production of the trucks and take a tour of the plant. Although the original purpose was to help selling efforts for prospective customers, the visits to the production line have also enhanced partnerships with long term customers.

Inseparability of product and production may now apply to goods as well as services. Firms should plan and design for transparency and the marketing appeal of production.

In this era most offerings are classified as services: 80+% of GDP in the USA (CIA 2022). Furthermore "servitization," which includes software as a service – SaaS, cars as a service – CaaS, IT as a service – ITaaS, and the on-demand trend is spreading in numerous goods markets.

Servitization is especially important in B2B markets but is attractive wherever customers focus on the use of a good but not on managing maintenance or repair. As companies are opening up the production processes and enabling consumers to influence these processes to a larger extent, there is also increased emphasis on servitization in B2C markets. With an increased focus on service, the marketing interest shifts from the sales of a good to the use of a service. The experience of using the product is the key for increasing the sales of goods. This has implications for the development of a good, since the producer is normally not present where the customer experience takes place.

These changed perspectives on goods push all offerings toward services. As a consequence, there is much to be learned from experience developing services that will benefit all NPD. When New Service Development (NSD) was initiated as a theoretical concept in the 1980s, it borrowed heavily from NPD (Johne and Storey 1998). In their literature review of NSD, Papastathopoulou and Hultink (2012), the authors emphasized that service and manufacturing are becoming more intertwined and there might be a need for a common framework for innovation activities instead of keeping NPD and NSD separate. Since then, theory on NSD has developed on its own, so the authors believe that *it may be time for NPD for goods to consider the lessons of NSD and service innovation.*

This chapter focuses on three trends observed in research on service innovation: (1) servitization, (2) innovation collaboration with customers, and (3) a focus on customer experiences. In the remainder of this chapter, we investigate these three trends and how they apply to product innovation when all products are services.

24.1 Servitization – Products Are Becoming Services

All products – goods or services – involve coproduction or co-creation between the manufacturer and the consumer. A user is not buying a hammer but a role in a job like hanging pictures, co-creating the experience of custom decorating a home. A key part of managing and innovating a product is understanding the "job" for the product (Christensen et al. 2007).

The development of a large installed base of products (i.e. products that are still in use on the market) provides an opportunity to more completely serve the customer. Associated goods and services may include repairing the good, maintaining the good, providing training of employees, and using data to improve the efficiency and performance of the good and services.

A single product is today often sold in a variety of ways, as a stand-alone product or good, as a part of a solution, and as a component of getting the job done through a pure service. A larger scope of the use of a product would be to help the customers better utilize their products; for instance, to build platforms of interconnected tools that share some expensive parts (e.g., batteries) or to enable it to be shared among neighbors and friends or part of the sharing economy (e.g., online solutions and the Internet of Things).

24.1.1 What Is Servitization and What Does It Mean for the Manufacturer?

Traditionally, servitization refers to the organizational changes of a firm caused by the emphasis on service (Forkmann et al. 2017), while service infusion refers to an increased emphasis on services in the offering. Servitization is "the innovation of an organization's capabilities and processes to shift from selling goods to selling integrated products and services that deliver value in use" (Baines et al. 2009). It is driven by user needs and producer opportunity. Users don't want or need to own and maintain every good they use to get their job done. Servitization appeals to manufacturers that value a stream of revenue from service and replacement parts, and potentially regular subscription income for a service.

Consider how often various products (e.g., drills, lawn mowers, chain saws, or cars) are actually used by a non-professional user. A car, for example, is parked 95% of the time on average (Schmitt 2016). If products were to be connected to various platform-based services, the usage of the products could potentially increase dramatically, and consumers would not have to own and maintain as many goods and use as much storage space. A solution could be a peer-to-peer service, connected to an online solution to create revenue for the owner. One B2C firm that is planning for such a development is Tesla. With the growing capability of autonomous vehicles, Tesla has plans for its electronic vehicles to operate as a robotaxis, generating income for owners on a ridesharing network (Thompson 2016). A firm must understand product use and opportunities for changes in the business model to design for the future.

Servitization of manufacturing firms will accelerate in B2C markets and continue in B2B markets. This trend is supported by increased complexity of products, solution selling, and enhanced connectivity of goods, based on the Internet of things (IoT). This evolution has created a debate on who really owns a product – and when a customer has the right to repair a product. Digitalization is increasing both efficiency and effectiveness of agriculture. But, with increased digitalization the question comes – what can a customer do with the product – and what is the customer allowed to do. Is it really John Deere's tractor? As a user are you simply driving it?

Through digitalization, a firm can provide both additional functionality and services through pressing the keyboard on a computer. These services are increasingly a key component of their business model – and therefore a key part of their innovation process. As an example, the CEO of Volvo Trucks announced a strategic direction that by 2030, 50% of revenue should come from associated services, an example of service innovation within a traditional manufacturer.

Goods need to be prepared for services – or to be a piece of a platform for service provision.

24.1.2 Implications of Servitization for NPD

Since many products are part of a solution or sold as a service or part of a platform, servitization influences how NPD needs to be performed. All stages of the NPD process should include NSD activities. There should be experimentation with different services and business models. When the longer cycle of activities for NPD are running, multiple shorter cycles of activities for NSD can be performed. This enables service innovation to progress and also provide input into the longer NPD cycles working on the physical design of the product.

First, in each stage of the NPD process there is a need to consider the development of the service content of the offering. Traditionally, a product was designed first and then service design took place after the product design or hardware was decided upon. With an increased share of the revenues coming from services, it is important to design services in the offering. That may involve including IoT sensors, contracting who owns the data, and who will have access to collected data.

> *Service developers must be included in the important stages in the product design. One of the authors participated in the design of digital services for a bus. Unfortunately, when he and the service team joined, it was already too late to change the design of the bus to track the needed key data to predict maintenance needs. Service designers had not been represented in the initial development team. Therefore significant potential revenues from digital services were never realized.*

Second, the increased emphasis on service provision and growth of good-as-a-service or pay-for-use means that *the business model, or often multiple models, become important to consider in the NPD process.* Tools such as service blueprinting and the business model canvas become important to map what service provision, now or in the future, means in practice and how it contributes to revenues. Test marketing of new business models can be performed. When Volvo introduced its first car sharing service Sunfleet, they introduced it only in the Swedish market. It enabled the company to learn about the car sharing business model and to understand what the implications are for the design of the car.

Third, the focus on digital service provision demands new competences to be represented in the development team. In addition to service developers, a team should have competence on IoT, digital data analysis, and business modeling. These competencies need to work together with the product designers so that all of the demands on the design of the product are considered. One example is the use of the installed base of products as a source of data, such as in what condition are the products being used, how they are treated, and how well they work in combination with competitor products.

24.2 Co-creation and Collaboration in Innovation

Customers are active and use products, goods and services, provided by one or more firms to perform a variety of tasks or jobs. Products are employed by customers to create value and meet their needs (Christensen et al. 2007) A washing machine can serve as an example:

> *The washing machine has limited value if the customer is unable or has not been trained to place the laundry in the machine and start it correctly. Its value may be forfeited altogether if clothes are washed at the wrong temperature setting or are mixed with clothes of a different color. The customers may not need a washing machine if clean clothes can be delivered some other way. (Gustafsson et al. 2016)*

This illustrates that the services provided by goods are coproduced/co-created and that the customer has a unique perspective on what matters to them. A provider of goods removed from where value occurs may know how the product is intended to be used but may lack information about how knowledgeable individual customers are, or how customers actually use the product.

24.2.1 Co-creation and the Curse of Knowledge

Many people, on first hearing a recording of themselves speaking, utter "is that me?" We don't sound as we think we do when we talk to others. In the innovation management literature, a similar phenomenon is typically labeled "over the wall." Users often perceive newly launched offerings differently than producers imagine. An unpublished research study for a dissertation at Stanford University in the early 1990s assigned participants to a simple game of tapping the rhythm of a well-known song (Newton 1990). The listener's job was to guess what song it was. The researcher asked the tapper to predict how many listeners would guess the song correctly. The average prediction was that around 50 % of the listeners would accurately get the song. However, less than 3 % of the listeners guessed the right answer.

In the management literature, the phenomenon illustrated in the previous paragraph is referred to as the curse of knowledge (Heath 2003). Tappers cannot avoid hearing the tunes within their own heads while tapping, leading to a grossly overestimated number of correct answers. Yet the listeners hear a few taps sounding as inexplicable Morse code. The Stanford experiment is thought of as a replica of what is taking place in the innovation. Product developers may assume that customers will use a product the way they would. However, training or transfer of knowledge affects the value that the customer is experiencing.

Virtually all partners in an ecosystem suffer from an information imbalance which makes understanding more difficult. Businesses need to *collaborate* with their customers to provide a foundation for future value-creation. Services must provide customers with resources needed in order to co-create value. Companies that provide service need to understand how their customers use products, how they perceive the experience and how they live their lives. By collaborating, companies can learn from interactions and shorten the distance between themselves and their customers.

24.2.2 Implications of Co-creation for NPD – Different Shades of Collaboration

The range of collaboration that takes place between a company and its customers is generally referred to with the popular label co-creation. Co-creation was coined by researchers (Prahalad and Ramaswamy 2004), to compromise multiple activities, including coproduction of the actual delivered service and collaboration in the innovation of a service.

Co-creation during the *value realization* phase – the use of the product – is also referred to as coproduction. The customer is an active participant in service delivery. Even a relatively simple service like banking with an ATM depends on a customer having a current card, remembering a password, and having patience to see the process through. More complex, customizable, or experience-dependent services require more of the user.

Collaboration takes many forms throughout the innovation process, for instance, responses to market research requests, direct input on what to produce (Kristensson et al. 2004), or active innovation by motivated users such as "lead users." (von Hippel et al. 2006) Customers are the true source of many innovations.

As evident from the co-creation framework, useful collaboration in innovation is more likely to occur (von Hippel 1986; von Hippel et al. 2006) when:

1. *Customers want or need something strongly.* The Lead User literature notes that Lead Users can be identified as either being the most anxious for innovation or so motivated that they are working on innovation of the product themselves.

2. *Customers are better at something.* This might entail expertise or other types of advantages that a customer might have, such as knowledge within certain application areas.
3. *Customers want to avoid something.* Service innovation processes are facilitated when customers are strong proponents of risk minimization. Patient participation in health care is more likely if it results in a lowered risk of becoming ill or experiencing negative consequences from a disease (Lusch et al. 1992).
4. *Customers have resources.* Customers who are in possession of resources, for example physical capital of some kind, are more likely to use these resources to co-create value for themselves. Tools, workspace, cooperation of a team, or other types of manifest resources can be made available to targeted customers to advance innovation.

Co-creation is present in some form in all services. Coproduction, one form of co-creation, can be a relatively simple exercise such as using an ATM for banking or far more complex like learning to sky-dive with an instructor. Similarly, user collaboration in the innovation process can range from sharing knowledge, to allowing a company to observe use, to actively working with the company during innovation, or even sharing innovations the user has already developed.

Customers collaborate with companies in innovation for several reasons. The overall motivation for companies and customers to collaborate is the information asymmetry that exists between companies and customers and that hinders the process of value-creation. As a result, customers are collaborating in product and service innovation projects. Both parties in the collaboration can be more confident that the process of value-creation should lead to desired results.

24.2.3 Example: Managing Collaboration in a Rapidly Changing Market

A new market is a fertile environment for active customer collaboration – experimentation is natural in such an environment and the users may be anxious for improved tools to use in the new space. Smart companies plan for such collaboration in their innovation process, customer service and service design. One of the authors studied the innovation process in a single financial technology ("fintech") firm as the market for its services matured. (Schirr 2021) The firm supported online trading in financial products.

As von Hippel predicted there were Lead Users anxious to provide input to firms supporting their trading and operations when trading originally moved online. There was a critical need for a variety of products and services, such as connectivity to exchanges where the online financial instruments traded, networks to multiple exchanges, intuitive order-entry devices, and real time tracking and risk management support. Participants were anxious to effectively utilize the new trading opportunities and therefore in some cases worked side by side with the firm's software engineers to develop enhanced products.

After a few years these users seemed to become more wary of providing intellectual property ("IP") to the firm as it and the other fintech companies in the market grew and became profitable. In addition, customers were now generally dealing with multiple products. The firm found that it needed to change the nature of the collaboration over time as the market matured to deal with these IP concerns and the new attitudes of customers.

The author doing this research identified five phases in the company's development and corresponding user involvement in each:

1. Pre-launch – *Consulting for ideas from potential online trading organizations*
Contract software development and custom products for firms moving trading online, with the explicit goal to identify a market and develop proprietary products.

2. Launch – *First Customers*
Meeting product development needs of initial customers. Fortunately, the early customers included far-sighted, demanding lead users. This is not surprising, as they were taking advantage of new opportunities, but needed support that was under development.

3. Rapid Change – *Lead Users*
Demanding customers drove early development. Most of them openly shared ideas and often their own prototypes for a product or service.

4. Rapid Growth – *Encourage Users to Collaborate*
Collaboration had to be encouraged. Engineers were assigned to demanding user work sites. Some were sent abroad for special applications. Some users were induced to participate in development by free use of the firm's products, extra support, or even rent-free workspace.

5. Maturing Growth – *Become the User!*
Rich information became harder to gather from users. Perhaps users noticed the firm and its revenues had grown, realized the value of the collaborative IP, or they were now working with multiple vendors. The firm created a separate profit center to participate in the online markets – to "become a customer themselves" (Griffin 2012).

As shown in the outlined five steps, the founder of the start-up in this case study actually started the business as a technical consultant and contract developer in a fast-growing market. His purpose was to find the products needed by the evolving marketing place. (Interviews with competitors and companies in related products uncovered several others who started in that manner.) Once the firm launched products it benefited from Lead User customers who not only provided them with innovative ideas, but pushed them into aggressive innovation.

However, as time passed, the market matured, and a handful of firms were perceived as leaders. Whether because of the perceived value of IP the collaborating users were providing, or simply because most users now used more than one solution in the space, or some other reason, collaboration was not as "natural." As the market matured, the firm became more proactive and aggressive in facilitating user collaboration for innovation, eventually offering incentives from free products to rent-free office space. Finally, the company actually set up a subsidiary to be a user (phase 5). This strategy, becoming a user, was a step discussed in the third edition of the *PDMA Handbook of New Product Development* (Griffin 2012).

Interviews with other fintech firms in these markets showed that they had also gone through multiple forms of collaboration including at least some of the five phases that emerged from this single case study. This firm's five phases of collaboration cannot be considered a definitive outline of the phases of collaboration at this time, despite the support for the phases from the experiences of other firms in the online trading market. Experiences may vary in markets that develop differently, but the common experiences of the firms in this diverse industry illustrates that:

- *Collaborative innovation processes may need to change as markets evolve, and*
- *Firms should be proactive and plan to facilitate user collaboration in changing and maturing markets.*

24.3 Designing the Product for User Experience

Service and goods offerings facilitate value creation with the goal of increasing customer well-being. Well-being comes from successfully completing the task the product was acquired to perform and leaving the customer with a positive feeling about the experience. Firms need product attributes that speak to both "needs" and "wants" of their customers. Much attention has been focused on uncovering elusive customer needs that may be tacit, sticky, or contextual, with a focus on observing the user in the process of using a product. How does an innovator uncover the wants and emotions within the experience of use?

24.3.1 Understanding Needs and Wants

The "needs" help customers achieve their goals – do the "job" discussed earlier. Firms produce drills to enable customers to make a hole and washing machines to clean clothing. In today's competitive situation all producers are aiming to meet user needs. Functions can be copied by other similar products, or the needs can be satisfied by another solution. It is doubtful that anyone really wants to own a drill or a washing machine apart from their function.

Many products fit into a system of linked products and services provided by a firm or should work with other products that a customer may have access to. An example of this is connected homes that make it possible to control everything from light and sound to who is in the house. Customer preferences can be discerned from collected data. This also means that devices from the same manufacturer or devices certified to be compatible with the platform may be favored.

24.3.2 The Internet of Things and Customization

When connecting more products either as part of a solution platform or part of the Internet of Things, firms move toward customization or adaptation. Products should get smarter and adapt to a customer's use. There are, of course, privacy issues – when do producers know too much about users? How do you load a product with customization abilities without saving too much information about customers? *A firm must learn to trade off customers' wishes for customization and demand for privacy.*

Another issue is that features can play different roles over the lifespan of a product. Attributes that are appealing to customers during purchase may not have the same appeal during usage. For instance, research has found that product personalization or anthropomorphization may be beneficial during a purchase phase while the same feature plays a different role during customer use (Hoyer et al. 2020). Human-like features help to create relationships and encourage customers to rely and depend on a product. However, when that service or support is disappointing to a customer, the anthropomorphization of a product may make the failure experienced by the customer more personal and damaging.

24.3.3 Implications of Customer Experience for NPD – Customization, Simulation, and Neuroscience

Many innovation theories and methods have focused on understanding and uncovering customer needs; there has been much less focus on understanding wants and emotions. Attributes that connect to emotions or attachment are desirable to firms and their customers. From a firm perspective these attributes may be more difficult to mimic and copy into a new offering. Small details may have a large impact on how a product is perceived. For instance, the temperature or texture of a product may impact the user experience and how long a product is used.

The Internet has facilitated new techniques to study user needs. Collaborating and observing the use of services and goods have been made easier; crowdsourcing has been made possible. The move online has made many services more like software or apps and therefore led to software development techniques and data analysis being employed in service innovation. Methods such as "agile" innovation and A/B testing are widely used. In similar fashion, developing technology will expand innovator's understanding of user experience, emotion, and wants.

The introduction of new technology-based services such as Internet of Things (IoT), Augmented Reality (AR), Virtual Reality (VR), Mixed Reality (MR), virtual assistants, chatbots, robots, and now the meta-verse will change product innovation. These services are typically powered by Machine Learning (ML) and Artificial Intelligence (AI), with the potential to simulate alternative user experiences, understand a user's feelings, and ultimately to dramatically transform the customer experience. The fact that they are powered by AI implies two important things; they can adopt to customer behavior during and after use, but more importantly they are designed to collect information about the users as the information occurs in an unobtrusive way.

Again, innovators using these technologies will face privacy issues. Capturing and analyzing information from everything a user does and acts and interacts with the environment without involving the actual interpretation of the situation by the user is the purpose. We may be used to this in the online space, getting suggestions for "useful" products based on our acts or what we have expressed, now these techniques are moving into the physical space. One example of application of these tools is Spotify's patent for a solution to read emotions from a user's voice as they speak on a phone to adapt the streaming music to the user's mood.

To understand wants and the emotional impact of co-creation by customers, a company needs to experiment and understand what customers are doing and experiencing with products in their own context (value-in-context). Tools, such as neuroscience measures, that have been around for a long time have been prohibitively expensive, both in terms of instruments and skilled personnel, to use widely during a development process. The cost of these tools, however, is dropping dramatically, new versions are being developed, and data analysis is more accessible. *Forward-looking product innovators should take the time to understand these tools, how to use them, and build a more solid understanding of customers' experience and emotions about products in use.*

A set of tools that can be categorized as neuroscience measures are being developed (Verhulst et al. 2019). We see that as a natural evolution given that service providers want to learn more from their users. These neuroscience tools are developed to detect changes in the human body at three vital levels – the brain, the peripheral system, and the neurotransmitters and hormonal system. Common tools include eye-tracking, galvanic skin response (GSR), and electroencephalogram (EEG). Spotify's voice patent, mentioned earlier, is a simple example of such tools.

What a company can learn from these tools is what really happens in a user's own context during the experience of actual use (Alvino et al. 2020). This contrasts with a user's later recall of the use. This is significant as a memory is inexact and not every action is cognitively processed; the latter typically is connected to routine behavior. Furthermore, tools such as VR or AR are potentially useful to simulate and test new solutions and their emotional impact before rollout.

The authors would suggest to product innovators that early experimentation with and adoption of technologies measuring the emotional experience of using a service and technologies simulating new or changed services will be a competitive advantage for organizations.

24.4 In Conclusion: Product Innovation Is Service Innovation

As discussed in the beginning of this chapter, early comparisons of innovation for services and goods focused on the factors distinguishing services from goods, such as the inseparability of production and consumption in services. These factors as well as practice in service and goods innovation helped define differing goods-dominant and service-dominant logics of innovation (Mele et al. 2014).

24.4.1 Goods-dominant and Service-dominant Logic for Product Innovation

A service logic for NPD can infuse new methods and activities that support product development. The key difference of a goods and service logic for NPD is in the view of the customer. A traditional product logic for goods implies developing new features, or a redesign of a good; whereas the service logic focuses on developing a new value-creating experience (Gustafsson et al. 2016).

Sometimes improving the value of customer experiences from using a good may not require a redesign of the good or even a new feature. Changes in how the company is carrying out activities, for instance a new business model or customer contact center, can impact the experience. Service logic views the customer as a co-creator of value while a goods logic is placing its emphasis on product features. The co-creation of value is addressed in the traditional NPD process, including the lead user methodology, but a service logic suggests that co-creation should be the focus through all activities in the innovation process. Table 24.1 illustrates how the service logic focus on the user value creating experience affects each phase of the innovation process.

Through the lens of the service-dominant logic, the NSD process is iterative and based on experimentation. Understanding how the customer co-creates value in the best possible way takes time to understand and may involve multiple experiments and iterations. Yet these activities can be performed in parallel, and experiments can accelerate choices in design and options, so the total innovation time could well be reduced. Parts of the transition from a product to a service logic for NPD is supported by three ongoing trends in service innovation, previously discussed in this chapter.

TABLE 24.1 LOGIC OF TRADITIONAL GOODS AND SERVICE INNOVATION FOR NPD.

Dimension	Goods Logic for NPD	Service Logic for NPD
Product	Develop the Goods with the best product features. (Value-in-exchange)	Develop the Goods/services that perform best from a customer perspective in use. (Value-in-use)
Customer role in product exchange	The customer purchases and uses developed goods.	The customer uses the goods and services as a co-producer of value.
Innovation process	A structured development process supports and controls development work. Focus on developing a tangible output.	Iterative and inspirational development process used to encourage, support, and steer development work. Focus on developing intangible outputs (such as a valuable experience for a user).
Customer/user role in NPD process	Users are utilized to gain knowledge about the use of the goods. (Passive/Active participation)	Users actively create knowledge about the design of the new goods/services. (Active participation)
Production of goods and services	Production plans for goods will adjust for planned features and pricing.	Production of services is inseparable from use and therefore production is an integral part of service innovation.
Development time	Development of goods takes time, based on planning, testing, and executing a plan.	Goods/services development takes time. Based on experimentation in use, the plan is reconfigured and changed.

24.4.2 All Product Innovation Is Service Innovation

The chapter started with an illustration of why one of the traditional key differentiators of service – the inseparability of production and product – is changing. From a discussion of the benefit of a traditional manufacturing firm making its production process visible to prospect and existing customers came the observation that:

- A product innovation process should include analysis and plans to make production attractive and transparent to customers.

The three trends of service and service innovation (1) servitization, (2) collaboration with customers, and (3) a focus on customer experiences, discussed in this chapter, were shown to have implications for product innovation as shown:

- *Servitization* – Design and plan for multiple product deliveries: stand-alone product, servitized product, part of a solution, and part of a platform. How will it fit in the IoT?
- *Co-creation of service and collaboration in innovation* – Study the product in use. Plan for training of users. Create an innovation process that involves users throughout and adapts to changing conditions.
- *Focus on customer experience* – Understand the importance of the emotions and experience of a user and designing a product accordingly.

Table 24.2 goes into more detail by summarizing how each of these three trends impact six facets of an innovation process and customer co-creation.

Each of the service trends has implications for how to perform product innovation and introduce new activities in the product development process. The service trends put additional demands on the NPD project suggesting new activities to add to the effort. Table 2 highlights additional implications from these service trends, including:

- The product will be consumed as a service or as part of a solution. This will influence the product design – it must be designed as a service through a focus on its use.
- New product development projects should use new tools to experiment with customers to better understand the user experience and feelings from the use of products/platforms.
- New competences on services, digitalization, and business models need to be added to the development team.
- A better knowledge about how the customer uses the product should result in shorter development time and reduce unnecessary development work.
- Experimentation becomes an important activity throughout the NPD process, including experimentation with business models, neuroscience tools, and designs of the service system.

TABLE 24.2 THREE SERVICE TRENDS – SIX FACETS OF INNOVATION AND CO-CREATION.

	Servitization	Co-creation	Customer Experience
Product (Goods)	Goods need to be prepared for service provision.	Understand how the customer uses the product.	View the product as a platform for experiences.
Offering (Goods and Services)	The design of goods and services need to consider that they will be provided through and as solutions.	Understand how the offering can help the customer to get the job done.	Understand how the product can create experiences, taking the users' service systems into account.
Development process	All stages of the NPD process should include service development.	Use co-creation methods to identify latent needs of the customer.	Understand how different attributes contribute to the process of usage.
Development projects	Shorter development time enables services to experiment with different offerings and business models.	Involve lead users in the development team to co-create the new product.	Use customer journeys to understand how products and services can be integrated into the life of the user.
Development Competence	The importance of competence on IoT, Digital Data analysts, Service development, and business models.	Importance of competence on co-creation methods, market research methods, and service design methods.	Importance of competence on tools (neuroscience) to understand customers emotions.
Development Time	Should not be influenced by an increased emphasis on service provision.	Can be shortened due to better understanding of latent customer needs.	Can be shortened due to understanding of how customers emotionally connect with the offering.

There is benefit in studying and focusing on each of NPD for (1) goods and (2) services. As predicted from a service-dominant logic, the theories and practices of service innovation increasingly apply to goods. Goods can be viewed as frozen services waiting to be put into a value-creating experience (Vargo and Lusch 2008).The importance of the service-dominant logic is the shift in focus, away from the offering to the process and experience of user value creation.

It may be time to start viewing both NDP for goods and service innovation as vital and interesting subsets of all "product innovation." *Goods manufacturers who have embraced the inseparability of services and goods, and other service innovation practices into their NPD gain a competitive advantage over firms that have not yet incorporated the insights of service innovation.*

References

Alvino, L., Luigi P., Abhishta, A., and Robben, H. (2020). Picking your brains: where and how neuroscience tools can enhance marketing research. *Frontiers in Neuroscience* 14.

Baines, T.S., Lightfoot, H.W., Benedettini, O., and Kay, J.M. (2009). The servitization of manufacturing: a review of literature and reflection on future challenges. *Journal of Manufacturing Technology Management* 20 (5): 10.

Christensen, C.M., Anthony, S.D., Berstell, G., and Nitterhouse, D. (2007). Finding the right job for your product. *MIT Sloan Management Review* 48 (3): 38.

CIA (2022). *GDP – composition, by sector of origin*. CIA.gov.

Forkmann, S., Henneberg, S.C., Witell, L., and Kindström, D. (2017). Driver configurations for successful service infusion. *Journal of Service Research* 20 (3): 275–291.

Griffin, A. (2012). Obtaining customer needs for product development. In: *The PDMA Handbook of New Product Development*, 3e (ed. K.B. Kahn), 504. NY: Wiley.

Gustafsson, A., Kristensson, P., Schirr, G.R., and Witell, L. (2016). *Service Innovation*. Business Expert Press.

Heath, C. (2003). Loud and clear. *Stanford Social Innovation Review* 2003 (Winter): 18–27.

Hoyer, W.D., Kroschke, M., Schmitt, B. et al. (2020). Transforming the customer experience through new technologies. *Journal of Interactive Marketing* 51: 57–71.

Johne, A. and Storey, C. (1998). New service development: a review of the literature and annotated bibliography. *European Journal of Marketing* 32 (3/4): 184–251.

Kristensson, P., Gustafsson, A., and Archer, T. (2004). Harnessing the creative potential among users. *Journal of Product Innovation Management* 21 (1): 4–14.

Lusch, R.F., Brown, S.W., and Brunswick, G.J. (1992). A general framework for explaining internal vs. external exchange. *Journal of the Academy of Marketing Science* 20 (2): 119–134.

Mele, C., Colurcio, M., and Russo-Spena, T. (2014). Research traditions of innovation: goods-dominant logic, the resource-based approach, and service-dominant logic. *Managing Service Quality*.

Newton, L. (1990). Overconfidence in the communication of intent: heard and unheard melodies. Unpublished doctoral dissertation. Stanford, CA: Stanford University.

Papastathopoulou, P. and Hultink, E.J. (2012). New service development: an analysis of 27 years of research. *Journal of Product Innovation Management* 29 (5): 705–714.

Page, A.L., and Schirr, G.R. (2008). Growth and development of a body of knowledge: 16 years of new product development research, *1989–2004. Journal of Product Innovation Management* 25 (3): 233–248.

Prahalad, C.K. and Ramaswamy, V. (2004). Co-creation experiences: the next practice in value creation. *Journal of Interactive Marketing* 18 (3): 5–14.

Schirr, G.R. (2021). Innovation from collaboration – start-up to growing firm. *34th Global Research Conference on Marketing and Entrepreneurship*. online.

Schmitt, A. (2016). It's true: the typical car is parked 95 percent of the time. Streetsblog USA. https://usa.streetsblog.org/2016/03/10/its-true-the-typical-car-is-parked-95-percent-of-the-time.

Thompson, C. (2016) Elon Musk wants to let Tesla owners make money on their cars when they are not using them. Business Insider Elon Musk Reveals Tesla Shared Fleet for Renting Out Your Car. businessinsider.com.

Vargo, S.L. and Lusch, R.F. (2004). Evolving to a new dominant logic for marketing. *Journal of Marketing* 68 (1): 1–17.

Vargo, S.L. and Lusch, R.F. (2008). From goods to service (s): divergences and convergences of logics. *Industrial Marketing Management* 37 (3): 254–259.

Verhulst, N., De Keyser, A., Gustafsson, A. et al. (2019). Neuroscience in service research: an overview and discussion of its possibilities. *Journal of Service Management*.

Von Hippel, E. (1986). Lead users: a source of novel product concepts. *Management Science* 32 (7): 791–805.

von Hippel, E., Franke, N., and Prugl, R. (2006). Efficient identification of leading-edge expertise: screening vs. pyramiding. *Technology Management for the Global Future – PICMET 2006 Conference*, 884–897.

Zeithaml, V.A., Bitner, M.J., and Gremler, D.D. (1985). Services marketing. *Journal of Marketing* 49 (2).

Anders Gustafsson is a professor of marketing at the BI – Norwegian Business School. He is a distinguished professorial fellow at the University of Manchester's Alliance Manchester Business School (AMBS). He is the first international president for AMA's (American Marketing Association) academic council and a recipient of the Christopher Lovelock Career Contributions to the Services Discipline Award. He has published articles in *Journal of Marketing, Journal of Marketing Research, Journal of Consumer Research, and Journal of the Academy of Marketing Science*.

Per Kristensson is a professor and the director at CTF, Service Research Center, Karlstad University, Karlstad, Sweden. He is also a senior fellow at Hanken, Svenska Handelshögskolan in Helsinki, Finland. Aside from his academic career, he is a member of several boards and is an experienced business speaker.

Gary R. Schirr was Associate Professor, Marketing at Radford University in Radford VA until his retirement at the end of 2021. He conducted research and taught on service innovation, service marketing, and direct marketing. He has published in the *Journal of Product Innovation Management*, and along with the other three authors of this chapter he co-authored *Service Innovation*, published by Business Expert Press. Before joining Radford, he had worked in financial derivatives marketing at financial firms in the US and Asia, and been a part of several financial technology startups.

Lars Witell is a professor at the CTF Service Research Center at Karlstad University, Sweden. He also holds a position as Professor in Business Administration at Linköping University, Sweden. He conducts research on service innovation, customer experience and service infusion in manufacturing firms. He has published in *Journal of the Academy of Marketing Science, Journal of Retailing, Journal of Service Research* and *Journal of Business Research*.

CHAPTER TWENTY-FIVE

NEW PRODUCT DEVELOPMENT BY EXTENDING THE BUSINESS MODEL

Christer Karlsson and Thomas Frandsen

25.1 Introduction

New product development can take several alternative or complementary strategic directions. The company can improve and develop existing products, develop a new product generation or new product platform, or perhaps create something entirely new – a technological breakthrough. However, along with technological development, it is vital to consider the application, customer, and market dimensions and what the company does for the customer or user. The company may go further and not only offer the product and/or service but also participate in part of the customer's process where its product and/or service is used within the customer's domain. The company may run part of the customer's operation in which it is an expert, and which is not the customer's key business. These are ways in which the company may extend its business model, defined, and described in terms of a concept that entails three consecutive parts: value proposition, value creation, and value capture (Osterwalder and Pigneur 2010).

This chapter describes and develops the business model extension strategy, including first services and product performance, and eventually solutions for the customer. This strategy is more than business model development of the products and processes; extending the business model means that the company does some of what the customer used to do as a service to the customer. This business model has become very effective in enabling customers to focus on

The PDMA Handbook of Innovation and New Product Development, Fourth Edition. Edited by Ludwig Bstieler and Charles H. Noble.

FIGURE 25.1 EXTENDING THE BUSINESS MODEL.

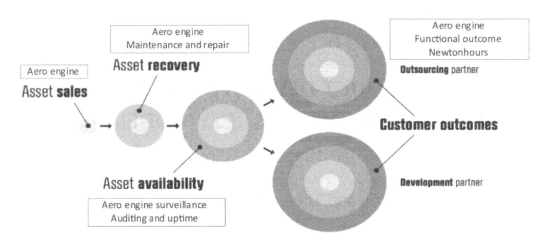

Source: Avlonitis et al. (2014)/with permission of Elsevier.

value creation for the end customer or user (Karlsson et al. 2018). Figure 25.1 shows how a manufacturer extends its business model from primarily focusing on selling a product. Extending the model includes first adding recovery services and second taking on responsibility for the availability of the equipment by selling uptime. Eventually, extending the model involves selling not the product but the performance of the equipment, and as a solution provider, carrying out a function in the customer's production system. In a real case, a manufacturer of aircraft engines moves from the product base of the engine, adds services and maintenance, then surveys engines in use to manage the uptime, and finally, sells not the engine but the engine's outcome in terms of power times time, Newton hours. The push on the aircraft, we call it.

Becoming and being a solution provider offers considerable potential gains, but they do not come easily, there are several developments that must be carried through. Potentials include creating stable revenue, higher margins, and closer customer relationships. Challenges involve changing the value proposition to include services and solutions, which requires a drastic change in the organizational culture, competencies, organizational structures, financial instruments, and objectives. Generally, the effects are well worth the effort. A more stable revenue stream comes from the continuous delivery of the services instead of the now and then sale of equipment. Higher margins are possible from the value creation in the customer processes, but also less competition from manufacturers who cannot offer complete solutions. A much stronger competitive position comes from closer relations with the customer, since switching between solution providers is much more complicated than switching between equipment suppliers.

In addition to the gains and other advantages for companies applying the solution provider business model, recent research has demonstrated potential positive effects on a macro level. Through specialization and higher volumes, the solution provider can achieve higher effectiveness and economy of scale than the customer, resulting in better resource utilization. This gives lower environmental footprints and lower negative impact on climate change, enabling customers to meet sustainability requirements. With increasing societal awareness and growing regulatory pressure, solution providers can leverage their capabilities to develop commercial sustainability offerings when expanding their business model (Frandsen et al. 2022a).

This chapter shows how to analyze the potentials and challenges of applying a "solution provider" business model (Baines et al. 2009; Storbacka et al. 2013).

25.2 Extending the Business Model: Vertical and Horizontal Integration

The business model can be extended in different directions that involve vertical and horizontal integration. Vertically can be either backward toward raw materials or base services or forward toward the customer or market. Horizontally includes widening the business to new products and services for existing and new markets. Vertical technologies such as chemicals, electronics, and mechanics build up to the final product, while horizontal application or feature technologies provide product functions such as safety or sustainability. The vertical backward arrow is crossed over as a warning since it is seldom profitable and only applied if there is shortage of supply and a not functioning supplier market. See Figure 25.2.

FIGURE 25.2 VERTICAL AND HORIZONTAL TECHNOLOGIES.

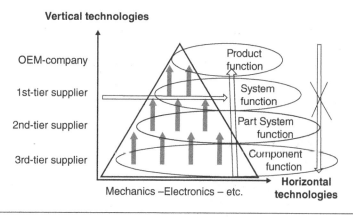

Source: Own presentation, partly in Karlsson and Sköld (2007)/Emerald Publishing Limited.

25.2.1 Analyze Sourcing

Sourcing is a crucial and strategic consideration in vertical integration into the business model. It is a matter of how the firm sources and how its customers source, and which options for service provision this opens up. Customers considering sourcing and shoring question whether they should make goods and perform activities themselves or outsource, and whether they should locate operations domestically or offshore. Customers' strategies present business opportunities, and the provider may influence them. To handle more complex offerings, many firms abandon lower levels of technology and allow suppliers to contribute to whole processes in addition to systems, sub-systems, and components (Chiesa et al. 2000). Production and product development are increasingly sourced externally from fewer and larger sources that supply complete production process solutions. External sourcing is not limited to manufacturers but is also valid for service providers and technology vendors. It can take place by leveraging existing partners and with new ventures.

25.2.2 A Customer's Sourcing Is a Business Opportunity

A customer's reason for outsourcing is a business opportunity to understand. In situations where suppliers have specialized knowledge, superior performance, or development capabilities, there are business opportunities. However, beyond this supplier perspective, there can be opportunities to perform better in a broader part of a customer's business and make this part of the firm's business. This presents opportunities for extending business models and creating an entire operation in which a component and/or service is the basis, offering, and new business.

Flexibility and responsiveness to technological and market changes are reasons for an original equipment manufacturer (OEM) to consider outsourcing. Committing to investing in technologies, which subsequently become obsolete with radical innovations, has proven dangerous and therefore affects structures and drives networking. Not having the best global production system creates competitive disadvantages compared to other actors that take advantage of the best sources. Pursuing the best sources yields a high probability that many components, systems, and entire processes are preferably sourced externally. The best sourcing strategy pushes a shift in perspective from the plant to the industrial network.

25.2.3 Take a Network Perspective in Managing Your Company and Its External Relations

Product functions and service offerings are increasingly important in a more competitive environment. It is not simply the product itself that is important but also the product function and the brand. Companies move from the product level to the level of selling functions that create customer value (Kotler 1976). Consider again the initial picture (see Figure 25.1) of selling the function an aircraft engine performs rather than selling the engine itself. This sale of functions involves an entirely different business model in which the engine is not sold as an asset.

Networks may be based on ownership as well as on more complex and varied types of integration, such as outsourcing, licensing, and joint ventures of various kinds. Specialized supplier networks offering solutions can enhance an OEM's competitiveness (Karlsson and Sköld 2007). As perspectives change or are extended, there is a need for new thinking and to escape previous paradigms.

With a network perspective and analysis, it is natural and expected to perform many or most activities outside the company in focus, simply because there are firms specializing in components or services at lower levels in the product's composition (Karlsson 2003). For a company at a higher product system level, drawing on a network to be an OEM will enhance competitiveness. For an individual company, it is essential to find a role in the network where a unique function can be offered, which makes customers even more competitive. In the analysis of extended business models, a key question is how to increase customer effectiveness by executing part of the customer's operation, an operation in which the provider is an expert. The question can be brought up to the OEM, which may perform operations for the end user with a product or service offering.

25.2.4 Information and Other Technologies as Enablers of Business Model Extension

Information technologies play very important roles in enabling the development of the business model, with internal development and horizontal and vertical extensions. Very important enablers include the overlapping phenomena of digitalization, the Internet of Things, connectivity, and artificial intelligence, all of which have revolutionized extensions of business models and are forecast to continue to do so.

Beyond the meaning of transforming data to digital, digitalization has come to mean the use of digital technologies to develop a business model to provide new revenue and value-producing opportunities, that is, to move to a digital business (Tronvoll et al. 2020). This is important to extend the business model, especially for communication with the new horizontal and vertical contacts.

A key factor, and often the start of digitalization is the Internet of Things (IoT). It describes physical objects (or groups of such objects) that have sensors, processing ability, software, and other technologies that connect and exchange data with other devices and systems over the Internet or other communication networks. Thus, the Internet of Things is a key factor in networking in the production system, for example, in enabling autonomous solutions (Frandsen et al. 2022b). Information technology (IT), meaning the use of computers to create, process, store, retrieve, and exchange all kinds of electronic data and information, is just the technological base.

At a higher level of technology, and for imagining the future, connectivity should be considered. The concept refers broadly to technological, information, and social connections integrated into mediated communications systems. Such systems have the ability to accumulate social and financial capital stemming from actors' connections and activities on social and technological platforms. The platform and connectivity concepts originate from the technology field but have been widely applied in social fields such as organization and management. We can foresee the increased importance of connectivity. Potentially, network effectiveness can be continuously improved by artificial intelligence in machine learning systems. These technologies can substantially increase the sophistication and automation of prediction and intervention in complex production systems.

25.2.5 The Strategic Considerations of Positioning in the Hierarchy

A key strategic question is where a company should be and move to in the pyramid. Should the company develop its OEM final product offering into solutions, as discussed? Different positions and alternative moves are illustrated in Figure 25.3. The company can become a part of its customer's processes, be a new player as a solution provider at the next level, or both. If the company does not, when digitalization occurs, there is a danger that the equipment becomes merely a component of a more comprehensive value system in which digital solution providers develop platforms that are less dependent on the company's particular equipment.

Should the company remain an OEM company that sells products? There are challenges in moving to a higher level in the value chain when different and new resources are needed. Moving up also means competing with customers, which may expose significant parts of the business to risk.

When choosing between strategic alternatives, there are often no correct answers, and they depend on an analysis of the competitive situation.

FIGURE 25.3 WHERE TO BE AND MOVE IN THE SYSTEM OF SYSTEMS.

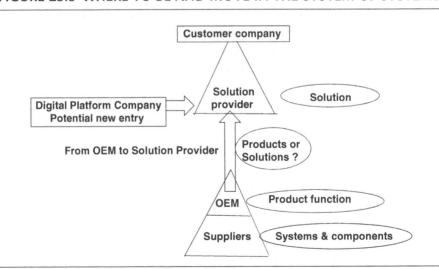

However, the extensive development of transformation and innovation platform companies (from the left in Figure 25.3) shows that pure product and component companies risk being pushed down the pyramid to lower power and profit margins.

25.2.6 The Business Model Changes Significantly When Extended

The logic of the extended business model will change entirely and be nearly the opposite of conventional business logic. Factors that previously generated income become costs, and other factors may also change. The analysis is summarized in Table 25.1.

The first and rather drastic change in the business model is that the key income factors from products become costs. The company used to profit from products, spare parts, and services. For the solution provider, these factors become cost drivers, while the functions provided to the customer generate income. The financial effects used to be immediate from income and costs, but now will be longer term, stemming from productivity and asset management. Risks can be important but were limited to the product's function. However, a solution provider is responsible for the outcome and its customer's operations. Depending on the contract, the provider may be responsible for operations

TABLE 25.1 CHANGES IN THE BUSINESS MODEL FROM MANUFACTURER TO SOLUTION PROVIDER.

Issue	Existing Business Model	Solution Provider
	Product	
Parts	Profit	Cost
Service	Profit	Cost
Equipment	Profit	Cost
	Financial Effects	
Profit and Loss Profit	Immediate effect	Deferred effect
Balance sheet	Optimize sales price and reduce production and distribution costs	Residual value, decrease cost, optimize productivity on site
	Stable	Focus on asset management
	Sensitivity	
Sales	Increase volume	Balance volume
Risks	Low	High
Potential	Low	High
		Finance solutions, adapt capacity
	Resources	
Development	Structured step by step	Iterative
IT governance	Harmonize and centralize	Agility
Human Resource	Reduce headcount	Balance headcount
Competence	Product and process technology	Services
		Subscriptions, no cure, no pay

downtime and the economic effects, although the income can be considerably higher. Stakes will be higher.

The solution provider performs part of what its customer used to perform itself. Thus, another significant request is organizational development to enable the company to perform part of a customer's operations. Providing solutions typically requires additional staff with new competencies. Such a transformation takes time. Contract negotiations are often challenging. Some OEM firms realize the advantage of outsourcing parts of their operations, while others have difficulty losing that part of their income. It is crucial to demonstrate the performance deliverables, such as the value of activities carried out for the customer. In summary, becoming a solution provider is usually a long-term project, not simply a change in written strategy. It involves multiple changes in the business model, as illustrated in Table 25.1.

25.3 From Manufacturing Equipment to Providing Their Outcomes and Effects

This section outlines how the business model is extended step by step from the manufacturer in principle and provides an example of the development of a business model into a solution provider. To illustrate the business model extension, two cases of equipment producers show the transformation of manufacturers into solution providers.

Becoming a solution provider presents an opportunity to become a stronger actor in a network. However, to be a solution provider, the supplier must develop its business model along with the manufactured product and basic services, offer the functional performance of the product, and eventually, run the process for the customer.

25.3.1 Extending the Business Model from Manufacturer to Solution Provider

OEMs may differentiate their products and develop new revenue streams, which reduces their dependence on core products. These include new services and innovative packaging solutions. Such developments involve the potential for creating stable revenue, higher margins, and closer relationships with customers, but pose risks to short-term performance.

Researchers have dealt with these issues under labels such as integrated solutions (Davies et al. 2007) and the servitization of manufacturing (Vandermerwe and Rada 1989). New service business models are partly enabled by technological developments and digitalization, where connectivity and intelligence enhance capital equipment by allowing remote diagnostics and intelligent solutions. The transformation from a manufacturer to a solution provider is complex, affecting operations and shifting the orientation from focusing on products to customers and services.

25.3.2 Why a Solution Provider

Three primary customer motives enable different future roles for an OEM. First, customers may have a clear desire to focus on their core business. Second, higher volatility in markets and technology increases the importance of adaptability. Third, customers emphasize their balance sheets and reduce the amount of capital they commit to assets.

25.3.3 What a Solution Provider Is

The business model concept entails three consecutive parts: value proposition, value creation, and value capture. In other words, the business model raises questions about value for the customer, the value created by products and services through transformation processes, and the fair share of the value offered to customers to be captured through revenue. The business model is extended by adding to the manufacturing more offerings of services and product performance and eventually, solutions for the customer. Figure 25.4 illustrates the five steps in the development process from manufacturer to solution provider. Two cases illustrating the process in practice follow. The cases are, although anonymous, real cases that the authors have been doing longitudinal collaborative research with.

CASE: Alpha Construction Equipment Company (ACEC): Acting as a process optimization consultant

Equipment sales are considered necessary among OEMs, and after-sales figures emphasize spare parts sales. The focus is on growing the volume of the spare parts business, while the risks and the potential are modest. However, OEMs are now expanding their business models into more advanced customer services. The services will challenge the reseller and the OEM with increased financial exposure, processes and systems, and skills.

FIGURE 25.4 EXTENDED VALUE PROPOSITION/BUSINESS MODEL.

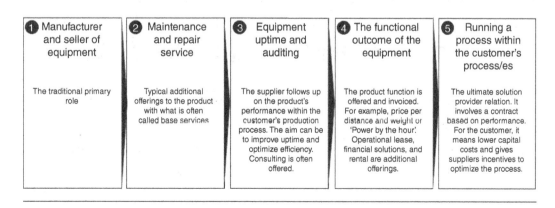

① Manufacturer and seller of equipment	② Maintenance and repair service	③ Equipment uptime and auditing	④ The functional outcome of the equipment	⑤ Running a process within the customer's process/es
The traditional primary role	Typical additional offerings to the product with what is often called base services	The supplier follows up on the product's performance within the customer's production process. The aim can be to improve uptime and optimize efficiency. Consulting is often offered.	The product function is offered and invoiced. For example, price per distance and weight or 'Power by the hour'. Operational lease, financial solutions, and rental are additional offerings.	The ultimate solution provider relation. It involves a contract based on performance. For the customer, it means lower capital costs and gives suppliers incentives to optimize the process.

ACEC is one of the biggest in the industry, a global company with a wide range of equipment, each specially designed and optimized for performing defined tasks in the value chain of road construction, landscaping, gravel production, etc. See Figure 25.5. As ACEC explores the options of expanding its business model into new types of customer services, such as production process optimization, a holistic and consultative approach has to be incorporated into the service offering. This offering requires objectivity and a focus on the total output across all equipment types, functionality, and brands. As the identified improvements could reduce or even exclude ACEC's products from the customer's production process, the potential risk of cannibalizing the OEM's products and decreasing parts and service sales must be managed.

A recent customer pilot project identified potential savings of around 4–5 % of the total production costs. Despite this solid value proposition, the project was challenged by the high complexity of identifying the financial baseline as an enabler for measuring progress and the goals achieved. Delivering this type of service requires digital platforms, connectivity, analytics, lean methodology, and industry skills that many traditional resellers struggle to perform. ACEC sees this as one of the main obstacles, which can be mitigated with extensive channel development and partnerships, and requires a radical shift in the firm's mindset, competencies, and business model.

CASE: Beta Forklifts Europe Company (BFEC): New digital platform for collecting data and securing collaboration between the stakeholders in the ecosystem.

Beta Forklifts is one of the biggest companies in that industry in the world while BFEC is a reseller that rents out the majority of its new material handling equipment. Consequently, BFEC has become more exposed to financial risks, as well as the risk of high repair costs and residual value degradation. Rental agreement profitability depends significantly on optimal cooperation of the

FIGURE 25.5 STEPS OF THE VALUE-ADDING PROCESS.

| Load & Haul | Crush & Screen | Load | Outbound Logistics | Construction Aggregate |

Construction equipment is used in various areas to produce outputs that extend the business model of the equipment manufacturer. This includes advanced production processes involving various sub-processes, external partners, and machines with great potential for alignment and improvements to avoid waste and inefficiency. The figure illustrates the value-adding process steps involved in creating the aggregate used in construction.

stakeholders in the ecosystem, from the service engineer and the operator to the warehouse manager. This presents opportunities to leverage technologies to facilitate coordination to manage risks of long-term agreements. Specifically, digitalization and connectivity enable real-time collection and sharing of information about the equipment's location, use, and condition.

In a co-creation project between BFEC and IT Provider Company (ITPC), a platform for collaboration between each of the stakeholders was developed with a focus on supporting the operator to perform efficient safety pre-checks and report damage to the machine, avoiding consequential damage. This digital collaboration tool combines the collection of machine sensor data (IoT) and data collected by each stakeholder via a mobile application. The tool presents the data in intuitive dashboards enabling BFEC (a reseller) to improve the first-time fix rate, reduce unnecessary travel, reduce uncertainty about in- and out-of-contract costs, and increase uptime and safety. The platform enables the reduction and risk control of repair costs, improving the rental agreement profitability for the reseller and increasing uptime for the customer.

25.4 Partnerships and System Providers Providing Solutions

In the previous section, manufacturers transformed from manufacturers to solution providers. In this section, an example of manufacturers transforming into system providers is provided. This section describes how an OEM and a digital platform provider with competencies in connectivity and data analysis enable the development of business models that utilize data for new customer services and improve internal efficiency. Connectivity and digitalization allow for data collection in the ecosystem around the machine, providing the basis for insights across business functions.

CASE: ITPC and OEMs: Deploying hardware and data connections to enter into partnerships

Previous telematics technology was characterized by simple data collection (e.g., location and hours), and the business model did not require harvesting any additional data. With developments in technology, many possibilities for collecting and utilizing data have emerged, enabling OEMs and customers to enhance the value of their equipment, particularly by developing partnerships with ITPCs and other service and solution providers. This first step in IT development based on digitalization is illustrated in the left half of Figure 25.6. Connectedness around the machine provides data and analysis of machine functions and performance for the gain of the operator and the direct customer.

In the next step, telematics embraces machine sensor data and data collected in the ecosystem around the machine. Data begin to flow across data silos at the OEM, reseller, customer, and operator, enabling a new wave of data-driven services and insights across business functions. This data collection expands value creation by including the broader value system around the equipment. This is illustrated to the right in Figure 25.6. A digital solution provider delivers

FIGURE 25.6 DIGITAL PARTNERSHIPS AND VALUE CREATION.

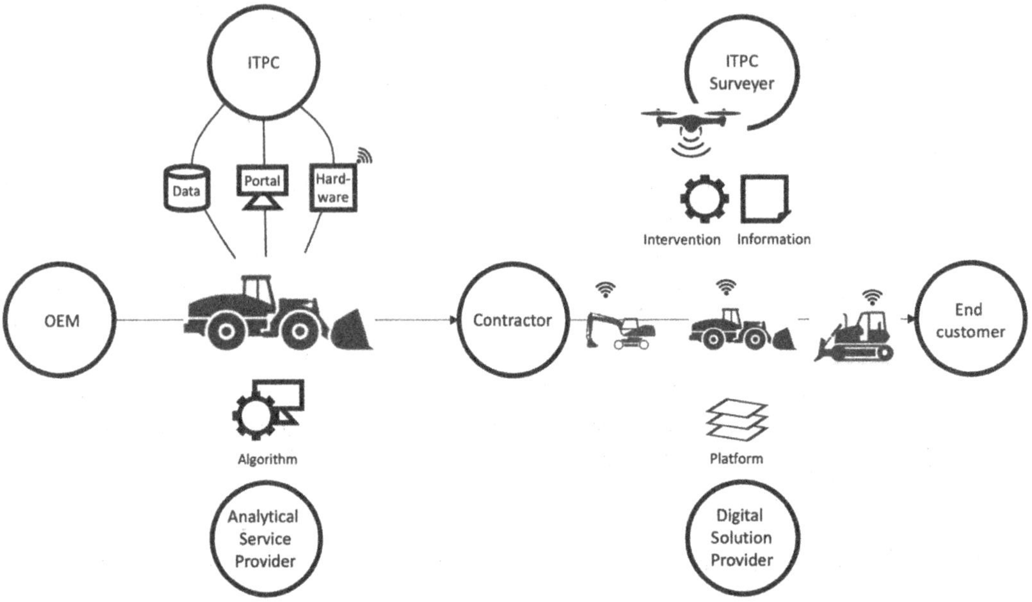

the end-to-end process outcome to the end customer with all equipment and activities surveyed and analyzed.

ITPC is a mid-size company which has specialized in developing technology to connect and monitor fleets of equipment. It has grown rapidly and has developed a platform which is now widely used within its focus industry. ITPC is at the forefront of the fast technological shift to collecting higher volume and varied data (CAN bus data) at a higher speed and lower prices, enabling low-cost and even non-powered equipment to be connected. In addition, ITPC has developed digital portals that allow fleet management and an enhanced digital interface between the operator and the equipment. Both offer possibilities for improved utilization of the equipment. Therefore, ITPC enables the OEM to create business models that utilize data for new customer services and to improve the internal efficiency and external effectiveness of the customer's processes.

Learnings from several global OEMs in the construction equipment industry demonstrate that transforming from product-centric to digital solution-centric is challenging. ITPC conceptualized the development and launch of digital services, introducing a new agile and lean start-up methodology in a modular offering including shaping, ideation, incubation, and development. Additionally, collecting data from equipment in use enables value creation for customers as part of their operating processes. For example, a collaboration by the ITPC, OEM, contractor, and surveyor employed drone technology and advanced simulation modeling to optimize the construction process to enhance the total value.

The OEM's foundation and legacy are machines and equipment. Moving from producing "iron" to digital services is extensive and requires revisiting and extending the current business model. This business transformation requires different skill sets, working across organizational structures, and in some cases, bypassing existing development and deployment processes at the OEM. The OEM's executive management acknowledges that external help is needed to start this business transformation.

ITPC has identified this gap and is now offering to bridge it with a proven model for launching digital services combined with a project and a consultative team that facilitates and supports this business transformation. A partnership model allows OEMs to think big but start small, fast, and agile. The transition is facilitated by OEMs actively seeking partnerships with digital solution providers to develop and integrate digital platforms when extending their business models. An essential part of developing digital services is constantly aligning all initiatives with the business model, ensuring that costs and revenue are synchronized. The OEM should not only consider their part of the hierarchy for equipment but also in the broader network where their equipment is part of a more comprehensive digital ecosystem.

25.5 Challenges and Other Issues in Becoming a Solution Provider

In this section, experiences from many companies that attempted to become solution providers are collected.

Section 25.3 introduced what it means and takes to extend a business model and become a solution provider. Studies of international experiences point to dilemmas resulting from changing a company's value proposition to services and solutions. Doing so requires significant changes in the culture, competence, organizational structures, financial instruments, and objectives. The following section provides insights into what managers of equipment manufacturers experienced when they transformed into solution providers.

25.5.1 Strategic Transition

The core of the business model must focus on customers' perceptions of value and the criteria for which customers are willing to pay, including opportunities for customization. Developing solution-focused offerings can challenge customers that focus on equipment to start thinking about how to build solutions around it. Providers need to offer industry insight and technology developments at a speed that their customers cannot keep up with themselves. Solutions should allow for more differentiation, flexibility, and scalability in the product offering and provide customized solutions. As solution providers, firms sell a solution that can improve the customer's business, which involves selling a return on investment. From the perspective of the reseller, service agreements can mean significant risks to warranty, which can be reduced through the design of the offerings. Collecting and analyzing data to enable proactive service interventions can avoid overload and abuse of equipment and reduce risks.

Solution provision typically enables higher profitability than product sales, thus increasing the price-to-earnings ratio. The most robust solution typically involves higher margins and is a more defensible position to differentiate, thus presenting a competitive advantage. However, investments in time, resources, and effort are required. For the reseller, benefits can be calculated by estimating residual value risk improvements and repair cost reductions. Furthermore, the total value of installing telematics solutions across all stakeholders can be calculated, including efficiency gains and risk mitigation. Pricing can be an issue, because solutions often involve indirect resources, such as overhead costs for support facilities.

25.5.2 Developing Markets and Customers

Solutions providers need to be close to their potential customers when selling solutions integrated into the customer's organization. People across regions and cultures may not understand each other sufficiently. Industries evolve at different paces. For example, the material handling industry developed service agreements, established call centers, and used telematics years before the construction equipment industry. Within an industry, OEMs can need to start the discussion at entirely different levels across segments. Adding services and digital solutions offers options for OEMs and resellers to extend the scope of their offerings, improve efficiency, and control risks. Solution providers must understand cultures and value systems because they will act within their customers' domain. Customers in different regions may have different levels of openness to digital services and solutions. Advanced customers may be open to utilizing digital services, while conservative customers and those less willing to change their business model need particular attention.

For solution providers, selling should be based on the value for the customer, with offerings tailored to customer needs. The customer should be the starting point for the solution provider to develop its offerings. Solution providers should expand from the idea of selling their products to selling a revenue stream to allow it to become a service. Selling solutions requires developing additional expertise in sales and technical and product issues.

25.5.3 Research and Development

A solution provider must first show competency and some success before seeing the benefit of growth. It is particularly useful to develop a solution that nobody knew they wanted and to have solved a problem that was not visible to the customer. Large and demanding customers can make the solution provider better as a business, as they may have a different perspective on their business and need more sophisticated solutions. Co-development enables growing the current business and leverages learnings to expand to other new customers and applications.

Creating an advanced R&D group may take time and commitment from the company but can provide critical components to the product development group, creating customized offerings. There may be different emphases on technologically advanced solutions across regions. Standardization is vital, and hardware allowing for scalability and flexibility is needed. Building prototypes and testing are essential functions. Software is required to provide quick solutions and instant value for the customer. The software can be changed, and the appearance of what the end customer sees with a differentiated solution. Digital services may become very equipment specific.

It is important to remember operators that have a substantial personal interest in the data, which can be leveraged to incentivize them to become better operators and improve their employment options. Digitalization can be a way to add value to customers and differentiate products. Digital services take many shapes, from elementary services to very advanced services that require organizational change. Digital services require controlling data to limit vulnerabilities. Partnerships with customers can enable access to data collection but are challenging, as many do not want to share such data.

25.5.4 Organizational Development

Changing from a hardware seller to a solution provider is about developing relationships within the customer organization, knowing who the influencers are, and having relationships with them. Showing what is happening in customer industries, what has happened previously, and storytelling are effective managerial measures. To sell solutions, the company will likely need to create a separate business unit as well as a complete engineering section that follows the processes and how they are managed and continuously developed.

Becoming a solution provider requires different skill sets and new capabilities and recruitment of employees, such as data analysts and program managers. The need for an advanced R&D group was discussed earlier. While production may be outsourced, control of hardware development throughout the whole chain must be maintained. In this way, customized solutions that are unique to the customer can be provided.

25.5.5 Obstacles to Change

How to organize the company to provide solutions is,and has been shown to be the most significant challenge. Obstacles usually exist because of past precedent and communication patterns, and a lack of understanding. Obstacles can mainly be overcome through examples, logic, and business cases. Remember that the cultural challenges OEMs have, their resellers mostly have as well.

25.6 Summary: How to Analyze, Develop, and Implement

The extension of the business model from selling equipment to providing solutions with vertical and horizontal expansion and integration was analyzed, and several cases were then discussed. The challenges of becoming a solution provider were summarized, and recommendations for how to overcome them and manage the transformation process are outlined (and shown in Figure 25.7).

Four main challenges facing OEMs in becoming a solution provider were identified, each of which includes several vital aspects. First, it is essential to recognize that a *strategic transition* is involved: The business model is changed by the development of solution offerings, which can have profound financial implications. Second, the transition involves understanding and *developing the market and customers* and recognizing differences in cultures. This consists of selling value, which places different requirements on the sales processes. Third, firms' experiences show the importance of *research and development* in finding solutions to unresolved customer problems. Technology plays an essential role in digitalization and the IoT, reshaping value creation and requiring new ways of thinking for OEMs. Fourth, the transition requires *organizational development* to define and implement the vision for solutions as well as recognize that new skillsets and capabilities are required for implementation.

Therefore, moving from manufacturing equipment to providing solutions is not trivial. There are many issues to consider, and planning and managing is complex and challenging. After analyses are conducted, a comprehensive checklist of things to consider and plan for, can be set up. When a company moves from being a manufacturer to becoming a solution provider, it must consider

FIGURE 25.7 CONSIDERATIONS WHEN BECOMING A SOLUTION PROVIDER.

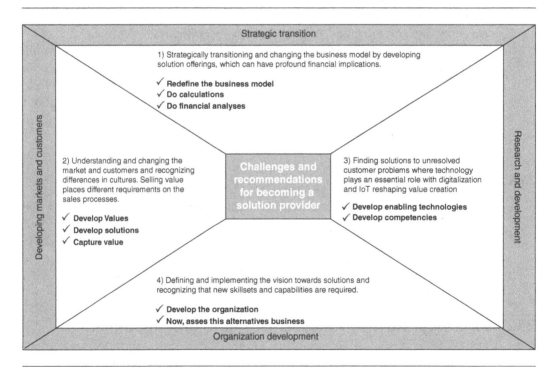

adding service offerings, performance outcomes, and eventually, solution provisions. The process does not apply to all companies, but it can serve as an inspiration to plan and initiate considerations within a business context.

- **Redefine the business model:** Carefully develop and analyze the intended business model and compare it with the current model. This involves developing new objectives and key performance indicators (KPIs). Analyze strengths, such as competitiveness, and weaknesses, such as ease of copying. What value can be created for the customer, and are they willing to pay for it?
- **Do calculations**: Try to estimate not only the potential revenues but also the higher risks. In the new business, sources of income, such as equipment and parts, become costs. Taking responsibility for the customer's process means taking high risks.
- **Conduct financial analyses:** Products that used to be sold are now owned, and the equipment manufacturer must make soft investments. Because of the timing, there are many development costs before the income materializes.
- **Develop enabling technologies:** An extension of the business model, internally but even more externally in horizontal and vertical relationships, is largely enabled by digitalization together with applying platform thinking in developing products and processes. The information system itself may have continuous improvements through machine learning.
- **Develop competencies:** If risks are accepted and efforts are made, start investing in, and developing service competencies. Two main kinds of new competencies are needed. One is service-focused individuals to develop and deliver services. The other is information system developers for digital platforms, connectivity, and support.
- **Develop the organization:** While new staff is acquired, develop the organization and its culture. Restructure to focus on operations that work closely and integrate with customers.
- **Develop values:** More complex than finding a new organizational structure is developing values in the organization and its staff members. Changing the thinking from physical products to a customer focus happens only during a long process. Values and perspectives are more persistent than objects.
- **Develop solutions:** Product development must change from a planned activity based on technological development and forecasts to an adaptive process of customer problem-solving. Having "application engineers" is one possible practice.
- **Capture value:** Going out there to meet potential customers implies a new customer interface with new participants. It is about making bigger deals at a higher organizational level. In addition to value proposition and value creation, ensure value is captured.
- **Assess this alternative business**: The fundamentals of the business change from production and distribution costs to paid-for effects in the customer's production process, meaning no pay when there are no effects.

References and Suggested Further Reading

Avlonitis, V., Frandsen, T., Hsuan, J., and Karlsson, C. (2014). *Driving Competitiveness through Servitization: A Guide for Practitioners.* Copenhagen, Denmark: Copenhagen Business School.

Baines, T.S., Lightfoot, H.W., Benedettini, O., and Kay, J.M. (2009). The servitization of manufacturing: a review of literature and reflection on future challenges. *Journal of Manufacturing Technology Management* 20 (5): 547–567.

Chiesa, V., Manzini, R., and Tecilla, F. (2000). Selecting sourcing strategies for technological innovation: an empirical case study. *International Journal of Operations & Production Management* 20 (9): 1017–1037.

Davies, A., Brady, T., and Hobday, M. (2007). Organizing for solutions: systems seller vs. systems integrator. *Industrial Marketing Management* 36 (2): 183–193.

Frandsen, T., Hsuan, J., Raja, J., and Ritter, T. (2022a). Sustainability as a service offering: Denmark as a leader in "sustainability solutions" business models. Copenhagen Business School, CBS. P+S+D Integration Booklet Series.

Frandsen, T., Raja, J.Z., and Neufang, I.F. (2022b). Moving toward autonomous solutions: exploring the spatial and temporal dimensions of business ecosystems. *Industrial Marketing Management* 103: 13–29.

Karlsson, C. (2003). The development of industrial networks: challenges to operations management in an extraprise. *International Journal of Operations & Production Management* 23 (1): 44–61.

Karlsson, C. and Sköld, M. (2007). The manufacturing extraprise: an emerging production network paradigm. *Journal of Manufacturing Technology Management* 18 (8): 912–932.

Karlsson, C., Stjernquist, P., and Frandsen, T. (2018). *Becoming a Solution Provider: The Case of Equipment Producers and Trackunit as Enabler.* Copenhagen, Denmark: Copenhagen Business School.

Kotler, P. (1976). *Marketing Management.* London, UK: Prentice-Hall International.

Oliva, R. and Kallenberg, R. (2003). Managing the transition from products to services. *International Journal of Service Industry Management* 14 (2): 160–172.

Osterwalder, A. and Pigneur, Y. (2010). *Business Model Generation.* Hoboken, New Jersey: John Wiley & Sons.

Storbacka, K., Windahl, C., Nenonen, S. and Kowalkowski, C. (2013). Solution business models: transformation along four continua. *Industrial Marketing Management* 42: 705–716.

Tronvoll, B., Sklyar, A., Sörhammar, D., and Salonen, A. (2020). Transformational shifts through digital servitization. *Industrial Marketing Management* 89: 293–305.

Vandermerwe, S. and Rada, J. (1989). Servitization of business: adding value by adding services. *European Management Journal* 6 (4): 314–324.

Visnjic, I., Wiengarten, F., and Neely, A. (2016). Only the brave: product innovation, service business model innovation, and their impact on performance. *Journal of Product Innovation Management* 33 (1): 36–52.

Christer Karlsson is Professor Emeritus of Innovation and Operations Management at Copenhagen Business School and Stockholm School of Economics, and Professor at the European Institute for Advanced Studies in Management (EIASM). He received his PhD degree from Chalmers University of Technology, Gothenburg. He is Founder, Fellow, and lifetime board member of the European Operations Management Association (EurOMA), Honorary Fellow of EIASM, and Fellow of the Product Development Management

Association (PDMA), He is a member of several editorial boards of professional journals, including *International Journal of Operations and Production Management, Journal of Product Innovation Management,* and *International Journal of Innovation Management.*

Thomas Frandsen is Associate Professor at Copenhagen Business School in the areas of Operations and Innovation Management. He received his PhD degree from Copenhagen Business School. He has researched business model extension focusing on topics such as servitization, digital innovation and pricing of industrial services. He has had the role of co-principal investigator on multiple applied and externally funded research projects. He has published his research in journals such as *International Journal of Operations and Production Management, Technological Forecasting and Social Change, Journal of Business Research, International Journal of Production Research, International Journal of Production Economics* and *Industrial Marketing Management.*

CHAPTER TWENTY-SIX

HOW TO BUILD SUBSCRIPTION BUSINESS MODELS

Charley Qianlei Chen, William C. Zhou, and Sunny Li Sun

26.1 The Rise of Subscription Business Models

The rise of subscription business models is significantly changing business practices related to product design, marketing, and consumer behavior (Aral and Dhillon 2021; Tzuo 2017). According to the Subscription Economy Index (SEI) published by Zuora, the SEI level has grown 4.6 times faster during 2021 than the S&P 500 index – a market index that represents traditional, product transaction-based businesses. More specifically, the 10-year compound annual growth rate (CAGR) of SEI is 17.5%; significantly higher than the 3.8% CAGR of the S&P 500 and the 4.1% CAGR of the retail sales sector.[1] In particular, subscription business models are gaining popularity in the software-as-a-service (SaaS) industry, which is the fastest developing sector of all industries in the SEI.[2]

Adobe is generally recognized as the first company in the software industry to make the complete transition from a product transaction-based to a subscription-based model (Tzuo 2017) with its *Adobe Creative Cloud Membership Plans*. In 2018, Gartner estimated that all startup software companies and 80% of incumbent software providers would shift to subscription-based models before 2020 (Tzuo 2017). Currently, many prominent software providers including Microsoft, HubSpot, and Salesforce are serving their customers utilizing subscription-based business models.

From 2017 to 2021, Netflix, HBO, Spotify, and other video on demand (SVOD) streaming service firms kept growing their subscription businesses with an average annual growth rate of 11.7% during the pandemic; much higher

The PDMA Handbook of Innovation and New Product Development, Fourth Edition. Edited by Ludwig Bstieler and Charles H. Noble.

than the benchmark index – Invesco Dynamic Media Index Sales which is less than 2%.[3] According to a recent survey by McKinsey & Company, 46% of US customers subscribe to at least one online streaming service, and 15% subscribe to an e-commerce service. Although the medium number of subscription services for an active subscriber is two, nearly 35% have three or more subscriptions (Chen et al. 2018).

The adoption and expansion of subscription business models in the manufacturing sector is somewhat more intriguing and counterintuitive. SEI's manufacturing sector recorded 6.7% annual growth from 2018 to 2021, but the S&P 500 Industrial Sector Sales only reported approximately 1% annual growth.[4] For example, Komatsu launched a new subscription service called *Smart Construction* that supports their on-site bulldozers by employing radar technology to map out 3-D construction sites helping customers improve work efficiency. Similarly, Caterpillar provides subscription-based plans to assist clients with dirt-removal solutions as a complementary, bundled service when selling or renting Caterpillar excavators. Since the pandemic broke out, Peloton has transitioned to the subscription business model leveraging the trend of at-home fitness to scale up in a fast manner, and its connected fitness subscribers quadrupled to 2.77 million.[5] As Peloton considers itself a service provider (vs. a hardware manufacturer), it carved out a niche from competitors by providing online training sessions and a dynamic interactive community for subscribers.

26.2 Value Proposition of Subscription Business Models

While we are impressed by the rise of subscription business models across multiple industries, we cannot help but ask what differentiates subscription business models from non-subscription business models, as these two models of business being in the same industry basically do not have differences in terms of human capital, technologies, products, corporate governance, and management. However, we believe the value delivered to customers by subscription business firms does make a difference – the value as sensed and appreciated by customers upon receiving goods both timely and conveniently, the value beyond the goods and services portfolio, and the value as embodied in continuous personal care and customization. Thus, the value propositions of all firms must answer one fundamental question: Why do customers buy from your firm over the many alternatives? The value propositions we observe in subscription businesses in the marketplace do have a key component in common – customer-centric value construction in the business model.

New product development (NPD) provides a sequence of activities to identify, create, build, and maintain customer-centric value in a business model which conceptualizes and delivers a product or service. Applying NPD in conceptualizing and delivering a new product or service, we offer four methods to help managers make the transition to subscriptions. At the same time, a business model is a "formal conceptual representation of how a business functions"

(Massa et al. 2017, p. 73). A business model might simplify the characteristics of a highly complex environment, but provided there is enough representation of reality, it can guide NPD managers to focus on key elements and components. We provide four tools: subscription business model canvas, customer journey analysis, marketing strategy, and performance metrics to assist in designing a new subsection product or service. An overall framework is presented in Figure 26.1. In this figure, while the NPD perspective provides four design methods to enable new subscription service and product, the business model perspective contributes four tools to guide the managers in developing a new service or product, or in making a transition to a subscription business model. The NPD methods facilitate the development of subscription business models. Finally, we discuss the impediments and challenges in the transition to a subscription business model. We suggest that the transition to subscription business models is critical for firms to enable them to capture and retain long-term customers. Managers should adopt a new mindset to develop new products and services based on subscription business models.

26.3 How to Design Subscription Business Models Based on NPD Methods

While NPD provides many methods to design and promote a customer-centric value proposition, we suggest four methods to explore how to properly design a subscription business model for startups or legacy firms in their transitions.

FIGURE 26.1 THE ROLES OF NPD AND BUSINESS MODELS IN ENABLING SUBSCRIPTION PRODUCTS/SERVICES.

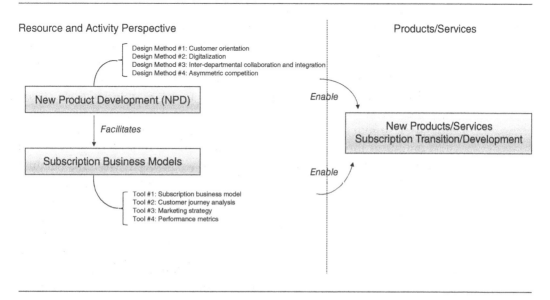

26.3.1 Design Method #1: Customer Orientation

- Subscription business models highlight customer orientation more than non-subscription business models.

Subscription business models shift the business logic from a product-centric mindset to a customer-centric one, treating "everything as a service" (Tzuo 2017). In a subscription business model, the product design and development, product quality and quantity, marketing, sales promotion, and channel management are not just measured by revenues or profits. Instead, the customers themselves, their identities, profiles, personas, individualized demands and sustainable consumptions, level of satisfaction and feedback, implicit or explicitly expressed expectations of products and services all become focal objectives and success metrics of the business. It is a shift from product to people, and from constantly manufacturing and selling goods to continuously servicing and satisfying customers.

Every consumer brand must take on the overwhelming challenges to know better and do more for all targeted customers. Why those phenomenally successful subscription box companies are doing well? They transform their customers' regular or periodic demand for personal goods into recurring offerings to serve and satisfy their customers' joys in daily life. Dollar Shave Club (razors and personal grooming products), Rent the Runway (women's fashion), Birchbox (cosmetics), Stitch Fix (clothing), Freshly (meals), Graze (snacks), Fabletics (active wear), and Stance (socks), and many others never stop studying customers' buying behaviors and feedback, and they are continuously providing customers with prompt and increasingly personalized products and services.

- Subscription business models commit to deliver continuous and prestigious service to customers.
  - In a subscription business model, no customer is treated as a generic buyer as subscription-based firms commit to continuously updating their knowledge about each customer and to supplying better goods and services to satisfy each set of personalized demands. SaaS-based software subscriptions have become the standard practice in the marketplace because subscription accounts in a forever-online mode enable software developers to quickly identify and continuously serve their clients with ongoing upgrades, urgent fixes, one-on-one troubleshooting, etc. Thanks to 4G/5G telecommunications and smart-home appliances, remote diagnostic and healthcare services as well as mental health advisory treatments become suitable for a subscription-based model where businesses can tailor services to customers who want health care in the privacy of their homes.
  - Subscribers can save money and mitigate risks by avoiding a large, lump sum payment instead spreading the cost over multiple installments especially where they prefer to convert the ownership of purchased good to the right to use the leased good for a shorter period. For example, Rent the Runway (RTR) is a firm that rents designer fashions and accessories to

customers who like increased selection and greater flexibility to wear eye-catching apparel for days, rather than buy and own clothing destined to eventually be forgotten somewhere in their closet. For those buyers who want to purchase large-value assets, the subscription-based model or similarly based rent-to-own solutions may significantly alleviate liquidity pressures that often prevent outright purchase of a product.

– Subscribers can save time and enjoy more services. For those frequently purchased and light-weight goods like men's razors, snacks, cosmetic, and health-care products, etc., more and more consumers are choosing to have the products regularly and timely delivered to their doorsteps rather than spending tedious amounts of time shopping in malls, supermarkets, and online. Coupled with the delivery service, subscribers may anticipate surprise bonus products and coupons, free market-testing merchandise, exposing broader range of items of personal interest. Moreover, one of the most attractive features of the subscription-based model, in software and streaming services for example, is the constant upgrades and addition of new functions and content that subscribers can access without delay or difficulties.

– Subscribers experience the benefits of engaging with like-minded people that actively participate in a growing community. Most subscribers are avid consumers of goods and services, and they often share opinions and suggestions that contribute to product development and service upgrades. These customers have a strong sense of community and a desire to discuss with like-minded peers, adding to and benefiting from the organic growth of subscription-based businesses.

– Subscribers may feel status, recognition, and prestige as diamond-class members: for most subscription-based businesses, the subscribers can feel recognition as high-profile customers under the personal care of services tailored to satisfy individual needs and contingent requests.

– Enable customers' participations into the value co-creation process. In subscription business model, the information exchange between customers and firms on daily usage of goods, and weekly and monthly renewal of services, shall take place more frequently than non-subscription business model. Therefore, the subscribers would feel less hassle and more engaged to supply their opinions to the re-design and upgrading of products and services. When the firms build up tight customer relationships and a close-knit community, they facilitate subscribers to participate into the co-creation with community members by sharing user experiences and feedbacks on product improvement and market development.

26.3.2 Design Method #2: Digitalization

In order to transition into a subscription business model which requires the service totally centered on customers and integrate all functional departments to optimize service provisions, the firms must make digital transformation across all departments, in and out, based on SaaS and Cloud Technology, through webs, apps, and all touchpoints over marketing and sales channels, by using emails, call center, messaging, linkages, to connect and serve all stakeholders, partners, and customers. Digitalization not only ensures the firms to be efficient, agile, and accurate on managing client orders and developing subscription-based business, but also contributes directly to social media marketing on exploring new customers through word-of-mouth. With digitalization in place, the firms may effectively transition themselves from depending on linear and one-way transactional channels to leveraging a dynamic relationship with customers and subscribers, because they have necessary powers and resources to carry on data collection and analysis on customer journey, improve service level on promptness, accuracy, and personal care, as well as offer individualized solutions after searching out customer's preferences, urgent needs, or unnoticed consumption patterns.

Since Adobe's textbook-case transition started in 2012, the subscription business model has been growing continuously across the industries which are immersed in digital supply chains. These industries mainly create, produce, and distribute the products of software, gaming, music, movies, animation, online education, news, reports, and media. They incubated the most influential subscription-based pioneering startups such as Netflix, Spotify, Amazon Prime, Linkedin Learning, and Zoom; they also witnessed more and more legacy corporations such as HBO, Microsoft 365, Apple Plus, and Disney Plus to encompass subscription business model to expand and compete.

For the subscribed non-digital goods like perishable but frequently purchased staple food or groceries (Freshly), fast-moving consumer goods (Dollar Shave Club), or even fashion and apparel (RTR, Stitch Fix), total digitalization plays the fundamental role for these firms to stay competitive and keep supplying data-driven services which differentiate and value more than traditional merchandise suppliers.

RTR, an entrepreneurial firm supplying subscription-based rental of designer apparel and accessories since 2009, reported in its 2021 prospectus that it has captured more than 6,200 unique data points per subscriber per year and up to 27 unique data points per item each time. These data were garnered from customer online activities (hearts, dislikes, shortlist, clicks, and browse lines, etc. on websites), post-wear feedbacks about the happiness and occasions of wearing experiences, and the customer data like photo reviews, professions, preferences, body sizes, life stages, and events. All data from customers are analyzed with algorithm to generate personalized recommendations which in return assist customers to make satisfactory choices over next round of leasing or even purchasing dress and accessories. Such personalized recommendations powered by total digitalization of business contributed to the 30% share in total rented items and help extend the tenure of those subscribers 2.7 times longer than the subscribers who do not use personalized recommendations.[6]

More and more manufacturing corporations rolled out subscription-based packages to grow the current customer base and deliver more valuable solutions. They sell hardware in bundles with service packages to help subscribers better use the hardware, and optimize the resource management of energy, consuming materials, and environment effects related to the hardware operation. For example, Peloton sells Bike and Tread in addition to the subscribed monthly packages which supply workout training services with live or on-demand streaming video classes.[7] For the past years, Apple has increasingly mobilized its resources to develop subscription-based businesses such as Apple Music, Apple TV+, Apple News+, Apple Fitness+, and Apple Arcade to generate new revenue streams and encompass all customers into Apple One scheme, hoping to create more value of a brand to secure stronger loyalty and greater lifetime value from subscribers.[8]

26.3.3 Design Method #3: Inter-departmental Collaboration and Integration

Subscription business model requires entrepreneurial transformation by coordinating, integrating, or restructuring most functional business units within a firm, and the partners beyond the firm. The management of customer relationships in subscription business model shall be positioned to achieve the paramount goals and metrics of business, therefore, these relationships are no longer just managed as supportive functions in a firm but become the baton in whole business re-setup to guide inter-departmental collaboration and integration as well as drive the generation of revenues and profits.

Therefore, the collaborations across departments in the subscription business model should be configured and fine-tuned with proper infrastructure, communicative channels, feedback loops, and powerful tools to drive information flow and decision process in a friction-free and fast-responsive way before the solid foundation of customer-centric relationship management is constructed. More importantly, the *soft* parts across a firm, such as open, trustful, and collaborative culture, departmental or individual key performance indicators (KPIs), and the corresponding managerial approaches, shall be aligned with the strategy on how to optimize customer value delivery.

26.3.4 Design Method #4: Asymmetric Competition

Amazon allows more than 200 million Prime-subscribers to enjoy cross-over benefits, including free shipping within two days, access to exclusive deals, free book reading, and streaming of movie, gaming, music. Though most e-tailers cannot compete against Amazon in the scale and variety of their offered goods and services, some subscription-based retailers are increasingly popular. For consumers, the subscription of goods in a box delivered home offers a convenient, personalized, and often lower-cost way to buy what they need and want. For subscription-based businesses, they can elaboratively design, improve, and then take advantage of the subscription model to win over an asymmetric competition in market. For example, because of recurring payments of monthly or yearly subscription fees from thousands or millions of customers, these businesses have more predictive and stabilized cash flows, better inventory control than those non-subscription-based business rivals. Additionally, the subscription model helps the firms establish closer customer relationships and keep up customer loyalty. The subscribers may never get lost on their journeys while traveling through the forever services, anytime and anywhere, connected and interacted, on the webs, Apps, social media channels of brands and firms, and touchscreens embedded in subscribed products.

The managers shall design the subscription business model to maximize the advantages of asymmetric competitive position, lock the customers with high satisfaction, and develop a recurring supply of products and services. The firms can also take preemptive adjustments to keep one step ahead of competitors on upgrading goods and services as well as entrenching their market shares if they build the continute advantages of asymmetric competition.

26.4 How to Deliver Subscription Business Models?

After discussing four methods to design a subscription business model, we elaborate on managerial tools about how to develop and deliver a subscription business model in this section.

The business model canvas (BMC) is a strategic tool to visually develop and display a business model to present the insights about what customers to serve, what value propositions to offer through what channels, and how a firm can make money, etc. Initially proposed in 2005 by Alexander Osterwalder, BMC can be used as a template to either develop a new business model or document the existing business model (Osterwalder and Pigneur 2010). The subscription business model canvas in Figure 26.2 documents the nine building blocks of a business model. Value proposition, as the most critical component of the model, is channeled through marketing strategy and subscription plan customization.

FIGURE 26.2 SUBSCRIPTION BUSINESS MODEL CANVAS.

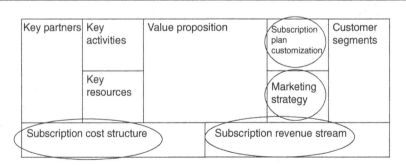

There are design and marketing tactics to encourage, attract the incumbent customers to subscribe services and implement the customized subscription plans. For example, the freemium strategy (free + premium = freemium), which prompts customer to choose either a free but very basic option or another paid but enhanced subscription plan, is a plug-and-play tactic for a firm to launch subscription business easily and quickly. However, whether the monthly recuring revenues can grow sustainably and steadily while keeping low the churn (cancellation of subscription within a period of week, month, or year) and high the customer lifetime value, is ultimately centered on true value proposition of a business model – why shall customers buy from you, not from others?

It is noteworthy the subscription-based businesses shall leverage social media and subscribers' community as best as they can to develop organic marketing (unpaid digital marketing) and referral marketing (word-of-mouth). Both subscribers and membership customers shall have customer-ID-based access seamlessly connecting to social media on every touchpoint where they communicate, interact, and make deals with peers and brands.

Since its business operation in 2009, RTR has spent less than 10% of its total revenues on marketing but organically acquired 88% of its total customers. Most of its customers were naturally generated through word-of-mouth channels like the conversations online and offline, rather than the paid ads on digital and non-digital media. As RTR community grows continuously, it benefits more and more from the powerful virality of word-of-mouth organic marketing. As stated in its IPO prospectus in 2021, 81% of RTR subscribers have shared rental experiences with at least five persons, and 32% of subscribers shared with more than 20 persons. Renting designer fashions from RTR became an inherently social behavior – 86% of subscribers rent apparel and accessories together with their friends or colleagues.[9]

26.4.1 Tool #2: Customer Journey Analysis

To identify the value proposition of subscription business model, we further apply the customer journey analytic tool which observes and analyzes the complete experiences a customer goes through and interacts with subscription business provider. Customer journey analytics can capture, study, predict, and enhance the customer experiences across every virtual (digital) and physical touchpoint. It is one of the most effective ways to increase customer lifetime value, improve customer loyalty, and drive revenue growth.

The value proposition of subscription-based businesses is fundamentally about why customers shall buy goods or services from firms, or what extra value, tangible or not, can be delivered to customers through the monthly or yearly subscriptions in comparison to non-subscription business competitors. A subscription-based business can transform an original linear and one-way transaction-focused model to a circular, two-way customer-relationship-centered model. With the customer journey analytics which monitor the customer experiencing process through all physical and digital touchpoints, the subscription business model is able to match all interactions between subscribers and firms with customer relation management system, from when customers are aware of a brand and buy the goods to their lifetime purchasing behaviors and activities in subscribers' community.

As the customer journey analytics demonstrated, the firm with subscription business model may no longer have to rely exclusively on customer feedbacks which rarely tell the whole story or true story through consumption process. Without customer journey analytics, the firms barely know about why customers go away without saying goodbye by just doing customer satisfaction surveys.

26.4.2 Tool #3: Marketing Strategy

In this section, we also discuss how the entrepreneurs use different marketing strategies to help promote their subscription business models. The roles of 4P marketing mix (product, price, place, and promotion) in a subscription business model may have structural differences with non-subscription business model in terms of the impact on business outcomes.

For example, for some subscription-based businesses who break up an expensive price total into affordable monthly plans, the proper marketing message shall be considered to clear up the confusion of customers who have believed the lofty brand position of products or services in accord with original pricing level. This type of marketing nuance in subscription-based business was confirmed in an interview in March 2022 when we had conversations with two executives from a high-tech startup in northeastern American which supplies the market-leading fitness and health-monitoring wearables for athletes and sports activists.

As to the product marketing, there has a little more nuance on marketing subscription boxes which contain the goods in different but appealing assortments from month to month. The marketers shall do great job to fundamentally understand their customers and clearly answer the questions – What are customers' regular shopping behaviors and pattern? What shall be put in the subscription boxes to transcend the total value for customers if they go shopping and pick goods by themselves?

The promotion part of subscription business marketing shall play on different battlefields, focusing on customer acquisition, engagement, retention, and co-creation which can help develop the community for consumers. For example, the marketers shall keep eyes on the recency, frequency, depth, and breadth of customer engagements to comprehend and visualize how customers deal with their firms and expect from the subscription.

26.4.3 Tool #4: Performance Metrics

As emphasized in the subscription model canvas, both revenue stream and cost structure merit the strategic attentions from business leaders. The metrics of a subscription-based business described in below paragraphs need to stand out as the primary instrument for decision-makers to observe revenues and profits, in addition to profit and loss financial analytics, to gauge and sharpen their earning abilities. These indicators are so important that they directly expose cash earning levels and growth trends in a dashboard, to pinpoint what shall be done to keep the subscription-based business in a promising track. These metrics are not only the key measurements on costs and revenues, but also the fulfilment indicators on customer relations to be observed by all departments.

- Customer acquisition cost (CAC) is the amount of money spent to acquire a new customer, including marketing costs plus any direct trial expenses to acquire a potential customer to be a subscriber of the firm.
- Customer lifetime value (CLV) is the total revenue generated by a customer over all time when he or she purchases goods and services from firm as a subscriber or non-subscriber. Since not all customers are equally valuable, this metric is important to identify, serve, retain, and grow the core customer segment, as well as being explored to expand the customer base.
- Monthly recurring revenue (MRR) and annual recurring revenue (ARR), simply speaking, are total billings from existing subscribers in a given month and a given year. They are predictable total revenue, fundamentally different with traditional business, to act as the most important figures of subscription-based business as noticed by the board and analyzed by investors.
- Average revenue per user (ARPU) or average revenue per account (ARPA) is the revenue dividing total monthly recurring revenues by total users or accounts of the subscription service. They assess the expected value to the monthly recurring revenue (MRR) of each additional subscriber and can be a tool to estimate and determine any changes of customer spending behaviors.
- Churn rate, also known as attrition rate or customer churn, is the proportion of subscribers who discontinue their subscriptions out of the total existing subscribers in a given time period. It equals to the number of subscription-canceled customers during a month divided by the total subscription customers at the month beginning. Factors of customer churn can be related to either passive antecedents like credit card expiration or active ones when the customer voluntarily calls to cancel subscription.

26.5 Impediments and Challenges of the Transition to Subscription Business Model

There are two main impediments to business model design and reconfiguration: structural and cognitive impediments. First, managers make sense of their (and others') ways of doing business by creating cognitive representations. Business models are a set of heuristics that "organize managerial understanding of the design of firms' value-creating activity systems" (Massa and Hacklin 2020, p. 21). The activities organized by business models may relate to assumptions about organizational identity, macroeconomics and target market, customer identification and engagement, value propositions, and use of technology. Those heuristics are greatly helpful for bounded-rational individuals to make quick and efficient decision. However, building the ideal heuristics in scalability is challenged to the entrepreneurs. For example, it took three years for John Foley, the founder of Peloton, to write up the playbook of business model until venture capital investors recognized that Peloton is not selling bikes, but "a subscription digital content business model" (Sherman 2020, p. 9) which is highly scalable.

However, as environment changes, heuristics may become a source of inertia. Heuristics are self-enforcing and the rigid ones become "dominant logic." Old heuristics may prevent the organization from identifying and exploiting emerging opportunities. Therefore, organizations must overcome cognitive impediments in order to redesign and reconfigure their business models.

Second, under a dynamic environment, exploiting new opportunities may require the focal organization to assemble resources of limited access, and conduct activities that are incompatible with the current configuration within the existing business model, thus structural impediment arises (Chesbrough 2010). Because business model is a system of organized and inter-dependent activities, changing activities may generate conflicts and chaos within the organization. Structural impediment threatens the complementarity among business model activities and the strategic fit between the business model and its environment. When the incompatibility between resources and activities occurs, the firms must overcome structural impediments to continue to reconfigure their business models.

In addition to the business model design impediments theoretically elaborated earlier, the subscription business planners and practitioners will confront more challenges and hardships in the execution process. Below are two exemplar questions to answer before transitioning to the subscription business model.

26.5.1 Inherently Loving Subscription?

According to the recent survey of McKinsey (Chen et al. 2018), consumers do not necessarily demonstrate an inherent love of subscriptions, but they like an end-to-end experience to be directly served through the customer journey and have strong willingness to subscribe to the packages only when the automated purchasing can offer them tangible benefits of either lower costs or increased personalization. As long as consumers like the subscribed goods and services, they will be used to keeping long-term and regular purchasing as sticky and stable customers. For example, the firms like Amazon Prime, Dollar Shave Club, and Loot Crate have higher rates of long-term subscription relative to many other emerging subscription e-commerce players. On the other hand, consumers are very prompt to cancel subscription services which cannot deliver superior consumer experiences because of poor product quality, their dissatisfaction with the assortment, or just lower-than-expectation perceived value (Chen et al. 2018).

26.5.2 Category Fit for Subscription

Researching, testing, and discovering the proper subscription category fit shall be the primary action to assess whether or not a business has the potential to take on a successful subscription model. For sure, the goods and services that can be digitalized have the potential to embark on subscription business development. The lightweight, quickly packaged, and frequently purchased goods with regular replenishment on cycle may also neatly fit the category. However, fashion and apparel, most high-value goods with hard-to-standardized services are difficult to apply subscription business model despite several promising cases like RTR, because customer-service infrastructure, personalized merchandising, data science, and cost management require too much investment and human capital to catch up the fast development of consumer needs.

26.6 Conclusion

The rapid expansion of subscription business models is changing the business landscape and competitive environment. In this chapter, we contribute to using a NPD perspective and methods on business model design practices, by discussing four methods to design subscription business model: customer orientation, digitalization, inter-departmental collaboration and integration, and asymmetric competition. We also provide a toolbox that is helpful for entrepreneurs to develop their subscription business models via subscription business model canvas, customer journey analysis, marketing strategy, and performance metrics. Finally, the impediments and challenges on the transition to subscription business model are generally discussed from theoretical and practical perspectives. Faced with those challenges, managers should take into consideration whether their business fits a subscription business model and make cautious decisions about the transition. The transition to subscription business models is critical in a service economy. Managers should adopt this transition in product/service design, marketing, and competition based on NPD and subscription business models.

Notes

1 https://www.amic.media/media/files/file_352_2844.pdf (The Subscription Economy Index, Subscribed Institute, Zuora, Feb. 2022).
2 https://www.amic.media/media/files/file_352_2844.pdf (The Subscription Economy Index, Subscribed Institute, Zuora, Feb. 2022).
3 https://www.amic.media/media/files/file_352_2844.pdf (The Subscription Economy Index, Subscribed Institute, Zuora, Feb. 2022).
4 https://www.amic.media/media/files/file_352_2844.pdf (The Subscription Economy Index, Subscribed Institute, Zuora, Feb. 2022).
5 https://www.citybiz.co/article/240255/why-pelotons-new-subscription-plan-is-a-big-red-flag.
6 http://www.sec.gov/Archives/edgar/data/0001468327/000119312521291103/d194411ds1.htm.

7 https://www.citybiz.co/article/240255/why-pelotons-new-subscription-plan-is-a-big-red-flag.

8 https://www.subscriptioninsider.com/type-of-subscription-business/magazines/apple-is-developing-apple-one-subscription-bundles-to-grow-recurring-revenue.

9 http://www.sec.gov/Archives/edgar/data/0001468327/000119312521291103/d194411ds1.htm.

References

Aral, S. and Dhillon, P.S. (2021). Digital paywall design: implications for content demand and subscriptions. *Management Science* 67 (4): 2381–2402.

Chen, T., Fenyo, K., Yang, S., Zhang, J., and McKinsey & Company. (2018). Thinking inside the subscription box: new research on e-commerce consumers. February https://www.mckinsey.com/industries/technology-media-and-telecommunications/our-insights/thinking-inside-the-subscription-box-new-research-on-ecommerce-consumers.

Chesbrough, H. (2010). Business model innovation: opportunities and barriers. *Long Range Planning* 43 (2–3): 354–363.

Massa, L. and Hacklin, F. (2020). Business model innovation in incumbent firms: cognition and visual representation. In: *Business Models and Cognition*, 203–232. Emerald Publishing Limited.

Massa, L., Tucci, C.L., and Afuah, A. (2017). A critical assessment of business model research. *Academy of Management Annals* 11 (1): 73–104.

Osterwalder, A. and Pigneur, Y. (2010). *Business Model Generation: A Handbook for Visionaries, Game Changers, and Challengers*. Netherlands: Wiley.

Sherman, L. (2020). How Peloton built the foundation for enduring success (A).

Tzuo, T. (2017). *Subscribed: Why the Subscription Model Will Be Your Company's Future and What to Do about It*. New York: Penguin Random House.

Charley Qianlei Chen is a PhD student at the Manning School of Business, University of Massachusetts Lowell. His research interests include marketing strategy, interorganizational relationships, international business, and climate-change-related CSR. He worked for over 15 years in business consulting, business development, and at the Chinese Mission to the European Union (Brussels) with research on EU-China business relationships.

William C. Zhou is a PhD candidate of entrepreneurship at the Manning School of Business, University of Massachusetts, Lowell. He will join the Sacred Heart University as an assistant professor in 2023. His research interests include international entrepreneurship, technology entrepreneurship, business models, and innovation. His research has appeared in leading academic journals like Research Policy.

Dr. Sunny Li Sun is an associate professor of entrepreneurship and innovation at the University of Massachusetts Lowell. His research interests cover entrepreneurship, corporate governance, venture capital, network, and institution. He has published over 60 papers in academic journals and over 150 practice articles, including in Harvard Business Review. Before joining the academia, Dr. Sun had 11 years' industrial experience in new venture creating, financing, and consulting.

SECTION SIX

APPLICATIONS IN NEW PRODUCT DEVELOPMENT

CHAPTER TWENTY-SEVEN

OBTAINING CUSTOMER NEEDS FOR PRODUCT DEVELOPMENT

Abbie Griffin

27.1 Why Your Development Team Must Understand Customer Needs in Depth

Products and services that don't solve people's problems, or don't solve them at a competitive cost, fail. Motorola discovered this with Iridium. Iridium's main function was to enable wireless communication worldwide. However, in developing a solution to this problem, potential users were not asked about the details of or the specifics for what that meant. Thus, the technology solution chosen – satellite delivery to a bulky phone requiring a large antenna that could only be used outdoors and was very expensive to buy and make phone calls – did not achieve the physical functionality customers wanted simultaneously with the communications functionality. The vast majority of customers were willing to give up coverage in some very remote places for smaller, lightweight phones. The result was predictable: the demise of Iridium, with an $8 billion technology development write-off.

> *The most successful new products match a set of fully understood consumer problems with a cost-competitive solution to those problems.*

The most successful product development efforts match a set of fully understood customer[1] problems with a cost-competitive solution to those problems.

The PDMA Handbook of Innovation and New Product Development, Fourth Edition. Edited by Ludwig Bstieler and Charles H. Noble.

"The devil is in the details," as they say. Palm's first Palm Pilot was so successful because the development team interacted extensively with potential users to understand the details of both form and function. Functionally, they identified that managers were using a combination of computer- and paper-based organizing and memory tools, with the mix changing depending on whether they were at or near their computer or away from it. The solution to these problems was to create an organizing system with both computer-based and remote capabilities rather than just a standalone product that did not connect to the computer. The development team determined the frequency of use of each of the organizing capabilities (address book, memos, calendar, etc.) and designed the product so that those most frequently used were the easiest to access. Finally, they thought about how people might want to carry the remote device around. As men frequently carry date books and other notes in pockets, especially shirt pockets, they designed the Palm to fit in the smallest shirt pocket made – which happens to be the pocket on Brooks Brothers shirts. The result of their development efforts was a wildly successful product that changed the way people organized their lives.

New product success can be obtained by two paths. On one path, the firm first captures a complete understanding of the complex set of needs surrounding a problem for which a set of customers would like a better solution. They then develop a product or service that solves the set of problem. This is the path the Palm team took. Alternatively, firms can develop products that do new things, are based on new technology, or have new features and then see if they solve enough problems for people to buy at the price the firm can afford to charge. Motorola developed Iridium using the second path. Although teams can be successful this way, it is a riskier path to success.

On the one hand, firms that ignore customers, or only talk to them in general terms, risk wasting money developing solutions to problems that do not exist or for which potential customers already have an adequate solution. On the other hand, interacting with and talking to customers can be misleading if firms ask them for information that they inherently are not able to provide. The key is to observe customers in context and talk to them using methods and asking questions in a form that customers can answer based on facts, and that can provide information useful for developing new products. This chapter presents information and qualitative market research methods to help product development teams understand customer needs.

27.1.1 Information Customers Cannot Provide

Although customers can easily give firms direction about how to evolve their products incrementally, they generally cannot tell firms exactly what products to develop, especially breakthrough products or radical innovations. They cannot provide the details of exactly what the future blockbuster product for your firm should look like, the features it should have, or the technologies it should contain. If you find someone who can, hire that person! He or she is doing the job your development team should be doing. That is, this person understands the customers' problems fully and has the technical capability to translate their needs into yet-to-be-developed technologies and forecast the features that will meet those needs effectively in the future.

Customers also cannot provide reliable information about anything they have not experienced or with which they are not personally familiar. By definition, therefore, customers are not familiar with technologies that have not been commercialized. They cannot be familiar with a new product the firm may be thinking of developing and thus generally cannot provide reliable information when asked to react to a concept or prototype. This is especially true for radical or new-to-the-world products. They will, of course, provide answers to questions (most people want very much to be helpful). In reacting to product concepts without experiencing them, some customers may try to imagine how they think they will feel. Others will just tell us what they think we want to hear. Information derived from unknowledgeable customers is at best inaccurate and at worst is an irrelevant fantasy. To act upon it is extremely risky.

27.1.2 Information Customers Can Provide

Customers can provide reliable information about the things with which they are familiar or which they directly have experienced. They can provide the subset of needs information that is relevant to them in an overall area of customer problems. They can articulate their own problems and needs. They can discuss the products and features they currently use to meet their needs, indicating where these products fall short of solving their problems and where they excel. The only way that a full set of customer needs for a product area can be obtained is by coming to understand in detail the needs of a number of customers, each of whom contributes a piece of the needs information.

27.2 Basic Principles for Obtaining Customer Needs

The objective of this chapter is to define and present techniques for obtaining the qualitative customer needs necessary to start product development. These needs can be used for quantitative market research later in the project (see Chapter 29). More importantly, this information provides the detailed understanding of the functional nuances to the development team that will dictate the engineering trade-offs they make during product development. The techniques presented focus on producing rich, detailed, context-specific information and ensuring that this information is transferred completely to those who need it: the development team.

> *Current customers and potential users can provide reliable information about the problems and needs they experience, those that are relevant to them. For each person, this is a subset of the full set of information needed for effective product development.*

27.2.1 Defining Customer Needs

Customer needs are the problems that a product or service solves and the functions it performs. They describe what products let you do, not how they let you do it. General needs and problems are fairly stable; they change only slowly, if at all, over time. For example, many people want to "read for enjoyment and pleasure wherever I am."

Features deliver the solutions to people's problems. Features are the ways in which products function. Nearly 1,000 years ago, the hand-copied Anglo-Saxon Chronicles and illuminated manuscripts provided the ability to read for pleasure. Although these were the solution in the distant past, they are not a particularly feasible solution today. Today, a printed book or a magazine may deliver a partial solution to the problem of being able to "read for enjoyment wherever I want." However, more complete solutions currently may be provided by e-readers or the iPad. As this example demonstrates, solutions and features change more rapidly than general needs.

Customers have general problems that need solutions and that relate to the overall product function. For example, an e-reader must "let me see what I am reading." Rather than having physical pages containing the printed content, as books have, e-readers have screens that allow me to read, changing the print that I see as I move through the book.

Customers also have very specific needs or details of the overall function that a successful product also must solve. Most detailed needs are specific to the particular contexts in which the product is used. E-readers are used in many different venues. Some of the detailed needs include "let me read ... "on my couch in the morning," "...in my bed," "...on an airplane at night," "...in a hotel room," and my personal favorite, "...while sitting next to the ocean." There is great controversy over which type of screen, the Kindle e-ink gray-tone screen or the iPad LED-backlit screen, is the better solution to the problem overall. The Kindle solution works very well in bright light by the pool or ocean. However, because the Kindle is not easily readable in the dark, its owners can be seen reading it with a portable book light illuminating their screen on airplanes at night. iPad screens, on the other hand, work very well in low light but tend to wash out in strong light – like next to a pool or the ocean.

Customer problems generally are very complex, and frequently different needs conflict. At the same time that I want to be able to read in all those different venues, I also want to be able to keep reading for a long time without recharging – such as on an overseas flight. The Kindle's e-ink solution sips battery power, giving days of power, compared to the iPad's LED-backlit screen, which may give 8 to 10 hours of use between charges. The development team thus needs to have a good understanding of the relative importance of all the contexts and ways in which their products will be used, misused, and abused to select the most appropriate feature sets for their product. It is first uncovering and understanding these detailed needs, and then providing a solution to them, that differentiates between product successes and failures.

As the Kindle and iPad examples clearly show, no product is perfect. Each product is a compromise, only partially meeting the complex set of customer needs for any function. Products consist of sets of features that deliver extremely well against some needs, adequately against others, and not at all against still others. I was an early adopter and happy Kindle user. However, as different firms develop new technologies and features, product compromises shift across the set of customer needs. While the Kindle was great for reading books, the graphic support for reading magazines like *Business Week* and *Fortune* was awful. Basically, there were no graphics. Thus, I was still carrying my hard-copy versions of these magazines with me when I traveled. In 2011, however, I shifted to an iPad for my reading, giving up reading easily in bright light and extended battery time for the ability to read my business magazines in full graphic glory electronically (and with many added audio bonuses).

Because of both technology and competitor evolution, customer needs tend to be far more stable than specific features offered in products. I still want to be able to read for pleasure and enjoyment wherever I am. Providing product development teams with a rich understanding of the complex and detailed customer needs and problems prepares them to select the best technology and feature set compromises in the future to continue delivering successful products for the firm.

So, how should the project team define what constitutes useful statements of customer needs? There are four C's of good statements of customer needs and problems:

- *Customer Words*: They are not the voice of the team and do not contain company-specific or technical jargon.
- *Clear*: They are easily understandable by all over time. Some teams even create dictionaries with detailed definitions of specific terms and phrases.

> *Understanding features leads to today's dominant products; understanding needs leads to tomorrow's dominant products.*

- *Concise*: They are not wordy. They contain only the words necessary to describe the need.
- *Contextually Specific*: They include all contextual references and provide situational details. For example, "let's me read when I'm by the pool."

27.2.2 From Whom to Obtain Needs

Many teams embarking on a voice of the customer (VoC) project call their salespeople and ask them for customer contacts. When this is how customers are identified to interact with, the team generally is put into contact with the "usual suspects" – the customers who have solid relationships with the salesperson.

However, if you only watch or talk to your firm's own customers, you are only interacting with people who already like the products you are marketing (at least somewhat). The real benefit in doing a VoC project comes from obtaining information from individuals who are not your customers but who still have or experience the general category of problems you are trying to understand. Some of these consumers will be customers of your competitors. Others may have jury-rigged their own solution. Still others may just be suffering along, with no solution at all. These are the people who can provide you with information you do not currently have.

Experts also may not be the best individuals to talk to or watch. Experts will use and interact with products in a much different manner than either the average user or (especially) the novice. One medical device company goes out of their way to avoid understanding the needs of experts. Their position is that if you make a device that only the experts can use, then the target market is much more limited than if you make one for the clumsy doctor or technician. Thus, in deciding who to talk to and watch, the development team should focus on understanding the needs of a heterogeneous set of people within the overall target market.

Only the people involved with the details of how a problem affects the day-to-day way they perform their job or live their life can provide you with their needs. And only the people who interact with, use, or are affected by the operation of a particular product can provide you with the details of how that product excels at and fails to solve their problems. A purchasing agent cannot identify the logistical and physical problems that a grocery clerk has in operating a point-of-sales scanner system. Nor can he or she help you understand the difficulty of the procedure the general manager of the grocery store must go through to produce a daily income statement or rectify the store's inventory position at the end of the month with the software associated with the scanner system. Similarly, a mother cannot provide adequate information about the athletic protector her son needs for playing baseball or concrete details about the feminine hygiene needs of her newly adolescent daughter. The details of customer needs and problems must be gathered directly from the people who have them.

27.2.3 Special Considerations in Business-to-Business Markets

Gathering detailed information generally is more difficult in business-to-business markets because most products affect multiple groups of people. Because people cannot provide accurate information about something they don't actually experience, several different groups must be investigated to obtain complete information about the detailed issues surrounding a function (McQuarrie and Coulter 1998). Grocery store general managers only partially understand the customer needs of their clerks. They have general information, but not the details that will help firms differentiate between acceptable and superior products. The need to investigate multiple groups' needs increases the cost and effort associated with obtaining good, complete product development market research for business-to-business markets. However, spending this money is worthwhile, as it will increase a new product's probability of success.

27.3 Techniques for Deeply Understanding Customer Needs

Firms can obtain a detailed understanding of customer needs through at least three market research techniques:

- Be an involved customer with those needs and problems.
- Critically observe and live with customers who have those needs.
- Talk to customers with needs.

> *Different techniques for understanding needs produce different kinds of information. No one technique is sufficient to produce a full understanding of customer and potential user needs.*

Table 27.1 summarizes the main aspects of these techniques.

27.3.1 Be a User: Be an Involved Customer of Your Own and Competitors' Goods and Services

27.3.1.1 What to Do and Keys to Success. An enormous amount of customer needs knowledge and understanding can be gained by putting all development team members in situations where they are customers actively involved with the problems your firm is trying to solve. Also, when your firm already has a product commercialized in a particular functional area, encourage team members to use your products and all competitive products routinely in everyday as well as extraordinary situations.

TABLE 27.1 SUMMARY OF NEEDS-OBTAINING TECHNIQUES.

Needs-Uncovering Techniques	Information Obtained	Major Benefits	Major Drawbacks
Be a user	Tacit knowledge	Obtain knowledge in depth	Hard to transfer knowledge to others
	Feature trade-off impacts on product function	Generate irrefutable belief in identified needs	Time and expense
Watch Users Critically	Process knowledge	Learn customer language	Time and expense
	Tacit knowledge	Find unarticulated needs	Must translate observations into words
Interview users for needs	Large number of details	Speedy information collection	Cannot elicit reliable tacit and process needs
	Context-specific needs	Information breadth	"Marketing's job"

At Procter & Gamble (P&G), both men and women work on the product development team for feminine hygiene pads. Teams at P&G are known for the lengths to which they go to try to fully understand and identify with customer problems. The entire team personally tests their own and competitors' current and new products. Male and female team members have worn pads underneath armpits and in shoes to test chafing and smell-elimination characteristics. They also have worn these pads in the anatomically appropriate area, with and without having doused the pads with liquid to simulate various normal-use conditions.

Team members at another firm in the point-of-sales scanner system market work full shifts as checkout clerks several days a year in different kinds of local stores. Store managers readily agree to cooperate because they do not have to compensate them and because they hope to get improved products. By working full shifts, development personnel learn about shift startup and close-out, as well as the effects of different payment modes, breakage, and fatigue, and they are exposed to a random day's worth of the strange things that can happen in a checkout line that can affect the operator and the system. Operating a system in a laboratory setting just does not provide the same breadth of interaction experience.

While routine continual personal gathering of customer information is not feasible for all product areas, with a little imagination it is possible to do far more than many firms encourage development teams to do.

27.3.1.2 What Kind of Information Is Obtained. Having employees become actively involved customers is the best way, sometimes the only way, to transfer "tacit" information to the product development team. Tacit information is knowledge someone has but cannot articulate, or cannot articulate easily. It is the intuitive aspect of the knowledge a person has about his or her needs. Becoming a routine customer for all of the various products in the category also may be the most efficient way to drive home to development teams the trade-offs firms have made in their products and the effects these trade-off decisions have had on product function.

General Motors (GM) misses out on an inexpensive way of imbuing their employees with a great deal of competitive and daily ownership information by some of their policies. GM requires any employee traveling on business for the firm and renting a car to rent a GM car. Development team members miss out on great opportunities to learn inexpensively how other firms' design differences affect performance. In addition, GM provides managers with new cars and then assumes responsibility for maintaining those cars. Because of this policy, senior people at GM lose an appreciation for how recalls and the need to maintain a car over time cause problems for customers – especially those who are very short on time.

27.3.1.3 Codicils. Although being a customer is a good technique to bring rich data to the product development team, it is only one of several techniques that should be used because of several inherent problems:

- The firm must learn how to transfer one person's experience and knowledge to others. A means of codifying experiences must he found.
- If experiences are not well documented, retaining personal knowledge becomes a critical problem if team members frequently shift product areas or end markets or leave the firm.
- Project management must ensure that individuals do not think that their own needs are representative of the market. They will differ from the average customer in both predictable and unexpected ways.
- Encouraging team members to be customers takes time, money, and personal effort. Obtaining cooperation from team members requires management support and example.

> *By actively using products, developers are exposed to needs that are tacit and difficult to extract verbally from customers.*

27.3.2 Critically Observe and Live with Customers

27.3.2.1 What to Do. Product developers who cannot become customers may be able to live with customers, observing and questioning them as they solve a set of problems. Developers of new medical devices for doctors usually cannot act as doctors and personally test devices on patients. However, they can observe operations, even videotape them, and then debrief doctors about what happened and why they took particular actions later, with or without viewing the videotape simultaneously.

Sometimes observing customers in their natural setting leads directly to new products or features (Lilien et al. 2002). Development team members at Chrysler observed that many pickup truck owners had built holders for 32-ounce drinks into their cabs. When asked, drivers told the team that they drank "big gulps, not Perrier water in tiny bottles." So, starting in 1995, Ram truck cup holders could accommodate 32-ounce drinks. In other instances, observation only points out the problem. The team must still determine whether the problem is specific to that person or applies across the entire target market, and if so, develop an appropriate solution. Another Chrysler engineer had watched the difficulty his petite wife had wrestling children's car seats around in the family minivan. It took him several years to convince the firm that his solution – integrating children's car seats into the car's seating system – would solve a major problem for a large number of customers. It did.

27.3.2.2 Keys to Success. Critical observation, rather than just casual viewing, is the key to obtaining information by watching customers. Critical observation involves questioning why someone is performing each action rather than just accepting what he or she is doing.

The best results are achieved when team members spend significant time with enough different customers to be exposed to the full breadth of problems people encounter. They must spend enough time observing customers to uncover both normal and abnormal operating conditions. In addition, using team members from different functional areas is important because people with different types of training and expertise see and pay attention to different things.

27.3.2.3 What Kind of Information Is Obtained. Living with customers is an effective way to identify tacit information and learn the customers' language. It is also the most effective means of gathering work-flow or process-related information. These customer needs are particularly important for firms marketing products to other firms. The products and services they develop must fit into the work flows of those firms, which means that the work flows must be understood fully. For example, the Palm team did not just talk to executives; they also watched how they worked while managing information and remaining organized. Even when questioned in detail, people frequently forget steps in a process or skip over them. Although forgotten or unimportant to the customer, these steps may be crucial to product design trade-offs.

> *When new products must fit into work flows or customer processes, critically observing customers is crucial to effective development.*

27.3.2.4 Codicils. Observing and living with customers is not especially efficient. Its problems include the following:

- Gathering information broadly requires significant team time and expense. Actions unfold slowly in real time.
- Observation or even unobtrusive videotaping may change people's behavior; natural actions are not captured.
- The team again has to turn actions into words, reliably capturing customer needs.

27.3.3 Talk to Customers to Get Needs Information: Capturing the Voice of the Customer

27.3.3.1 What to Do. By talking to customers, development teams can gather their needs faster and more efficiently than by emulating or observing them. A structured, in-depth probing, one-on-one situational interview technique called voice of the customer (VoC) can uncover both general and very detailed customer needs (Griffin and Hauser 1993; Zaltman and Coulter 1996). The way questions are asked in this method differs significantly from standard focus group qualitative techniques in four ways. VoC:

- Is grounded in reality. Customers only talk about situations and experiences they actually have had. This keeps customers from fantasizing inaccurately about things they know nothing about.

- Uses indirect rather than direct questions. Thus, rather than asking customers "What do you want?" directly (as happens in focus groups), VoC indirectly discovers their wants and needs by leading customers through the methods they currently use to find and utilize products and services to fulfill particular needs.
- Asks questions from a functional orientation, not a product or feature orientation. For example, one study asked customers about the various ways they transported food they had prepared at home to another place and stored it for some period of time before later consuming it. This is the general function that picnic baskets, coolers, and ice chests fulfill. Asking about the function rather than a product yields information about many different and unexpected products that customers use to perform this function, including knapsacks and grocery store bags. Detailed probing draws out the specific functions, needs, drawbacks, and benefits of each product. Most important is delving into why various features of the products are good and bad. What problem does each of these features solve? At the same time, does a particular feature cause any other problems? Probing the reasons why uncovers the needs.
- Inquires about multiple situations or contexts in which the customer faced a particular problem, because the information desired is the breadth and depth of needs details.

In addition to these differences, research has found that one-on-one in-depth interviews are more cost effective than using focus groups.

One advantage of interviewing is that many different use situations can be investigated in a short period of time, including a range of both normal and abnormal situations. Each different use situation provides information about additional dimensions of functional performance that a customer expects. A good way to get started is to ask each customer to describe the last time he or she used a product that fulfilled the function. The food transporter study began: "Please tell me about the most recent time you prepared food in your home, to be shared by you and others, then took the food outside your home and ate it somewhere else later." By asking customers to relate what they did, why they did it, and what did and did not work well, both detailed and general customer needs are obtained indirectly.

After customers relate their most recent experience, they are asked how they fulfilled the function in a series of other potential use situations. These use situations are constructed by the team to attempt to cover all the performance dimensions within which customers will expect the product to function. For example, customers were asked about the last time they took food with them:

- On a car trip
- To a football or baseball game
- On a bike trip
- Canoeing or fishing
- To the beach
- On a romantic picnic
- Hiking or backpacking

> *Buried in consumers' stories about specific use instances are the nuggets of detailed needs that a superior product must deliver.*

Customers also were asked to relate the most disastrous and marvelous times they ever took food with them. Although no customer had experienced all situations, the food transporting and storing needs resulting from each situation were also fully uncovered by the time 20 people had been interviewed (Griffin and Hauser 1993).

27.3.3.2 Keys to Success. Although VoC is not difficult, it gathers needs differently than traditional focus group or other qualitative market research techniques. It results in a much larger list of far more detailed and context- or situation-specific needs, because the objective is to obtain a level of detail that enables teams to make engineering trade-offs during product development. There are several keys to being successful in obtaining the VoC.

First, it is critical to ask customers about functions (what they want to do), not features (how it is done), because only by understanding functional needs can teams make the appropriate trade-offs in technologies and features as they become feasible in the future. It is the continual probing about why something is wanted or works well that uncovers underlying needs. Second, the VoC should cover only reality. If someone has never been on a romantic picnic, she cannot be asked about what she would like in this situation, because she does not know. What she would relate is pure fantasy.

Finally, it is vital to ask detailed questions about specific use instances. General questions produce general needs. General needs are not as useful in designing products and making trade-offs over features as are the details of problems. Customers are capable of providing an excruciating amount of detail when they are asked to relate the story of specific situations that occurred during the last year.

27.3.3.3 What Kind of Information Is Obtained. Both the details of customer problems as well as more general functional needs are obtained with VoC. Through indirect questioning, customer needs that relate to technical design aspects can be obtained, even from nontechnical customers. For example, by relating how her car behaves in various driving situations (flooring the accelerator at a stop sign, traveling at city speeds of around 35 mph, and traveling at interstate speeds), an elderly woman can provide information that helps a car company determine the gear ratios governing the speeds at which an automatic transmission shifts gears, even though she may have no idea of how the company's transmission works.

27.3.3.4 Codicils. The development team obtains a better understanding of a full set of detailed needs if the team interviews customers personally rather than outsourcing this function to a market research group.

- Some customers are completely inarticulate. Getting them to converse is like pulling teeth. Indeed, it always seems that one of the first two customers the team interviews will be inarticulate.
- Extreme care must be taken to maintain the words of the customer and not immediately translate one problem into a solution before understanding the full set of needs.
- Tacit and process-related needs may not be complete.

27.4 Practical Aspects of Gathering Customer Needs

Regardless of which technique is used to gather customer needs, the development team will be interacting with customers, which always involves some risk. By structuring and planning the interactions carefully, firms can increase the probability that both the team and the customer will benefit. This will increase the likelihood that a particular customer will agree to work with the firm in the future.

27.4.1 How Best to Work with Customers

The most basic principle behind working with customers is that they should be involved only so that the firm can learn from them. If they are involved for any other reason, such as to provide an excuse to delay decisions about a project, the firm is probably wasting its money. If product features have already been defined, and customer needs are gathered after that has been done to "prove" that the team has specified the "right" product, the firm is also wasting money. Gathering customer needs makes sense only if the task is completed before the product is specified.

Customers will be most willing to interact with the development team if they see how they can benefit. For most household markets, this generally means that customers receive money for interviews or observation periods. Development teams investigating business-to-business markets may find that they can provide benefit to customers by helping them gain an understanding of their own end customers. Gathering customer needs proceeds more smoothly when the interaction becomes a two-way conversation rather than a grilling.

Most firms have a portfolio of products that they have already commercialized. If the product development team collects customer needs themselves rather than contracting with a market research firm to gather the data anonymously, most or all of the customers interviewed will be familiar with at least some of their current product line. Some customers, especially in business-to-business markets, may spend the first 10 to 15 minutes of an interview venting their anger and frustration at current products. The team needs to be careful not to get defensive during this tirade but to listen to what the customers say and try to find out why these items bother them. Once customers understand that the team is talking to them to try to serve them better in the future by developing better products, and once they have vented their immediate anger, they generally calm down and gladly answer questions.

27.4.2 Pitfalls to Avoid When Interacting with Customers

There are several pitfalls to avoid when gathering information from customers. The first is to avoid selling the company's products. The interview team is not there to sell, even if it includes a salesperson. They are strictly on a fact-finding mission. Selling will both use up the limited time scheduled with customers and erode their willingness to interact.

The second pitfall to avoid is not talking to enough customers to obtain a complete set of needs. Observing only one firm's business processes or talking only to your firm's people as surrogates for actual customers is almost more dangerous than not interacting with any customers. No one customer provides a full set of customer needs for any product area. Interaction with about 20 customers is required to obtain about 90 percent of customer needs (Griffin and Hauser 1993). Interviewing 30 customers produces about 97 percent of customer needs. Unfortunately, the most important needs are not always those customers cover first; they trickle out of customer mouths at the same rate as the unimportant needs.

The third pitfall is to avoid ignoring the results the team has obtained. Several steps must be taken to ensure that these results are used because information that does not affect the product development effort wastes the time and energy of the team as well as the firm's money. Results are more likely to be used when the information users were involved in the data gathering. Both technical specialists and managers find the data more believable if they have assisted in collecting the information. Data that are in a usable form also are more likely to be used in the product development effort. Data that are buried in a report are less likely to be used than those that are pasted all over the walls of the development area. Reminders of what was learned can never hurt.

27.5 Summary

No one technique provides all the customer needs that product developers seek. Tacit needs are best conveyed by being a customer. Process-related needs are best identified by critical observation of customers. In-depth interviewing is the most efficient means to obtain masses of detailed needs, but it may not provide the tacit and process-related information required. Unfortunately, few projects can afford the time and expense of fully implementing all these processes. When personnel are fairly stable, management may be able to implement an ongoing customer need-generating process that works to provide product developers continuously with customer interactions. Otherwise, it is best for development teams to use the most appropriate customer need-generating technique(s), given the informational requirements, budget, and time frame for their project.

Note

1 *Customer* in this chapter refers to current customers, competitors' customers, potential customers, and all others who have unsolved problems and unmet needs. Customers can be individuals interested in solving their own needs or people in firms trying to solve business needs. They can be seeking either product or service solutions to their problems.

References

Griffin, A. and Hauser, J.R. (1993). The voice of the customer. *Marketing Science* 12 (1): 1–27.

Lilien, G.L., Morrison, P.D., Searls, K. et al. (2002). Performance assessment of the lead user idea-generation process for new product development. *Management Science* 48 (8): 1042–1059.

McQuarrie, E.F. and Coulter, R.H. (1998). *Customer Visits: Building a Better Market Focus*, 2e. Sage Publications.

Zaltman, G. and Coulter, R.H. (1996). Seeing the voice of the customer: metaphor-based advertising research. *Journal of Advertising Research* 35 (4): 35–24.

Abbie Griffin holds the Royal L. Garff Presidential Chair in Marketing at the David Eccles School of Business and is the Associate Dean of Business Innovation in the Medical School at the University of Utah. Her research focuses on measuring and improving the process of new product development, including the marketing techniques associated with developing new products. Her latest research can be found in the book *Serial Innovators: How Individuals in Mature Firms Create Breakthrough New Products*. She is the former editor of the *Journal of Product Innovation Management*. She holds a PhD from MIT, an MBA from Harvard Business School and a B.S. in Chemical Engineering from Purdue University. She is an avid quilter, a scuba diver, and a hiker.

CHAPTER TWENTY-EIGHT

THE EVOLVING INFLUENCE OF CUSTOMER NEEDS ON PRODUCT DEVELOPMENT

Kristyn Corrigan

28.1 Why Understanding Customer Needs Remains Relevant

The most successful products and services incorporate a complete and detailed understanding of customer needs, which requires systematic Voice of the Customer (VOC). VOC is traditionally defined as a complete list of customer needs, organized and prioritized by customers. Increasingly, teams are employing VOC needs assessment throughout the product development funnel. Customer needs are fundamentally different from solutions or features. A customer need is a description, in the customer's own words, of the benefit to be fulfilled by a product or service (Griffin and Hauser 1993). This differs from the solution, which is the way in which the product will ultimately deliver those benefits. By having a foundational understanding of customer needs, companies can develop creative solutions that meet those needs.

Whether an organization seeks incremental innovation or is designing a breakthrough new product, VOC methodologies reduce the risk of new product failure, and they optimize product launch. We uncover customer needs by examining customer experiences and understanding customers to develop a keen sense of empathy. A recent survey of product development professionals done by PDMA (Product Development Management Association) measured the innovation success rates of companies across a variety of industries (Lee and Markham 2016). The objective was to determine what separates the "best" or more successful companies from the "rest" or less successful companies. VOC activities were cited as a top indicator of new product success. In other words, companies that engaged in VOC activities to enhance their understanding of customer needs were more successful at launching new products than those that did not.

Having consulted with hundreds of leading global organizations for nearly two decades, I have observed that many companies do the reverse: they create solutions, then search for a need to fulfill, rather than letting customer needs drive the process and developing solutions to meet those needs. Google Glass is one example of building a solution before identifying a critical need. Although the first version of Google Glass was a very elegant and advanced technological solution, it did not meet any important unmet needs, and it was extremely expensive. Google Glass had two basic functions: to quickly capture images and to have a real-time Internet feed a glance away. However, Google targeted the wrong customer. The technology and its benefits were well suited for a professional market. Instead, Google pursued the consumer market, first targeting "hip" early adopters and celebrities who were mostly uninterested in taking photos from their glasses or navigating the Internet with eye movements. Because Google did not have a clear understanding of market needs and dynamics, they made dangerous assumptions, and many consider Google Glass a failure.

28.1.1 Types of Customer Needs

There are two categories of needs, which are both critical to understand:

- **Stated needs,** which customers can articulate through discussion
- **Latent needs**, which are unspoken, and customers may not even realize they have.

Latent or unspoken needs are typically more difficult to uncover. Best practice is to observe customers using a product or experiencing a particular service. By using ethnographic research, researchers often observe issues or pain points that customers do not know they have. For example, someone may state that their last trip to the grocery store was fast and easy. When observing the shopping process, however, researchers may notice that shoppers often return to aisles they had already visited, typically because they forgot to grab an item from their shopping list. By observing these customers, researchers could uncover needs customers either could not recall or did not realize they had.

Both stated and latent needs can be functional, emotional, or experiential:

- **Functional needs** are the most basic needs that products or services must satisfy. When customers purchase a new car, for example, they typically want to satisfy several functional needs, including efficient gas mileage, comfortable seating, and pleasing interior features and color.
- **Emotional or psychological needs** are needs that evoke a deeper feeling beyond the surface. Customers often choose products for the emotions they induce. Luxury car brands, for example, may evoke feelings of wealth, trendiness, uniqueness, and environmental consciousness.

- **Experiential needs** are the benefits that customers seek while interacting with a service, product, or brand. For example, many customers place a great deal of weight on the quality of service and maintenance provided by the dealer after their automobile purchase. This is often a central driver of repurchase, likelihood to recommend, and brand loyalty. In response to critical customer needs, automobile companies have introduced concierge services, incorporated premium food and beverages in their waiting areas, and installed working stations with Wi-Fi for those waiting for their car to be serviced.

28.1.2 Voice of the Customer Drives Innovation at Every Level of the Business

Letting customer needs drive the new product development and innovation process offers both focus and efficiency in the product development process. It allows product managers to spend time solving the right problems. It creates a common understanding between engineering, marketing, and sales.

Voice of the Customer drives innovation in three key areas:

- **Product Engineering**
  An understanding of customer needs allows for more focused and effective product development and helps companies to avoid over engineering. Several years ago, I consulted with an organization developing a sophisticated virtual reality training system for dental students. When we began our partnership, the engineering team was focused on building capabilities to allow students to practice as many dental procedures as possible. However, VOC research revealed that it was not the *number* of procedures driving potential adoption, but realism. Institutions valued realistic haptic feedback over the quantity of procedures. More procedures could be released over time, in a modularized format. By having a clear picture of the market's prioritized needs the company was able to change course and develop a successful new product, now commonly used by many higher education institutions in the US and globally.
- **Service Design and Experience Creation**
  The typical customer completes several tasks along their purchase journey. They also access touchpoints (e.g., a brand's website, employees, social media) while completing these tasks. Underlying these tasks and touchpoints, of course, are customer needs, or the benefits customers seek during their experience. A recent study of knee-replacement patients identified that the biggest barrier patients faced during their patient journal was scheduling their surgery. Patients referenced "putting off" the procedure and uncertainty about recovery time. In this circumstance, pinpointing this pain point in the journey allows a hospital to focus its innovation efforts on the most critical area of risk.
- **Business Model Definition**
  VOC activities often provide a deeper glimpse into how an organization creates and delivers value. Through understanding customer needs and pain

points, new business models can emerge. In 2015, Starbucks launched "Mobile Order and Pay," allowing customers to order ahead and pick-up in stores. The app was developed in response to several important, unmet customer needs, including not wanting to wait in a long line, streamlined customization of beverages, and earning rewards for loyalty. Through the successful development of this app, Starbucks was able to establish a leading mobile payment app, ahead of competitors.

28.2 Voice of the Customer and Customer Needs in the Product Development Process

28.2.1 Exploratory Voice of the Customer

Traditionally, Voice of the Customer activities happen in the fuzzy front end of innovation, when an organization is exploring important, unmet needs on which to innovate. The activity is exploratory in nature because the goal is to hear all types of customer needs. Exploratory VOC is often one of the first activities in the product development funnel.

Figure 28.1 depicts five stages of product development:

FIGURE 28.1 THE FIVE STAGES OF PRODUCT DEVELOPMENT.

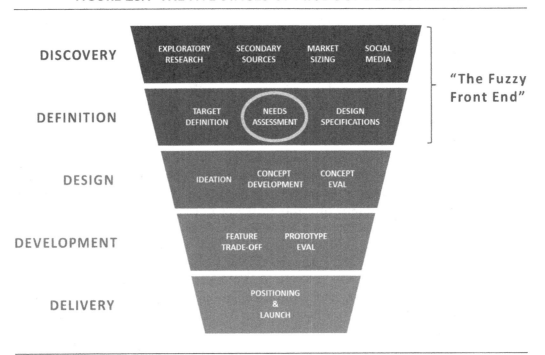

- **Discovery:** What are the potential opportunities of interest? What customer segments exist? How large are they?
- **Definition:** What detailed needs must be satisfied? How well is our company satisfying the needs relative to market alternatives?
- **Design:** How does our company come up with better features and solutions that address key, unmet needs?
- **Development:** Which features or solutions are worth developing? How much are our customers willing to pay for them?
- **Delivery:** Can our company reliably produce the product or service? Is the innovation worth the investment?

Griffin and Hauser (1993) define VOC as the process of:

- Identifying a complete set of customer needs, in the customer's language
- Organizing the needs into an actionable hierarchy or affinity diagram
- Having customers rate needs on importance and performance, to identify opportunity areas for innovation.

Under this definition, traditional VOC activities exist in the "definition" phase of the product development process. The activity is investigative, in that the goal is to unearth all the needs that exist in a given area. Years ago, my consulting firm, Applied Marketing Science, partnered with General Cinemas Corporation, a US-based movie chain that has since merged with AMC. General Cinemas sought to understand the comprehensive list of movie-goer needs and prioritize those needs for innovation. The needs they uncovered encompassed the entire experience of going to the movies including ticket purchase, seat selection, concessions and the movie watching experience. From this VOC effort, two successful innovations emerged. General Cinemas Corporation uncovered the customers' need to have more food options available, beyond the typical candy and popcorn. They were the first theater to partner with external chains like Godiva and Starbucks to bring in more varied and premium food options. Additionally, the theater chain was the first to introduce the concept of stadium seating, in reaction to customer needs related to seat comfort and wanting a view from anywhere in the theater.

28.2.2 Validation Voice of the Customer

Increasingly, organizations are incorporating Voice of the Customer activities further down in the product development funnel. These activities often happen following a traditional, exploratory VOC effort or in place of that effort.

Validation VOC efforts are synonymous with activities like concept and prototype testing. The effort is focused on determining whether a product concept or prototype at hand will satisfy key, unmet customer needs better than alternatives. During traditional qualitative concept or prototype testing, teams seek to understand overall concept appeal and the appeal of specific product features. Additionally, teams may test things like believability, clarity, and barriers to purchase. When teams incorporate VOC and needs assessment into this stage of the product development process, they seek to understand if the concept addresses key, unmet customer needs. A key element of this process is presenting customers with a minimally viable stimulus (MVS). An MVS is 'good enough' and often imperfect, which makes it easy to change. It can be a displayed in a variety of ways, spanning from a written description, a series of images, or even an early prototype. Presenting customers with an MVS, either virtually or in person, allows the product development team to remain agile and continuously make changes.

By way of example, I recently partnered with a medical technology company to design a new line of consumables for an enteral feeding pump. Both home and health-care users interacted with a new set of consumables for the pump. When gathering customer reactions to the prototype, health-care respondents shared the need for a set of consumables that instilled confidence in users. In their feedback, they expressed that the bag was noisy and increased the risk of air bubbles. This "cheapness" diminished their confidence that the product would succeed in the market. These findings changed the direction of the product concept and informed the team's path forward with product development

28.3 Effective Voice of the Customer Planning

28.3.1 Whose Needs Must Be Satisfied for a Product or Service to Be Successful?

On the surface, the question "who is the customer?" seems simple, but it is almost always more complex than teams imagine. A well thought-out and systematic sample plan is an important first step in any Voice of the Customer effort. In both consumer and B2B markets, there are purchase decision-makers, influencers, buyers, users and many in between. Layered on top of this may be segmenting variables such as geography, company size or other demographics and firmographics.

Critical elements for the team to consider:

- What are the segments of interest? Will needs differ by segment?
- Who in the value chain is it necessary to speak with to optimize or improve the product? Whose needs must be met to be successful?

Companies may be quick to call their biggest and best accounts or "The A List" to learn about their needs. While there is certainly nothing wrong with including these kinds of people in VOC research, the danger is in *only* talking with these customers. These "best" accounts or customers have a bias that may not match the opinions of the broader marketplace (McQuarrie 2008). There is much to be learned from speaking with non-customers – those customers who do not currently do business with your organization and those who used to be customers and have since migrated to a competitor. In many circumstances, understanding the perspective of non-customers is one of the most critical lenses for innovation and VOC. Understanding these customers' motivation to buy, reasons for switching, and product usage tells us a lot about what they value.

28.3.2 Considerations for Agile Teams

As organizations adopt agile frameworks, there is often a struggle to understand how systematic Voice of the Customer can fit within iterative, fast-paced processes. Traditional VOC methods can be adapted to fit with quicker timelines, increased iterations and an ever-changing backlog.

While traditional qualitative VOC efforts seek a sample size of $N = 30$ to uncover the universe of customer needs (Griffin and Hauser 1993), Agile teams can implement "interview sprints" with a smaller subset of customers, typically $N = 5–8$. As the team collects needs and reactions to product concept or MVS, they can choose to add additional customers to the sample or continuously go back to the same group of 5–8 for additional insights. Teams adjust their interview questions and product backlog as the VOC progresses. This methodology allows teams to collect VOC in a fast-paced, iterative way.

28.4 Traditional and Emerging Methods for Understanding Customer Needs

28.4.1 Traditional

28.4.1.1 One-on-One Interviews. Traditionally, exploratory and validation Voice of the Customer insights are gathered through one-on-one conversations with customers. Griffin and Hauser (1993) identify one-on-one conversations as the most effective and efficient way to unearth detailed, unique customer needs. These 45–60-minute interviews give customers the opportunity to go into detail about their unique experiences and needs. Additionally, each customer has more airtime than they would in a traditional 90-minute focus group.

These one-on-one discussions are qualitative in nature, meaning the interview is meant to investigate the *why* and *how* rather than count *how many*. Qualitative interviewing is an effective way to understand issues and experiences, and to explore nuances. These interviews are detailed conversations around a set of topics to hear customers' thoughts and experiences in their own words. The goal is to uncover needs and explore in detail why people behave the way they do.

The most important aspect of one-on-one interviewing is probing or asking "why." Researchers ask why to understand customers' underlying needs and to get customers to explain generalities. For example, if a customer claims something is "good" or "bad", researchers should not accept that answer and stop there. Instead, they strive to understand what makes a product or service particularly good or bad and why.

Storytelling is another great technique because it allows the interviewer to hear about the customer's experience firsthand, in their unique situation, and in detail. It reduces the likelihood of a customer filtering what they think the researcher wants to hear. Storytelling also helps prevent socially desirable responses. Once the customer becomes more invested in telling their story, the truth emerges more easily.

While in-depth interviews have historically been conducted in-person, either at a research facility or on-site with a customer, one-on-one interviews are increasingly conducted over the telephone or through video conferencing. Video conferencing allows the interviewer to understand another nuance of customer emotion. Customer body language and facial expressions can provide helpful context behind the needs customers express. It also unearths contradictions and areas to probe for deeper understanding.

28.4.1.2 Ethnographic Observation.

Ethnographic or observational research can augment Voice of the Customer in-depth interviews. Ethnography involves observing people as they live and work, watching what they actually do, and not just what they say they do. Sometimes what people say and what they do are different and observing customers through ethnography provides a check on what people say.

Ethnography involves the study of culture, including the beliefs, rituals, symbols, language, gender roles, social divisions, and other aspects of the population under study. The study of these cultural issues goes back to ancient times, but the modern emergence and popularity of ethnographic research is commonly associated with people like Margaret Mead. Mead studied the development of adolescent girls in native populations in Papua New Guinea in the 1920s and 30s. She became well-known in the field and continued to write and speak throughout the middle of the 20th century.

While ethnography comes from the academic field of anthropology, marketers and product developers can use the technique in a variety of disciplines. Applied to research and product development, the idea is to observe a customer rather than ask questions. By observing how customers use a product or service, latent or unspoken needs may emerge.

Coke Fridge Pack is an example of how a raw materials company – in this case Alcoa, used observational research all the way downstream with consumers to get insights to drive the market. Alcoa wanted to figure out how to drive demand for beverages sold in aluminum cans. They did ethnography with consumers including shop-alongs and home visits. They found that the riskiest link in the supply chain was between the consumers' trunk and the refrigerator. If canned beverages were placed directly in the fridge, they would be consumed very quickly. If, on the other hand, cans went in the pantry or the basement, they tended to sit there a long time. The unstated need was to make it easier for consumers to get canned beverages right into the fridge. Working with Coca-Cola and a packaging supplier, Alcoa invented the Fridge-Pack, which is the now familiar format of a slender carton that holds 12 cans in a 2×6 configuration. It fits easily on a refrigerator shelf and is designed with a built-in dispenser. It was a smashing success.

Although very popular today, ethnography is more effective in some situations than others. It works best when you have something tangible to observe, such as customers using a physical product or completing a task. It is less effective with businesses that do not have a tangible product, such as understanding decision-making criteria and processes among business managers. Furthermore, ethnography does not tell us why customers act the way they do. This requires talking to customers in an in-depth interview. As a result, product developers typically combine ethnographic observation and interviewing. For example, researchers might observe a customer doing something with a product and say: "I noticed you did this. Tell me more about why." The idea is that researchers might not have known to ask about that behavior if we had not observed customers acting a particular way, and we would not know why they acted that way if we had not asked.

28.4.2 Emerging

28.4.2.1 Machine Learning Review of User Generated Content.
User generated content (UGC) is an emerging, alternative source of Voice of the Customer insights. Machine learning (ML) and artificial intelligence (AI), when coupled with professional, human analysis, allows for the discovery of customer needs from UGC and can identify needs that are either infrequently mentioned or may be overlooked completely in more traditional qualitative research.

ML algorithms can utilize a variety of content types, so it is critical to identify the best sources of user-generated text content given your product innovation questions. In some cases, useful content exists on product review sites, social media pages, customer forums, or in blog posts. In other instances, it exists in call center data, transcripts of customer interviews from prior studies, or answers to open-ended survey questions. Regardless of the source, the goal is to find places where customers discuss brands and share their opinions and stories. These data can be a ripe source for ML analyses.

Timoshenko and Hauser (2019) developed an effective algorithm for identifying customer needs from UGC, which uses a convolutional neural network (CNN) to filter out noninformative content and cluster dense sentence embeddings to avoid sampling repetitive content. In their article, they tested their algorithm on an oral care product. They successfully uncovered nearly every unique customer need identified with traditional research. Furthermore, they identified several insights that were unlikely to appear in a random sample of UGC of equal size.

Recently, a major manufacturer of snowplows and snow spreaders was looking to unearth customer insights to spur innovation and strengthen their marketing. While they had done primary research in the past, they weren't sure they had uncovered all possible needs in the category. They partnered with my consulting firm, Applied Marketing Science, and piloted Timoshenko and Hauser's (2019) algorithm for needs identification. Figure 28.2 depicts the process of using said algorithm for needs identification in user generated content.

- The effort started by mining UGC from websites where snow equipment operators talk about the products they use and the snow removal process. On these forums, community members rely on each other to answer questions,

FIGURE 28.2 THE PROCESS OF USING TIMOSHENKO AND HAUSER'S (2019) ALGORITHM FOR NEEDS IDENTIFICATION.

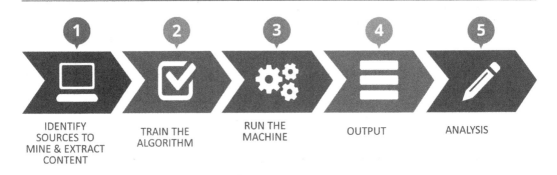

Source: Adapted from Timoshenko and Hauser's (2019).

troubleshoot issues, and provide general advice. These forums are full of rich insights, as oftentimes operators post about their experience right after key moments of truth – for example, when there is an equipment failure in the middle of a large storm.

- After compiling the dataset, the machine learning algorithm was trained by human analysts to distinguish informative content, that which provided insights into customer wants and needs, from uninformative content that did not.
- Next, researchers ran the algorithm. In a matter of minutes, the machine assessed all the records of data and uncovered 1,600 statements that contained unique customer insights related to the snow plowing category. Trained research analysts then reviewed these statements and formulated a comprehensive list of the operators' wants and needs in a matter of hours.
- In total, the process was able to identify 107 unique customer needs related to snowplows. The needs covered a variety of areas including the lights on the vehicles, the challenges of removing snow from sidewalks, de-icing, maneuverability, and much more.
- Ultimately, our research enabled the manufacturer to quickly (within 3–4 weeks) and inexpensively develop a detailed view of the entire category. The insights uncovered ultimately led to the company's release of their next generation plowing blade system. It has been one of their most successful product launches to date.

UGC continues to be an effective source for Voice of the Customer insights when sufficient text data are available. The machine learning methodology has been applied across a vast number of categories from consumer goods, industrial products, and medical devices.

28.5 The Continued Impact of Understanding Customer Needs on Product Development

28.5.1 Which Customer Needs Must Be Incorporated into New Offerings?

Following a qualitative or ML needs-gathering effort, some teams have the insights they need to move directly to engineering or re-engineering. Many teams, however, require quantitative measurement to understand the greatest opportunity areas for innovation. This interim quantitative survey research can allow teams to:

- Reach a statistically valid sample of customers and prospects, overall and by key segment or geography
- Measure the appeal, likelihood to adopt and willingness to pay for a product concept
- Measure customer importance or criticality of customer needs
- Measure how satisfied customers are with the ability of the market or a specific brand to deliver on that need

Quantitative Voice of the Customer prioritization allows teams to identify the focus areas (high importance, low performance) of the market as well as potential areas for disruption. See Figure 28.3, which plots need importance vs. need performance, with a goal of pinpointing opportunity areas for innovation.

28.5.2 Transforming Needs into Specifications and Solutions

Voice of the Customer efforts often fuel in-person or virtual team ideation sessions. In these sessions, the team brainstorm features or solutions that meet the most important, unmet needs of the market. Instead of starting from a more general state, teams can be more efficient in their efforts.

VOC can also feed QFD (Quality Function Deployment) or "The House of Quality" sessions where a team collaborates to correlate engineering capabilities and metrics to meet key, unmet customer needs (Hauser and Clausing 1988). QFD is used by teams who desire a high level of precision in selecting product attributes and engineering metrics. Figure 28.4 depicts the QFD house of quality. Customer needs reside on the y axis and related product attributes and engineering metrics on the x axis. Teams complete an in-depth correlation exercise to prioritize product attributes that meet the most critical, unmet customer needs.

FIGURE 28.3 MARKET OPPORTUNITY MAP.

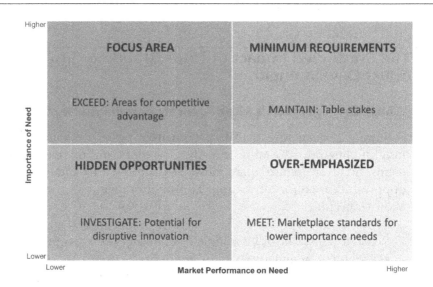

FIGURE 28.4 THE HOUSE OF QUALITY.

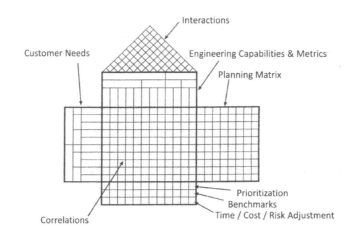

28.6 Summary

A systematic understanding of customer needs remains essential to effective product development, feature selection, and marketing/sales strategy. While companies may not have the time or resources to conduct in-depth Voice of the Customer for all product releases, customer insights can be incorporated at some level of rigor throughout the product development process using both traditional and emerging techniques.

References

Griffin, A. and Hauser, J.R. (1993). The voice of the customer. *Marketing Science* 12: 1–27.

Hauser, J.R. and Clausing, D. (1988). The house of quality. *Harvard Business Review* 66 (3): 63–73.

Lee, A. and Markham, S. (2016). PDMA Comparative Performance Assessment Study (CPAS): methods and future research directions. *Journal of Product Innovation Management* 33: 3–19.

McQuarrie, E. (2008). *Customer Visits*. New York: Routledge.

Timoshenko, A. and Hauser, J.R. (2019). Identifying customer needs from user-generated content. *Marketing Science* 38: 1–20.

Kristyn Corrigan is a principal and co-owner of Applied Marketing Science, a Boston based market research consultancy that helps companies develop better products and services through harnessing the power of customer insights.

In her nearly two decades of product development consulting experience, Kristyn has led and designed customer insights engagements for leading global B2B and consumer brands. She specializes in helping companies understand stated and latent customer and how to measure their criticality to innovation.

Kristyn's published work has been featured in Fast Company and Marketing News. She has been a longtime supporter of PDMA, holding various leadership positions, and most recently was elected to the organization's Board of Directors.

CHAPTER TWENTY-NINE

CHOICE-BASED CONJOINT ANALYSIS: REVEAL CUSTOMER PREFERENCES TO INCREASE PRODUCT-MARKET FIT

Garth V. Brown

The intellectual appeal for many professionals in New Product Development (NPD) is that the field is simultaneously art and science. Proficient Product Managers are comfortable being artists who envision innovative products and services for customers to buy and use. These professionals are also increasingly required to be comfortable with gathering and analyzing data that drives decisions. In other words, a new product manager is frequently an analytical role. In this role, awareness of the experimental and statistical tools used in measuring the utility and desirability of new products gives one an advantage in the market. For the professional already comfortable with the art of designing products and measuring sales performance, this chapter introduces you to experimental and statistical tools that provide deeper insights into the customers' wants and needs long before the design of the product is complete. Collectively, these tools are called *conjoint analysis* (CA) and, in the right hands, can provide deep insight into customers' desires for both physical products and services in almost any market.

29.1 What Is Conjoint Analysis (CA)?

Understanding conjoint analysis is helped by reviewing a commonly accepted economic principle. Consumers make purchasing decisions based on a belief in

The PDMA Handbook of Innovation and New Product Development, Fourth Edition. Edited by Ludwig Bstieler and Charles H. Noble.
© 2023 John Wiley & Sons, Inc. Published 2023 by John Wiley & Sons, Inc.

the value of various parts of a product or service. We refer to a product or service as *a product* for simplicity throughout the rest of this chapter. Given that one has money to spend, a product whose price is below the value we assign to it is an attractive purchase. For example, if the product's price exceeds our assessed value, we consider it overpriced and do not purchase it. When a product has more beneficial attributes than its alternative, we find it more valuable and are willing to pay more.

Similar products may be differentiated by having or not having specific attributes. Attributes are product characteristics that purchasers consider when comparing similar products. To better orient you to conjoint analysis, we will use the standard terminology specific to this field. Product attributes, or characteristics we wish to measure, are known in conjoint methods as *features*. Examples of features include the battery life of a laptop computer, the distance between charges of an electric vehicle, or the cabin class of an international plane ticket. Let us not forget one of the most important product features: *price*. John R. Hauser, a professor at MIT's Sloan School of Management, summarizes this by writing: "The goal of conjoint analysis is to determine how much each feature contributes to the overall preference" (2007, p. 3).

Once we recognize that consumer purchasing decisions are based on the collective value of a product's attributes, we appreciate the significance of measuring the relative importance of each feature. By presenting potential customers with differing combinations of features, then asking questions via a survey, we can deconstruct preferences into constituent parts and statistically measure the worth of each element. A group of techniques for this discovery is *conjoint analysis*.

In the following pages, we will explore concepts to help you understand the principles of a potent type of conjoint analysis (CA). Choice-based conjoint (CBC) analysis is a research method that asks respondents to choose between different product concepts. By adjusting the attributes of each concept, we can statistically determine the importance of each product characteristic and uncover the product design, price, or configuration which best fits a given market. The power of CBC relies, in part, on the substantial size of supported choices that consumers can make. The field of CBC has matured to the extent that survey techniques, mathematical formulas, and analysis models make for a vast field of research choices. For these reasons, software tools quickly become necessary to help us manage the design, data collection, and reporting of CBC surveys. This chapter does not have the space to explore every choice you can make as a market researcher. Undertaking a CBC marketing research project will soon require using one of the many suitable software tools available as online services. Although academic scholars have cited reliance on the expertise of software solutions (Wittink and Cattin 1989), you may still wish to compare the service options available. The use of software solutions should be a practical extension of your proficiency in CBC. However, its application is beyond the scope of this text.

29.1.1 Conjoint Analysis: The Fundamentals

Conjoint analysis is a statistical tool popularized by marketing professionals revealing preferences between attributes (also called features) of a given product. These features cause a prospective customer to choose one product over another but were difficult to measure before applying statistical models. Even the most basic of products is an aggregation of many attributes. Beyond the physical characteristics of a new product (e.g., color, storage capacity, weight), this technique is one of the few validated statistical tools to measure price influence on buying decisions (Kahn 2005). New product managers have many options to consider when pursuing their competition. Production materials, design, quality, availability, packaging, and price are just a few choices. When choices are made that combine undesirable product features or attributes, the product's future may be at risk. However, we can use surveys, and statistics, to understand the most desired combination of attributes customers wish to have in a single product design. In this way, we can design the specific products that customers find most attractive. Many of these experiments can be done through surveys before physical prototypes are made.

29.2 What Does CA Reveal?

Consider how the choices between products are at once complex and straightforward. When options are provided to a potential customer looking to purchase a specific product, the "best choice" is frequently, but not always, clear. Simultaneously, the variation between features makes a choice complicated. This paradox illustrates how designing new products is complicated, but a well-designed product makes choosing easier for buyers. Reducing the near-infinite combinations of features into a small set compared to each other is one of the values of conjoint analysis. Beyond lowering the possible choices of product configurations, these statistical tools help us to understand which feature – in which combination – adds value to a given product design. Dissecting these features and arrangements into the worth of the individual parts is called *partworth estimation* and is the foundation of conjoint analysis.

29.2.1 Understanding Customer Desire

What makes a product desirable is often attributed to the deep and broad experiences that come from a long career in a specific industry or market, especially in the field of new product development. Experience is invaluable when designing a new product. However, we should not dismiss that an ongoing conversation with customers yields valuable insight into how their needs and desires are constantly evolving. Even the keenest product managers must estimate how their regular customers will accept new features, delivery models, or marketing messages.

To complicate this ongoing customer conversation, customers often do not know what they desire until presented with a representation of a novel product option. We can show prototypical representations of the new products we have in mind, including models, mockups, photorealistic renderings, and often 3D-printed examples. However, soliciting feedback on one or two examples does not clarify which feature differences capture the highest desirability.

To gather more information, surveys, in the form of experiments, allow customers to share their known and unknown preferences methodically. The data from these experiments are available for analysis through sophisticated statistical techniques. Later in this chapter, we will illustrate this concept using an experiment based on product research of a consumer webcam.

There are a few concepts that are beneficial to review at this point. First, the value of a given product is the sum of the value of its features (Karniouchina 2011). Second, your company can afford to add only so much value to your product at any proposed price. Third, not all features have an equal value to every customer. Finally, to reconcile these limitations, product managers make trade-off decisions between one feature and another to achieve a product that best fits customers' desires. Conjoint analysis provides a way to dissect the value of each part, that is, *partworth* value. Knowing the customer value of each product feature allows us to make informed decisions about which features stay or go. One of the most reliable types of conjoint analysis is Choice-based Conjoint (CBC) analysis. We will focus on this technique for much of this chapter.

29.2.2 Partworth and Total Value

Choice-based Conjoint (CBC) analysis is one of many techniques used to determine consumer preference, also called *utility*. Alternate techniques attempt to measure the individual desirability of individual attributes. The sum of these separate utilities is considered the total value of the product. We expect buyers to choose the product configuration with the highest *total utility*. We call these *composable* approaches because they sum or compose the value of each measured partworth utility.

CBC is different from other forms of CA in that it asks survey respondents simply to choose between profiles. By sampling a sufficient number of people and applying a statistical model (*discussed later*), we can *decompose* – into individual parts – the total utilities of options (profiles) into feature-level utilities. As a result, we refer to all CBC approaches as *decompositional*. It is difficult or impossible to describe potential products fully. For the purposes of New Product Development, we acknowledge this theoretical limitation, yet we move forward regardless. Do not be disappointed by this news. Academic researchers have concluded that conjoint analysis (CBC specifically) are reliable predictors of real-world marketplace choices (Hauser 2007; Steiner and Meißner 2018).

To illustrate that the sum of partworth utilities equals the total utility visually, consider the value diagram in Figure 29.1: *Example Sum of Partworth Utilities.* First, we must introduce a CBC scenario and return to this example throughout

FIGURE 29.1 EXAMPLE SUM OF PARTWORTH UTILITIES.

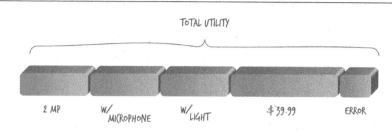

the chapter. It is a relatively simple but sufficient example to use to explain concepts. Imagine you are the product manager of a line of webcams used for videoconferencing. The design and feature details are yours to decide. You have narrowed the configuration options (features and feature levels) to *camera resolution (2 MP vs. 8.5 MP)*, with (or without) *onboard microphone*, and with (or without) *onboard lighting*, and at two *prices ($39.99 vs. $59.99)*. By presenting product configuration options to survey respondents, CBC allows us to decompose the partworth utility of each option as summarized in this illustration.

29.2.2.1 Stochastic Error. Figure 29.1: Example Sum of Partworth Utilities a theoretical illustration of the utility (preference) for a webcam with 2 MP resolution, an onboard microphone, and an onboard light, which costs $39.99. Additionally, you notice an amount of *error*. This is not an error in the sense that something is incorrect. Instead, the stochastic error represents all unmeasured nuances that the survey respondent considered when choosing this profile. This nuance (error) can include the respondent being tired, missing but relevant variables, measurement bias in the survey, and other subtleties (Louviere and Woodworth 1983). Eggers and his coauthors capture the ingredients of this type of error this way, "The stochastic error term catches all effects that are not accounted for" (Eggers et al. 2022, p. 787).

29.2.3 Predicting New Product Performance

Before we move on to the steps involved in a practical choice-based conjoint analysis exercise, you may want to predict how your new product will perform in the marketplace. Until now, we have discussed how CBC provides customer preference insights. As you begin to think about how you will conduct CBC market research, consider including profiles representing competitive products or services. Discovering the partworth values of competitive products, including their brand name and current prices, will provide you with the necessary data to create a "what if" simulator. These market simulators help product managers compare the estimated sales performance of a new product configuration with that of existing products (Green and Krieger 1988).

Suppose you know the size of an existing market in revenue or units sold. You include choice profiles in your CBC market research that describe the current products in that market. In that case, you have all the statistical ingredients needed to simulate changes in your product design. These changes have a predictive value for your product design choices. Keep in mind that the predictive reliability of any statistics simulation is only as good as the data on which it is built. Simulator tools are available as part of many online CBC tools and add-ins for typical spreadsheet applications.

29.3 Steps of a Choice-based Conjoint Analysis

After understanding choice-based conjoint analysis and what it may reveal about customers, let us place this powerful technique chronologically into a new product development process. You may already have a process framework that prescribes sequential or iterative steps (i.e., phases) for new product development. CBC does not favor one process framework over another. However, knowing when to start your analysis is helpful.

Whether your NPD organization relies on a sequential process such as a phase-and-gate framework or an iterative approach such as agile, features and feature levels can be collected early and throughout the ideation (discovery) phases/cycles. Your preferred sequential process likely has a scoping phase. This may be the right time to conduct your CBC market analysis. If your preferred sequential process suggests a "market analysis" phase, your CBC research should be conducted then. However, suppose your sequential process is unclear about when to conduct market studies. In that case, we recommend that you conduct the CBC just before risk analysis (e.g., SWOT). Using an agile process, the CBC is most helpful during roadmap planning (epic and release planning) and can be reconducted after significant milestones.

29.3.1 Choose a Calculation Tool

Sawtooth Software, Qualtrics, XLSTAT, Conjointly, and many other software vendors offer applications to help you design CBC experiments, conduct surveys, analyze results, and "what if" simulators. When you are ready to run your first CBC research experiment, you may consider one of these vendors. Still, an Internet search will reveal many additional choices with varying abilities and prices. Understanding the difference between tools is helpful before making a purchasing decision. However, you may want to wait until considering (or completing) the remaining steps explained shortly. What you learn in the following steps may provide valuable insight into choosing a CBC tool. *For notes on each of the four tools mentioned earlier, see Appendix,* Table 29.3.

29.3.2 Determine Product Features

Depending on where you are in a new product design lifecycle, you are likely to have already considered how your product will differ from existing products.

While practically any difference that distinguishes two products is an excellent candidate to be a CBC *feature*, there are some limitations.

We are often conducting CBC research virtually. Therefore, some features are differentiators while not lending themselves to online research methods. For example, olfactory cues (smells) have changed consumers' purchasing behaviors in many situations (Purdy et al. 2021). In the current state of remote research, product smell may be a poor CBC feature, even if it is essential to your product design. Taste, tactile texture, and similar attributes are also tricky. However, CBC research can be conducted in field studies in person. While the later step of gathering responses will be different, much of the rest of your research is unchanged.

29.3.3 Define Feature Levels

Once you have determined the potential features of your product, you need to select the variation of each. When choosing between smartphones, color is often a deciding factor for consumers. If product color is one of your identified *features*, the *feature levels* may be black or silver. Likewise, laptops can be differentiated by battery life, a *feature* where *feature levels* are 4.5 hours vs. 7 hours. Bicycles can be distinguished by frame material (a *feature*) where *feature levels* are carbon fiber vs. titanium. Theoretically, there is no limit to the number of features or feature levels you can imagine. However, in practice, the complexity of choices will proliferate Suppose you choose to conduct CBC market research. In that case, you are advised to keep the research design as straightforward as possible while still answering the most critical design questions. Three to eight features with two levels for each is manageable. You will learn to deal with many features and levels by becoming proficient in this research technique. However, keeping your first CBC analysis simple will reveal customer opinions and grow your knowledge of the research process.

29.3.4 Feature Matrix

A feature matrix is a table containing all the permutations of features and feature levels. A matrix with four (4) features, each having two (2) possible levels, would contain sixteen (16) distinct profiles. To measure each feature's part-worth utility (i.e., value), you need not intend to manufacture or offer every profile for your customers. However, the analysis will likely become confusing if illogical options are available. After designing potential product choices, consider if any of the choices are not physically practical and reevaluate your features and feature levels.

New Product Managers must analyze the consumer market to understand which features and feature levels are most attractive to buyers in the market, that is, a webcam used for videoconferencing. After discussions with salespeople within your company, you discover four features that customers consider before purchasing a webcam: *price ($), camera resolution*

(MP), onboard microphone (Y/N), and *onboard lighting (Y/N)*. After consulting with production engineers, you understand that several attribute levels are practically possible for your new webcam. Hence, you choose to study combinations of these four features using two feature levels each. These levels are camera resolution 2 megapixels (MP) vs. 8.5 MP, with and without an onboard microphone, with and without onboard light, and finally, retail price of $39.99 vs. $59.99. These are summarized in the matrix in Table 29.1. Four (4) features and two (2) feature levels result in 16 profiles ($2^4 = 16$).

29.3.5 Profiles

A profile is a single instance of one of the combinations of attributes created by the feature matrix. Our webcam analysis has sixteen profiles which is a relatively simple example. Because of the multiplying effect of adding features or feature levels, the number of profiles can become very large. While the process and arithmetic of CBC support large matrices with many profiles, the cognitive burden on experiment participants is quickly overwhelming. Therefore, we use the matrix as a theoretical set of choices but only offer a subset of profiles when collecting preferences. These subsets are called *design choice sets*.

TABLE 29.1 16 POSSIBLE PROFILES OF EXAMPLE WEBCAM EXPERIMENT.

Profile #	Resolution	Microphone	Light	Price
1	2 MP	Yes	Yes	39.99
2	2 MP	Yes	No	39.99
3	2 MP	No	Yes	39.99
4	2 MP	No	No	39.99
5	8.5 MP	Yes	Yes	39.99
6	8.5 MP	Yes	No	39.99
7	8.5 MP	No	Yes	39.99
8	8.5 MP	No	No	39.99
9	2 MP	Yes	Yes	59.99
10	2 MP	Yes	No	59.99
11	2 MP	No	Yes	59.99
12	2 MP	No	No	59.99
13	8.5 MP	Yes	Yes	59.99
14	8.5 MP	Yes	No	59.99
15	8.5 MP	No	Yes	59.99
16	8.5 MP	No	No	59.99

29.3.6 Design Choice Sets

Even with smaller feature matrices, as in our webcam example, 16 profiles become challenging to keep separate in the minds of survey respondents. CBC allows us to reduce this cognitive difficulty by limiting the number of profiles within *design choice sets*. For comparison, a set can contain two, three, four, or more profiles. Experts debate the recommended number of profiles within a single choice set. Leading scholar Vithala R. Rao suggests a rule-of-thumb that a person cannot efficiently process more than 20 pieces of information as part of a decision (2014). These twenty pieces could be four profiles (product variations) for a choice set, each with five features. Alternatively, two profiles, each with ten features, would also contain twenty pieces of information. Earlier scholars paid little attention to the number of feature levels; they simply recommended no more than six features as part of your CBC study (Green and Srinivasan 1990).

Figure 29.2 is an example of a design choice set presented to a survey respondent. Respondents are asked to review the project features and feature levels of three product options. Respondents are also invited to consider not purchasing any of the alternatives. Multiple screens such as this are presented, with different production options, including all possible profiles (full factorial design) or a subset of profiles (fractional factorial design).

29.3.7 Gather Responses

CBC used to be conducted with in-person interviews or recorded on paper forms. Fortunately for New Product Managers, most data gathering is automated using an online service specializing in conjoint analysis surveys. Currently,

FIGURE 29.2 ONE OF SEVERAL CHOICE SETS SHOWN TO RESPONDENTS.

Which of the following Webcams would you choose?

Camera Resolution	8.5 MP	2 MP	2 MP	✖ None of the above
Onboard Microphone	No	No	Yes	
Onboard Light	No	Yes	No	
Price	$39.99	$39.99	$59.99	

Go back

several online services allow one to design an analysis study, send survey invitations, record survey responses, and analyze the gathered data. A quick query with your favorite search engine will provide a list of tool options. If you are part of a product design group, consider democratizing the decision of which tools to use. As in CBC itself, there are trade-offs to make; not the smallest is price. You will find inexpensive options, but inexpensive is relative to your budget.

After choosing your software or online service and designing your study, it is time to gather responses. As with most research designs, you are advised to conduct a small pilot, or test, survey on the people around you. By asking friends, family, and work colleagues to take your survey, you can find and correct any issues you have overlooked. This pilot may delay your analysis, but it is usually worth it; unresolved confusion in your main study may question the results. It is better to accept critique about your service design early than to be faced with critique about the survey results later.

29.3.7.1 How Many People Should You Survey? 385 There are some magic-like numbers in statistics that are accepted as trustworthy. One of these numbers is 385. The instructions accompanying your survey tool (or online service) will likely provide advice on the number of respondents you need to be confident in your statistical analysis. As you may expect, you should survey more prospective customers to reduce error and increase confidence in your research. However, it may seem less evident that the relationship between more survey takers and statistical surety is not linear. In other words, once we have about 385 survey respondents, having more provides no additional information. The number 385 assumes that you feel comfortable with a statistical confidence level of 95% and a margin of error (confidence interval) of 5%. These statistical quality metrics are widely accepted in marketing research. Keep in mind that you do not need 385 respondents if you are willing to live with lower confidence. For example, suppose you can tolerate a confidence level of 90% and a margin of error of 10%. In that case, you can have this with only 68 completed surveys.

29.3.7.2 Finding Survey Respondents Until this point in your CBC research project, you have been able to work alone or with trusted members of your product teams. Next, you can connect with potential users or customers. Fortunately, CBC (and other survey types) are much more accessible than they once were. Because of the Internet, we can find respondents worldwide (if you wish). Or, it is possible you already have a list of research subjects willing to take your CBC survey. If you do not yet have a list of survey takers, there are a few ways to find them. If your firm has a marketing department, you may already have access to a list. Suppose you have a marketing department, but your company does not have a pool of research contacts. In that case, you may find a new ally with a marketing background with whom to work jointly on this task. Some CBC software service providers offer paid access to their own pool of respondents. Still, you can also source these using services like Prolific (www.prolific.co) or Amazon Mechanical Turk, also known as MTurk (www.mturk.com).

Since most CDC researchers use an online service to handle surveys, responses, and data collection, the rest of this chapter will assume that you are doing the same. A point of caution is in order – sending emails to conduct marketing research can be complicated. You are advised to research legal compliance regulations such as the GDPR in Europe and other jurisdictions' respective rules.

To conduct your CBC research, you will need willing research subjects. Since your survey is online, the most direct way to connect with potential survey takers is to send an email to each of them. Most online survey services can send templated emails on your behalf. Additionally, some services have a pool of respondents, although you will likely pay extra for this access. A general-purpose survey respondent pool is often sufficient to understand the preferences of a broad population. However, your new product may be sufficiently specialized and is best served by bringing your own study subjects.

If your new product fills a niche market, consider some study participant options you may already have available. These may include previous customers, newsletter subscribers, and trade group email lists. If you think you will draw on a pool of research subjects in the future, consider working with your marketing colleagues to create an ongoing *research panel*. A research panel is a group of customers or buyers who agree to provide input to your organization. They need not be *your* customers; competitors' customers often provide insight into how your product compares to their alternative options. Beyond your CBC and other conjoint analysis needs, a well-maintained research panel can help with qualitative-type research, such as through focus groups and discussion panels. Treat the members of your research panel like the VIPs they are, and they will provide you with market insights you did not know existed.

29.3.8 Response Data Analysis

Because so much of the statistical work required in CBC research is done for the researcher by the software tools, this introduction chapter need not dive deep into the mathematical models. However, being familiar with a few concepts will help you understand CBC better. While CBC has matured significantly over the previous two decades, it remains a fertile area of science, and improvements are discussed frequently (Eggers et al. 2022). The following are general models used in conjoint analysis and common in choice-based conjoint (CBC).

Before the rise of CBC as a favored preference-measuring method, conjoint analysis was usually based on surveys that asked respondents to rate (on a scale, e.g., 1–10) products or attributes. This rating system is helpful to product researchers as it is simple for participants. In addition to being practical, the statistical model used to calculate partworth preference is relatively simple. Predominantly, marketers used Ordinary Least Squares (OLS) as their statistical model. However, because multiple products or attributes could be evaluated as similarly poor or good, rating = 1 or rating = 10, this conjoint analysis does not replicate buyers' real-life decisions. Since CBC avoids choice ties by forcing one option (profile) to be selected over an alternative in a choice set, the choice ties were resolved. Still, new statistical models are required to calculate preferences when using CBC.

29.3.8.1 Ranked Preference of Profiles Statistical tools often can use the choices made by respondents to calculate a ranked order of profile preference. This is particularly helpful to product managers who offer a limited number of products but want to understand which profile (product configuration) is preferred above others. An example – still using our proposed webcam profiles – is found in Table 29.2, which includes the most and least preferred options.

TABLE 29.2 RANKED LIST OF PREFERRED PRODUCT PROFILES.

Camera Resolution	Onboard Microphone	Onboard Light	Price	Value	Rank
8.5 MP	Yes	Yes	$39.99	30.3	1
8.5 MP	Yes	No	$39.99	27.1	2
8.5 MP	Yes	Yes	$59.99	15.7	3
8.5 MP	No	Yes	$39.99	15.6	4
8.5 MP	Yes	No	$59.99	12.5	5
8.5 MP	No	No	$39.99	12.4	6
2 MP	Yes	Yes	$39.99	2.2	7
8.5 MP	No	Yes	$59.99	1.0	8
2 MP	Yes	No	$39.99	−1.0	9
8.5 MP	No	No	$59.99	−2.2	10
2 MP	Yes	Yes	$59.99	−12.4	11
2 MP	No	Yes	$39.99	−12.5	12
2 MP	Yes	No	$59.99	−15.6	13
2 MP	No	No	$39.99	−15.7	14
2 MP	No	Yes	$59.99	−27.1	15
2 MP	No	No	$59.99	−30.3	16

29.3.8.2 Relative Importance of Features. A key output of a choice-based conjoint analysis is the mathematically rigorous insight into the value of one product feature over others. As managers of product development projects, you probably understand that feature decisions significantly influence the success or failure of new products. It would be best if you also recognized that not each feature is equally important compared to the others. CBC experiments allow us to understand which features matter and which matter less. However, an analysis output is referred to as *relative importance.* Relative importance is represented as a percentage for each feature. Since each feature contributes something to the total value of a product, the percentage of each feature will sum to 100%. For easy viewing, many CBC tools present the relative importance as a chart, like the example shown in Figure 29.3. As a new product manager, spending more of your limited resources on features with higher relative importance will frequently yield the most significant improvement in customer desire.

29.3.8.3 Preference for Levels. Upon understanding which features matter more to customers than others, we must understand which feature levels are most preferred. Often this is intuitive (e.g., customers prefer higher camera resolution),

FIGURE 29.3 RELATIVE IMPORTANCE OF FEATURES IN WEBCAM EXAMPLE.

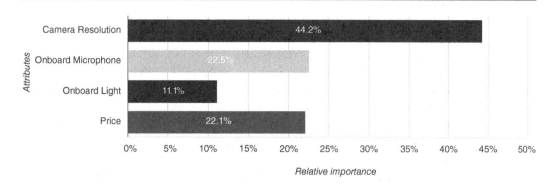

FIGURE 29.4 FEATURE LEVEL PREFERENCE.

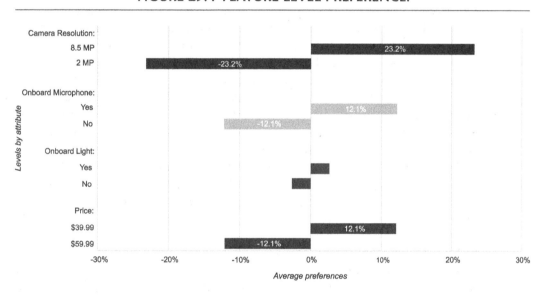

but surprising results often occur. It is also essential to understand if given multiple options (feature levels), customers greatly or just slightly prefer one over the others. Like relative importance, feature level preference is also presented as a percentage. Unlike relative importance, which must sum to 100%, level preferences include unpreferred levels (represented as a negative value) and preferred levels (represented as a positive value). Thus, each level must average to 0%, as illustrated in the image shown in Figure 29.4.

What to make of relative importance and preferred feature levels? Consider each of these analytical outputs for what they are. Both are averages across all survey participants. If your development project has resources, spend more of them on features with high relative importance. Within each feature, look for opportunities to make more significant distinctions between levels. In the example of the webcam, the feature of onboard light has the most negligible relative importance. This is reinforced by the fact that there is a modest difference between preferences of having or not having the onboard light.

29.3.8.4 Multinomial Logit (MNL) Model. The logit models are the most commonly used statistical models in CBC research (Raghavarao et al. 2010). Logit models are also referred to as logistical regression models. Unlike linear regression models, which attempt to predict outcomes based upon a continuous spectrum (e.g., price, weight), logistical regression models deal with discrete outcomes such as a specific product feature's presence (or absence). Since CBC depends on choices, logit models fit nicely into our survey-based *choosing* experiments. However, logit models are not perfect, and there is one case where you should be aware of the drawbacks.

The challenge with multinomial logit (MNL) models is that they cannot anticipate unforeseen options correctly. In statistics, this is referred to as the *Independence of Irrelevant Alternatives* (IIA) assumption. It is illustrated by the red-bus/blue-bus problem, which imagines a transportation market made up of red buses and cars. Consider a CBC survey conducted in which 50% of the consumer chose to ride the (red) bus. In comparison, the other 50% decided to commute by car. In this experiment, there are two features with two feature levels. Still, the red bus color is only measured when the vehicle is a bus. Suppose we simulated the entry of a new profile, perhaps a blue bus. In that case, an MNL model may predict that blue buses would capture 33% of the total transportation market. However, it seems illogical that introducing a blue bus would cannibalize the market share of red buses and cars equally. We should assume that today's red bus riders are more likely to choose a blue bus, versus a car, on tomorrow's commute. The IIA problem can largely be avoided in your consumer surveys through careful design of features and feature levels, as described earlier in this chapter.

29.3.8.5 Multinomial Probit (MNP) Model An alternative to multinomial logit models are *multinomial probit models* (MNP). Probit, less commonly used than logit, is not encumbered by the IIA issue; therefore, it can avoid illogical simulation conclusions such as the red-bus/blue-bus problem. While this may seem like a superior model choice, MNP is much more difficult to calculate because of the prework required to estimate the importance of features. If you assume that the sampling errors, such as subject bias, are normally distributed – rather than logistically distributed – then using a probit model may be advantageous (Eggers et al. 2022).

29.3.8.6 Advanced and Emerging Models MNL and MNP are not the only statistical models useful in discrete-choice experiments such as CBC. As your CBC skills advance, alternative statistical models to investigate may include the *polyhedral method, support vector machines* (SVMs), and *hierarchical Bayesian.* Not all software applications support these statistical models; however, as new models become popular in CBC studies, we should expect the most popular new models to be available in your application of choice.

29.3.8.7 Simulating Product Performance As demonstrated in the tables and illustrations earlier, CBC studies provide detailed insight into the preferences of potential buyers. Once these insights are quantified, these studies can also forecast new product performance, making CBC a favorite tool among new product managers and marketers (Baier and Gaul 2001; Moore et al. 1999). Let us explore two ideas that provide additional strength to CBC studies.

First, consider that your design choice sets can include products sold by your firm or competitors. Until this point, we only considered the desirability of variations of a webcam design. However, we can easily add profiles of existing products, and this technique adds additional insights. We can simulate the new product's market share by comparing existing products (with known revenue) to our novel product. Many commercial CBC software tools include such ability and usually refer to this ability as a *simulation.*

Secondly, in CBC studies that consider both profiles of competitor products and price as a feature, conjoint software applications can assist in price optimization (Marn et al. 2004). A CBC software tool that includes simulation ability allows one to run *what-if* analyses to predict which price maximizes total sales revenue, knowing that underpricing and overpricing both causes reduce revenue. While most conjoint software applications include simulation functionality, if price optimization or market share forecasting is important to you, confirm this before selecting an application.

29.4 Overlooked Attributes in Practice

We have covered several examples of product features. However, there are attributes of the purchase which can provide as much value to the buyer as some of the attributes of the product. For product managers from the sales or marketing departments, the following examples of overlooked attributes may not be surprising. New product managers who come from engineering or manufacturing may be interested in learning that there are features of the buying experience that affect consumer choices. Non-physical features often make helpful additions to our discrete-choice experiments.

29.4.1 Brand Names

One of the most insightful discrete choices consumers make, and we can measure, is that of a brand name. Brand names are like meta-features that contain positive or negative feelings based on the consumer's past experiences. For this reason, some authors refer to brand names as an *enriched feature.* Researchers have recognized the embedded package of useful information revealed by brand names for decades. The following example reflects this preference package: "Brand name provides substantial information about likely performance, expected benefits, and problems in use, as well as the relative price of the item" (Huber et al. 1992, p. 276). While testing for brand name preference may be valuable, we must recognize the challenge of brand names as preferences.

Because well-known brand names are packed with feature-like information, we must be careful when measuring them with choice-based conjoint analysis. CBC attempts to decompose the relative importance of individual features of a product. Our study steps backward, not forward, if a brand name or brand recognition recombines these disparate features together. Suppose product managing researchers do not assume any implicit meaning of a famous brand name. In this case, CBC can help measure the overall importance of a brand name in the preference of one profile over another within a choice set.

29.4.2 Seals of Approval

As brands and products have proliferated in the marketplace, many consumers rely on trusted seals of approval. Examples of seals of approval include *Good Housekeeping* and *Consumer Reports* which lend their trusted names and logos to individual products after meeting specific quality standards or tests. In the earliest days of the development of CBC, these seals of approval were of interest to researchers (Green and Wind 1975). The popularity of endorsement seals was common among household products; however, the concept of trusted seals has also had success in helping manufacturers differentiate themselves from competitors. The *International Organization for Standardization* (ISO) has developed and popularized many standards of quality proudly brandished by companies that, after auditing and confirmation, can claim one or more achievements, such as ISO 9000.

When using brand names or seals of approval as features in your product profile design, be careful not to make assumptions about the qualitative meaning of a trusted name association. Recognize that brands contain ambiguous features. For example, consider that an ultra-premium brand of automobile or clothing is desired by some consumers while considered pretentious by others. In the same way, a discount brand name may simultaneously signal low price and lowered quality. Be cautious in concluding why a brand increase or decreases a profile's preference in your choice sets.

29.4.3 Refurbishment

One of the field of interest that has evolved recently is the *circular economy*. This is the study of recycling, reuse, and refurbishment of goods. As consumers become more aware of the extraneous costs of manufacturing or products, they have become more aware of how their choices can impact the environment. Academic researchers have used CBC to investigate the impact of refurbished products on the choice market (Wallner et al. 2022). Researchers have also found that CBC reveals differences between consumer preferences that depend upon the types of studied products (Hunka et al. 2021). Therefore, understanding the influence of a reused, recycled, or refurbished product option on new product sales may reveal new sales channels (reselling) without developing a new product.

29.4.4 Money-back Guarantees

Money-back guarantees (MBG) have long been favorite features in consumer product CBC. In one of the earliest mainstream publications of the concepts of choice-based conjoint analysis, an article in *Harvard Business Review* described, in part, the measurement of MBG as a feature of a household cleaner in 1975 (Green & Wind). Later, MBG as a product feature was validated when it was found to influence purchasing choices among delivery pizza options (Verma et al. 1999). Like other non-physical product features, money-back guarantees vary depending on buyer preference. Each product category and consumer type are different. Therefore, as you conduct discrete choice experiments, avoid assuming that partworth results persist across products and customers. You are advised to remeasure features when you change product design, resale model (circular economy), and target market (demographic).

29.5 Other Types of Conjoint Analysis

CBC is of particular value during the development of new products. The frequency of choice-based conjoint demonstrates this value as the preferred method in many commercial – real-world – research studies. One reason that explains the popularity of CBC over other methods is that it most closely replicates the actual shopping behaviors of consumers (Hauser 2007; Natter and Feurstein 2002; Rao 2014). However, to better understand CBC, we review alternative conjoint analysis methods to understand better the trade-offs that occur when selecting one technique over another.

29.5.1 Ratings-based Conjoint Analysis

Before the rise in popularity of CBC, conjoint analysis was often conducted by asking survey respondents to *rate* the preferability of product profiles. As an example, a profile would be measured, by a survey taker, between 0 ("Do not prefer it") to 10 ("Prefer it greatly"). This contrasts with CBC, where we ask respondents to choose between profiles. The fundamental drawback to this rating-based conjoint method is that respondents are not precluded from assigning a rating of 10 to multiple profiles, which obscures the actual preference. When choosing a webcam, or other product, consumers rarely find that two options are so close in desirability that they decide to buy both. In other words, rating-based conjoint "do not mimic consumers behavior in the marketplace" (Eggers et al. 2022, p. 784).

29.5.2 Ranking-based Conjoint Analysis

Conjoint Analysis can also be conducted when survey respondents (or interviewees) place alternative choices in preference order. With this technique, and unlike rating-based CA, there can be no tie between options. As with CBC, this forced choice mechanism matches the behavior expected in real shopping decisions. However, unlike CBC, measuring a respondent's desire to choose none becomes challenging. In other words, ranking choices will tell us relative preference, but it still does not tell us if consumers would prefer to buy. Finally, ranking choices cause a cognitive load on the respondent, unlike other preference-measuring methods. Maintaining a mental model of previous value judgments is strenuous and potentially unreliable, especially in larger choice sets.

29.5.3 Best–Worst Scaling (BWS)

Product designers, marketers, and researchers have recognized that surveying the "best choice" leaves valuable data untapped. With this rise in interest, software applications have begun to address the desire to capture a respondent's "worst choice" as well. Since we know that we understand preferences by discovering the most preferred profile, we can uncover additional information by asking about the least-favorite profile. The principle of surveying for the worst choice (least preferred) profile is that this choice should represent the opposite, and supported statistical models can compensate for this mirror effect.

BWS is commonly used interchangeably with the underlying concept called *MaxDiff*, which is much older than BWS. For clarity, BWS is a discrete-choice experiment type. At the same time, *MaxDiff* refers to the choice theory on which best-worst scaling is based.

29.5.4 Adaptive Conjoint Analysis

By now, you have likely recognized that the conjoint analysis methods discussed thus far have at least two characteristics that limit their ability to scale. First, repetitive comparing of profiles within choice sets can fatigue survey respondents. Secondly, not every respondent cares about the same product features (attributes). Adaptive conjoint analysis (ACA), which is almost exclusively used in computer-delivered surveys, helps by "detecting" which features are more critical to a respondent's choices. Once an important feature is recognized, the software can replace uninformative features with other features helpful to the product manager. This way, the number of features included in the product profile increases without further fatiguing the respondent. For a deeper technical explanation of ACA, see Green et al. 1991.

29.5.5 Self-explicated Method (SEM)

Compared to the other models, the self-explicated method (SEM) is a *compositional*, rather than *decompositional*, preference measurement technique. In short, an SEM survey asks respondents to evaluate their preference for features directly, without being presented with a profile. While this model is not as popular as CBC, you may wish to consider it if your new product development project deals with a novel product or product category. SEM may be a technique that allows you to understand customer preferences before designing product candidates. Because of this model's focus on the features in isolation, SEM better handles large numbers of features, although at the sacrifice of discovering relative importance. Considering a large set of potential features and difficulty narrowing down the feature list for a CBC study, consider an SEM survey as the first part of a two-part analysis.

29.6 Where CBC Disappoints

Introducing you to choice-based conjoint (CBC) analysis and other preference measuring techniques, this chapter is incomplete until you become aware of some weaknesses of CBC. This is an appropriate place to remind ourselves that CBC and other conjoint methods are powerful in many ways. However, as with any powerful tool, it carries dangers when misapplied (McCullough 2002). The risk in market research is to present a misunderstood or misinterpreted result on which your organization bases expensive decisions.

One drawback of CBC is that it is susceptible to misleading results, not by the methodology itself, but by innocent-looking changes made between similar experiments. Expert marketing researcher, Dick McCullough, warns of three types of methodological errors that can cause problems with research results (2002). Assume you conduct a CBC survey with two price levels as in the webcam example. In a follow-up experiment, you expanded the number of price options to four to gain more specific preferences. Comparing the results of the two experiments would be misleading due to the *number of levels effect* (NOL), which would make the second survey over-represent the *importance* of price as a feature.

Similarly, increasing the variance among numeric choices (feature levels) in subsequent studies causes problems. In our survey of webcams, we compared two prices (feature levels): $39.99 vs. 59.99. If we later changed the options to $49.99 and 79.99, this seemingly modest change may exaggerate the relative importance of the price (compared to our initial study). This is referred to as the *attribute range* (AR) problem. Finally, the *attribute additivity* (AA) problem occurs when many less important features crowd out the smaller number but more relevant features.

Another challenge for CBC is that "Unlike traditional CA, with DCE [discrete choice experiment], respondents do not provide sufficient data so that it may be analyzed at the individual level" (Raghavarao et al. 2010, p. 8). This is not a significant problem in surveys of homogenous respondents such as electronics consumers in our webcam example. However, suppose your new product is targeting a small market where you want to understand individual differences between customers. In that case, CBC is not the best methodology for your market research. CBC relies upon the sum of averages and not the individual profile selection within a choice set.

29.6.1 Gestalt-like Products or Services

While CBC and other conjoint analysis techniques work well for mass-produced items, we consider these gestalt-like products when the product is bespoke and not comparable to the alternatives (Karniouchina 2011). Gestalt-like products are not suitable for CBC research. Products with subjective features suffer similar challenges (Creusen 2011). Imagine the difficulty in conducting a discrete-choice experiment on two films playing at the theater. Not only would the description of features and levels be difficult, but the way multiple attributes combine to become more than the sum of their parts would not be captured. Consider comedic actors Abbot and Costello, who arguably delivered more entertainment value (total utility) when performing together rather than apart. Novel products may also suffer from difficult comparisons with a state-of-the-art alternative. Comparing the first smartphone to its alternative would be problematic. The number of new abilities available in a smartphone, for example, touch screen, apps, GPS were not available in its alternative (feature phone); thus, choices between features are not possible.

29.7 Conclusion

This chapter introduced the concepts of conjoint analysis in general and choice-based conjoint (CBC) analysis in particular. As New Product Development has become a better-understood profession, the tools required for success have also matured. As organizations grow, they often become aware of market risks and adopt systematic research to reduce these risks. Product managers are well-positioned to introduce research methods such as CBC in their companies. Others often welcome efforts when they develop a deeper understanding of customers, their preferences, and how current or proposed products fit customers' desires.

For product managers expecting scientific research to impede creative thinking, we hope this chapter demonstrates that awareness of quantitative research methods improves new product development. Statistics cannot replace innovative and design thinking. Instead, consider CBC another tool that will allow you and your organization to have deeper – and more scientific – conversations with colleagues, executives, and customers.

References

Baier, D. and Gaul, W. (2001). Market simulation using a probabilistic ideal vector model for conjoint data. In: *Conjoint Measurement* (ed. A. Gustafsson, A. Herrmann, and F. Huber), 97–120. Springer, Berlin, Heidelberg.

Creusen, M.E. (2011). Research opportunities related to consumer response to product design. *Journal of Product Innovation Management* 28 (3): 405–408.

Eggers, F., Sattler, H., Teichert, T., and Völckner, F. (2022). Choice-based conjoint analysis. In: *Handbook of Market Research* (ed. C. Homburg, M. Klarmann, and A. Vomberg), 781–819. Springer.

Green, P.E. and Krieger, A.M. (1988). Choice rules and sensitivity analysis in conjoint simulators. *Journal of the Academy of Marketing Science* 16 (1): 114–127.

Green, P.E., Krieger, A.M., and Agarwal, M.K. (1991). Adaptive conjoint analysis: some caveats and suggestions. *Journal of Marketing Research* 28 (2): 215–222.

Green, P.E. and Srinivasan, V. (1990). Conjoint analysis in marketing: new developments with implications for research and practice. *Journal of Marketing* 54 (4): 3–19.

Green, P.E. and Wind, Y. (1975). New way to measure consumers' judgements. *Harvard Business Review* 53: 107–117.

Hauser, J.R. (2007). A note on conjoint analysis. MIT Sloan Management. http://www.mit.edu/~hauser/Papers/NoteonConjointAnalysis.pdf.

Huber, J., Wittink, D.R., Johnson, R.M., and Miller, R. (1992). *Learning effects in preference tasks: choice-based versus standard conjoint.* Paper presented at the Sawtooth Software Conference Proceedings.

Hunka, A.D., Linder, M., and Habibi, S. (2021). Determinants of consumer demand for circular economy products. A case for reuse and remanufacturing for sustainable development. *Business Strategy and the Environment* 30 (1): 535–550.

Kahn, K.B. (2005). *The PDMA Handbook of New Product Development*, 2e (ed. K.B. Kahn). Hoboken, NJ: John Wiley & Sons, Inc.

Karniouchina, E.V. (2011). Are virtual markets efficient predictors of new product success? The case of the Hollywood stock exchange. *Journal of Product Innovation Management* 28 (4): 470–484.

Louviere, J.J. and Woodworth, G. (1983). Design and analysis of simulated consumer choice or allocation experiments: an approach based on aggregate data. *Journal of Marketing Research* 20 (4): 350–367.

Marn, M.V., Roegner, E.V., and Zawada, C.C. (2004). *The Price Advantage*. John Wiley & Sons.

McCullough, D. (2002). A user's guide to conjoint analysis. *Marketing Research* 14 (2).

Moore, W.L., Louviere, J.J., and Verma, R. (1999). Using conjoint analysis to help design product platforms. *Journal of Product Innovation Management: An International Publication of the Product Development & Management Association* 16 (1): 27–39.

Natter, M. and Feurstein, M. (2002). Real world performance of choice-based conjoint models. *European Journal of Operational Research* 137 (2): 448–458.

Purdy, M., Klymenko, M., and Purdy, M. (2021). Business scents: the rise of digital olfaction. *MIT Sloan Management Review* 62 (4): 1–5. https://sloanreview.mit.edu/article/business-scents-the-rise-of-digital-olfaction.

Raghavarao, D., Wiley, J.B., and Chitturi, P. (2010). *Choice-based Conjoint Analysis: Models and Designs.* Chapman and Hall/CRC.

Rao, V.R. (2014). *Applied Conjoint Analysis.* Springer.

Steiner, M. and Meißner, M. (2018). A user's guide to the galaxy of conjoint analysis and compositional preference measurement. *Marketing: ZFP–Journal of Research and Management* 40 (2): 3–25.

Verma, R., Thompson, G.M., and Louviere, J.J. (1999). Configuring service operations in accordance with customer needs and preferences. *Journal of Service Research* 1 (3): 262–274.

Wallner, T., Magnier, L., and Mugge, R. (2022). Do consumers mind contamination by previous users? A choice-based conjoint analysis to explore strategies that improve consumers' choice for refurbished products. *Resources, Conservation and Recycling* 177: 105998.

Wittink, D.R. and Cattin, P. (1989). Commercial use of conjoint analysis: an update. *Journal of Marketing* 53 (3): 91–96.

Garth Brown is a technology development advisor whose clients have included retail, semiconductor, aerospace, energy, and health-care companies. He holds bachelor's and master's degrees in Leadership as well as multiple industry certifications. He is also a PhD candidate at the Amsterdam Business Research Institute (ABRI), which is part of the School of Business and Economics (SBE) at Vrije Universiteit Amsterdam. His academic research focuses on the creation and productivity of temporary and self-managing product development teams. Garth and his wife, Alevtina, live in Dallas, Texas, USA.

APPENDIX

TABLE 29.3 SAMPLE OF CONJOINT ANALYSIS SOFTWARE TOOLS.

Software	Website	Notes
Conjoint.ly	conjointly.com	A relatively new entrant into the market. Introduced in 2016, the software has experienced brisk adoption due to its ease of use. * Many screen captures in this chapter were created using this tool.
Qualtrics DesignXM	www.qualtrics.com/design-xm/	Qualtrics is a well-regarded name in consumer research. Their DesignXM tool focuses on the needs of product design decision-makers.
Sawtooth Software	sawtoothsoftware.com/	Founded in 1983 by respected conjoint researcher Richard M. Johnson, this is considered a field stalwart. Popular with practitioners and academic users.
XLSTAT	www.xlstat.com/en/	Notable for its integration with Microsoft Excel® as an installed add-in. May be an attractive choice for those who desire to work without an Internet connection or prefer a fixed-price perpetual software license.

CHAPTER THIRTY

CREATIVITY TOOLS FOR NEW PRODUCT DEVELOPMENT

Teresa Jurgens-Kowal

Creativity is intelligence having fun.

— ALBERT EINSTEIN

Even those working in the field of innovation shy away from creativity. Many, if not most, adult professionals express doubts of their own creativity. Yet, everyone can be creative. Product development practitioners, especially, must harness creativity for success.

Creativity is not limited to the arts, though it is prominently recognized in literature, music, and painting. Children are deemed creative and imaginative, creating small worlds and stories for playtime. Somehow, over time, we begin to view creativity as a specialized function for the performing arts or story time for toddlers. School systems and working environments often reward serious behavior and accurate answers over creative play and exploration.

The fall-out from conforming to pre-determined solutions is that people take fewer risks, and we end up constraining product development to easy tweaks of existing features. New products then fail to intrigue customers, and companies become trapped in an endless cycle of incremental improvement. Product launches are met with yawns from the marketplace rather than delighted consumers purchasing exciting products that generate profit for the firm. For example, Jamie Dimon of JPMorgan Chase lamented the lack of "creative combustion" and collaboration necessary for innovation during the coronavirus when employees worked individually and remotely (Zarroli 2020).

The PDMA Handbook of Innovation and New Product Development, Fourth Edition. Edited by Ludwig Bstieler and Charles H. Noble.
© 2023 John Wiley & Sons, Inc. Published 2023 by John Wiley & Sons, Inc.

In this chapter, you will learn to define *creativity* through the *design thinking* process. We then suggest several tools to empower creative exploration for product development teams. We will focus on the front-end of innovation and early stages in the design thinking process. More details on creativity for the back-end of innovation (Jurgens-Kowal 2022) and in-depth application of tools can be found by reviewing the references. The chapter closes with a few tips to construct a creative working space. Finally, the appendix includes a reference table of selected creativity tools and techniques corresponding to the various phases of the design thinking process.

30.1 Part 1 – the Creative Process

30.1.1 Creativity Is a Process

Unlike the popular perception of a lone creative genius inventing a game-changing new product in the garage, creativity is a process. It does not come through a lightning strike, nor does it materialize instantaneously. Creating valuable new products and services requires the necessary learning and effort to ensure delivery of features to benefit customers and end-users while generating profit for the producer.

One common process (Stillman 2014) for creativity is shown in Figure 30.1.

30.1.1.1 Preparation. Because creative outcomes do not instantaneously materialize in eureka moments, we prepare our minds for creativity by learning. Both deep learning and broad-based learning covering many different topics stimulate creativity. Often creative ideas for new products are transferred between knowledge domains. So, preparing for creativity means acquiring lots of information from a wide variety of sources.

30.1.1.2 Incubation. Of course, many people report a flash of inspiration when they are in the shower or doing a task completely unrelated to the problem they are trying to solve. Incubation is the period during which the subconscious works on the challenge and links the apparently disparate ideas into a novel proposal. Stepping away from the challenge allows the relaxed brain to generate connections among varying pieces of information. This is an important, but often overlooked, step in the creative process. For example, Bill Gates, Jr., regularly took a "think week" away from Microsoft to think creatively and strategically (Clifford 2019).

FIGURE 30.1 CREATIVITY PROCESS.

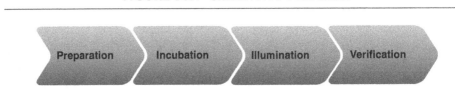

30.1.1.3 Illumination. When several nuggets of information collide and connect to generate a unique idea, the creative process is realized. Cartoons and comics show this moment with the proverbial light bulb illuminating the character's new insights. The light bulb serves as appropriate imagery for the illumination step in which the unique solution to the problem is born. There is no retreat from the creative process at this point as the new idea has taken form.

30.1.1.4 Verification. Naturally, creative ideas formed in the shower may fail to see production. Verification is a step in the creative process to address and formulate a new concept for an appropriate audience and package it in a way that users will accept the idea. For instance, Beethoven transformed his deep and dark emotions into stunning symphonies that are consumed by musical patrons in the creative arts. As a creative application, ride-sharing services (like Uber and Lyft) connect people with cars to those that need a ride. Pleasure and satisfaction come from enjoying a creative new symphony while profit results from generating a creative new product or application for a company.

30.1.2 Design Thinking

The most common creative process we encounter in product and service development is **design thinking**. Design thinking is both a process and a set of tools used to explore customer interfaces with products and services. We define design thinking as a "systematic and collaborative approach for identifying and creatively solving problems" (Luchs 2016). This framework is illustrated in Figure 30.2. Note the similarities among the steps of the standard creative process and those in design thinking.

30.1.2.1 Discover The focus of the **discovery stage** in design thinking is to uncover new customer insights. Breakthrough ideas begin with a deep understanding of customer needs and challenges. Learning the qualitative problems facing customers allows an innovation team to build empathy for the

FIGURE 30.2 DESIGN THINKING MODEL.

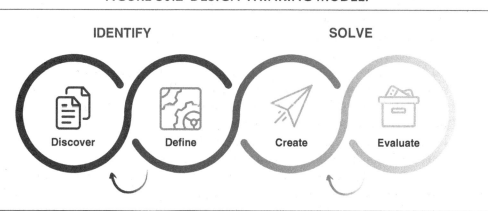

context of opportunities. Like preparation in the creative process, discovery requires gathering data and collecting knowledge. For instance, product managers investigate different segments of a population to learn specific pain points as well as to understand drivers for customer behaviors. These drivers may include market trends, technology advances, and competition.

30.1.2.2 Define. Distilling customer insights into a specific, actionable problem statement parallels the individual creator's step of incubating an idea. An innovation team may approach a customer need from many different points of view, so that **defining** the correct problem to solve is an iterative process. Understanding customer needs in the design thinking process links information and insights such as type of customer, market needs, consumer activities, and commercial opportunities. Once an innovation team defines a problem statement, they validate that it explains the opportunities identified in the discovery phase and then move to generating potential solutions.

30.1.2.3 Create. The purpose of the **create stage** in design thinking is to develop a concept that will be shared with the target market. Innovation teams need this feedback to understand if the potential product solution addresses the previously identified customer needs. Concepts vary in fidelity throughout the product development process so idea generation and prototyping in the create stage involves several different tools and techniques. Many prototyping tools are industry specific. However, the outcome remains consistent with a goal of designing a feature set and/or full product that addresses customer needs identified in the discovery stage, and focusing on solutions to the specific problem statement generated in the definition phase. Prototype testing further minimizes commercial investment risk.

30.1.2.4 Evaluate. Ideas, concepts, and prototypes validate solutions to the defined problem. In most cases, prototypes are tested and **evaluated** by real customers who suffer the identified pain point. However, in early stage iterations, the ideas and concepts may assess specific features or elements of the final product. In these situations, internal trials validate the subset of features or technical functionality. All final products must be presented to real customers to gather authentic end-user feedback for the new product launch. The evaluate stage ends with ensuring that a product solution meets customer needs as defined throughout the cyclical design thinking process.

30.1.2.5 Iterative Process. *Design thinking is an iterative process.* As potential solutions are proposed, they are matched with discovery to ensure that customer needs are met. Iteration makes the process more agile and flexible. With increased customer interactions, design thinking also lends itself to higher levels of product success. As customers are consulted throughout each phase, the end product results in greater user satisfaction.

In the first phase of design thinking in which the problem is identified, creativity processes focus on *divergent thinking* (Rodriguez 2020). Divergent thinking is often called "outside the box" thinking in which all ideas are captured – no matter how crazy they might sound. There is no filtering of ideas, such as technical feasibility to implement the idea in a physical solution. The goal of divergent thinking is to generate as many ideas as possible with very few constraints on the problem statement.

Creativity also requires applying the best potential solution to the problem at hand. Not all ideas are appropriate to apply at any given point in time, so we use *convergent thinking* to categorize ideas into specific themes and applications that can be transformed into relevant product solutions. In the design thinking model, convergent thinking is utilized more during the solve phase.

30.1.2.6 Collaboration. Compared to the creative process implemented by an individual problem-solver (Figure 30.1), design thinking succeeds only when developers seek collaboration among customers and across disciplines. Collaboration allows a broader view of customer needs as well as blending various sources of knowledge and information into potential solutions. Collaborative creativity expands the sphere of ideas simply by bringing together people of different experiences and skills. It is important to build collaborative innovation teams of cross-functional team members to maximize creativity in product development. Diversity enhances the numbers and types of creative ideas generated by any innovation team.

30.2 Part 2 – Creativity Tools and Techniques

In designing creative solutions to meet customer needs, a creativity toolkit includes techniques to explore problems from a broad perspective, empathize with the customer, gather unique insights, build prototypes, trial products in the marketplace, and validate data for continuous improvement. The set of creative tools at the disposal of product managers is truly only limited by their imagination. We will describe a few of the more common creativity tools that repeatedly elicit valuable results in product development. For reference, the appendix maps the most-used creativity tools to the design thinking phases discussed here.

30.2.1 Creativity Tools for Discovery

Many of the creativity tools used in the identification stage of the design thinking model focus on understanding customer needs and pain points. Thus, product developers utilize divergent thinking to generate as many ideas as possible. You can use these techniques with internal product development teams, but they are especially valuable when applied to consumers and end-users, lead customer panels, or focus groups, including both users and non-users of your product and/or of competitive products.

30.2.1.1 Word Association Sometimes called *free association*, **word association** is a quick way to generate novel connections among ideas (Markov 2019). The technique can be used by an individual or in a small group setting. A word, perhaps associated with a feature of a competitor's product or a customer's problem, is presented to a group. Under a set time limit, participants record any words that come to mind associated with the provided guide word. Ideas are shared with the larger group for further divergent thinking and/or categorization to move into the definition stage of design thinking.

This tool complements mind-mapping (see Section 30.2.1.2) with free association (Gross et al. 2012) across several categories or may be produced in a liner fashion. An example of linear word association is shown in the figure below where two participants started with the word "coffee" to generate new product and service ideas. In the first example, the association of the guide word "coffee" with "door" may encourage the developers to find a new way of door-to-door selling or home delivery of coffee products. In the second example, you may consider an outdoor coffee kiosk serving cyclists on a bike path or trailway.

You can try word association to stimulate creative ideas for especially sticky problems. Require participants to record a minimum number of words within a set period of time (e.g., one to two minutes) to generate even more "outside the box" thinking.

30.2.1.2 *Mind-mapping* **Mind-mapping** builds on free association but involves images and sketches in addition to words. In Figure 30.3, a linear word association is presented and mimics the childhood game of "telephone" where words are whispered to individuals in a line. The result at the end of the line is often very different than the starting word. Word association is typically a verbal or written exercise.

In contrast, mind-mapping involves more random idea generation and is a visual technique. Typically mind-mapping is better suited to individual creativity but can be extended to group ideation if the participants are grounded in trusted relationships. Mind-mapping starts with several minutes to an hour of time in which ideas, words, sketches, and other pieces of information are documented on large sheets of paper or a whiteboard. Connections are noted between and among the various factors, often resulting in convergent thinking in which categories or themes are developed further. In many cases, mind-mapping starts with

FIGURE 30.3 EXAMPLE OF LINEAR WORD ASSOCIATION.

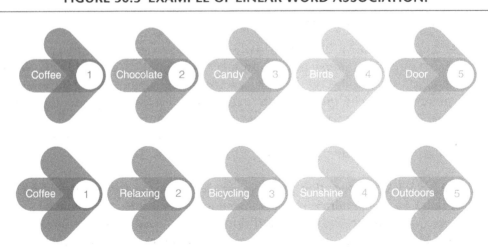

a problem statement expressed in the middle of a blank page resulting in concentric elements of ideas that add granularity to identifying or expressing the problem statement.

As an example, the author has applied mind-mapping with a software development team to build a new product taxonomy. Mind-mapping revealed product and platform dependencies as well as identifying independent product lines. Ultimately, defining products with associated product lines and portfolios led to restructuring the organization for higher effectiveness in product development.

Like many creativity tools, mind-mapping is not reserved solely for the discovery phase and often serves well to resolve the definition of a product challenge in the definition stage as well. Software tools are available to add aesthetics and formal organization to mind-maps. However, we encourage you to draw your mind-maps by hand to enhance creative inspiration.

30.2.1.3 Brainwriting While the technique of brainstorming is widely used in many industries and trouble-shooting situations, there are several drawbacks to traditional brainstorming. Often, less vocal individuals do not openly share their ideas while groupthink can follow an influential person's callout of an idea in the brainstorming session. An alternative to brainstorming is **brainwriting** (Mattimore 2012).

Brainwriting takes advantage of both individual and group creativity in a short ideation session. The problem statement is presented to the group. Individually, under a strict time limit, each person records an idea on a sheet of paper. Time limits are typically 30 to 60 seconds. After recording a novel idea, the participant passes the sheet of paper to the person sitting to the right. This individual can either record a second new idea or build on the previously recorded idea.

The process repeats for a few minutes, typically with about a half dozen ideas recorded on each sheet of paper. At this point, the paper is returned to the originator who quickly scans the ideas for the "best" idea. This idea is then shared with the group via a traditional brainstorming session where ideas are called out and recorded for the group to view collectively. Benefits of brainwriting include capturing divergent thinking individually and collaboration among team members.

Note that brainwriting is easily adapted to a virtual setting by using an online whiteboard, shared documents (e.g., Google sheets), or instant messaging/texting applications. An example showing the flow of brainwriting is shown in Figure 30.4. The author often uses brainwriting as a warm-up creativity exercise for product discovery starting with a well-known but unrelated product. For example, one exercise was to find alternate uses of water bottles for an automotive parts manufacturer. As the team discovered their creativity with a standard item (the water bottle), they quickly expanded to new ideas for their own product lines (auto parts).

30.2.1.4 Collage The word **collage** is derived from the French word meaning "to glue." Literally, collages involve various print media cut and glued into a single image. From the perspective of encouraging divergent and creative thinking,

FIGURE 30.4 EXAMPLE OF FLOW IN BRAINWRITING.

collages lend insight into connections that customers make between the problem they are trying to solve, products and features that address these issues, as well as other items of concern. For instance, a collage that includes images of cash registers or money may indicate that customers are concerned about the cost of a product. Thus, having users share and describe their collage with the innovation team is the most important element for product development practitioners to discover customer needs with this technique.

Internal product development teams can also use collages in the early phases of product design. Products are typically composites of various feature sets. Discovering how features might fit together by encouraging team members to paste together various elements lends insights for further product development. Collages emphasize creative combinations of features while also underscoring potential design complexities for the development team. In a specific example, the author encouraged a digital retail client to create collages from fashion magazines to help identify target customers and end-users.

***30.2.1.5 SCAMPER* SCAMPER.** is an acronym for substitute, combine, adapt, modify, put to another use, eliminate, and reverse (Rodriguez 2020). Innovation teams use SCAMPER in the discovery phase to probe customer needs and to identify new product opportunities. By viewing a customer need, competitive product, or existing product through the lens of each verb in SCAMPER, new ideas are revealed.

- **Substitute** – Can we use a different raw material or packaging for the product? Can we substitute another product for this one? What other products and services could we use?

- **Combine** – If green is generated by mixing blue and yellow, what features can we mix to create a new product? Can we combine this product with another product, service, or application to generate a new offering?
- **Adapt** – How can we approach different markets? What can we learn from accessories used with our product? Can we borrow a technology from another industry and apply it for a unique product solution?
- **Modify** – How can we change the look, shape, or feel of the product? Are there existing features to highlight in a new way? Can we use an element of this product in another one?
- **Put to Another Use** – Can this product be transferred to a different purpose or industry? Are there other markets that can use an existing product? Is there another use for the product within an existing category?
- **Eliminate** – Can we remove features to simplify the product or application? How could you take away a portion of the product and allow it to continue to function in the same way? Can we make it faster, cheaper, or lighter?
- **Reverse** – Just as looking through the wrong end of binoculars gives an extended view instead of a close-up view, can we look at the product from an opposite perspective? Can we reverse steps in the process? What if we sequenced features in a different way?

30.2.1.6 Tell Me About … Open discussions with customers and potential customers are invaluable to build empathy and gather insights. As problem-solvers, many product development professionals naturally gravitate toward closed-end questions with yes/no answers. However, to gain the most qualitative information in the discovery phase, we want to explore customer needs with open-ended and broad-ranging questions. To facilitate open conversations, product innovation teams can ask a potential customer to describe a situation in which they faced the problem, used a particular product, or experienced the need for a different product solution. "Tell Me About …" is a starting phrase that encourages open and broad-based informational responses.

"Tell Me About …" can be used in conjunction with many of the other creative discovery tools. For instance, if a customer has created a collage, ask them to explain it. You can gain further insights by asking the customer to "Tell Me About…" specific images or arrangements of images in the collage.

30.2.2 Creativity Tools for Define

As an innovation team generates a list of unmet customer needs and opportunities during discovery, we must carefully define the problem statement for the product development project. The problem statement should not suggest a solution, yet should stimulate further creative ideas in crafting the ultimate product solution. For example, a problem statement asking how to make a sharper kitchen knife suggests a solution – a sharper knife. On the other hand, a problem statement asking how to facilitate chopping vegetables opens the solution space wider. In this case, a potential solution is an electric food processor. An even broader problem statement would inquire how to serve individual portions of vegetables at home. With this open question, a team might identify how a home chef could purchase pre-cut vegetables at the supermarket or buy a prepared mixture.

Again, many of the creativity tools discussed in this chapter find application across the design thinking process. Certainly, there is much overlap to effectively apply tools and techniques between the discover and define stages. You may also find utility in applying the definition tools as you begin to create prototypes in the solve stage of the design thinking process.

The goal of the define stage is to establish a list of product features and requirements. Specific product design components are then embodied in concepts and prototypes in the subsequent create phase.

30.2.2.1 Interview. As indicated in the "Tell Me About …" technique, structured interviews with potential customers can refine the description of their needs into clear problem statements. As an outcome of the discovery stage, the innovation team highlights specific categories and themes of customer problems. **Interviews** with open-ended dialogue gather information regarding the accuracy of the problem statement. Because the identification stage of design thinking is iterative, product managers typically recycle and iterate between the discovery and definition phases multiple times before narrowing the specific problem definition statement. Interviews not only suggest the issue facing customers but also validate very specific problem(s) they encounter.

Interviews may be coupled with observations, such as the customer using a product in a real-world situation or actively comparing features of competitive products. One of the greatest challenges common to innovation teams is the ability to accurately capture information from interviews. A best practice involves having one person asking questions and another acting as a scribe. The interview technique works equally well to gather information in-person or remotely. Observation of tangible product use is typically better in-person while software use can be easily captured virtually. Be sure to check local laws and to ask a potential customer for permission before recording the interview.

30.2.2.2 Five Whys. Borrowed from quality management practices (Five Whys and Five Hows) and root cause analysis, the **Five Whys** technique uses subsequent, telescoping questions to better define the problem facing a customer. It is frequently difficult for customers to describe their unmet needs clearly and concisely. Further, asking a customer to define a potential solution results in incremental designs since most consumers are unable to ideate outside of known product boundaries (e.g., thinking "inside the box").

For instance, many are familiar with the dilemma faced by Henry Ford to market the brand-new automobile. He is famously quoted as saying that if he had asked people what they wanted, the response would have been "a faster horse." Thus, repeated questioning to target a root cause through Five Whys drives product definition to a data-based problem statement.

An example of Five Whys is shown in the following conversation in which a boss might subjectively determine employees are lazy because they are late to

TABLE 30.1 EXAMPLE OF 5 WHYS.

Boss	Employee
Why were you late for work today?	*My car wouldn't start.*
Why didn't your car start?	*It was out of gas.*
Why was your car out of gas?	*I didn't fill the tank last night.*
Why didn't you fill the tank?	*I didn't have any cash.*
Why didn't you have any money?	*I forgot to cash my paycheck.*

work. Initially, the issue is assumed to be a time-related delay, but the questioning technique reveals a very different problem, as illustrated by a fictional conversation in Table 30.1.

In this situation, the *actual* problem statement is "How might we ensure employees receive their pay automatically" and not related to lazy employees oversleeping, as the boss might have assumed without root cause analysis.

While Five Whys is a great technique to help reveal the root cause problem, be sure to change the wording of the questions as you drill down through the subsequent levels. A customer might feel as if they are being interrogated if someone is constantly asking "why, why, why." Additional appropriate phrases include "Tell me more about what caused that situation," or "Can you explain what happened before/after that step."

In addition to using the Five Whys in the define phase of design thinking, it is a highly useful tool in product retrospectives. Team members can identify new ways of working and specific features for next generation products when applying root cause analysis in lessons learned reviews. In one such example, the author learned of significant project overload for team members during a Five Whys exercise as part of a retrospective. Late projects were a result of poor portfolio management rather than inefficiencies in design and development. Better prioritization of new product ideas reduced team member stress as well as introducing better effectiveness in delivering new products to the customer(s).

30.2.2.3 *Card Sorting.*
Because potential customers and end-users of new products often have a difficult time visualizing product features, the **card sorting** method helps innovation teams better define the necessary features and attributes for the product design. Consumers are presented with two choices for several different features and asked to choose the one they prefer. Several features are presented in rapid succession, and features are repeated often to validate the data.

Card sorting overlaps traditional market research tools such as A/B testing, morphological analysis, and conjoint testing (Rodriguez 2020). The innovation team will prepare 20 to 50 cards (such as 5 × 7-inch index cards) with feature comparisons as in Figure 30.5. (Note that images and sketches are often included as well as short narrative descriptions of features. The figure here is highly simplified for presentation purposes.) Each pair of cards is presented to customers, and they are asked to choose the feature that they like the best or is most important to them for the overall product functionality. The test is repeated multiple

FIGURE 30.5 CARD SORTING EXAMPLE.

times with several different target customer segments to fully refine the product definition.

Product managers also apply card sorting to determine which features to offer for an initial product launch and to create the product roadmap for future releases. In this way, card sorting helps to minimize new product investment and reduce risk.

30.2.2.4 Customer Journey Map. Often used to define customer experience as much as specific product features, the **customer journey map** helps to refine problems that customers face. This tool is especially useful for innovation teams seeking to understand alternate product uses, competitive products, and situations where there may be no existing product(s) in the marketplace.

The customer journey map (see Figure 30.6) traces the experiences customers have once they identify they have a problem (*awareness*) through the decision-making process (*consideration*), *purchase*, and post-purchase period (*warranty*). An important element of the customer journey map is to capture the customer's emotional response at each step. In the example shown, the customer had a positive experience in learning about the product, researching various products and features, visiting a retail location, and purchasing the product. However, when the customer faced an issue in using the product, they had a negative emotional response during the warranty period. These emotional indicators along the customer's journey of finding and using a product give greater clarity to the design and development of new products and next generation products and services.

FIGURE 30.6 EXAMPLE OF CUSTOMER JOURNEY MAP.

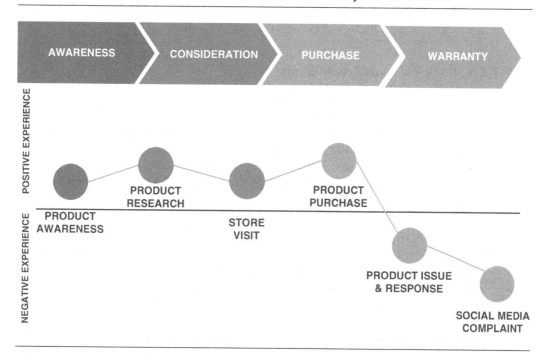

After a new product is launched, continued interactions with customers to track their experience with the product and service package encourage creative ideas for discovering and defining needs. As an example, the author worked with an IT team to craft their customer journey. Previously, the team had looked only at internal customers and when challenged to identify the end-user's customer journey, the IT team discovered several new products requiring investment (e.g., smartphone payment apps) while also recognizing a set of products to retire (e.g., checks and money order payments).

30.2.2.5 Affinity Diagrams.
A common tool for convergent thinking is the **affinity diagram**, also known as *clustering*. During the discovery phase of design thinking, the innovation team has likely collected hundreds of ideas. Clustering the ideas into like themes and categories minimizes the number and types of concepts that move into the next stage of the process. This technique is effective for in-person ideation, remote teams, and hybrid teams.

Each idea is recorded on a single sticky note, index card, or piece of paper by individuals. The ideas are then shared on a whiteboard and similar ideas are brought together into blocks using the wisdom and collaboration of the team. Typically, several clusters are identified, and hundreds of ideas are quickly sorted into six or eight potential product concepts. These concepts then move into the next phase of the design thinking process for concept and prototype testing.

You have probably applied concepts of affinity diagrams in how you organize your closet. Do you cluster all your formal clothes in one section and casual attire in another? Or do you cluster clothes by type (shirts, pants, dresses)? Or by color? The types of clusters identified by a product development team drive the feature sets that are built into a new product design.

30.2.3 Creativity Tools for Create Stage

The design thinking model represents the simplest two steps of problem-solving: first identify the problem, then solve the problem. In the solve stage, innovation teams create prototypes and evaluate them. Prototypes involve varying degrees of fidelity and functionality linked to the specific question or problem statement to be addressed. Thus, the evaluation tools are connected to the type of prototype created, especially whether the testing is internal or external to the innovation team.

The primary goal of the create and evaluate stages is to generate concepts and prototypes that will be tested and evaluated by the innovation teams and potential customers. Concepts and prototypes include components and elements to model or simulate specific features as well as whole products for full customer testing. In this way, form and function can be tested independently to speed time-to-market for new products.

30.2.3.1 Looks Like … As with other creativity tools, many of the tools used in the create stage find utility in other stages, particularly the define stage. **Looks Like …** is a technique that helps to refine product definition as well as building functions and features for prototype testing. Conceptually, the team identifies another product (or service) that offers the gold standard and compares the new product under development to the reference product.

Looks Like … works when comparing products within a category as well as to compare to the best-in-class of another category. For example, if the innovation team is designing a new wine refrigerator, the stimulus input is "*It looks like the Cadillac of wine refrigerators.*" This aids in framing the design to include luxury, comfort, and dependability into the functionality and feature sets of the refrigerator, which is otherwise a utilitarian item. Similarly, the team could be designing a traditional car and compare it to Cadillac but at a different price point, for a different market, or using a modified manufacturing process.

30.2.3.2 What Can It Not Do? Similar to *Looks Like …*, asking the innovation team what an existing product cannot do stimulates the design of novel features and functions. The question of **What Can It Not Do?** must be constrained within boundaries to address specific design or development needs. This tool is useful when contrasting existing products with those of competitors, creating next generation products, and matching customer needs with iterations of the product definition.

For example, in developing an insulated on-the-go coffee cup, asking "*What can a traditional insulated beverage holder not do?*" resulted in several design insights. A traditional insulated beverage holder did not fit into a car's cup holder due to its large diameter. Likewise, unscrewing the lid to pour the beverage was not a task most drivers could safely undertake during their morning commute. Standard coffee cups have handles and no lid, making them inappropriate to address the problem statement. These insights led to development of tapered designs for the body of the cup to fit into a cup holder and a locking, but easy-to-open, lid to prevent spills.

30.2.3.3 Sketching.

One of the most underutilized but most powerful creativity tools is simple **sketching**. Many professionals are hindered by their perceived lack of artistic ability. However, sketching is an important addition to verbal communication. There are no special tools needed for team members to use this method and with iteration, sketching can be incorporated into *storyboards* used in defining and creating service solutions (Rodriguez 2020). Storyboards show a simplified flow of a customer's interaction with a product (or service), serving to illustrate elements of the customer journey map, for instance.

Individuals should be given plenty of blank paper, colored pencils, or markers, and set free to sketch product or feature ideas. Assuring individuals that "ugly is okay" (Knapp 2016), they will be given an opportunity to describe their sketch to the group (especially with the addition of the "Tell Me About …" technique). However, adding descriptors to the sketch is helpful when the team sorts ideas and concepts throughout the iterative design thinking stages or when the originator is not available to describe the sketch. When individuals share their solution sketches, the conversation should be restricted to facts and evidence, with questions designed to clarify functionality only. Further refining of the design should wait until all team members have shared their sketches and the team reaches consensus on the most impactful components to include in the new product design.

30.2.3.4 Paper Prototype.

Also known as a *wire frame*, **paper prototypes** are non-working, low-fidelity prototypes that are used to test end-user interactions with a proposed product, feature, or function. Often used to illustrate computer or mobile phone screen interactions, the paper prototype is an efficient and cost-effective way to gain customer feedback later in the evaluate stage. Paper prototypes are an extension of sketching but with a specific product design in mind.

In the evaluation phase of design thinking, paper prototyping also offers a quick way to test various layouts of product features for tangible products, such as buttons, knobs, handles, and levers. By creating a simple diagram of the product, whether a mobile app or a toaster oven, the innovation team generates different options for the new product. Designers use paper prototypes to establish spacing of different features as well as to ensure logos and color schemes are appropriate for the product.

Specific to software and mobile applications, *wire frames* are paper prototypes illustrating the progression through various user screens. This blueprint of website screens is a visual guide and framework for developers as the product advances through successive design stages. Depending on the fidelity of the wire frame layout, the innovation team may test layouts with end-users or simply identify options for further refining during development. Paper prototypes may be combined with various evaluation tools or combined with the Wizard of Oz technique to further elicit customer feedback during product design.

A key benefit to paper prototypes in the create phase of design thinking is to draw on both individual ideas and team collaboration. When individuals sketch a concept independently, there are often features common among all participants. As a team, these sets of features for the new product are naturally revealed by combining the various concepts. The fidelity of the prototype will increase with iteration and as the teams further align ideas and concepts with customer inputs.

30.2.3.5 Wizard of Oz. The **Wizard of Oz** technique, also known as *the man behind the curtain*, refers to the famous movie in which one man controlled the entire land of Oz with a set of controls hidden behind a curtain. While visitors to Oz heard a loud, threatening voice, the man behind the curtain was, in actuality, neither large nor intimidating. In product development, the Wizard of Oz technique is used to simulate automatic responses of a product's features to a user's input, even though the "product" is incomplete and non-functional.

In software development, for instance, PowerPoint™ slides are built to simulate actions taken by a user without writing code (building from the wire frame or paper prototype). If a potential customer indicates they would click on a certain icon or section of the screen, a manual operation is executed behind the scenes to advance to the next "screen" or "action" required. Often the Wizard of Oz technique is combined with interviews and observation to better understand customer needs and challenges. Users are encouraged to explain why they are taking certain actions and what they expect as a result of that action. An expected outcome of this creativity tool is an understanding of potential market share and customer acceptance.

30.2.4 Creativity Tools for Evaluate Stage

In the second step of problem-solving in the design thinking model (solve), new products are tested and evaluated against customer criteria established in the identify stage. Evaluations include both internal testing (often called *alpha-testing*) as well as external feature and product testing with potential customers. In *beta-testing*, a ready-to-launch product is evaluated with ancillaries and accessories by actual customers to ensure the product functions as expected under different user circumstances. With *gamma-testing*, the product is assessed under real conditions to ensure that no unexpected side effects or secondary influences occur.

Many of the creativity tools used to test a feature, function, or product are tightly linked to the type of prototype or product that has been created at that stage. Recalling that the design thinking process is iterative, creating and testing prototypes in tandem ultimately helps to deliver the highest quality product to the customer. Some of the testing tools and techniques are more appropriate for internal evaluations, while others are applied equally well to external customer testing.

The primary outcome of the evaluation phase is a product ready for commercial launch.

FIGURE 30.7 EXAMPLE OF DOT VOTING.

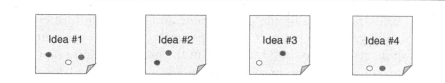

30.2.4.1 Dot Voting. Applied as an additional element to affinity diagrams or in selecting specific features of a prototype to advance, **dot voting** is an internal evaluation by the product development team. The tool works well in either a face-to-face environment or in remote workspaces using electronic whiteboards.

Features, functions, or other products and components are presented to the cross-functional team. As illustrated in Figure 30.7, these ideas are listed on sticky notes or note cards. For functional prototypes, the actual product may be presented. Team members are each given a fixed number of "votes." A typical vote for an in-person evaluation session is a colored sticky dot, thus the name of the technique. Participants affix the dots to the indicated idea, feature, or product that they feel will most successfully meet the required criteria. In the example above, three individuals were each given three votes and Idea #1 "wins" since it has the most "votes."

Typically, a facilitator will set a limit on both the time allotted for voting as well as the number of dots each individual is allowed. For online evaluations, the voting may be confidential with only the final vote tally revealed, as desired. Other rules for dot voting include individuals cannot vote for their own ideas or they cannot place multiple votes on a single idea.

Note that dot voting is primarily a qualitative method of evaluation by internal teams. Ideas, concepts, and product features must be vetted for financial viability, commercial feasibility, and market acceptance. However, depending upon the stage of product development and number of cycles in the design thinking loop, team members typically have the most insights toward customer needs and ability to achieve results from a technical standpoint. Therefore, dot voting is a good indicator of potential longer term product success.

30.2.4.2 Focus Groups. Focus groups are a stalwart of traditional market research but can be extended to several other situations to evaluate concepts and features throughout the design thinking process. Like many other tools and techniques discussed in this chapter, focus groups can be used in different phases of design thinking (e.g., the identify stage and the solve stage). For instance, discovering customer needs using small groups and lead customer panels is one implementation of focus groups in the identify phase of design thinking. Certainly, evaluating high-fidelity prototypes with groups of users and non-users is another situation to apply focus groups in the solve stage.

Formally, focus groups are defined as a group of people convened to test a product under the guidance of a trained facilitator (Rodriguez 2020). Collaboration among the group is encouraged to gain further product understanding. Often a product, or competitor's product, is presented for use to the whole group or to each individual in the group. A typical practice is to record the focus group session so that the innovation team can later study both audio and video elements to broaden their insights into product usage.

Focus groups are usually small with no more than eight or ten participants. Today, focus groups are conducted via digital meeting rooms as well as in-person. If the focus group is held remotely, each individual will need to receive the physical prototype to be tested in advance of the session. Depending on the scale of the product, such as testing a tractor or industrial equipment, cost-savings are incurred when the focus group is convened at the company's development center instead.

It is important to guide the focus group with a set of pre-planned questions and activities. A trained moderator, or facilitator, is also essential to successful outcomes in a focus group. The facilitator is typically not vested in the company or in the product ideas and can remain neutral throughout the session. Trained moderators are also able to navigate new ideas or to rein in the group as appropriate. Normally, multiple focus groups are planned with a variety of participants (users, non-users, etc.) to gain the most information. Moreover, focus group participants often create new uses for products or reveal alternate solutions pathways to a problem when they interact with a product or prototype. The focus group facilitator should investigate these alternatives and document them for next generation product ideas.

For example, a food and beverage client of the author discovered an alternate new product during focus group testing. Instead of offering a dip for snacks, a focus group participant suggested a seasoning salt. This simple suggestion improved product handling throughout the supply chain and reduced manufacturing costs.

While focus groups are often considered a conventional market testing tool, the innovation team can tap into a number of creative opportunities by closely observing the participants. For instance, a focus group on wine corks revealed that elderly participants had difficulty to obtain the necessary physical strength to pull the cork. When a synthetic cork was substituted for natural cork material, the physical challenges were lessened.

30.2.4.3 3D Printing. Another tool that crosses the line between create and evaluate is **3D Printing**. Designers want to test the form and function of products before expensive production or tooling. With 3D printing, an inexpensive, non-working prototype is produced that allows internal and external testing and evaluation. Especially for tangible products with long-lead times or complicated manufacturing constraints, 3D printing helps the innovation team evaluate feature placement and other product aesthetics.

For example, a client of the author's manufactures kitchen appliances. Due to supply chain issues during COVID, prototypes from Asia were significantly delayed. Rather than postponing the product launch or assembling expensive full metal dies for molding of the appliance base, the company began using 3D printing to produce a variety of prototype bases. Placement of switches, levers, and dials are now evaluated on the three-dimensional, non-working models with much better clarity than on a sketch or CAD drawing. This further saved costs of full-scale, imported prototypes in future product development efforts.

Industrial designers and engineers can easily adapt 3D printed models to interface with final manufactured tools and dies. 3D printing enhances creativity by generating very inexpensive prototypes. Some of these concepts may push the boundaries of expected designs and can introduce radical innovation. When innovation teams and potential customers can hold and handle a 3D form model, these disruptive designs are easier to understand and accept then when presented as a 2D sketch. Therefore, 3D printing can offer a competitive advantage in product development.

30.2.4.4 Assumptions Testing. Similar to testing a prototype, **assumptions testing** is designed to eliminate the highest risk ideas quickly (Mattimore 2012). New ideas, features, and concepts are ranked using any of the mentioned tools (such as affinity diagrams or dot voting). However, instead of prioritizing on the *best* idea, the selection ranks the function with the most technical or market challenges. It is this function that is tested internally with the innovation team or externally with a focus group of interested potential users. This is also a great technique to combine with the Wizard of Oz tool.

Suppose that in the design thinking process, a software team believes the best idea to move forward is a subscription model. Rather than generating code and creating a functioning prototype, the idea can be tested by simply surveying potential customers in the target market. Using a specific solution statement that addresses the needs discovered early in the design thinking process, the innovation team might ask, "Would you pay $30 per month to subscribe to this new service?" If a sufficient percentage of potential customers replies affirmatively, the team will continue to create prototypes and iterate within the solve stage.

However, if the response is negative, the team cycles back to the create step and generates a new prototype, perhaps a different set of features with a lower subscription price. Since the design thinking process is iterative, any information generated in the evaluation stage provides valuable insights for the other steps. An element of caution is needed, however, to determine the number of recycle loops or failed prototypes before the new product is deemed unworkable. Be sure to continually test the problem statement defined in the identify stage again the prototype concepts.

A summary of the creativity tools recommended for each phase of the design thinking process is included in the appendix.

30.3 Part 3 – Designing a Creative Space

Creativity is a skill that requires nurturing and growth as does any business or innovation skill. Many organizations today recognize the need for innovation and have appointed executives as Chief Innovation Officers. Organizations should also encourage creativity through collaboration and structured "play." To achieve this end, designing an appropriate **creative space** allows individuals and teams to generate more and better new product ideas.

Creative spaces are uncluttered yet supply a lot of different materials for the innovation teams to use creatively. Paper of all sizes and colored pens, markers, and pencils are necessary for sketching and recording ideas. Sticky notes and large whiteboards allow innovation teams to brainstorm ideas with a high degree of flexibility. Comfortable furniture, reference books, and sample products should be available to the team. Simple materials, such as magazines, tape, and scissors, are provided to teams to contemplate current trends and for tools such as *Collage* and *Looks Like ...* as discussed previously. Small snacks and beverages are frequently located in a central location for the innovation teams using the creative space as they may become engrossed in an activity for hours. Fidget toys are also useful to stimulate embodiment of designs, activity, and team energy.

Additionally, creative spaces should be isolated from the primary working areas of the company. Product development teams might get noisy as they are act out product ideas or debate key design points. Moreover, when people are in a new or different environment, creativity is inspired. The room should have a higher-than-average number of electrical outlets, computer stations, and printers. Moveable desks, chairs, and whiteboards are used to subdivide the room when individuals collaborate in smaller groups or need to focus on different elements of the product design and development. In short, the creative space should be inspiring, colorful, and fun.

For remote and hybrid teams, at least one face-to-face kick-off meeting is encouraged (Hardenbrook and Jurgens-Kowal 2018). Remote teams make extensive use of online whiteboards, breakout rooms, and other collaboration software tools. While in-person teams might spend a day or a week in a creative space together working through the steps in a design thinking model, remote and hybrid teams are unable to do so.

Typically, online creativity sessions should be limited between two to four hours with multiple breaks. A 15-minute break every hour and half to two hours allows participants to check email, stretch, get a snack, and check on other personal matters. Including scavenger hunts, contests, and small games during the online sessions and breaks encourages new creative thoughts. In facilitating innovation sessions for remote clients, the author often uses one 15-minute break for team members to find and display (on their video camera) a company logo item (such as a hat or pen). These activities simultaneously enhance collaboration and creativity, and people report having more fun in the online meetings!

30.4 Summary

Everyone is creative. Structured tools stimulate creativity to support new product development. The design thinking process guides an innovation team with both steps and tools to move from discovery and problem definition to creating and evaluating potential product solutions. Continuous customer interaction and feedback validate assumptions of the innovation team, providing speed-to-market and high levels of customer satisfaction. Creativity is at the heart of all we do as innovation professionals.

References

Clifford, C. (2019 July 28). *Bill Gates took solo 'think weeks' in a cabin in the woods – why it's a great strategy.* Retrieved April 12, 2022, from cnbc.com: https://www.cnbc.com/2019/07/26/bill-gates-took-solo-think-weeks-in-a-cabin-in-the-woods.html#:~:text=As%20the%20boss%20of%20Microsoft%2C%20Bill%20Gates%20would,It%20was%20what%20he%20called%20his%20%E2%80%9CThink%20Week.%E2%80%9D.

Five Whys and Five Hows. (n.d.). Retrieved June 29, 2022, from ASQ.org: https://asq.org/quality-resources/five-whys.

Gross, O., Toivonen, H., Toivanon, J.M., and Valitutti, A. (2012). Lexical creativity from word associations. *Seventh International Conference on Knowledge, Information, and Creativity Support Systems,* 35–42. IEE Computer Society.

Hardenbrook, D. and Jurgens-Kowal, T. (2018). Bridging communication gaps in virtual teams. In: *Leveraging Constraints for Innovation,* Vol. PDMA Essentials Volume 3 (ed. S. Gurtner, J. Spanjol, and A. Griffin), 95–117. Hoboken, NJ: Wiley.

Jurgens-Kowal, T. (2022 August 8). *Creativity in the Back-End of Innovation.* GNPS Press AIN B0B9228JK4.

Knapp, J. (2016). *Sprint.* New York: Simon & Schuster.

Luchs, M.G. (2016). A brief introduction to design thinking. In: *New Product Development Essentials: Design Thinking* (ed. M.G. Luchs, K. Swan, and A. Griffin), 1–11. Hoboken, NJ: Wiley.

Markov, S. (2019, May 31). *Free association – creative tecnique.* Retrieved March 10, 2022, from Genvive: https://geniusrevive.com/en/free-association-creative-technique.

Mattimore, B.W. (2012). *Idea Stormers: How to Lead and Inspire Creative Breakthroughs.* San Francisco, CA: Jossy-Bass.

Rodriguez, C.M. (2020). Product design and development tools. In: *Product Development and Management Body of Knowledge,* 2e (ed. A. Anderson and T. Jurgens-Kowal), 121–162. The Product Development and Management Association.

Stillman, J. (2014, October 1). *The 4 stages of creativity.* Retrieved March 7, 2022, from Inc. com: https://www.inc.com/jessica-stillman/the-4-stages-of-creativity.html.

Zarroli, J. (2020, September 24). *When everybody's working at home and the magic is gone.* Retrieved August 4, 2022, from npr.org: https://www.npr.org/2020/09/24/916211900/as-more-americans-work-from-home-some-ceos-reopen-offices-to-find-that-missing-s#:~:text=JPMorgan%20Chase%20CEO%20Jamie%20Dimon%20calls%20it%20%22creative,coming%20up%20with%20innovative%20ways%20to%20address%20pro.

Teresa Jurgens-Kowal is a speaker, writer, and facilitator for innovation leadership. She helps organizations improve time-to-market for new products by delivering enhancements across the innovation ecosystem. Teresa has worked with innovation teams around the world in a variety of industries, including food and beverage, medical devices, publishing, petrochemicals, non-profits, and more. She is the author of *The Innovation ANSWER Book,* currently in the 2nd edition and the companion book, *The Innovation QUESTION Book.* Teresa founded Global NP Solutions in 2009, specializing in product development consulting and product management coaching. She is NPDP certified and has been a long-time supporter of PDMA in a variety of roles.

APPENDIX

APPENDIX: CREATIVITY TOOLS

Discover	Define	Create	Evaluate
Word Association	Interview	Looks Like...	Dot Voting
Mind-Mapping	Five Whys	What Can It Not Do?	Focus Group
Brainwriting	Card Sorting	Sketch	3D Printing
Collage	Customer Journey Map	Paper Prototype	Assumptions Testing
SCAMPER	Affinity Diagrams	Wizard of Oz	
Tell Me About ...			

CHAPTER THIRTY-ONE

FORECASTING NEW PRODUCTS

Kenneth B. Kahn

31.1 Introduction

Forecasting is an elemental part of the new product development process because most, if not all, go/no-go decisions during the process require some kind of forecast on which to base these decisions. However, many companies tend to overlook the requirements leading to better new product forecasting and instead simply accept as fact that any new product forecast will be characteristically inaccurate due to uncertainties related to market acceptance, technical feasibility, and company capability to bring the new product to fruition. Successful new product forecasting is possible if the respective company has the proper understanding. Indeed, those companies successful at new product forecasting manifest such success by establishing forecasting parameters; applying appropriate techniques; using a cross-functional, systematic process; and focusing more on generating meaningful new product forecasts versus a preoccupation with accuracy. New product forecasting cannot be avoided nor ignored.

In the present chapter, the topic of new product forecasting is outlined along with how a company may achieve a better, more meaningful new product forecast. The chapter begins by establishing and defining the parameters of the new product forecasting effort. Next, the linkage between new product forecasting and the product development process is discussed. Various techniques available and considerations surrounding the decision to use a particular technique versus another are then presented. The chapter concludes with new product forecasting strategies and considerations surrounding new product forecast accuracy and a new product forecasting process.

The PDMA Handbook of Innovation and New Product Development, Fourth Edition. Edited by Ludwig Bstieler and Charles H. Noble.

31.2 Forecasting Parameters

Successful new product forecasting requires establishing forecasting parameters to set a scope of work. Among the more important parameters, clear definition is needed around the forecasting objective, type of forecast desired, forecasting level, forecasting time horizon, forecasting interval, and form in which the forecast will be provided. Table 31.1 summarizes these parameters.

The forecasting objective should be determined outright to clarify the questions that need to be answered. The type of forecast to be made, the forecasting level, the forecasting time horizon, the forecasting interval, and the forecasting form can then be properly derived. The forecasting objective thus serves to clarify the purpose and intent of the forecast so that a meaningful forecast can be made – meaningful in the sense that the forecast is presented in a usable, understandable form and addresses the needs of the company. Without a clear objective, an innumerable set of forecasts can be developed, leading to confusion over which forecast(s) should be employed. For example, a manufacturing company may wish to emphasize superior customer service at reasonable cost and initiate a corresponding forecasting objective. New product forecasts would align with this objective in that the type of forecast, forecasting level, forecasting time horizon, forecasting interval, and forecasting form, all would be oriented to help the company drive customer service and maintain/reduce cost.

Choosing the type of forecast to be generated is a second important consideration. Several types of new product forecasts are possible and can be broken down in terms of *potential* versus *forecast*, and *market* versus *sales*. *Potential* represents a maximum attainable estimate, whereas *forecast* represents a likely attainable estimate. *Market* represents all companies within a given industry

TABLE 31.1 PARAMETERS TO CONSIDER WHEN BEGINNING TO FORECAST A NEW PRODUCT.

Forecasting Parameter	Examples
Forecasting objective	- Enable the company to maintain superior customer service at reasonable cost
	- Help the company grow market share through competitive advantage
	- Provide market insights that allow the company to remain a technology leader
Type of forecast desired	Market potential, sales potential, market forecast, sales forecast
Forecasting level	Industry, company, strategic business unit, product line, stock keeping unit (SKU), stock keeping unit per location (SKUL)
Forecasting time horizon	One year, two years, five years, ten years
Forecast interval	Weekly, monthly, quarterly, annually
Forecasting form	Financial ($, €, £, ¥), unit volume, market share

marketplace, whereas *sales* pertains to only the respective focal company. The following new product forecast definitions are provided:

- *Market Potential*: the maximum estimate of total market volume reasonably attainable under a given set of conditions.
- *Sales Potential*: the maximum estimate of company sales reasonably attainable within a given market under a given set of conditions.
- *Market Forecast*: a reasonable estimate of market volume attainable by firms in that market under a given set of conditions.
- *Sales Forecast*: a reasonable estimate of company sales attainable within a given market under a given set of conditions.

During the new product forecasting effort, one or all of the above may be of interest. The key is to clarify through the objective what is needed for decision-making at particular points during the new product development (NPD) process.

Once what is to be forecast is established, the forecast needs to be further defined in terms of level, time horizon, interval, and form. *Forecasting Level* refers the focal point in the corporate hierarchy where the forecast applies. Common levels include the stock keeping unit (SKU) level, stock keeping unit per location (SKUL) level, product line, strategic business unit (SBU) level, company level, and/or industry level. *Forecasting Time Horizon* refers to the time frame for how far out into the future one should forecast. New product forecasts could correspond to a single point in the future or a series of forecasts extending out for a length of time (the latter is more common). Examples include a one- to two-year time horizon, which is typical for most fashion products; two to five years for most consumer product goods; and ten plus years for pharmaceutical products. One reason for the longer time horizon in the case of pharmaceuticals is the consideration toward patented technology and length of term for the respective patent. *Forecasting Time Interval* refers to the granularity of the forecast with respect to the time bucket as well as how often the forecast might be updated. For example, a series of forecasts can be provided on a weekly, monthly, quarterly, or annual basis. *Forecasting Form* refers to the unit of measure for the forecast. Typically, new product forecasts early on are provided in a monetary form (e.g., $, €, £, ¥) or market share and later provided in terms of unit volume for production purposes. Some new product forecasts also can be in the form of narrative scenarios that describe a future event.

31.3 Forecasts during the New Product Development Process

Because decisions differ across the different stages of the new product development process, forecasts too will differ to support proper decision-making across product development process stages. Early in the process the forecasting focus will be market potential. Such forecasts are normally in dollars and used to answer the question of whether this is a good opportunity to pursue? Marketing and finance departments would play key roles in generating forecasts at this early stage.

During concept generation and pretechnical evaluation stages, forecasts investigate sales potential in answering the question of whether this is a good idea for our company to pursue? Again, forecasts at this point would be normally in the form of dollars, and under the auspices of marketing and finance departments.

Entering technical development and launch stages, unit sales forecasts would become critical in order to plan for the launch and ensure adequate supply through the channel. At this point, operations and supply chain department would play a key role in developing these sales forecasts to drive operational decisions. Specific testing like product testing during technical development and market testing as part of pre-launch activities would help to qualify key assumptions and better estimate unit demand and sales revenues from such unit demand.

31.4 Forecasting Techniques

One will find that there are numerous forecasting techniques available. Among the multiple ways in which to categorize these techniques, one way is to organize new product forecasting techniques into the three categories of judgmental techniques, quantitative techniques, and customer/market research techniques. Albeit showing only a sample of techniques, Figure 31.1 presents the more popular techniques associated with each of these three categories.

Judgmental techniques represent techniques that attempt to turn experience, judgments, and intuition into formal forecasts. Six popular techniques within this category include Jury of Executive Opinion, Sales Force Composite, Scenario Analysis, Delphi Method, Decision Trees, and Assumptions-Based Modeling (see Kahn 2006 for more information on these techniques):

FIGURE 31.1 A SAMPLE OF NEW PRODUCT FORECASTING TECHNIQUES.

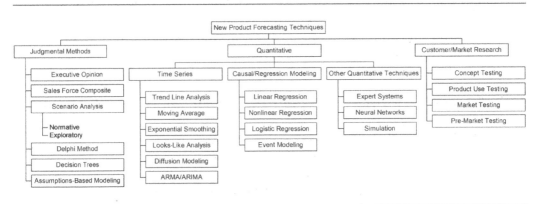

- *Jury of Executive Opinion*: a top-down forecasting technique where the forecast is arrived at through the ad hoc combination of opinions and predictions made by informed executives and experts.
- *Sales Force Composite*: a bottoms-up forecasting technique where individuals (typically salespeople) provide their forecasts. These forecasts are then aggregated to calculate a higher-level forecast.
- *Scenario Analysis*: an analysis involving the development of scenarios to predict the future. Two types of scenario analysis include exploratory and normative approaches. Exploratory scenario analysis starts in the present and moves out to the future based on current trends. Normative scenario analysis leaps out to the future and works back to determine what should be done to achieve what is expected to occur (see Wright et al. 2019).
- *Delphi Method*: a technique based on subjective expert opinion gathered through several structured anonymous rounds of data collection. Each successive round provides consolidated feedback to the respondents, and the forecast is further refined. The objective of the Delphi method is to capture the advantages of multiple experts in a committee, while minimizing the effects of social pressure to agree with the majority, ego pressure to stick with your original forecast despite new information, the influence of a repetitive argument, and the influence of a dominant individual (see Belton et al. 2019).
- *Decision Trees*: a probabilistic approach to forecasting where various contingencies and their associated probability of occurring are determined – typically in a subjective fashion. Conditional probabilities are then calculated, and the most probable events are identified. The example in Figure 31.2 shows two scenarios under consideration, option A and option B. A has two demand scenarios with their associated probabilities of occurrence; B has three demand scenarios with their associated probabilities. Using a decision tree approach, option A looks more attractive because the forecast for expected revenue is $2 million, versus no revenue in the case of option B.
- *Assumption-Based Modeling*: a technique that attempts to model the behavior of the relevant market environment by breaking the market down into market drivers. Then by assuming values for these drivers, forecasts are generated. These models are also referred to as chain models or market breakdown models. As illustrated in Figure 31.3, an assumptions-based model begins with an overall market size and uses these drivers to break down the market size proportionally (see Kahn 2007; Kahn and Mohan 2021).

Quantitative techniques are broken into the three subcategories of time series, "causal"/regression modeling, and other quantitative techniques. Time series techniques analyze sales data to detect historical "sales" patterns and construct a representative graph or formula to project sales into the future. Time series techniques used in association with new product forecasting include:

FIGURE 31.2 FORECASTING USING A DECISION TREE APPROACH.

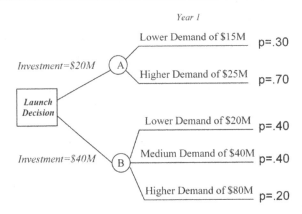

Expected Return After Year 1

E[A] = (.30*$15M) + (.70*$25M) - $20M = $22M - $20M = **$2M**

E[B] = (.40*$20M) + (.40*$40M) + (.20*$80M) - $40M = $40M - $40M = **$0**

FIGURE 31.3 GENERAL STRUCTURE OF AN ASSUMPTIONS-BASED FORECASTING.

- *Trend Line Analysis*: a line is fit to a set of data. This is done either graphically or mathematically (see Hyndman and Athanasopoulos 2021).
- *Moving Average*: a technique which averages only a specified number of previous sales periods (see Hyndman and Athanasopoulos 2021).
- *Exponential Smoothing Techniques*: a set of techniques that develop forecasts by addressing the forecast components of level, trend, seasonality, and cycle. Weights or smoothing coefficients for each of these components are determined statistically and applied to "smooth" previous period information (see Hyndman and Athanasopoulos 2021).

- *Looks-Like Analysis (Analogous Forecasting)*: a technique that attempts to map sales of other products onto the product being forecast. Looks-Like Analysis is a popular technique applied to line extensions by using sales of previous product line introductions to profile sales of the new product. Figure 31.4 shows a product line's sales curves for the prior two product launches, proportioned by month across the initial 12 months of sales. An average sales curve would be extrapolated from these data and used to forecast sales for the next line extension (see Kahn 2006).

FIGURE 31.4 LOOKS LIKE ANALYSIS.

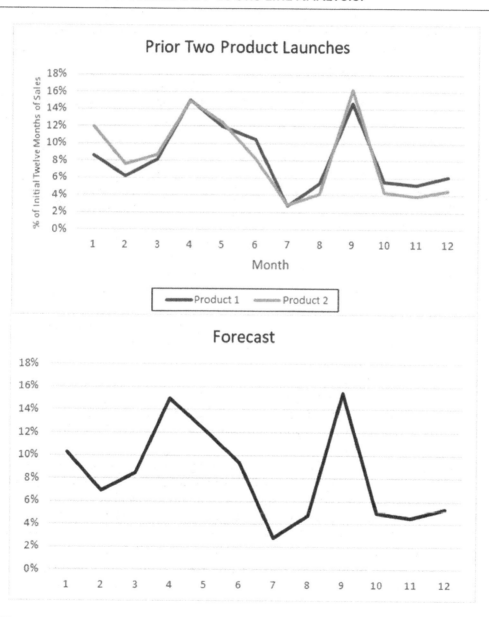

- *Diffusion Models*: models that estimate the growth rate of product sales by considering various factors influencing the consumer adoption process. Considerations taken into account include the rate at which mass media (the coefficient of innovation) and word of mouth (the coefficient of imitation) affect lead users, early adopter, early majority, late majority, and laggard customer segments. Different types of diffusion models exist including the Bass Model, Gompertz Curve, and Logistic Curve are examples of popular diffusion models. Diffusion models are also referred to as technology S-curves (see Chandrasekaran and Tellis 2018; Mahajan et al. 2000).
- *Autoregressive Moving Average (ARMA)/Autoregressive Integrated Moving Average (ARIMA) Models*: a set of advanced statistical approaches to forecasting, which incorporate key elements of both time series and regression model building. Three basic activities (or stages) are considered: (1) identifying the model, (2) determining the model's parameters, and (3) testing/applying the model. Critical in using any of these techniques is understanding the concepts of autocorrelation and differencing. ARMA/ARIMA models are also referred to as Box-Jenkins Techniques (see Hyndman and Athanasopoulos 2021).

"Causal"/regression modeling techniques use exogenous or independent variables and through statistical methods, develop formula correlating these with a dependent variable. The term "causal" is very loosely used because these models are predicated on correlational relationships and not true cause and effect relationships. Four popular techniques within this subcategory include:

- *Linear Regression*: a statistical methodology that assesses the relation between one or more managerial variables and a dependent variable (sales), strictly assuming that these relationships are linear in nature. For example, price may be an important driver of new product sales. The relationship between price and the quantity sold would be determined from prior data of other products within the product line and then used to predict sales for the forthcoming product (see Hair et al. 2018).
- *Nonlinear Regression*: a statistical methodology that assesses the relation between one or more managerial variables and a dependent variable (sales), but these relationships are NOT necessarily assumed to be linear in nature (see Hair et al. 2018).
- *Logistic Regression*: a statistical methodology that assesses the relation between one or more managerial variables and a binary outcome, such as purchase versus non-purchase. A logistic regression model calculates the probability of an event occurring or not occurring (see Hair et al. 2018).
- *Event Modeling*: often a regression-based methodology that assess the relation between one or more events, whether company-initiated or non-affiliated with the company, and a dependent variable (sales). For example, a promotion used with prior product launches would be analyzed and the bump in

sales cause by this promotion statistically determined. The expected bump in sales would be correspondingly mapped to the sales of the new product.

The other category contains those techniques which employ unique methodologies or represent a hybrid of time series and regression techniques. A sample of these forecasting techniques include:

- *Expert Systems*: typically, computer-based heuristics or rules for forecasting. These rules are determined by interviewing forecasting experts and then constructing "if-then" statements. Forecasts are generated by going through various applicable "if-then" statements until all statements have been considered (see Liebowitz 1998).
- *Neural Networks/Artificial Intelligence*: advanced statistical models that attempt to decipher patterns in a particular sales time-series. These models can be time-consuming to build and difficult to explain. In most cases, these models are proprietary.
- Simulation: an approach to incorporate market forces into a decision model. "What-if" scenarios are then considered. Normally, simulation is computer-based. A typical simulation model is Monte Carlo simulation, which employs randomly generated events to drive the model and assess outcomes (for a detailed discussion of Monte Carlo simulation, see Thomopoulos 2013).

Customer/market research techniques include those approaches which collect data on the customer/market and then systematically analyze these data to draw inferences on which to make forecasts. Four general classes of customer/market research techniques include (see Kahn and Mohan 2020 for more information on the following techniques):

- *Concept Testing*: a process by which customers (current and/or potential customers) evaluate a new product concept and give their opinions on whether the concept is something that they might have interest in and would likely buy. The purpose of concept testing is to proof the new product concept.
- *Product Use Testing*: a process by which customers (current and/or potential customers) evaluate a product's functional characteristics and performance. The purpose of product use testing is to proof the product's function.
- *Market Testing*: a process by which targeted customers evaluate the marketing plan for a new product in a market setting. The purpose of market testing is to proof the proposed marketing plan and the "final" new product.
- *Pre-Market Testing*: a procedure that uses syndicated data and primary consumer research to estimate the sales potential of new product initiatives. Nielsen BASES is an example of a proprietary new product forecasting model associated with pre-market testing and commonly employed in the consumer products goods industry.

31.5 New Product Forecasting Strategy

While there are a number of forecasting techniques available, it is important to realize that not all of them are appropriate for every forecasting situation. Qualitative techniques are quite adaptable, but very time-consuming; they would not be appropriate in situations where a severe time constraint exists. Quantitative techniques require data and rely on the critical assumption that current data will correspond to future states; if this cannot be assumed, quantitative techniques would not be appropriate. Customer/market research tools are time-consuming and expensive to perform. Budget constraints could seriously hamper what degree of customer/market research may be applied. A toolbox approach is therefore recommended for applying new product forecasting techniques.

To assist in decisions related to new product forecasting, a variation of the Product-Market Matrix is tailored to reveal four new product forecasting situations (refer to Figure 31.5). Mapping market uncertainty and product technology uncertainty on the two dimensions of current and new reveals four cells, each of which is represented by one of the following new product forecasting strategies: sales analysis, life cycle analysis, customer and market analysis, and scenario analysis.

Sales analysis is associated with the situation of current market and current product technology, where the uncertainties of market and product technology are lowest. Cost reductions and product improvements would populate this cell. The nature of these products would mean that sales data would be available because the product has previously existed. Analysis would focus on looking for deviations and deflections in sales patterns based on previous cost reductions and improvements in the product. Quantitative techniques such as times series and regression could be quite useful and manifest objective forecasts.

Product life cycle analysis is associated with the situation of current market and new technology. Line extensions are associated with this cell and represent higher product technology uncertainty. Because of understanding with the

FIGURE 31.5 NEW PRODUCT FORECASTING STRATEGY.

	Product Technology	
	Current	New
Market — Current	*Cost Reductions / Product Improvements* **Sales Analysis**	*Line Extensions* **Product Line/ Life Cycle Analysis**
Market — New	*New Markets/New Uses* **Customer and Market Analysis**	*New-to-the Company/ New-to-the-World* **Scenario Analysis (What-If)**

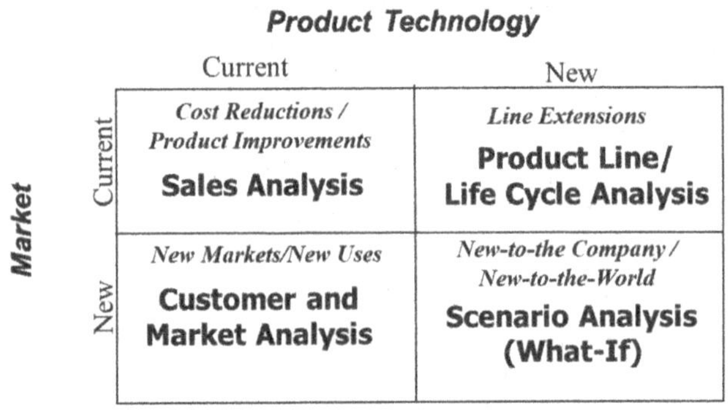

current marketplace, analyses would attempt to overlay patterns of previously launched products in the product line onto the new line extensions. These patterns would characterize a launch curve or life cycle curve by way of looks-like analysis or analogous forecasting.

Customer/market analysis would be necessary in the case of current technology and a new market due to higher market uncertainty. The purpose of this forecasting strategy would be to understand the new market and therefore reduce such uncertainty and manifest greater understanding about the new market. Various customer/market research studies might be engaged along with the use of assumption-based models in an attempt to specify market drivers, which would be validated by the customer/market research performed. Products in this cell would include new use and new market products.

A scenario analysis strategy would correspond to the situation of new market and new product technology, representing high market and product technology uncertainties akin to new-to-the-company (new category entries) and new-to-the-world products. Scenario analysis would be employed to paint a picture of the future and future directions to be taken. Note that a scenario analysis strategy should not be confused with just the use of scenario analysis; rather the intent of forecasting in this situation is to develop various scenarios on which to base the NPD decision. Given a lack of data, potential difficulty in identifying the specific target market, and questions regarding technology acceptance, subjective assessment techniques would play a major role here.

It should be recognized that forecasting techniques have applicability in each of the cells, depending on the specific situation. Customer/market research could greatly benefit market understanding related to cost reductions, product improvements, and line extensions. Subjective assessment techniques can be readily applied to all types of new products. The issue is to recognize the resources necessary and outcome desired. For example, subjective assessment techniques may not provide enough detail to forecast next year's sales of a product improvement. Hence, the intent of the proposed framework is to suggest a strategy that encompasses those techniques appearing to be most appropriate so as to facilitate the new product forecasting effort. In no way should techniques be viewed as exclusive to only those cells indicated.

31.6 New Product Forecasting Benchmarks

Even with a plethora of techniques and keen strategy, new product forecasting is characteristically associated with low accuracy (high forecast error). In fact, comparing results of research studies published in 2002 and 2018 indicate that new product forecasting accuracy has remained low. Referring to the more recent study by Kahn and Chase (2018) and as shown in Table 31.2, the overall average accuracy across the six types of new products is approximately 58%, with cost improvements generally 68% accurate; product improvement forecasts 68% accurate; line extension forecasts 60% accurate; new market forecasts

TABLE 31.2 % FORECAST ACCURACY.

Type of New Product	Accuracy (Kahn 2002)	Accuracy (Kahn and Chase 2018)
Cost Reduction	72%	68%
Product Improvement	65%	68%
Line Extension	63%	60%
New Use/New Market	54%	58%
New Category Entry	47%	51%
New-to the-World	40%	46%
Average Across Categories	57%	58%

Sources: Kahn 2002; Kahn and Chase 2018.

58% accurate; new category entry (new-to-the-company) forecasts 51%; and new-to-the-world products 46% accurate. Note that these mean values of new product forecasting accuracy were collected by asking respondents to indicate the average forecast accuracy achieved one year after launch (refer to Kahn and Chase 2018). The nature of these accuracies reaffirms that newer markets are more troublesome to forecast (i.e., new markets, new category entries, and new-to-the-world products), than those situations where a current market is being served (i.e., cost improvements, product improvements, line extensions).

Related to accuracy is the nature of the bias associated with new product forecasting. In other words, is there a greater tendency to under- or over-forecast? Research finds that new product forecasting is typically optimistic in nature due to the need to garner support and approvals for the respective NPD project. Such optimism corresponds to higher forecasts and over-forecasting, which if not properly managed, can lead to excess inventory and overstock situations post-launch.

In terms of process characteristics, benchmarking research suggests that in almost two-thirds of companies the marketing department has responsibility for the new product forecasting effort. Even if not responsible, the marketing department is heavily involved in the new product forecasting effort, with the departments of forecasting and sales also having an appreciable level of involvement in the new product forecasting effort. In terms of technique usage, new product forecasting relies heavily on judgmental forecasting techniques and customer research; time series techniques and other statistical techniques are used, but to a lesser extent (see Kahn and Chase 2018).

31.7 The New Product Forecasting Process

While applying an appropriate forecasting technique will benefit the new product forecasting effort, there are further considerations. Various uncertainties inherent in the new product should be accounted for, including potential cannibalization effects and market penetration to be achieved, along with finding pertinent data and having the time to perform the necessary analyses to address these and other uncertainties. The way in which successful companies have done this is through a process perspective, specifically creating a structured, systematic new product forecasting process. Such a process builds on experiences from prior new product forecasts, cross-functional communication (especially with marketing), and customer feedback. Integrated within a process enables organizational learning and understanding on which to make a credible and realistic forecast.

Assumptions management is an important part of the new product forecasting process. The need exists to clearly specify assumptions and make them transparent so that there is company understanding of what underlies these assumptions. After launch, successful forecasting companies would implement tracking systems that closely monitor and control these assumptions to determine if forecasts are coming to fruition or whether a deviation is occurring. Transparency of assumptions is particularly valuable for clarifying whether the forecast is based on sound rationales or just optimism.

Companies successful at managing the new product forecasting endeavor also realize that new product forecasts should be range forecasts, not point forecasts. These ranges typically become more narrowed as the product approaches and enters the launch phase. For example, pessimistic, likely, and optimistic cases could be connected with the monitor and control of assumptions to determine which scenario is playing out.

Successful companies further embrace the notion of forecast meaningfulness versus forecast accuracy. Meaningfulness recognizes and accepts that error will exist. Doing so, the company focuses on developing various forecast scenarios, understanding the assumptions that underlie these scenarios, reaching agreement regarding the most likely scenario, and planning contingencies should other scenarios come to pass. Meaningfulness emphasizes usability of the forecast in helping the company plan properly around expected error. In other words, a forecast accuracy orientation implicitly seeks to avoid error and penalizes when error occurs; a forecast meaningfulness orientation accepts that error will occur and mandates that the company plan for and put in place contingencies to manage such error.

31.8 Summary

New product forecasting is certainly not easy, and there is no silver bullet when it comes to new product forecasting. However, companies who employ appropriate techniques coupled with a well-structured new product forecasting process show a greater propensity for new product forecast success. Techniques play the key role of establishing a sound initial baseline forecast. The new product forecasting process then refines and augments this baseline forecast. Together these elements help to manifest the best possible, most meaningful new product forecast on which to support new product development decision-making and eventual new product launches.

References

Belton, I., MacDonald, A., Wright, G., and Hamlin, I. (2019). Improving the practical application of the Delphi method in group-based judgment: a six-step prescription for a well-founded and defensible process. *Technological Forecasting and Social Change* 147: 72–82.

Chandrasekaran, D. and Tellis, G.J. (2018). A summary and review of new product diffusion models and key findings. In: *Handbook of Research on New Product Development* (ed. P.N. Golder and D. Mitra), 291–312. Northampton, MA: Edward Elgar Publishing, Inc.

Hair, J.F., Babin, B.J., Anderson, R.E., and Black, W.C. (2018). *Multivariate Data Analysis*, 8e. Andover, Hampshire, UK: Cengage Learning.

Hyndman, R. and Athanasopoulos, G. (2021). *Forecasting: Principles and Practice*, 3e. OTexts. https://otexts.com/fpp3.

Kahn, K.B. (2002). An exploratory investigation of new product forecasting practices. *Journal of Product Innovation Management* 19: 133–143.

Kahn, K.B. (2006). *New Product Forecasting: An Applied Approach*. Armonk, NY: M.E. Sharpe.

Kahn, K.B. (2007). Using assumptions-based models to forecast new product introductions. In: *The PDMA Toolbook 3 for New Product Development*, 257–272. ed. A. Griffin and S. Somermeyer, Hoboken, NJ: John Wiley and Sons.

Kahn, K.B. and Chase, C.W. (2018). The state of new-product forecasting. *Foresight: The International Journal of Applied Forecasting* 51 (Fall): 24–31.

Kahn, K.B. and Mohan, M. (2021). *Innovation and New Product Planning*. New York, NY: Routledge.

Liebowitz, J. (1998). *The Handbook of Applied Expert Systems*. Boca Raton, Florida: CRC Press.

Mahajan, V., Muller, E., and Wind, Y. (2000). *New-Product Diffusion Models*, International Series in Quantitative Marketing Volume 11. Boston, MA: Kluwer Academic Publishers.

Thomopoulos, N.T. (2013). *Essentials of Monte Carlo Simulation: Statistical Methods for Building Simulation Models*. New York, NY: Springer.

Wright, G., Cairns, G., O'Brien, F.A., and Goodwin, P. (2019). Scenario analysis to support decision making in addressing wicked problems: pitfalls and potential. *European Journal of Operational Research* 278 (1): 3–19.

Kenneth B. Kahn, PhD, is a Professor and Dean at Old Dominion University's Strome College of Business. His teaching and research interests include product innovation, product management and demand forecasting of current and new products. Dr. Kahn has published over 50 articles, authored the books *Product Planning Essentials* (Sage Publications, 2000; M.E. Sharpe, 2012), *New Product Forecasting: An Applied Approach* (M.E. Sharpe, 2006), *Innovation and New Product Planning* (Routledge, 2021); and served as editor of the 2nd and 3rd editions of the *PDMA Handbook on New Product Development* (Wiley & Sons, 2004, 2013).

CHAPTER THIRTY-TWO

A PRACTICAL GUIDE TO FACILITATING A DESIGN THINKING WORKSHOP

Wayne Fisher

This chapter outlines a practical approach to facilitating teams through the Design Thinking process to develop new products and services. PDMA's comprehensive handbook on Design Thinking (Luchs et al. 2015) clearly outlines the use of Design Thinking in implementing a systematic, more creative, and human-centered approach to innovation. This chapter assumes a basic understanding of Design Thinking principles and focuses on the role of the Design Thinking facilitator in the days leading up to, during, and following an immersive Design Thinking workshop. These best practices were developed by the author during his career at Procter & Gamble and are primarily within the context of the Consumer Products industry. However, these best practices have held up well in over 10 years of consulting practice in Business-to-Business, Health Care, and Non-Profit organizations.

This chapter describes:

- A typical design and commonly used tools for a Design Thinking workshop, including problem definition, idea generation, and idea selection/ evaluation
- Techniques for actively engaging target customers throughout the Design Thinking process
- Examples of how Design Thinking was used to solve a broad range of business challenges

The PDMA Handbook of Innovation and New Product Development, Fourth Edition. Edited by Ludwig Bstieler and Charles H. Noble.

32.1 Design Thinking and Innovation

One helpful definition of innovation is a continuous cycle of proactively seeking important new problems to solve, identifying new approaches to solve existing problems, and finding new ways to implement business-building ideas. This definition reinforces the critical role of the Creative Problem Solving (CPS) process in innovation.

Min Basadur's Simplexity™ Framework (Figure 32.1) is an innovative thinking and creative problem solving process that separates innovation into clearly defined steps, to take you from initial problem-finding right through to implementing the solutions you've created. The process has three distinct phases – Problem Formulation, Solution Finding, Planning and Execution (Basadur 2021).

One helpful framing of Design Thinking is actively involving target customers in the execution of a Creative Problem Solving workshop. Specifically, customers are involved in three of the eight steps shown in Figure 32.1:

- **Step 2 – Fact Finding**: Customer interviews and observational studies are critical to engender deep empathy and uncover important unmet needs. This is a cornerstone principle of the human-centered design approach.

FIGURE 32.1 THE CREATIVE PROBLEM-SOLVING PROCESS.

- **Step 3 – Problem Definition**: Problems are defined and articulated from the customer's point of view. Top problems are identified that most closely align with the insights derived from fact finding.
- **Step 5 – Idea Evaluation and Selection**: A. G. Lafley led the revitalization of Procter & Gamble in the 2000s in part with his mantra, "consumer is boss" (Lafley and Charan 2008). Nowhere was this more apparent than during Design Thinking workshops, where consumers were brought back into the studio to provide immediate feedback and idea strengthening suggestions while workshop sponsors observed.

32.2 Lather, Rinse, Repeat – The 2-DAY Design Thinking Workshop

Hundreds of design thinking workshops have been held at The GYM at Procter & Gamble (Anthony 2011). The majority of these workshops followed a repeatable design and flow as shown in Figure 32.2. Each of these steps will be described in more detail in the balance of the chapter.

Day 1 begins with a brief Sponsor kick-off and declaration of the workshop objective. Next the Problem Owner gives a 30–60 minute overview of the project goals, target customers, enabling technologies, and other key project details. The balance of the morning focuses on conducting and documenting in-context customer interviews. After lunch, the interviewees share their findings with the broader team and the team works together to explore possible problem statements. In general, it is best to have individuals write problem statements silently before sharing with the group to leverage the diverse perspectives of the participants. The group then converges on the top problem statements before brainstorming possible solutions (Ideation Round 1).

Day 2 begins with two rounds of ideation using advanced idea generation tools from the Creative Problem Solving toolbox. Before lunch, the team agrees

FIGURE 32.2 TYPICAL PRODUCT INNOVATION WORKSHOP FLOW.

Day 1	Day 2
Session Objective Project Background Customer Interviews	Ideation Round 2 Ideation Round 3 Top Ideas
Customer Storytelling Problem Definition Ideation Round 1	Rapid Prototyping Customer Feedback Action Planning

on the top ideas for prototyping prior to sharing with customers for evaluation and strengthening. After lunch, the top ideas are brought to life using 2D or 3D prototyping. Then, the customers are brought into the workshop space to give their feedback on the ideas. Based on the customer feedback, the solving team converges further and decides on the best ideas to explore further in qualitative testing with a larger pool of customers.

Here are some statistics for a typical Design Thinking workshop at The GYM:

- 20 workshop participants
- 10 consumers, interviewed in pairs
- 200 highly granular problem statements resulting from the consumer interviews
- 5 key problem themes leading to 10 top problem statements
- 200 ideas from three rounds of idea generation
- 20 top ideas for 2D/3D prototyping for consumer feedback
- 5 refined concepts for further testing
- Rough 90-day action plan to maintain momentum coming out of the workshop.

The most common Design Thinking application was to build the product pipeline for existing billion dollar brands. Other applications that benefited from the two-day design include:

- Designing products that leverage a breakthrough technology under development
- Creating affordable products for low income markets
- Improving supply chain efficiencies with key customers and suppliers
- Streamlining internal innovation and business work processes
- Designing a company intranet that enables both occasional and power users to find the information they need.

32.3 Pre-workshop Preparation

32.3.1 Assembling the Solving Team

The first step in planning a Design Thinking workshop is to assemble a solving team. The key players include:

- The **Problem Sponsor,** who helps define a clear workshop objective aligned with the organization's stated business and innovation goals. The Sponsor is also responsible for securing the time/money/resources needed to execute the workshop and evaluate the ideas that emerge during the workshop through qualitative testing.

- The **Problem Owner,** who is primarily responsible for ensuring that any product or service ideas from the workshop are aligned with the stated workshop objectives. The Owner also typically leads the post-workshop qualitative evaluation of top ideas.
- The **Planning Team**, who are responsible for designing the actual design thinking experience. The Planning Team usually consists of an experienced Facilitator, the Problem Owner, and a few core members of the project team.
- The **Solving Team**, which can include both project team members and external "creatives." Their role is to conduct customer interviews, define their needs, generate ideas and prototypes, and conduct customer evaluations.
- The **Customers**, who represent the (hypothesized) target audience for the product or service offering under consideration.
- The **Facilitator**, whose job is to ensure that the team has the tools and training needed to execute the workshop as planned. An experienced facilitator will also know when the workshop design needs to be adjusted to address challenges and learnings in real time. Importantly, the Facilitator is leading the Design Thinking process, but is not actively participating in the process.

32.3.2 Setting the Workshop Level of Ambition

The next step in planning a Design Thinking workshop is to establish the sponsor's Level of Ambition for the engagement. Simply stated, how much of a change is desired and how quickly is the change needed? This corresponds to "Step 1 – Opportunity Finding" in the Creative Problem Solving process.

For example, imagine you are the manager of a popular Starbucks location close to a college campus. You would like to grow year-on-year sales, of course, but what do you imagine is the biggest opportunity area for growth? Encouraging more consumption from existing customers represents a modest Level of Ambition, while attracting new customers (especially non-coffee drinkers) would represent a higher Level of Ambition. It is also important to have some sense of how urgently any solution would need to be implemented. Is it important to start driving sales in the next 6–12 months, or will longer term solutions be considered? Like every step in the Creative Problem Solving process, defining the workshop objective should be a deliberate divergent/convergent process. The Level of Ambition template from the Organizational Design for Innovation chapter is one helpful tool used by the Facilitator and Problem Owner to brainstorm a wide range of potential workshop objectives (long-term vs. short-term, breakthrough vs. incremental). Then, working with the Problem Sponsor, craft the final workshop objective statement to guide the workshop design and execution.

32.3.3 Preparing a Customer Interview Guide

The primary purpose of conducting customer interview guides during a Design Thinking workshop is to engender empathy, discover unmet needs, and identify new problems to be solved. A well-crafted interview guide is critical to uncovering these insights. Again, preparing an interview guide (Step 2 – Fact Finding) is a divergent/convergent process.

The Planning Team is primarily responsible for the development of an interview guide. The process begins with brainstorming a list of 20–30 potential interview questions based on the workshop objective. Next, the Planning Team identifies about 10 questions most likely to generate inspiring insights and problems to be solved. Finally, the questions are developed into an interview guide using the following flow:

- Easy "Warm-up" questions
- Narrow –> Broad –> Narrow –> Broad
- Reflection –> Projection
- Open ended "Wrap-up" questions

Warm-up questions are designed to relax the customer and set the broad context for the interview. For our Starbucks example, it would be ideal to meet the customer at the target location and conduct an in-context interview. In this case, example warm-up questions might be, "what brings you to Starbucks today?" and "what do you typically order when you come to Starbucks?"

Narrow –> Broad questions are used to transition between different topics for further exploration. Narrow questions have direct and concrete answers:

> *How often do you go to Starbucks?*
> *Is this the Starbucks you usually visit?*

Broad questions are open-ended and allow the customer to elaborate on the specific topic you just introduced. For example, if the customer above indicated that the target location is their favorite, you might ask questions like:

> *What do you like about this Starbucks location?*
> *Why do you usually come to Starbucks?*

Reflection –> Projection questions are used to help customers imagine a new future and understand what would motivate them to change current habits and beliefs. It is often difficult for people to imagine a better future without comparing it directly to something they have already experienced. Reflection questions help the customer recall past experiences and articulate likes and dislikes with the current offerings:

When did you start drinking coffee on a regular basis?
Do you drink more or less coffee than you did in the past?
Do you always order the same thing, or do you mix it up?

Reflection questions can be probed further to uncover likes and dislikes, but they are primarily set-ups for the Projection questions that imagine a better future:

Are there any items on our menu that you've never tried, but might like to?
Are there any new items you'd like to see on our menu?

Open ended **Wrap-up** questions signal that the interview is coming to a close and give permission to the customer to share whatever else came up for them during the interview. They give control of the conversation back to the customer to speak whatever is on their mind.

If there was one thing you could change about this Starbucks location, what would it be?
"Is there anything else I should have asked you about but didn't?

32.4 In-Workshop Facilitation

As shown in Figure 32.2, Day 1 of a typical Design Thinking workshop focuses on grounding the solving team on the business challenge being addressed, conducting in-context customer interviews, customer storytelling, customer problem definition, and capturing top-of-mind ideas. Day 2 focuses on advanced idea generation tools, rapid prototyping of top ideas, and obtaining immediate customer feedback to prioritize and strengthen top ideas for future study.

32.4.1 Session Objective and Project Background

Often, the solving team includes a core team of problem owners complemented by internal and external subject matter experts. In this case, it is helpful to ground the solving team in the session objective, ideal outcomes, and fit with the company's business and innovation strategy. Other examples of useful background for Solving Team members include:

- Target customer and known challenges
- Current product offerings vs. competitive products
- New product pipeline and lighthouse vision
- Market landscape and important trends

32.4.2 Customer Interviews

One of the most rewarding aspects of facilitating a Design Thinking workshop is sending out participants to conduct their first customer interview. They often feel some nervousness at first about conducting qualitative research ("isn't that someone else's job?"). However, they universally return from the interviews excited about what they learned and anxious to tell their customer's "story."

Here are some best practices for conducting customer interviews in the context of a Design Thinking workshop:

- Use a 2-on-1 interview format. One person leads the conversation while the other person listens and actively takes notes.
- Introduce yourselves briefly and explain why you're there. ("we're looking for ways to improve the customer experience when visiting our Starbucks location.")
- Confirm timing ("we'll be done by 2:00 pm, is that still ok?")
- Use the interview template as a discussion guide, not a checklist (modify/adapt as appropriate for your customer).
- Be conversational. Use your own natural voice and avoid industry jargon.
- Pay particular attention to body language, contradictions, compensating behaviors, pain points (often the role of the note taker).
- Allow for pauses; probe further with neutral questions, "How do you mean?"
- Avoid the temptation to "correct" or "problem solve." Simply note the fact that the customer is confused, frustrated, etc. – along with any compensating behaviors.
- Thank them for their time.

32.4.3 Customer Storytelling and Problem Definition

Empathy Maps, as illustrated in Figure 32.3, are commonly used to distill a 30–60 minute customer interview into 10–12 key insights to be shared with the solving team. These insights are then transformed into initial problem statements using the "How Might We" technique.

The left-hand side of the Empathy Map captures surprising quotes and behaviors uncovered during the interview. These should reflect words the interviewee literally said or actions literally described during the interview. The right hand side of the Empathy Map are often insights inferred from the conversation with the interviewee. "If they say and do these things, what must they be thinking? What emotions must they be experiencing?" They interviewee may tell you these things directly, or they may be elicited with probes like, "that must be frustrating …" If your interpretation is wrong, the interviewee will likely correct you and provide the insight you are seeking.

FIGURE 32.3 EMPATHY MAP TEMPLATE.

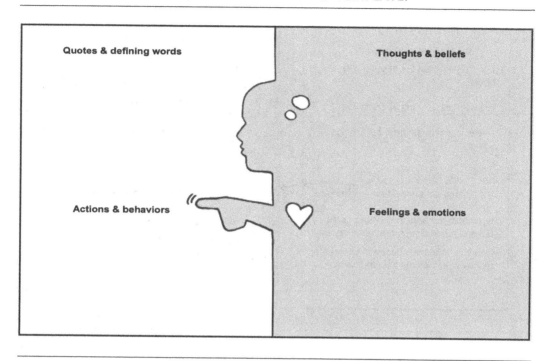

Let's look at an example Empathy Map from the Starbucks case study. Figure 32.4 is an Empathy Map from a conversation with "Alice," a frequent customer of Starbucks. The interesting thing to note is that Alice wasn't that interested in coffee! Her reasons for going to Starbucks were largely social, and new beverage offerings weren't likely to motivate her to visit more often or consume more per visit.

During Customer Storytelling, the interview pairs debrief their Empathy Maps and the facilitator encourages the solving team to capture the problems they are hearing on Post-it notes in the form of a "How Might We?" statement. Much has been written about the power of a How Might We statement to turn negative "facts" uncovered during the interview into more positively stated problem statements that stimulate idea generation (Berger 2012). Asking solving team members to capture How Might We statements silently leverages their unique perspective and uncovers a wider range of potential solution approaches.

Returning to the conversation with Alice, here are some potential How Might We (HMW) statements that the location manager may not have considered previously:

- HMW make our store location a more inviting "third place" where our customers feel immediately welcome?
- HMW help our customers (who want to) make meaningful connections with each other during their short visits with us?

FIGURE 32.4 EXAMPLE EMPATHY MAP FROM THE STARBUCKS CASE STUDY.

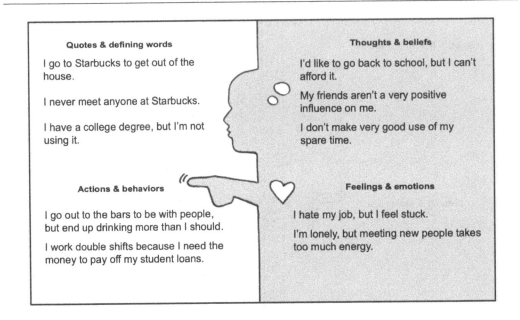

It's easy to imagine, with 10 customer stories and 20 solving team members sharing their individual HMWs, the total number of problem statements to consider will easily exceed 200. Affinity Diagrams, as illustrated in Figure 32.5, are

FIGURE 32.5 USING AFFINITY DIAGRAMS TO ORGANIZE HMW STATEMENTS.

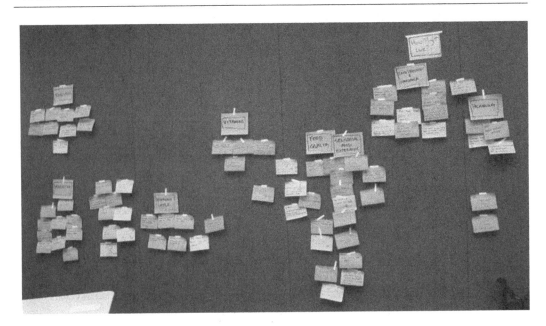

used to organize the individual problem statements into common themes uncovered during customer storytelling.

"Power Dotting" is one common technique for converging on top problems for further exploration and idea generation. Each member of the solving team is given five power dot stickers and asked to identify the individual Post-its that satisfy the "4I" convergence criteria:

- **Importance**: Will solving this HMW help address our workshop objective?
- **Influence**: Will we be able to implement potential solutions to this HMW?
- **Interest**: Are you personally interested in spending time and energy solving this HMW?
- **Imagination**: Does this HMW stimulate new ways of thinking about our workshop objective, or is it something we can simply execute today?

Inevitably, there will be overlap among the HMWs, and much time can be wasted debating which ones can be combined, etc. Far more productive is to simply have participants pair up and pick an individual HMW with many power dots that they have mutual passion for solving!

One tool that greatly facilitates the transition from problem definition to idea generation is the Why-Why-Why analysis, as illustrated in Figure 32.6. Working in pairs, participants explore their HMW by first imagining the ideal outcome if that HMW were completely solved. It's not important that the pair have a complete solution in mind (the HOW), but instead are describing for the other participants the characteristics of their ideal outcome (the WHAT). Comparing the current situation to their ideal outcome, the pair then identifies potential barriers currently preventing them from achieving the desired result. Finally, for the most important barriers, the pair hypothesize potential causes for each barrier.

FIGURE 32.6 USING THE WHY-WHY-WHY TECHNIQUE TO DIG DEEPER ON TOP PROBLEMS.

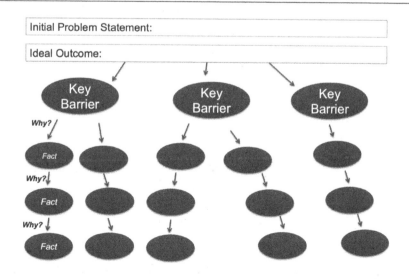

32.3.4 Idea Generation

As time permits, idea generation takes place in three rounds:

- Round 1 – Brainstorming
- Round 2 – Targeted Stimulus
- Round 3 – Non-Targeted Stimulus

Brainstorming naturally takes place during in-depth problem definition. Having an "idea catcher" board where participants can capture ideas throughout Day 1 helps free up brain space for exploring new ways of thinking.

When it is time for group brainstorming, here are some best practices to keep in mind:

- Have each pair share their detailed Why-Why-Why analysis and ask for help generating ideas to overcome their barriers and root causes (don't waste time debating the details of the analysis).
- Have the solving team members capture their starting ideas *silently* on Post-it notes before any ideas are shared with the group. The principle here is "write, then talk."
- Have the solving team members share their ideas and ask for help strengthening the idea further. As facilitator, vigorously shut down any criticism of the ideas as they are being shared.
- When all of the Why-Why-Why analyses have been reviewed, use the Affinity Diagram technique to group and theme the ideas.

Round 2 of idea generation uses Targeted Stimulus. One universally applicable technique is Innovation by Analogy (Markman 2012), shown in Figure 32.7.

FIGURE 32.7 INNOVATION BY ANALOGY IDEA GENERATION APPROACH.

2. Who Else Has This Problem?	3. How Have They Solved It?
1. What Problem Are You Trying To Solve?	4. How Might You Reapply Their Solution?

The mindset here is that "someone, somewhere, has already solved this problem." In our Starbucks case, one problem statement was "How Might We make our store location a more inviting 'third place' where our customers feel immediately welcome?" The Wikipedia (2022) article on Third Place is a treasure trove of best practices and resources for designing an inviting third place. Other potential sources of Targeted Stimulus include:

- A survey of relevant technologies readily available or under development
- Presentations by subject matter experts, bloggers, or influencers
- Presentations by trends experts or industry analysts

Round 3 of idea generation, time permitting, uses Non-Targeted Stimulus. Michael Michalko (2006) has assembled an extensive toolbox of creativity tools that are applicable to a Design Thinking workshop. Some techniques worth consideration include SCAMPER, Morphological Analysis, and a variety of forced connection techniques (Random Word, Random Picture, Famous Quotes).

As Rounds 2 and 3 of idea generation are completed, new ideas are added to the Affinity Diagram. Again, a typical Design Thinking workshop will generate over 200 "thought starters" that need to be narrowed down for prototyping and customer feedback. The same power dotting technique can be used to identify the top 20 or so ideas for further development.

32.4.5 Rapid Prototyping

Rapid prototyping is used to bring ideas to life quickly and inexpensively to facilitate meaningful interactions with customers during the customer feedback session. One universally applicable technique is the "60 Second Elevator Pitch" using the Idea Card template shown in Figure 32.8. The purpose of the Idea

FIGURE 32.8 60-SECOND ELEVATOR PITCH IDEA CARD TEMPLATE.

Ideas Start Here

Idea Name:

Problem It Addresses:

The Idea Is:

How it would work:

Card is to more fully describe the idea so the customer has a better understanding of the idea before providing feedback.

In practice, an Idea Card should be supplemented with one or more additional rapid prototyping techniques. Common techniques include:

- **Quick Sketches** – we often employ professional sketch artists who can quickly bring ideas to life as they hear them (analogous to caricature artists you might encounter at an amusement park).
- **Storyboarding** – a series of sketches or images used to describe a sequence of activities that might take place as the idea is executed.
- **Skits or Role Play** – used to emphasize new human interactions associated with a new product or service idea.
- **Low Resolution Prototypes** – especially helpful for those who don't feel comfortable with their artistic ability, 3D prototypes of an idea can be created using simple art supplies.
- **Flow Diagrams** – particularly useful for describing proposed new work processes or supply chain designs.

32.4.6 Customer Feedback

Reflecting on A. G. Lafley's mantra "consumer is boss," it is important to remind the solving team that their opinion of their own ideas doesn't matter. Only the customer's opinion matters. The most common approach for soliciting customer feedback is "Speed Dating." The top ideas are stationed around the room along with their prototypes. Customers visit each station in turn (usually five minutes per station) and share their feedback. After reviewing all of the ideas, the customers are invited to share their favorite ideas and why.

As the participants are sharing their ideas, some important principles to keep in mind:

- It's not about you, it's about the idea. Don't try to sell the idea, just present it.
- After sharing the idea and prototype, ask probing questions such as, "what did you like about this idea?", "what was unclear about the idea?", "what would you do to improve the idea further?", "how does this compare to similar products you might use today?"
- Avoid directly answering any questions from the customer. Questions reveal which part of the idea is unclear. For example, if a customer asks if a product or service idea provides a certain benefit, don't answer yes or no. Instead, ask probing questions such as, "what makes you unsure?" or "how might we make that benefit more clear to you?"

32.4.7 Action Planning

At this late stage of the workshop, participants will be mentally exhausted. Rapid fire Action Planning techniques are used to quickly identify key next steps to maintain momentum going forward. If more robust action planning is required, it is best to schedule a follow up session soon after the workshop.

Some rapid fire action planning techniques used with good success that take about 60 minutes to complete:

- **I Will/I Need Commitment Card** – the simplest technique, participants are handed a card with the words "I Will" on one side and "I Need" on the other. Participants take a few minutes reflecting on what they can do to continue delivering on the workshop objective and what help they need going forward.
- **What/Who/When Action Plan** – have participants brainstorm on all of the potential action steps going forward (the WHAT), prioritize and arrange in some logical order, and have individuals volunteer to complete the work by a certain date (WHO, WHEN).
- **Displayed Thinking/Immersive Critical Path Schedule** – this works well for multifunctional teams who need to coordinate their activities. Have the functional *subteams* identify their key milestones and deliverables, and estimate when they might be delivered. Create a giant wall chart as shown in Figure 32.9 and identify any potential conflicts that may need to be addressed in a future working session.

FIGURE 32.9 DISPLAYED THINKING/IMMERSIVE CRITICAL PATH SCHEDULE.

	January	February	March	April	May	June
Marketing	Milestone M-1		Milestone M-2			Milestone M-3
Finance		Milestone F-1			Milestone F-2	
Products Research	Milestone PR-1			Milestone PR-2		
Process Development		Milestone PD-1	Milestone PD-2			Milestone PD-3
Packaging			Milestone P-1		Milestone P-2	
Engineering			Milestone E-1	Milestone E-2		Milestone E-3

32.5 After the Workshop

Step 7 of the Creative Problem Solving is Acceptance Finding (Basadur 2021). There is not much written in the literature on this topic, but this is a critical step toward the commercialization of any new product or service idea. The Stakeholder Commitment Chart (see Figure 32.10) is a tool to be used with the workshop sponsor and planning team soon after the workshop. The key stakeholders are listed, along with the team's assessment of their current and required level of commitment. The inevitable mismatches reflect new problems to be solved by the project team. Reflecting on what motivates the key stakeholders (e.g., rewards) allows the team and their sponsor to think through how to overcome their likely objections.

It is also good to schedule a reconnect with the team in 2–3 months to check on progress and see if any additional facilitation help is required. Likewise, be sure to ask the team to keep you posted on progress toward commercialization.

32.6 In Summary

Design Thinking is a special case of Creative Problem Solving where target customers are actively engaged at critical points in the process. The Design Thinking skills step-up card shown in Figure 32.11 summarizes the work process for planning and executing a Design Thinking workshop. While this chapter provides some practical guidelines, facilitators should seek a Level 3 proficiency across all of these skill areas, or partner with colleagues who can step in and support key activities during workshops.

FIGURE 32.10 STAKEHOLDER COMMITMENT CHART.

Key Players	No Commitment	Let It Happen	Help It Happen	Make it Happen
A	X			→ O
B		X → O		
C			(X O)	
D		O ←		X
E	X		→ O	

Legend:
X = Present State
O = Desired State

FIGURE 32.11 DESIGN THINKING FACILITATOR STEP-UP CARD.

Design Thinking Facilitator Step-up Card

Proficiency Level:

1 - No capability

2 - Knows theory, can do with help

3 - Can do without help

4 - Very experienced, can coach and consult others on the subject matters.

5 - Expert in the subject matter, can design and implement new techniques

	Current Capability Assessment		Personal Capability Development Plan
	Current Level	Target Level	Action Step
Client Level of Ambition / Session Objective			
Workshop Design			
Consumer Discussion Guide			
Empathy Maps			
Insight Generation			
Effective Problem Definition			
Advanced Ideation Techniques			
Rapid Prototyping			
Concept Development			
Consumer Feedback			
Action Planning			

References

Anthony, S. (2011). Two rules for unleashing your company's innovation energy. https://hbr.org/2011/08/two-rules-for-unleashing-your.

Basadur, M. (2021). The power of innovation. http://www.basadur.com.

Berger, W. (2012). https://hbr.org/2012/09/the-secret-phrase-top-innovato.

Lafley, A.G. and Charan, R. (2008). The game-changer: how you can drive revenue and profit growth with innovation.

Luchs, M.G., Swan, S, and Griffin, A. (2015). Design thinking: new product development essentials from the PDMA.

Markman, A. (2012). Smart thinking.

Michalko, M. (2006). Thinkertoys: a handbook of creative-thinking techniques.

Wikipedia (2022). https://en.wikipedia.org/wiki/Third_place.

Wayne Fisher, PhD, founded Rockdale Innovation after spending 27 years as an R&D manager at Procter & Gamble. While at P&G, Wayne created a series of popular innovation workshops for all phases of new product development. He has trained thousands of managers, providing a common language and framework for innovation, fostering collaboration across P&G's diverse business units. He served as a full-time innovation consultant and Creative Problem Solving facilitator at The GYM, an IDEO-inspired design studio that remains a key enabler of P&G's innovation capability. Wayne serves on PDMA's Board of Directors and on the Outstanding Corporate Innovator award committee.

ABOUT THE PRODUCT DEVELOPMENT AND MANAGEMENT ASSOCIATION (PDMA)

The Product Development and Management Association (PDMA) is a global nonprofit professional association seeking to accelerate the contribution that innovation makes to the economic and professional growth of people, businesses, and societies around the world. Founded in 1976, the PDMA has grown to become recognized as the leading advocate for innovation and product development knowledge, practice, and community.

Innovation has become a buzzword in recent years. It appears in virtually every corporate vision statement and is touted as the key to success in virtually every annual report. PDMA takes a specific view of what constitutes innovation: it is the process of discovering, developing, and successfully delivering novel ideas for goods and services to the marketplace. The main purpose of the PDMA is to help bring more innovations to fruition by providing resources for professional development, information, collaboration, and promotion of new product development and management.

PDMA is a community of people passionate about developing innovative new goods and services, which combines the expertise of three groups:

- Academics – faculty and students who study, research, and report on new product development from a research perspective.
- Practitioners – professionals who work in firms of various sizes, from sole proprietorships to multinational corporations, trying to deliver new goods and services to the market.
- Solution Providers – firms and consultants who provide tools and/or guidance on how to manage innovation systems and new product development processes.

The PDMA Handbook of Innovation and New Product Development, Fourth Edition. Edited by Ludwig Bstieler and Charles H. Noble.
© 2023 John Wiley & Sons, Inc. Published 2023 by John Wiley & Sons, Inc.

PDMA brings these constituencies together into one community, where thousands of members can form, expand, test, modify, analyze, systematize, and evolve ideas and practices. Established best practices are collected in a body of knowledge (BoK) that is accessible to those who would benefit from it. Further, in the rapidly evolving and highly competitive environment that is the global marketplace, PDMA alerts members to *next practices* in innovation, even before they can be established as *best practices*. PDMA helps product developers and managers to stay informed about what works to accelerate innovation, what acts to inhibit it, and what emerging ideas might lead to the next breakthrough. The PDMA aims to capture, organize, codify, and disseminate such knowledge to people, businesses, and societies around the world that seek to grow professionally and economically through new product and service development.

The development of new goods and services involves an integrated set of unique activities. The PDMA is the only organization that focuses on addressing this challenge by providing:

- Communities of innovators through local chapters
- International conferences
- Practitioner publications
- The *Journal of Product Innovation Management*, the leading academic research journal in the field and a free benefit to all members
- Certification as a New Product Development Professional (NPDP)
- Awards and scholarships to support rigorous academic research
- Online access to a global virtual community and knowledge repository
- Volunteer opportunities for personal and professional growth

The PDMA brings together a diverse membership. Whereas most professional associations are focused on a particular industry or job function, the PDMA brings together those who share a common passion and aptitude for innovation. Members come from all industries, from aircraft manufacturing to commercial banking to candy distribution. Members come from all levels, from students to CEOs. Members come from all functions, from marketing to R&D to engineering to sales to finance and from over 30 countries around the globe. The recognition of bringing innovative and valuable new goods and services to market is what creates jobs, fuels growth, and improves standards of living, and is the common tie that binds the members of the organization. This unique diversity of the PDMA facilitates fertile cross-pollination of ideas and perspectives.

PDMA fills a void in the business world. Innovators can become isolated within their profession, within their industry, or within their geographic location. PDMA serves as an access point and as a conduit of communication between global innovators. The interaction among this diverse group accelerates the rate of learning. Fostering collaboration among different people in the innovation space enables higher quality, more valuable, and more successful innovation efforts. Connection builds knowledge, and knowledge facilitates connection.

Chapters

As a global community of innovators, PDMA offers a large number of chapters and affiliates that build product development networks and create professional learning opportunities at a local community level. Participation in a local chapter organization builds professional relationships that enable product developers and researchers to grow professionally, build strong development skills, and navigate the professional world. Events held by chapters provide opportunities for both networking and learning. These local organizations produce a variety of events, from meetings offering an insightful and interactive presentation by an expert in one of the many fields of product development to full-day summits featuring keynotes and breakouts that explore product development in depth.

Because these events are locally produced, they can be customized to the particular interests of a geographic region while still operating in a global context. This creates a focused network of innovators who are especially knowledgeable about the opportunities and challenges unique to their area while encouraging worldwide connections and learning. Combining the local community and the global vision of PDMA helps prepare professionals to effectively manage the constantly changing conditions of the global marketplace.

International Conferences

Annual Global Conference on Product Innovation Management. At the PDMA's Annual Global Conference, executives, practitioners, and solution providers gather to learn from one another and from innovation thought leaders about best practices, emerging practices, research findings, and case studies. From dynamic keynote speakers to deep-dive workshops, from executive panel discussions to interactive breakout sessions, product development professionals will be challenged and educated in a way that they can immediately apply to their work, becoming more effective and efficient at profitably delivering new goods and services to market. The Annual Global Conference is a great opportunity to build a global network of product development professionals while learning from some of the brightest minds in the field.

Annual JPIM Research Forum. Held in conjunction with the Annual Global Conference, the Annual JPIM Research Forum delivers the latest research findings in new product development and management. Researchers from around the world gather for interactive presentations and discussions. Practitioners benefit from the conventional-wisdom-defying knowledge that is presented, giving them a competitive edge on the job. Academic researchers have the opportunity to learn new methodologies, understand and debate challenging new findings, and gain practical insight from practitioners. Relationships of great value are established for future collaboration and exchange.

Jointly Sponsored Conferences. PDMA collaborates with external organizations to jointly sponsor conferences on product development and innovation. Some of these conferences are live face-to-face events that allow for interactive learning and networking. Others are virtual online events that maximize scheduling flexibility. All are excellent means of building greater skills in product development that will reap beneficial returns on innovation investment.

Webcasts. PDMA enables members and nonmembers alike to attend virtual meetings and presentations through regular webcasts. With diverse topics and dynamic speakers, product development professionals can interact with world-class thought leaders to increase their skill base, gain new ideas, and build on new insights.

Publications

The Journal of Product Innovation Management (JPIM). PDMA's prestigious academic research journal is recognized as a top-tier journal and the leading academic outlet focused on innovation and new product development issues. Rigorously researched and meticulously reviewed, each article provides insights for academics and practitioners alike, establishing proven strategies, tactics, and practices that deliver on innovation efforts.

Connections. PDMA's monthly e-newsletter communicates the latest happenings within the organization on both local and international levels. Featuring condensed articles and relevant news in a convenient electronic format, *Connections* helps keep members connected with the product development community while managing busy schedules.

Books. This fourth edition of *The PDMA Handbook of New Product Development and Innovation* is a comprehensive collection of information that spans the many different topics included in good product development practices. PDMA also publishes three topical deep-dive *ToolBooks*, all of which explore in detail specific tools used by product developers in all phases of the development process. In addition, PDMA offers a unique overview of the product development field with *New Product Development for Dummies*, which provides an excellent overview of product development knowledge. This collection of books will help get those new to the field off to a fast and effective start while serving as an ongoing reference for those who are more experienced. Whether readers are engaged in day-to-day management or long-term strategic planning, these publications should be of great value.

NPDP Certification

New Product Development Professional (NPDP) certification, a credential that confirms mastery of new product and service development principles and best practices, sets one apart as an expert and leader in the field. PDMA certifies individuals as being New Product Development Professionals, recognizing successful product development education, experience, and competence in the following key areas of product development:

- Strategy
- Portfolio Management
- Product Innovation Process
- Product Design & Development Tools
- Market Research in Product Innovation
- Culture, Teams, & Leadership
- Product Innovation Management

Candidates with appropriate professional experience can become certified by passing the NPDP Exam, which demonstrates competence in all of the above areas. PDMA and its chapters provide preparatory courses while the NPDP Body of Knowledge (BoK) combines the training materials needed for the exam.

PDMA Foundation Best Practices Survey

The PDMA Foundation Best Practices Survey on new product development is, as of 2022, in its fifth generation as the longest-running and most in-depth survey of its kind. Since 1990, the PDMA Research Foundation has captured new product development best practices from several thousand business units. This rich source of data is analyzed by leading academic professionals to identify trends in new product development best practices and to definitively identify those practices that separate the most successful innovation organizations from the rest.

The survey provides product developers with insight into where their organizations may make improvements. It delivers benchmarking data that identify performance gaps and define improvement metrics. Knowing the practices used by the most successful firms, identifying emerging practices that offer competitive edges, and understanding performance differences among industries, firm sizes, and technology bases all enable product developers to improve their performance and gain a competitive edge in their innovation efforts.

Awards

Outstanding Corporate Innovator (OCI) Award. This award is the only innovation award that recognizes organizations for sustained, quantifiable business results from new products and services. Since 1988, using a rigorous review process, PDMA has identified organizations that demonstrate the ability to consistently create and capture value through innovation based on:

- Sustained success in launching new products over a five-year time frame
- Significant company growth from new products
- Well-defined new product development practices and processes
- Distinctive innovative characteristics and intangibles

Winners are presented the prestigious award at the Annual Global Conference, where the winning firms share their best practices with the PDMA membership. Presentations from many past winners are archived on the PDMA website for ongoing access to the best historical approaches to sustained and successful new product development.

Crawford Fellows. The PDMA confers honorary recognition to a very select group of individuals who have made unique and significant contributions to advancing the field of new product development. This honor is named after the founder of the PDMA, Professor C. Merle Crawford, formerly of the University of Michigan.

JPIM **Albert L. Page Award for Outstanding Professional *Contribution.*** This award recognizes articles (original research or Catalyst) in JPIM that provide an outstanding contribution specifically to innovation management practice. The award is named after Professor Al Page, a major contributor to the PDMA, formerly of the University of Illinois at Chicago.

JPIM **Abbie Griffin High Impact Award.** This award recognizes the *Journal of Product Innovation Management* article published five years prior to the award date that has made the most significant (i.e., fundamental, far-reaching) contribution to the theory and practice of innovation management. The award is named after Professor Abbie Griffin, former editor of the Journal of Product Innovation Management, a major contributor to the PDMA, and Presidential Professor at the University of Utah.

JPIM **Best Reviewers.** The *Journal of Product Innovation Management* annually recognizes reviewers who have made exceptional contributions to the journal's editorial process, reflecting constructive, insightful, and timely evaluations of submitted manuscripts.

Thomas P. Hustad *JPIM* **Best Paper Award.** Each year, the Board of Directors of the PDMA and the Editorial Board of the *Journal of Product Innovation Management* award the article they feel most advances the state of the art of new product development and management the Thomas P. Hustad JPIM Best Paper Award. The award winner is announced at the PDMA Annual Global Conference and is named for the founder and first editor of *JPIM*, Thomas P. Hustad, Professor Emeritus, Indiana University.

Allan Anderson Ambassador Award. This award recognizes a member of the PDMA community who demonstrates exceptional commitment to the mission and ideals set forth by PDMA and as such is a true ambassador of the Association. The award, created in 2021, honors Allan Anderson, Professor Emeritus, Massey University, NZ, Chief Executive of NZ Dairy Research Institute, successful entrepreneur, and life-long advocate of sustainable product innovation practices. Over many years, Allan served in many capacities within PDMA, but it was Allan's dogged determination and his skills at identifying, building, and growing international relationships that inspired the creation of this award.

PDMA Research Competition. PDMA provides two grants to conduct academic research on the topic of product development and innovation management. Proposals are reviewed in a competitive, double-blind fashion on topic importance, conceptual development, methodological rigor, originality, and fit with the PDMA mission. Grants are awarded to those whose research will contribute significantly to the field of product development and/or to the practice of product development management. The grants encourage collaboration between academics and practitioners by supporting voluntary survey research among members. A summary of the findings is reported to PDMA respondents, providing early insights into what may work and what may not work in the practice of new product development.

Global Student Innovation Challenge. The Global Student Innovation Challenge is unique in that it encourages and supports students developing novel, real-world offerings, that include products, software, or services.

Website

PDMA hosts an interactive website, www.pdma.org, to help disseminate knowledge and build global collaboration networks. Through resources such as blogs, electronic publications, references, white papers, interactive forums, and an online store, PDMA keeps members up to date on the latest information in new product development knowledge.

Volunteer Opportunities

Volunteer opportunities exist within PDMA at both the local and global levels. Whether they are interested in providing leadership for a local organization, planning and producing a conference, writing articles for a publication, or serving on a committee, there are options available for those who would like to promote the field of new product development to others. By contacting a local chapter, or the international headquarters, product developers can discover ways to help accelerate the contribution that innovation makes to the economic and professional growth of people, businesses, and societies around the world.

Membership and Further Information

Interested in learning more about PDMA or in joining the organization? Please feel free to contact the association with any questions or comments.

Product Development and Management Association

www.pdma.org

1000 Westgate Drive

Saint Paul, MN 55114

Local: (651) 290-6280

Fax: (651) 290-2266

PDMA GLOSSARY OF NEW PRODUCT DEVELOPMENT TERMS

A/B testing: A form of multivariate research designed to test and compare two samples or variables. Other forms of multivariate testing, such as conjoint analysis, involve two or more variations and variables.

Agile product development: An iterative approach to product development that is performed in a collaborative environment by self-organizing teams.

Alliance: Formal arrangement with a separate organization for purposes of development, and involving exchange of information, hardware, intellectual property, or enabling technology. Alliances involve shared risk and reward (e.g., co- development projects).

Alpha test: Pre-production product testing to find and eliminate the most obvious design defects or deficiencies, usually in a laboratory setting or in some part of the developing organization's regular operations, although in some cases it may be done in controlled settings with lead customers. See also beta test and gamma test.

Analyzer: An organization that follows an imitative innovation strategy, where the goal is to get to market with an equivalent or slightly better product very quickly once someone else opens up the market, rather than to be first to market with new products or technologies. Sometimes called an imitator or a "fast follower."

The PDMA Handbook of Innovation and New Product Development, Fourth Edition. Edited by Ludwig Bstieler and Charles H. Noble.
© 2023 John Wiley & Sons, Inc. Published 2023 by John Wiley & Sons, Inc.

Applications development: The iterative process through which software is designed and written to meet the needs and requirements of the user base, or the process of enhancing or developing new products.

Acquisition effort: The extent to which your product or service is accessible to your customer.

Adjourning: The stage of a project team's work on the project is complete. In product innovation projects, the product is launched and turned over to standard business operations.

Architectural innovation: Combines technological and business disruptions. A well-quoted example is digital photography, which caused significant disruption for companies such as Kodak and Polaroid.

Architecture: See product architecture.

ATAR (Awareness-Trial-Availability-Repeat): A forecasting tool that attempts to mathematically model the diffusion of an innovation or new product.

Attribute testing: A quantitative market research technique in which respondents are asked to rate a detailed list of product or category attributes on one or more types of scales (such as relative importance, current performance, and current satisfaction with a particular product or service) for the purpose of ascertaining customer preferences for some attributes over others, to help guide the design and development process. Great care and rigor should be taken in developing the list of attributes, and it must be neither too long for the respondent to answer comfortably nor too short such that it lumps too many ideas together at too high a level.

Audit: When applied to new product development, an audit is an appraisal of the effectiveness of the processes by which the new product was developed and brought to market.

Augmented product: The core product, plus all other sources of product benefits, such as service, warranty, and image.

Augmented reality (AR): Similar to VR; whereas VR replaces the participant's real world with an entirely separate reality, AR overlays elements of a new reality into the participant's present environment.

Autonomous team: A completely self-sufficient project team with very little, if any, link to the funding organization. Frequently used as an organizational model to bring a radical innovation to the marketplace. Sometimes called a tiger team.

Awareness: A measure of the percent of target customers who are aware that the new product exists. Awareness is variously defined, including recall of brand, recognition of brand, recall of key features or positioning.

Balanced portfolio: A collection of projects where the proportion of projects in specific categories is selected according to strategic priorities.

Balanced Scorecard: A strategic management performance metric used to identify and improve various internal business functions and their resulting external outcomes.

Bass model: A tool used to forecast sales of an innovation, new technology, or a durable good.

Benchmarking: A process of collecting process performance data, generally in a confidential, blinded fashion, from a number of organizations to allow them to assess their performance individually and as a whole.

Benefit: A product attribute expressed in terms of what the user gets from the product rather than its physical characteristics or features. Benefits are often paired with specific features, but they need not be.

Best practice: Methods, tools, or techniques that are associated with improved performance. In new product development, no one tool or technique assures success; however, a number of them are associated with higher probabilities of achieving success. Best practices likely are at least somewhat context specific. Sometimes called effective practice.

Best practice study: A process of studying successful organizations and selecting the best of their actions or processes for emulation. In new product development, it means finding the best process practices, adapting them, and adopting them for internal use.

Beta test: A more extensive test than the alpha test, performed by real users and customers. The purpose of beta testing is to determine how the product performs in an actual user environment. It is critical that real customers perform this evaluation, not the organization developing the product or a contracted testing organization. As with the alpha test, results of the beta test should be carefully evaluated with an eye toward any needed modifications or corrections.

Big data: A collection of large and complex data from different instruments at all stages of the process which go from acquisition, storage, and sharing, to analysis and visualization.

Bottom-up portfolio selection: Starts first with a list of individual projects, and through a process of strict project evaluation and screening, ends up with a portfolio of strategically aligned projects.

Brainstorming: A group method of creative problem-solving frequently used in product concept generation. There are many modifications in format, each variation with its own name. The basis of all of these methods uses a group of people to creatively generate a list of ideas related to a particular topic. As many ideas as possible are listed before any critical evaluation is performed.

Brand: A name, term, design, symbol, or any other feature that identifies one seller's good or service as distinct from those of other sellers. The legal term for brand is trademark. A brand may identify one item, a family of items, or all items of that seller.

Brand development index: Sales of your brand compared with its average performance in all markets.

Break-even point: The point in the commercial life of a product when cumulative development costs are recovered through accrued profits from sales.

Breakthrough projects: These projects strive to bring a new product to the market with new technologies, depart significantly from existing organizational practices, and have a high level of risk.

Bubble diagram: Visual representation of a product portfolio. Typically, a bubble diagram shows projects on a two-dimensional X-Y plot. The X and Y dimensions relate to specific criteria of interest (for example, risk and reward).

Business analysis: An analysis of the business situation surrounding a proposed project. Usually includes financial forecasts in terms of discounted cash flows, net present values, or internal rates of returns.

Business case: The results of the market, technical, and financial analyses, or up-front homework. Ideally defined just prior to the "go to development" decision (gate), the case defines the product and project, including the project justification and the action or business plan.

Business Model Canvas (BMC): A strategic management and Lean startup template for developing new or documenting existing business models. It is a visual chart with elements describing an organization's or product's value proposition, infrastructure, customers, and finances.

Business-to-business: Transactions with non-consumer purchasers such as manufacturers, resellers (distributors, wholesalers, jobbers, and retailers, for example), institutional, professional, and governmental organizations. Frequently referred to as industrial businesses in the past.

Buyer: The purchaser of a product, whether or not they will be the ultimate user. Especially in business-to-business markets, a purchasing agent may contract for the actual purchase of a good or service, yet never benefit from the function(s) purchased.

Cannibalization: That portion of the demand for a new product that comes from the erosion of the demand for (sales of) a current product the organization markets.

Capacity planning: A forward-looking activity that monitors the skill sets and effective resource capacity of the organization. For product development, the objective is to manage the flow of projects through development such that none of the functions (skill sets) creates a bottleneck to timely completion. Necessary in optimizing the project portfolio.

Carbon credits: By a simple cost of goods calculation, the indirect cost from externalities (effects of a product or service on other people than the producer and user) are not reflected. This can be CO_2, but also social impact. Integrating all externalities in your (shadow) price gives the real price.

Cash cows: Products that have a high share of a market and low overall growth.

Centers of excellence: A geographic or organizational group with an acknowledged technical, business, or competitive competency.

Certification: A process for formally acknowledging that someone has mastered a body of knowledge on a subject. In new product development, the PDMA has created and manages a certification process to become a New Product Development Professional (NPDP).

Champion: A person who takes a passionate interest in seeing that a particular process or product is fully developed and marketed. This informal role varies from situations calling for little more than stimulating awareness of the opportunity to extreme cases where the champion tries to force a project past the strongly entrenched internal resistance of organization policy or that of objecting parties.

Charter: A project team document defining the context, specific details, and plans of a project. It includes the initial business case, problem and goal statements, constraints and assumptions, and preliminary plan and scope. Periodic reviews with the sponsor ensure alignment with business strategies. See also Product Innovation Charter.

Chasm: A critical part of the product life cycle occurs near the beginning – after introduction, but before growth has fully kicked in.

Checklist: A list of items used to remind an analyst to think of all relevant aspects. It finds frequent use as a tool of creativity in concept generation, as a factor consideration list in concept screening, and to ensure that all appropriate tasks have been completed in any stage of the product development process.

Chief innovation officer (or CINO): A senior executive devoted to overseeing innovation and new product development.

Circular economy: An economy that is restorative and regenerative by design, and which aims to keep products, components, and materials at their highest utility and value at all times, distinguishing between technical and biological cycles.

Cluster sampling: The population is divided into clusters and a sample of clusters is taken.

Collaborative product development: When two organizations work together to develop and commercialize a specialized product.

Co-creation/co-development: A collaborative NPD activity in which customers are invited by a company to actively contribute to the development or select the content of a new product offering.

Co-creation platform: A technology-based social platform that allows a company to engage with external partners for co-developing new products.

Co-location: Physically locating project personnel in one area, enabling more rapid and frequent decision- making and communication among them.

Commercialization: The process of taking a new product from development to market. It generally includes production launch and ramp-up, marketing materials and program development, supply chain development, sales channel development, training development, training, and service and support development.

Competitive intelligence: Methods and activities for transforming disaggregated public competitor information into relevant and strategic knowledge about competitors' position, size, efforts, and trends. The term refers to the broad practice of collecting, analyzing, and communicating the best available information on competitive trends occurring outside one's own organization.

Concept: A clearly written and possibly visual description of a new product idea that includes its primary features and consumer benefits, combined with a broad understanding of the technology needed.

Concept generation: The processes by which new concepts, or product ideas, are generated. Sometimes also called idea generation or ideation.

Concept screening: The evaluation of potential new product concepts during the discovery phase of a product development project. Potential concepts are evaluated for their fit with business strategy, technical feasibility, manufacturability, and potential for financial success.

Concept statement: A verbal or pictorial statement of a concept that is prepared for presentation to consumers to get their reaction prior to development.

Concept engineering: A customer-centered process that clarifies the "fuzzy front end" of the product development process with the purpose of developing product concepts. The method determines the customer's key requirements to be included in the design and proposes several alternative product concepts that satisfy these requirements.

Concept testing: The process by which a concept statement is presented to consumers for their reactions. These reactions can either be used to permit the developer to estimate the sales value of the concept or to make changes to the concept to enhance its potential sales value.

Concurrent engineering (CE): When product design and manufacturing process development occur concurrently in an integrated fashion, using a cross-functional team, rather than sequentially by separate functions. CE is intended to cause the development team to consider all elements of the product life cycle from conception through disposal, including quality, cost, and maintenance, from the project's outset. Also called simultaneous engineering.

Conjoint analysis: A market research technique in which respondents are systematically presented with a rotating set of product descriptions, each containing a rotating set of attributes and levels of those attributes. By asking respondents to choose their preferred product and/or to indicate their degree of preference from within each set of options, conjoint analysis can determine the relative contribution to overall preference of each variable and each level. The two key advantages of conjoint analysis over other methods of determining importance are (1) the variables and levels can be either continuous (e.g., weight) or discrete (e.g., color), and (2) it is just about the only valid market research method for evaluating the role of price, that is, how much someone would pay for a given feature.

Consumer: The most generic and all-encompassing term for an organization's targets. The term is used in either the business-to-business or household context and may refer to the organization's current customers, competitors' customers, or current non-purchasers with similar needs or demographic characteristics. The term does not differentiate between whether the person is a buyer or a user target. Only a fraction of consumers will become customers.

Consumer market: The purchasing of goods and services by individuals and for household use (rather than for use in business settings). Consumer purchases are generally made by individual decision-makers, either for themselves or others in the family.

Consumer need: A problem the consumer would like to have solved. What a consumer would like a product to do for them.

Consumer panels: Groups of consumers in specific sectors, recruited by research companies and agencies, who are used as respondents to answer specific research questions relating to product testing, taste testing, or other areas. Most often, they are a specialist panel who take part in numerous projects. Consumer panels are particularly useful for short, quick surveys, where the emphasis is on a sample of those with specialist knowledge rather than a representative sample of the general population.

Contingency plan: A plan to cope with events whose occurrence, timing, and severity cannot be predicted.

Continuous improvement: The review, analysis, and rework directed at incrementally improving practices and processes. Also called Kaizen.

Continuous innovation: A product alteration that allows improved performance and benefits without changing either consumption patterns or behavior. The product's general appearance and basic performance do not functionally change. Examples include fluoride toothpaste and higher computer speeds.

Convergent thinking: Associated with analysis, judgment, and decision-making. It is the process of taking a lot of ideas and sorting, evaluating, analyzing the pros and cons, and making decisions.

Cooperation (team cooperation): The extent to which team members actively work together in reaching team level objectives.

Copyright: The exclusive and assignable legal right, given to the originator for a fixed number of years, to print, publish, perform, film, or record literary, artistic, or musical material.

Core Benefit Proposition (CBP): The central benefit or purpose for which a consumer buys a product. The CBP may come either from the physical good or service, or from augmented dimensions of the product. See also value proposition.

Core competence: That capability at which an organization does better than other organizations, which provides them with a distinctive competitive advantage and contributes to acquiring and retaining customers. Something that an organization does better than other organizations. The purest definition adds, "and is also the lowest cost provider."

Corporate culture: The "feel" of an organization. Culture arises from the belief system through which an organization operates. Corporate cultures are variously described as being authoritative, bureaucratic, and entrepreneurial. The organization's culture frequently impacts the organizational appropriateness for getting things done.

Corporate strategy: The overarching strategy of a diversified organization. It answers the questions of "in which businesses should we compete?" and "how does bringing in these businesses create synergy and/or add to the competitive advantage of the organization as a whole?"

Creativity: "An arbitrary harmony, an expected astonishment, a habitual revelation, a familiar surprise, a generous selfishness, an unexpected certainty, a formable stubbornness, a vital triviality, a disciplined freedom, an intoxicating steadiness, a repeated initiation, a difficult delight, a predictable gamble, an ephemeral solidity, a unifying difference, a demanding satisfier, a miraculous expectation, and accustomed amazement." (George M. Prince, *The Practice of Creativity*, 1970.) Creativity is the ability to produce work that is both novel and appropriate.

Criteria: Statements of standards used by decision-makers at decision gates. The dimensions of performance necessary to achieve or surpass for product development projects to continue in development. In the aggregate, these criteria reflect a business unit's new product strategy.

Critical path: The set of interrelated activities that must be completed for the project to be finished successfully can be mapped into a chart showing how long each task takes, and which tasks cannot be started before which other tasks are completed. The critical path is the set of linkages through the chart that is the longest. It determines how long a project will take.

Critical path scheduling: A project management technique, frequently incorporated into various software programs, which puts all important steps of a given new product project into a sequential network based on task interdependencies.

Critical success factors: Those critical few factors that are necessary for, but don't guarantee, commercial success.

Cross-functional team: A team consisting of representatives from the various functions involved in product development, usually including members from all key functions required to deliver a successful product, typically including marketing, engineering, manufacturing/operations, finance, purchasing, customer support, and quality. The team is empowered by the departments to represent each function's perspective in the development process.

Crossing the chasm: Making the transition to a mainstream market from an early market dominated by a few visionary customers (sometimes also called innovators or lead adopters). This concept typically applies to the adoption of new, market-creating, technology-based products and services.

Crowdfunding: A particular type of crowdsourcing that enables individuals and companies to test the viability of their ideas or product concepts by seeking to raise funds from online communities in exchange for monetary or non-monetary rewards.

Crowdsourcing: The practice and use of a collection of tools for obtaining information, goods, services, ideas, funding, or other input into a specific task or project from a large and relatively open group of people, either paid or unpaid, most commonly via technology platforms, social media channels, or the Internet.

Culture: The shared beliefs, core values, assumptions, and expectations of people in the organization.

Customer: One who purchases or uses an organization's products or services.

Customer needs: Problems to be solved. These needs, either expressed or yet to be articulated, provide new product development opportunities for the organization.

Customer site visits: A qualitative market research technique for uncovering customer needs. The method involves going to a customer's work site, watching as a person performs functions associated with the customer needs your organization wants to solve, and then debriefing that person about what they did, why they did those things, the problems encountered as they were trying to perform the function, and what worked well.

Cycle time: The length of time for any operation, from start to completion. In the new product development sense, it is the length of time to develop a new product from an early initial idea for a new product to initial market sales. Precise definitions of the start and end point vary from one organization to another and may vary from one project to another within the organization.

Dashboard: A typically color-coded graphical presentation of a project's status or a portfolio's status by project resembling a vehicle's dashboard. Typically, red is used to flag urgent problems, yellow to flag impending problems, and green to signal projects on track.

Data: Measurements taken at the source of a business process.

Database: An electronic gathering of information organized in some way to make it easy to search, discover, analyze, and manipulate.

Decision tree: A diagram used for making decisions in business or computer programming. The "branches" of the tree diagram represent choices with associated risks, costs, results, and outcome probabilities. By calculating outcomes (profits) for each of the branches, the best decision for the organization can be determined.

Decline stage: The fourth and last stage of the product life cycle. Entry into this stage is generally caused by technology advancements, consumer or user preference changes, global competition, or environmental or regulatory changes.

Defenders: Organizations that stake out a product turf and protect it by whatever means, not necessarily through developing new products.

Deliverable: The output (such as test reports, regulatory approvals, working prototypes, or marketing research reports) that shows a project has achieved a result. Deliverables may be specified for the commercial launch of the product or at the end of a development stage.

Delphi: A technique that uses iterative rounds of consensus development across a group of experts to arrive at a forecast of the most probable outcome for some future state.

Demographic: The statistical description of a human population. Characteristics included in the description may include gender, age, education level, and marital status, as well as various behavioral and psychological characteristics.

Derivative projects: Spin-offs from other existing products or platforms. They may fill a gap in an existing product line, offer more cost-competitive manufacturing, or offer enhancements and features based on core organization technology. Generally, they are relatively low risk.

Design for assembly: Simplifies the product design to reduce the cost of assembly in the manufacturing process. An assembly is defined as a combination of parts and components needed to manufacture a product. This view includes all the working activities during and after production.

Design for the environment: The systematic consideration of environmental safety and health issues over the product's projected life cycle in the design and development process.

Design for excellence: The systematic consideration of all relevant life cycle factors, such as manufacturability, reliability, maintainability, affordability, testability, etc., in the design and development process.

Design for functionality: Functionality determines the final performance of a product. It allows for the intended behavior of the elements of design or their combination. DFF implies considerations such as design for safety (coffeemakers), design for simplicity (platform design), and design for redesign (product variants or derivatives).

Design for maintenance: Design decisions regarding the selection of materials, assemblies, parts, devices, and components determine the maintainability or the capacity of the system to be inspected, restored, and serviced when components fail as they reach their operational life. During the design stages, design for maintenance should facilitate executing corrective and preventive maintenance.

Design for production: Aims to minimize product costs and manufacture times while maintaining specified quality standards. A successful manufacturing process depends on the following factors:

- Rate: flow of materials, parts, components through the system.
- Cost: materials, labor, machines, equipment, tooling.
- Time: supply times, inventory flows, processing times, machine set-up times.
- Quality: Lost functions and deviations from the target.

Design for recycling: Design for recycling (DFR) accounts for the use of materials that allow for reusing or reprocessing production waste, products, and parts of products. The methods for DFR center on reusing products and reprocessing products.

Design for serviceability: Focuses on the ability to diagnose, remove, or replace any part, component, assembly, or subassembly of a product while performing service repairs and troubleshooting.

Design for Six Sigma: The aim of DFSS is to create designs that are resource efficient, capable of exceptionally high yields, and are robust to process variations.

Design for usability: Evaluates the functionality, serviceability, maintainability, ease of operation, reliability, safety, aesthetics, operating context and environment, and customizability of the concept system. Overall, DFU should be intrinsically connected to the product and manufacture design process.

Design patents: Protect new, non-obvious, and ornamental aspect of an invention for a limited term.

Design specifications: Where the concept statement provides a qualitative presentation of the product concept's benefits and features, the product design specifications provide the quantitative basis for further design and manufacture.

Design thinking: A creative solving approach – or more completely, a systematic and collaborative approach to identify and creatively solve problems.

Design validation: Product tests to ensure that the product or service conforms to defined user needs and requirements. These may be performed on working prototypes or using computer simulations of the finished product.

Development: The functional part of the organization responsible for converting product requirements into a working product. Also, a phase in the overall concept-to-market cycle where the new product or service is developed for the first time.

Development teams: Teams formed to take one or more new products from concept through development, testing, and launch.

Discontinuous innovation: Previously unknown products that establish new consumption patterns and behavior changes. Examples include microwave ovens and cellular phones.

Discounted Cash-Flow (DCF) analysis: One method for providing an estimate of the current value of future incomes and expenses projected for a project. Future cash flows for a number of years are estimated for the project, and then discounted back to the present using forecast interest rates.

Dispersed teams: Product development teams that have members working at different locations, across time zones, and perhaps even in different countries.

Disruptive innovation: Requires a new business model but not necessarily new technology. So, for example, Google's Android operating system potentially disrupts companies like Apple.

Distribution (physical and channels): The method and partners used to get the product (or service) from where it is produced to where the end user can buy it.

Divergent thinking: The process of coming up with new ideas and possibilities without judgment, without analysis, and without discussion. It is the type of thinking that allows for free-association, "stretching the boundaries," and thinking of new ways to solve difficult challenges that have no single, right, or known answer.

Early adopters: For new products, these are customers who, relying on their own intuition and vision, buy into new product concepts very early in the life cycle. For new processes, these are organizational entities that were willing to try out new processes rather than just maintaining the old.

Embodiment design: The stage of the design process that starts from the concept definition and continues to develop the design based on technical and economic criteria to reach the detail design stage, which leads to manufacturability.

Emotional design: Is based on eliciting the moods and feelings of consumers that allow for designing positive emotional associations and a feeling of trust in the product, and thus improve its usability.

Emotional intelligence: Comprised of self-management components and of elements directed toward managing relationships.

Empathy analysis: Involves the capacity to connect with and understand customers deeply and have a direct emotional connection with them.

Enhanced new product: A form of derivative product. Enhanced products include additional features not previously found on the base platform, which provide increased value to consumers.

Entrepreneur: A person who initiates, organizes, operates, assumes the risk, and reaps the potential reward for a new business venture.

Ethnography: A descriptive, qualitative market research methodology for studying the customer in relation to his or her environment. Researchers spend time in the field observing customers and their environment to acquire a deep understanding of the lifestyles or cultures as a basis for better understanding their needs and problems. See customer site visits.

Eye tracking: A specialized form of sensory testing that uses specialized tools, including connected headsets or goggles, measuring where people look and for how long. The equipment tracks and reports where participants look first, second, third, etc., and provides a visual scan overlaid on an image of the object tested. It is used to answer questions on consumers' reactions to various stimuli, online products and services, websites, apps, images of products, packaging, and messaging. It is widely used in software, retail product packaging, marketing, and advertising.

Factor analysis: A process in which the values of observed data are expressed as functions of a number of possible causes in order to find which are the most important.

Factory cost: The cost of producing the product in the production location including materials, labor, and overhead.

Failure rate: The percentage of an organization's new products that make it to full market commercialization, but which fail to achieve the objectives set for them.

Feasibility analysis: The process of analyzing the likely success of a project or a new product.

Feature: The solution to a consumer need or problem. Features provide benefits to consumers. A handle (feature) allows a laptop computer to be carried easily (benefit). Usually, any one of several different features will be chosen to meet a customer need. For example, a carrying case with shoulder straps is another feature that allows a laptop computer to be carried easily.

Feature creep: The tendency for designers or engineers to add more capability, functions, and features to a product as development proceeds than were originally intended. These additions frequently cause schedule slip, development cost increases, and product cost increases.

Feature roadmap: The evolution over time of the performance attributes associated with a product. Defines the specific features associated with each iteration/generation of a product over its lifetime, grouped into releases (sets of features that are commercialized).

Field testing: Product use testing with users from the target market in the actual context in which the product will be used.

Financial success: The extent to which a new product meets its profit, margin, and return on investment goals.

First-to-market: The first product to create a new product category or a substantial subdivision of a category.

Focus groups: A qualitative market research technique where 8 to 12 market participants are gathered in one room for a discussion under the leadership of a trained moderator. Discussion focuses on a consumer problem, product, or potential solution to a problem. The results of these discussions are not projectable to the general market.

Forecast: A prediction, over some defined time, of the success or failure of implementing a business plan's decisions derived from an existing strategy.

Forming: The first stage in team formation, where most team members are positive and polite. Some are anxious, as they haven't fully understood what the team will do.

Front End of Innovation (FEI): See Fuzzy Front End

Function: (1) An abstracted description of work that a product must perform to meet customer needs. A function is something the product or service must do. (2) Term describing an internal group within which resides a basic business capability such as engineering.

Function Analysis System Technique (FAST) A technique that builds on the results of a functional analysis. The purpose of FAST is to illustrate and provide insights on how the product system works in order to identify malfunctions, incoherence in the sequencing of operations, or operational flaws. The technique allows visualization of the cause–effect relationship among the functions in a product to enhance understanding of how it works.

Functional team: The project is divided into functional components with each component assigned to its own appropriate functional manager. Coordination is either handled by the functional manager or by senior management.

Fuzzy front end: The messy "getting started" period of product development, when the product concept is still very fuzzy. Preceding the more formal product development process, it generally consists of three tasks strategic planning, concept generation, and, especially, pre-technical evaluation. These activities are often chaotic, unpredictable, and unstructured. In comparison, the subsequent new product development process is typically structured, predictable, and formal, with prescribed sets of activities, questions to be answered, and decisions to be made.

Gamma test: A product use test in which the developers measure the extent to which the item meets the needs of the target customers, solves the problem(s) targeted during development, and leaves the customer satisfied.

Gantt chart: A horizontal bar chart used in project scheduling and management that shows the start date, end date, and duration of tasks within the project.

Gap analysis: Carried out in product development to determine the difference between expected and desired revenues or profits from currently planned new products if the corporation is to meet its objectives.

Gate: The point at which a management decision is made to allow the product development project to proceed to the next stage, to recycle back into the current stage to better complete some of the tasks, or to terminate. The number of gates varies by organization.

Gatekeepers: The group of managers who serve as advisors, decision-makers, and investors in a Stage-Gate® process. Using established business criteria, this multifunctional group reviews new product opportunities and project progress and allocates resources accordingly at each gate. This group is also commonly called a product approval committee or portfolio management team.

Greenwashing: When an organization or organization spends more time and money claiming to be "green" through advertising and marketing than actually implementing business practices that minimize environmental impact.

Growth stage: The second stage of the product life cycle, marked by a rapid surge in sales and market acceptance for the good or service. Products that reach the growth stage have successfully "crossed the chasm."

Heavyweight team: An empowered project team with adequate resourcing to complete the project. Personnel report to the team leader and are co-located as practical.

Hurdle rate: The minimum return on investment or internal rate of return percentage a new product must meet or exceed as it goes through development.

Idea: The most embryonic form of a new product or service. It often consists of a high-level view of the envisioned solution needed to solve the problem identified by a person, team, or organization.

Idea generation (ideation): All of those activities and processes that lead to creating broad sets of solutions to consumer problems. These techniques may be used in the early stages of product development to generate initial product concepts, in the intermediate stages for overcoming implementation issues, in the later stages for planning launch, and in the post-mortem stage to better understand success and failure in the marketplace.

Implementation team: A team that converts the concepts and good intentions of the "should-be" process into practical reality.

Implicit product requirement: What the customer expects in a product, but does not ask for, and may not even be able to articulate.

Incremental improvement: A small change made to an existing product that serves to keep the product fresh in the eyes of customers.

In-depth interviews: A qualitative research method that involves conducting longer intensive interviews probing and exploring a specific topic, one on one, with individual participants. The research gathers detailed insights, perspectives, attitudes, thoughts, behaviors, and viewpoints on a problem, idea, program, situation, etc.

Information: Knowledge and insight, often gained by examining data.

Initial screening: The first decision to spend resources (time or money) on a project. The project is born at this point. Sometimes called idea screening.

In-licensed: The acquisition from external sources of novel product concepts or technologies for inclusion in the aggregate NPD portfolio.

Innovation: A new idea, method, or device. The act of creating a new product or process. The act includes invention as well as the work required to bring an idea or concept into final form.

Innovation-based culture: A corporate culture where senior management teams and employees work habitually to reinforce best practices that systematically and continuously churn out valued new products to customers.

Innovation challenge or contest: A primarily online based approach via a social platform open to individuals for ideating, designing, or solving innovation or new product development related challenges, with financial rewards for winning submissions.

Innovation governance: A holistic system or high-level approach by top management to align goals, allocate resources, and assign decision-making authority for innovation across the company and sometimes with external parties.

Innovation steering committee: The senior management team or a subset of it responsible for gaining alignment on the strategic and financial goals for new product development, as well as setting expectations for portfolio and development teams.

Innovation strategy: Provides the goals, direction, and framework for innovation across the organization. Individual business units and functions may have their own strategies to achieve specific innovation goals, but it is imperative that these individual strategies are tightly connected with the overarching organizational innovation strategy.

Integrated Product Development (IPD): A philosophy that systematically employs an integrated team effort from multiple functional disciplines to effectively and efficiently develop new products that satisfy customer needs.

Intellectual property (IP): Information, including proprietary knowledge, technical competencies, and design information, which provides commercially exploitable competitive benefit to an organization.

Internal rate of return (IRR): The discount rate at which the present value of the future cash flows of an investment equals the cost of the investment. The discount rate with a net present value of 0.

Internet of Things (IoT): The interconnection via the Internet of computing devices embedded in everyday objects, enabling them to send and receive data.

Intrapreneur: The large-organization equivalent of an entrepreneur. Someone who develops new enterprises within the confines of a large corporation.

Introduction stage: The first stage of a product's commercial launch and the product life cycle. This stage is generally seen as the point of market entry, user trial, and product adoption.

ISO 9000: A set of five auditable standards of the International Organization for Standardization that establishes the role of a quality system in an organization and which is used to assess whether the organization can be certified as compliant to the standards. ISO 9001 deals specifically with new products.

Journal of Product Innovation Management: The premier academic journal in the field of innovation, new product development, and management of technology. The journal, which is owned by the PDMA, is dedicated to the advancement of management practice in all of the functions involved in the total process of product innovation. Its purpose is to bring to managers and students of product innovation the theoretical structures and the practical techniques that will enable them to operate at the cutting edge of effective management practice. Website: www.pdma.org/journal.

Journey maps: A representation as a flowchart of all the actions and behaviors consumers take when interacting with the product or service.

Kaizen: A Japanese term meaning "change for the better" or "continuous improvement." It is a Japanese business philosophy regarding the processes that continuously improve operations and involve all employees.

Kano method: Used to identify customer needs and latent demands, determining functional requirements, developing concepts as candidates for further product definition, and analyzing competitive product or services within a product category.

Kansei engineering: Used to identify the relevant design elements (color, size, and shape) embedded in a product as determinant of user preference.

Launch: The process by which a new product is introduced into the market for initial sale.

Lead users: Users for whom finding a solution to one of their consumer needs is so important that they have modified a current product or invented a new product to solve the need themselves because they have not found a supplier who can solve it for them. When these consumers' needs are portents of needs that the center of the market will have in the future, their solutions are new product opportunities.

Lean product development (LPD): The Lean approach to meet the challenges of product development. Lean product development is founded on the fundamental Lean methodology initially developed by Toyota (the Toyota Production System, or TPS).

Lean startup: An approach to building new businesses based on the belief that entrepreneurs must investigate, experiment, test, and iterate as they develop products.

Learning organization: An organization that continuously tests and updates the experience of those in the organization and transforms that experience into improved work processes and knowledge that is accessible to the whole organization and relevant to its core purpose.

Life cycle assessment: A scientific method for analysis of environmental impacts (CO_2 footprint, Water footprint, etc.).

Lightweight team: New product team charged with successfully developing a product concept and delivering to the marketplace. Resources are, for the most part, not dedicated, and the team depends on the technical functions for resources necessary to get the work accomplished.

Line extension: A form of derivative product that adds or modifies features without significantly changing the product functionality.

Manufacturability: The extent to which a new product can be easily and effectively manufactured at minimum cost and with maximum reliability.

Manufacturing design: The process of determining the manufacturing process that will be used to make a new product.

Manufacturing test specification and procedure: Documents prepared by development and manufacturing personnel that describe the performance specifications of a component, subassembly, or system that will be met during the manufacturing process, and that describe the procedure by which the specifications will be assessed.

Market penetration: The percentage of your target market that you have reached at least once in a specific period of time.

Market research: Information about the organization's customers, competitors, or markets. Information may be from secondary sources (already published and publicly available) or primary sources (from customers themselves). Market research may be qualitative in nature, or quantitative. See entries for these two types of market research.

Market segmentation: Market segmentation is defined as a framework by which to subdivide a larger heterogeneous market into smaller, more homogeneous parts. These segments can be defined in many ways demographic (men vs. women, young vs. old, or richer vs. poorer), behavioral (those who buy on the phone vs. the Internet vs. retail, or those who pay with cash vs. credit cards), or attitudinal (those who believe that store brands are just as good as national brands vs. those who don't). There are many analytical techniques used to identify segments – such as cluster analysis, factor analysis, or discriminate analysis. But the most common method is simply to hypothesize a potential segmentation definition and then to test whether any differences that are observed are statistically significant.

Market share: An organization's sales in a product area as a percent of the total market sales in that area. Sales may be for the organization, a brand, a product, etc.

Marketing mix: Comprises the basic tools that are available to market a product. The market mix is often referred to as the 4 Ps – Product, Price, Promotion, and Place.

Marketing strategy: A process or model to allow an organization to focus limited resources on the best opportunities to increase sales and thereby achieve a unique competitive advantage.

Maturity stage: The third stage of the product life cycle, where sales begin to level off due to market saturation. It is a time when heavy competition, alternative product options, and (possibly) changing buyer or user preferences start to make it difficult to achieve profitability.

Metrics: A set of measurements to track product development and allow an organization to measure the impact of process improvements over time. These measures generally vary by organization, but may include measures characterizing both aspects of the process, such as time to market and duration of particular process stages, as well as outcomes from product development such as the number of products commercialized per year and percentage of sales due to new products.

Mindmapping: A graphical technique for imagining connections between various pieces of information or ideas. The participant starts with a key phrase or word in the middle of a page, then works out from this point to connect to new ideas in multiple directions – building a web of relationships.

Mission: The statement of an organization's creed, philosophy, purpose, business principles, and corporate beliefs. The purpose of the mission is to focus the energy and resources of the organization.

Multidimensional scaling (MDS): A means of visualizing the level of similarity of individual cases of a dataset (for example, products or markets).

Multifunctional team: A group of individuals brought together from the different functional areas of a business to work on a problem or process that requires the knowledge, training, and capabilities across the areas to successfully complete the work. See also cross-functional team.

Multiple regression analysis: Often used in product innovation to analyze survey-based data. It provides detailed insight that can be applied to new products or improve products or services when there are any number of factors, key drivers, and product attributes that can impact the product's value proposition from the customer's point of view. It can be used to identify which variables have an impact on the topic of interest and is used to predict the value of a variable based on the known value of two or more other variables (predictors).

Multivariate analysis: Explores the association between one outcome variable (referred to as the dependent variable) and one or more predictor variables (referred to as independent variables).

Needs statement: Summary of consumer needs and wants, described in customer terms, to be addressed by a new product.

Net present value (NPV): The difference between the present value of cash inflows and the present value of cash outflows. NPV is used in capital budgeting to analyze the profitability of a projected investment or project.

Net promoter score: The likelihood someone would recommend your product or service to a friend.

Netnography: A qualitative research methodology utilizing adapted ethnographic research techniques to collect, observe, and analyze consumer interactions in social media (see Ethnography).

Network diagram: A graphical diagram with boxes connected by lines that shows the sequence of development activities and the interrelationship of each task with another. Often used in conjunction with a Gantt chart.

New product: A term of many opinions and practices, but most generally defined as a product (either a good or service) new to the organization marketing it. Excludes products that are only changed in promotion.

New Product Development (NPD): The overall process of strategy, organization, concept generation, product and marketing plan creation and evaluation, and commercialization of a new product. Also frequently referred to as product development.

New product introduction (NPI): The launch or commercialization of a new product into the marketplace. Takes place at the end of a successful product development project.

New Product Development process (NPD process): A disciplined and defined set of tasks and steps that describe the normal means by which an organization repetitively converts embryonic ideas into salable products or services.

New Product Development Professional (NPDP): A New Product Development Professional is certified by the PDMA as having mastered the body of knowledge in new product development, as proven by performance on the certification test. To qualify for the NPDP certification examination, a candidate must hold a Bachelor's or higher university degree (or an equivalent degree) from an accredited institution and have spent a minimum of two years working in the new product development field.

New-to-the-world product: A good or service that has never before been available to either consumers or producers. The automobile was new-to-the-world when it was introduced, as were microwave ovens and pet rocks.

Non-product advantage: Elements of the marketing mix that create competitive advantage other than the product itself. These elements can include marketing communications, distribution, organization reputation, technical support, and associated services.

Norming: The third stage of team formation, when people start to resolve their differences, appreciate colleagues' strengths, and respect the leader's authority.

Open Innovation (OI): The strategy adopted by an organization whereby it actively seeks knowledge from external sources, through alliances, partnerships, and contractual arrangements, to complement and enhance its internal capability in pursuit of improved innovation outcomes. These innovation outcomes may be commercialized internally, through new business entities, or through external licensing arrangements.

Operations: A term that includes manufacturing but is much broader, usually including procurement, physical distribution, and, for services, management of the offices or other areas where the services are provided.

Opportunity: A business or technology gap that an organization or individual realizes, by design or accident, exists between the current situation and an envisioned future in order to capture competitive advantage, respond to a threat, solve a problem, or ameliorate a difficulty.

Organizational identity: Fundamental to the long-term success of an organization is a clear definition and understanding of what the organization stands for, why it exists.

Outsourcing: The process of procuring a good or service from someone else, rather than the organization producing it themselves.

Outstanding Corporate Innovator Award: An annual PDMA award given to organizations acknowledged through a formal vetting process as being outstanding innovators. The basic requirements for receiving this award are: (1) Sustained success in launching new products over a five-year time frame; (2) Significant organization growth from new product success; (3) A defined new product development process, that can be described to others; (4) Distinctive innovative characteristics and intangibles.

Payback: The time, usually in years, from some point in the development process until the commercialized product or service has recovered its costs of development and marketing. While some organizations take the point of full-scale market introduction of a new product as the starting point, others begin the clock at the start of development expense.

Perceptual mapping: A quantitative market research tool used to understand how customers think of current and future products. Perceptual maps are visual representations of the positions that sets of products hold in consumers' minds.

Performance measurement system: The system that enables an organization to monitor the relevant performance indicators of new products in the appropriate time frame.

Performance metrics: A set of measurements to track product development and to allow an organization to measure the impact of process improvement over time. These measures generally vary by organization but may include measures characterizing both aspects of process, such as time to market and duration of particular process stages, as well as outcomes from product development, such as the number of products commercialized per year and percentage sales due to new products.

Performing: The fourth stage of team formation when hard work leads, without friction, to the achievement of the team's goals. The team structures and processes, established by the leader, are working well.

Personas: Fictional characters built based on objective and direct observations of groups of users. These characters become "typical" users or archetypes, enabling developers to envision specific attitudes and behaviors toward product features.

PERT (Program Evaluation and Review Technique): An event-oriented network analysis technique used to estimate project duration when there is a high degree of uncertainty in estimates of duration times for individual activities.

PESTLE: A structured tool based on the analysis of Political, Economic, Social, Technological, Legal, and Environmental factors. It is particularly useful as a strategic framework for seeking a better understanding of trends in factors that will directly influence the future of an organization – such as demographics, political barriers, disruptive technologies, competitive pressures, etc.

Phase review process: A staged product development process in which first one function completes a set of tasks, then passes the information generated sequentially to another function, which in turn completes the next set of tasks, and then passes everything along to the next function. Multifunctional teamwork is largely absent in these types of product development processes, which may also be called baton-passing processes. Most organizations have moved from these processes to Stage-Gate® processes using multifunctional teams.

Pipeline (product pipeline): The scheduled stream of products in development for release to the market.

Pipeline management: A process that integrates product strategy, project management, and functional management to continually optimize the cross-project management of all development-related activities.

Plant variety rights: An exclusive right to produce for sale and sell propagating material of a plant variety.

Platform product: The design and components that are shared by a set of products in a product family. From this platform, numerous derivative products can be designed. See also product platform.

Portfolio: Commonly referred to as a set of projects or products that an organization is investing in and making strategic trade-offs against. See also project portfolio and product portfolio.

Portfolio criteria: The set of criteria against which the business judges both proposed and currently active product development projects to create a balanced and diverse mix of ongoing efforts.

Portfolio management: A business process by which a business unit decides on the mix of active projects, staffing, and dollar budget allocated to each project currently being undertaken. See also pipeline management.

Portfolio management team: See gatekeeper.

Portfolio rollout scenarios: Hypothetical illustrations of the number and magnitude of new products that would need to be launched over a certain time frame to reach the desired financial goals; accounts for success/failure rates and considers organization and competitive benchmarks.

Primary market research: Original research conducted by you (or someone you hire) to collect data specifically for your current objective.

Process champion: The person responsible for the daily promotion of and encouragement to use a formal business process throughout the organization. They are also responsible for the ongoing training, innovation input, and continuous improvement of the process.

Process managers: The operational managers responsible for ensuring the orderly and timely flow of ideas and projects through the process.

Process owner: The executive manager responsible for the strategic results of the NPD process. This includes process throughput, quality of output, and participation within the organization.

Product: All goods, services, or knowledge sold. Products are bundles of attributes (features, functions, benefits, and uses) and can be tangible, as in the case of physical goods; intangible, as in the case of those associated with service benefits; or can be a combination of the two.

Product and process performance success: The extent to which a new product meets its technical performance and product development process performance criteria.

Product approval committee: See gatekeeper.

Product architecture: The way in which functional elements are assigned to the physical chunks of a product and the way in which those physical chunks interact to perform the overall function of the product.

Product backlog: A basis of agile product development. The requirements for a system, expressed as a prioritized list of product backlog items. These include both functional and non-functional customer requirements, as well as technical team-generated requirements.

Product definition: Defines the product, including the target market, product concept, benefits to be delivered, positioning strategy, price point, and even product requirements and design specifications.

Product design specifications: All necessary drawings, dimensions, environmental factors, ergonomic factors, aesthetic factors, cost, maintenance that will be needed, quality, safety, documentation, and description. It also gives specific examples of how the design of the project should be executed, helping others work properly.

Product development: The overall process of strategy, organization, concept generation, product and marketing plan creation and evaluation, and commercialization of a new product.

Product Development and Management Association (PDMA): A not-for-profit professional organization whose purpose is to seek out, develop, organize, and disseminate leading-edge information on the theory and practice of product development and product development processes. See the Appendix for more detail. Website: www.pdma.org.

Product development portfolio: The collection of new product concepts and projects that are within the organization's ability to develop, are most attractive to the organization's customers, and deliver short- and long-term corporate objectives, spreading risk and diversifying investments.

Product development process: A disciplined and defined set of tasks, steps, and phases that describe the normal means by which an organization repetitively converts embryonic ideas into salable products or services.

Product development team: That group of persons who participate in a product development project. Frequently each team member represents a function, department, or specialty. Together they represent the full set of capabilities needed to complete the project.

Product discontinuation: A product or service that is withdrawn or removed from the market because it no longer provides an economic, strategic, or competitive advantage in the organization's portfolio of offerings.

Product failure: A product that does not meet the objective of its charter or marketplace.

Product family: The set of products that have been derived from a common product platform. Members of a product family normally have many common parts and assemblies.

Product Innovation Charter (PIC): A critical strategic document, the Product Innovation Charter (PIC) is the heart of any organized effort to commercialize a new product. It contains the reasons the project has been started, the goals, objectives, guidelines, and boundaries of the project. It is the "who, what, where, when, and why" of the product development project. In the Discovery phase, the charter may contain assumptions about market preferences, customer needs, and sales and profit potential. As the project enters the

Development phase, these assumptions are challenged through prototype development and in-market testing. While business needs and market conditions can and will change as the project progresses, one must resist the strong tendency for projects to wander off as the development work takes place. The PIC must be constantly referenced during the Development phase to make sure it is still valid, that the project is still within the defined arena, and that the opportunity envisioned in the Discovery phase still exists.

Product life cycle: The four stages that a new product is thought to go through from birth to death introduction, growth, maturity, and decline. Controversy surrounds whether products go through this cycle in any predictable way.

Product life cycle management: Changing the features and benefits of the product, elements of the marketing mix, and manufacturing operations over time to maximize the profits obtainable from the product over its life cycle.

Product line: A group of products marketed by an organization to one general market. The products have some characteristics, customers, and uses in common and may also share technologies, distribution channels, prices, services, and other elements of the marketing mix.

Product management: Ensuring over time that a product or service profitably meets the needs of customers by continually monitoring and modifying the elements of the marketing mix, including the product and its features, the communications strategy, distribution channels, and price.

Product manager: The person assigned responsibility for overseeing all of the various activities that concern a particular product. Sometimes called a brand manager in consumer-packaged goods organizations.

Product owner: Commonly used in agile product development. The product owner is the single person who must have final authority representing the customer's interests in backlog prioritization and requirements questions.

Product platform: Underlying structures or basic architectures that are common across a group of products or that will be the basis of a series of products commercialized over a number of years.

Product portfolio: The set of products and product lines the organization has placed in the market.

Product positioning: How a product will be marketed to customers. Product positioning refers to the set of features and benefits that are valued by (and therefore defined by) the target customer audience, relative to competing products.

Product rejuvenation: The process by which a mature or declining product is altered, updated, repackaged, or redesigned to lengthen the product life cycle and in turn extend sales demand.

Product requirements document: The contract between, at a minimum, marketing and development, describing completely and unambiguously the necessary attributes (functional performance requirements) of the product to be developed, as well as information about how achievement of the attributes will be verified (i.e., through testing).

Product roadmap: Illustrates high-level product strategy and demonstrates how a product will evolve over time. It is essential to organizational alignment.

Product superiority: Differentiation of an organization's products from those of competitors, achieved by providing consumers with greater benefits and value. This is one of the critical success factors in commercializing new products.

Program manager: The organizational leader charged with responsibility of executing a portfolio of NPD projects.

Project: A temporary endeavor undertaken to create a unique product, service, or result. (See PMI-PMBOK Guide).

Project decision making and gate reviews: A series of Go/No-Go decisions about the viability of a project that ensure the completion of the project provides a product that meets the marketing and financial objectives of the organization. Includes a systematic review of the viability of a project as it moves through the various phase Stage-Gates in the development process. These periodic checks validate that the project is still close enough to the original plan to deliver against the business case.

Project leader: The person responsible for managing an individual new product development project through to completion. They are responsible for ensuring that milestones and deliverables are achieved and that resources are utilized effectively. See also team leader.

Project management: The set of people, tools, techniques, and processes used to define the project's goal, plan all the work necessary to reach that goal, lead the project and support teams, monitor progress, and ensure that the project is completed in a satisfactory way.

Project pipeline management: Fine-tuning resource deployment smoothly for projects during ramp-up, ramp-down, and mid-course adjustments.

Project plan: A formal, approved document used to guide both project execution and control. Documents planning assumptions and decisions, facilitates communication among stakeholders, and documents approved scope, cost, and schedule deadlines.

Project portfolio: The set of projects in development at any point in time. These will vary in the extent of newness or innovativeness.

Project resource estimation: This activity provides one of the major contributions to the project cost calculation. Turning functional requirements into a realistic cost estimate is a key factor in the success of a product delivering against the business plan.

Project sponsor: The authorization and funding source of the project. The person who defines the project goals and to whom the final results are presented. Typically, a senior manager.

Project strategy: The goals and objectives for an individual product development project. It includes how that project fits into the organization's product portfolio, who the target market is, and what problems the product will solve for those customers.

Project team: A multifunctional group of individuals chartered to plan and execute a new product development project.

Prospectors: Organizations that lead in technology, product and market development, and commercialization, even though an individual product may not lead to profits. Their general goal is to be first to market with any particular innovation.

Prototype: A tangible or non-tangible model of a new product/service concept. Depending upon the purpose, prototypes may be non-working, functionally working, or both functionally and aesthetically complete.

Prototyping: An iterative process to test alternative new product concepts.

Psychographics: Characteristics of consumers that, rather than being purely demographic, measure their attitudes, interests, opinions, and lifestyles.

Qualitative market research: Research conducted with a very small number of respondents, either in groups or individually, to gain an impression of their beliefs, motivations, perceptions, and opinions. Frequently used to gather initial consumer needs and obtain initial reactions to ideas and concepts. Results are not representative of the market in general nor projectable. Qualitative marketing research is used to show why people buy a particular product, whereas quantitative marketing research reveals how many people buy it.

Quality: The collection of attributes, which when present in a product, means a product has conformed to or exceeded customer expectations.

Quality assurance/compliance: Function responsible for monitoring and evaluating development policies and practices, to ensure they meet organization and applicable regulatory standards.

Quality-by-design: The process used to design quality into the product, service, or process from the inception of product development.

Quality control specification and procedure: Documents that describe the specifications and the procedures by which they will be measured which a finished subassembly or system must meet before judged ready for shipment.

Quality Function Deployment (QFD): A structured method employing matrix analysis for linking what the market requires to how it will be accomplished in the development effort. This method is most frequently used during the stage of development when a multifunctional team agrees on how customer needs relate to product specifications and the features that deliver those needs. By explicitly linking these aspects of product design, QFD minimizes the possibility of omitting important design characteristics or interactions across design characteristics. QFD is also an important mechanism in promoting multifunctional teamwork. Developed and introduced by Japanese auto manufacturers, QFD is widely used in the automotive industry.

Quantitative market research: Consumer research, often surveys, conducted with a large enough sample of consumers to produce statistically reliable results that can be used to project outcomes to the general consumer population. Used to determine importance levels of different customer needs, performance ratings of, and satisfaction with current products, probability of trial, repurchase rate, and product preferences. These techniques are used to reduce the uncertainty associated with many other aspects of product development.

Radical innovation: A new product, generally containing new technologies, that significantly changes behaviors and consumption patterns in the marketplace.

Random sample: A subset of a statistical population in which each member of the subset has an equal probability of being chosen.

Reactors: Organizations that have no coherent innovation strategy. They only develop new products when absolutely forced to by the competitive situation.

Reposition: To change the position of the product in the minds of customers, either on failure of the original positioning or to react to changes in the marketplace. Most frequently accomplished through changing the marketing mix rather than redeveloping the product.

Repurposing: A strategy of reusing of a product or its parts for a new functionality that differs from the originally intended purpose or that may serve a new market.

Resource matrix: An array that shows the percentage of each non-managerial person's time that is to be devoted to each of the current projects in the organization's portfolio.

Resource plan: Detailed summary of all forms of resources required to complete a product development project, including personnel, equipment, time, and finances.

Return on investment (ROI):   A standard measure of project profitability, this is the discounted profits over the life of the project expressed as a percentage of initial investment.

Reverse engineering: The implementation of value analysis (VA) tear-down processes to formulate ideas for product improvement.

Risk: An event or condition that may or may not occur, but if it does occur will impact the ability to achieve a project's objectives. In new product development, risks may take the form of market, technical, or organizational issues.

Risk acceptance: An uncertain event or condition for which the project team has decided not to change the project plan. A team may be forced to accept an identified risk when they are unable to identify any other suitable response to the risk.

Risk avoidance: Changing the project plan to eliminate a risk or to protect the project objectives from any potential impact due to the risk.

Risk management: The process of identifying, measuring, and mitigating the business risk in a product development project.

Risk mitigation: Actions taken to reduce the probability and/or impact of a risk to below some threshold of acceptability.

Risk tolerance: The level of risk that a project stakeholder is willing to accept. Tolerance levels are context specific. That is, stakeholders may be willing to accept different levels of risk for different types of risk, such as risks of project delay, price realization, and technical potential.

Risk transference: Actions taken to shift the impact of a risk and the ownership of the risk response actions to a third party.

Roadmapping: A graphical multistep process to forecast future market and/or technology changes, and then plan the products to address these changes.

Routine innovation: Builds on an organization's existing technological competencies and fits with its existing business models. Innovation is focused on feature improvement and new versions or models.

S-Curve (Technology S-Curve): Technology performance improvements tend to progress over time in the form of an "S" curve. When first invented, technology performance improves slowly and incrementally. Then, as experience with a new technology accrues, the rate of performance increase grows and technology performance increases by leaps and bounds. Finally, some of the performance limits of a new technology start to be reached and performance growth slows. At some point, the limits of the technology may be reached and further improvements are not made. Frequently, the technology then becomes vulnerable to a substitute technology that is capable of making additional performance improvements. The substitute technology is usually on the lower, slower portion of its own "S" curve and quickly overtakes the original technology when performance accelerates during the middle (vertical) portion of the "S."

Sales forecasting: Predicting the sales potential for a new product using techniques such as the ATAR (Awareness-Trial-Availability-Repeat) model.

Sales wave research: Customers who are initially offered the product at no cost are re-offered it, or a competitor's product, at slightly reduced prices. The offer may be made as many as five times. The number of customers continuing to select the product and their level of satisfaction is recorded.

Scamper: An ideation tool that utilizes actions verbs as stimuli. S – Substitute; C – Combine; A – Adapt; M – Modify; P – Put to another use; E – Eliminate; R – Reverse.

Scenario analysis: A tool for envisioning alternate futures so that a strategy can be formulated to respond to future opportunities and challenges.

Screening: The process of evaluating and selecting new ideas or concepts to put into the project portfolio. Most organizations now use a formal screening process with evaluation criteria that span customer, strategy, market, profitability, and feasibility dimensions.

Scrum: A term used in agile product development. Arguably it is the most popular framework for implementing Agile. With scrum, the product is built in a series of fixed-length iterations, giving teams a framework for shipping software on a regular cadence.

Scrum-master: Commonly used in agile product development. The facilitator for the team and product owner. Rather than manage the team, the scrum-master works to assist both the team and the product owner.

Scrum team: Commonly used in agile product development. Usually made up of seven, plus or minus two, members. The team usually comprises a mix of functions or disciplines required to successfully complete the sprint goals (cross-functional team).

Secondary market research: Research that involves searching for existing data originally collected by someone else.

Segmentation: The process of dividing a large and heterogeneous market into more homogeneous subgroups. Each subgroup, or segment, holds similar views about the product, and values, purchases, and uses the product in similar ways.

Senior management: That level of executive or operational management above the product development team that has approval authority or controls resources important to the development effort.

Sensitivity analysis: A calculation of the impact that an uncertainty might have on the new product business case. It is conducted by setting upper and lower ranges on the assumptions involved and calculating the expected outcomes.

Sensory testing: A quantitative research method that evaluates products in terms of the human sensory response (sight, taste, smell, touch, hearing) to the products tested.

Services: Products, such as an airline flight or insurance policy, which are intangible or at least substantially so. If totally intangible, they are exchanged directly from producer to user, cannot be transported or stored, and are instantly perishable. Service delivery usually involves customer participation in some important way. Services cannot be sold in the sense of ownership transfer, and they have no title of ownership.

Simulated test market: A form of quantitative market research and pre-test marketing in which consumers are exposed to new products and to their claims in a staged advertising and purchase situation. Output of the test is an early forecast of expected sales or market share, based on mathematical forecasting models, management assumptions, and input of specific measurements from the simulation.

Six Sigma: A level of process performance that produces only 3.4 defects for every one million operations.

Six thinking hats: A tool developed by Edward de Bono which encourages team members to separate thinking into six clear functions and roles. Each role is identified with a color-symbolic "thinking hat."

Social media: Computer-mediated tools that allow people, companies, and other organizations to create, share, or exchange information, ideas, and pictures/videos in virtual communities and networks.

Specification: A detailed description of the features and performance characteristics of a product. For example, a laptop computer's specification may read as a 90 megahertz Pentium, with 16 megabytes of RAM and 720 megabytes of hard disk space, 3.5 hours of battery life, weight of 4.5 pounds, with an active matrix 256 color screen.

Sponsor: An informal role in a product development project, usually performed by a higher-ranking person in the organization who is not directly involved in the project, but who is ready to extend a helping hand if needed or provide a barrier to interference by others.

Sprint: A term used in agile product development. A set period of time during which specific work has to be completed and made ready for review.

Stage: One group of concurrently accomplished tasks, with specified outcomes and deliverables, of the overall product development process.

Stage-Gate® process: A widely employed product development process that divides the effort into distinct time-sequenced stages separated by management decision gates. Multi-functional teams must successfully complete a prescribed set of related cross-functional tasks in each stage prior to obtaining management approval to proceed to the next stage of product development. The framework of the Stage-Gate® process includes workflow and decision-flow paths and defines the supporting systems and practices necessary to ensure the process's ongoing smooth operation.

Staged product development activity: The set of product development tasks commencing when it is believed there are no major unknowns and that result in initial production of salable product, carried out in stages.

Standard cost: See factory cost.

Star products: Products that command a significant market share in a growing overall market.

Storming: The stage in team formation where people start to push against the boundaries established. This is where many teams fail. Storming often starts where there is a conflict between team members' natural working styles.

Storyboarding: Focuses on the development of a story, possibly about a consumer's use of a product, to better understand the problems or issues that may lead to specific product design attributes.

Strategic balance: Balancing the portfolio of development projects along one or more of many dimensions such as focus vs. diversification, short vs. long term, high vs. low risk, extending platforms vs. development of new platforms.

Strategic fit: Ensures projects are consistent with the articulated strategy. For example, if certain technologies or markets are specified as areas of strategic focus, do the projects fit into these areas?

Strategic partnering: An alliance or partnership between two organizations (frequently one large corporation and one smaller, entrepreneurial organization) to create a specialized new product. Typically, the large organization supplies capital and the necessary product development, marketing, manufacturing, and distribution capabilities, while the small organization supplies specialized technical or creative expertise.

Strategic priorities: Ensures the investment across the portfolio reflects the strategic priorities. For example, if the organization is seeking technology leadership, then the balance of projects in the portfolio should reflect this focus.

Strategy: The organization's vision, mission, and values. One subset of the organization's overall strategy is its innovation strategy.

Stratified sampling: The population is divided into strata according to some variables that are thought to be related to the variables that we are interested in. A sample is taken from each stratum.

Support projects: Can be incremental improvements in existing products or improvements in manufacturing efficiency of an existing product. Generally, they are low risk.

Sustainable development: Development which meets the needs of current generations without compromising the ability of future generations to meet their own needs.

Sustainable innovation: The process in which new products or services are developed and brought to commercialization and in which the characteristics of sustainable development are respected from the economic, environmental, and social angle, in the sourcing, production, use, and end-of-service stages of the product life cycle.

Sustaining innovation: Does not create new markets or value networks but only develops existing ones with better value, allowing companies to compete against each other's sustaining improvements.

SWOT Analysis: Strengths, Weaknesses, Opportunities, and Threats Analysis. A SWOT analysis evaluates an organization in terms of its advantages and disadvantages vs. competitors, customer requirements, and market/economic environmental conditions.

Tangible product: The physical and aesthetic design features that give the product its appearance and functionality.

Target market: The group of consumers or potential customers selected for marketing. This market segment is most likely to buy the products within a given category. These are sometimes called prime prospects.

Task: The smallest describable unit of accomplishment in completing a deliverable.

Team: A small number of people with complementary skills who are committed to a common purpose, with a clear set of performance goals and approach, for which they hold themselves mutually accountable.

Team leader: The person leading the new product team. Responsible for ensuring that milestones and deliverables are achieved but may not have any authority over project participants.

Technology-driven: A new product or new product strategy based on the strength of a technical capability. Sometimes called solutions in search of problems.

Technology foresighting: A process for looking into the future to predict technology trends and the potential impact on an organization.

Technology roadmap: A graphic representation of technology evolution or technology plans mapped against time. It is used to guide new technology development or technology selection in developing new products.

Technology S-curve: The life cycle that applies to most technologies – embryonic, growth, and maturity stage.

Technology strategy: A plan for the maintenance and development of technologies that supports the future growth of the organization and aids the achievement of its strategic goals.

Technology transfer: The process of converting scientific findings from research laboratories into useful products by the commercial sector. May also be referred to as the process of transferring technology between alliance partners.

Test marketing and market testing: Market testing in its most general definition covers the research methods for all products, new or existing, tested under in-market conditions for the purpose of reducing the risks of launch or expansion failure, and includes test marketing methods. Where test marketing focuses on reducing risks for new product launches, market testing can also be defined more narrowly to mean testing the expansion of an existing product to a new market for the purpose of reducing the risk of a failed expansion strategy.

Time to market: The length of time it takes to develop a new product from an early initial idea for a new product to initial market sales. Precise definitions of the start and end point vary from one organization to another and may vary from one project to another within the organization. (also called Speed to Market).

Top down portfolio selection: Also known as the strategic bucket method, relies on starting with strategy and placing significant emphasis on project selection according to this strategy.

Total Quality Management (TQM): A business improvement philosophy that comprehensively and continuously involves all of an organization's functions in improvement activities.

Trade dress rights: Protect a products' total image or appearance, including the shape, color, or any other non-functional product features.

Trade secrets: Information related to IP that is retained confidentially within an organization.

Trademark: A symbol, word, or words legally registered or established by use as representing an organization or product.

Triple constraint: The combination of the three most significant restrictions on any project scope, schedule, and cost. The triple constraint is sometimes referred to as the project management triangle or the iron triangle.

Triple bottom line: Reports an organization's performance against three dimensions Financial, social, environmental.

TRIZ: The acronym for the Theory of Inventive Problem Solving, which is a Russian systematic method of solving problems and creating multiple-alternative solutions. It is based on an analysis and codification of technology solutions from millions of patents. The method enhances creativity by getting individuals to think beyond their own experience and to reach across disciplines to solve problems using solutions from other areas of science.

TURF analysis: Total Unduplicated Reach and Frequency (TURF) analysis has its roots in media scheduling and is used in product innovation and product management to understand and maximize the market potential of product lines and product platforms, especially when multiple choices and repeat purchases are involved over a product's life cycle.

Unarticulated customer needs: Those needs that a customer is either unwilling or unable to explain.

Usage and purchase intent: The extent to which someone says they will use or purchase your product or service.

User: Any person who uses a product or service to solve a problem or obtain a benefit, whether or not they purchase it. Users may consume a product, as in the case of a person using shampoo to clean their hair or eating a potato chip to assuage hunger between meals. Users may not directly consume a product, but may interact with it over a longer period of time, like a family owning a car, with multiple family members using it for many purposes over a number of years. Products also are employed in the production of other products or services, where the users may be the manufacturing personnel who operate the equipment.

User experience (UX): In current vernacular, UX is often associated with interface design, human factor design, etc., and while those are definitely a part of the user experience, UX ultimately comes down to understanding the customer.

User-generated content (UGC): A form of rich and unstructured textual data provided by users and harvested from online reviews, social media, and blogs. UCG is a promising data source from which to identify customer needs or sentiments quickly and at a low incremental cost to the company.

Utility patent rights: Protect new, non-obvious, and functional aspects of an invention for a limited term.

Value: Any principle to which a person or organization adheres with some degree of emotion. It is one of the elements that enter into formulating a strategy.

Value-added: The act or process by which tangible product features or intangible service attributes are bundled, combined, or packaged with other features and attributes to create a competitive advantage, reposition a product, or increase sales.

Value proposition: A short, clear, and simple statement of how and on what dimensions a product concept will deliver value to prospective customers. The essence of "value" is embedded in the trade-off between the benefits a customer receives from a new product and the price a customer pays for it.

Virtual reality (VR) testing: A growing segment of the market research field, conducted using specialized equipment including a headset and/ or gloves with tracking sensors that create three-dimensional (3D) simulations and enable participants to interact in a realistic environment.

Virtual team: Dispersed teams that communicate and work primarily electronically may be called virtual teams.

Vision: An act of imagining, guided by both foresight and informed discernment, that reveals the possibilities as well as the practical limits in new product development. It depicts the most desirable future state of a product or organization.

Vitality index: The sales of new products as defined by the business divided by the sales of all products for a given product line or department during a designated period.

Voice of the Customer (VOC): A process for eliciting needs from consumers that uses structured in-depth interviews to lead interviewees through a series of situations in which they have experienced and found solutions to the set of problems being investigated. Needs are obtained through indirect questioning by coming to understand how the consumers found ways to meet their needs, and, more important, why they chose the particular solutions they found.

Waste: Any activity that utilizes equipment, materials, parts, space, employee time, or other corporate resource beyond minimum amount required for value-added operations to ensure manufacturability. These activities could include waiting, accumulating semi-processed parts, reloading, passing materials from one hand to the other, and other non-productive processes. The seven basic categories of waste that a business should strive to eliminate are overproduction, waiting for machines, transportation time, process time, excess inventory, excess motion, and defects.

Waterfall process: A sequential design process used in software development processes, in which progress is seen as flowing steadily downward (like a waterfall) through the phases of conception, initiation, analysis, design, construction, testing, production/implementation, and maintenance.

Willingness to pay: The highest price a customer says they will definitely buy your product or service.

Whole product: A product definition concept that emphasizes delivering all aspects of a product which are required for it to deliver its full value. This would include training materials, support systems, cables, how-to recipes, additional hardware/software, standards and procedures, implementation, applications consulting – any constitutive elements necessary to assure the customer will have a successful experience and achieve at least minimum required value from the product.

Workplan: Detailed plan for executing the project, identifying each phase of the project, the major steps associated with them, and the specific tasks to be performed along the way. Best practice work plans identify the specific functional resources assigned to each task, the planned task duration, and the dependencies between tasks. See also Gantt chart.

INDEX

Page locators in **bold** indicate tables. Page locators in *italics* indicate figures. This index uses letter-by-letter alphabetization.

A

AA *see* attribute additivity
Abbie Griffin High Impact Award, 624
absorptive capacity, 10
ACA *see* adaptive conjoint analysis
accelerated development, 31–32
Accelerator Hub, **393**, 394–402, *395–397*
acceptance finding, 616
accountability
 performance, 25, 28
 Portfolio Management Framework, 160, 177
 really new products, 261
action planning, 614–615, *615*
adaptation, 470, 573
adaptive conjoint analysis (ACA), 560
Adobe, 497, 502
adoption
 customer needs, 530
 opportunistic new product development, 241–242
 really new products, 248–249, *249*, 253, 259–261

user-centric innovation, 320–321, 328–329
affect *see* emotion/affect
affinity diagrams
 creativity, 577–578
 facilitating a design thinking workshop, 610–611, *610*, 613
agile development
 customer needs, 534–535
 design thinking, 449
 digital transformation, 392–393, 399
 performance, 14, 16, 32–33, *33*
 user-centric innovation, 316
Agile-Stage-Gate hybrid model, 33, *33*
AI *see* artificial intelligence
Albert L. Page Award for Outstanding Professional Contribution, 624
Alfa Romeo, 451
Allan Anderson Ambassador Award, 624–625
AlphaGo, 411
alpha-testing, 580
Amazon, 371, 503 *504*

analogy
 facilitating a design thinking workshop, 612–613, *612*
 forecasting, 593, *593*
 front end of innovation, 208, *208–209*
 really new products, 251
Analytical Network Process (ANP), 434
annual recurring revenue (ARR), 507
anonymity, 359
ANP *see* Analytical Network Process
anthropomorphization, 470
Apple, 191, 250, 373, 503
AR *see* attribute range; augmented reality
ARIMA *see* autoregressive integrated moving average
ARMA *see* autoregressive moving average
ARPA/ARPU *see* average revenue per account/user
ARR *see* annual recurring revenue
artificial intelligence (AI)
 concepts and definitions, 425–426

The PDMA Handbook of Innovation and New Product Development, Fourth Edition. Edited by Ludwig Bstieler and Charles H. Noble.
© 2023 John Wiley & Sons, Inc. Published 2023 by John Wiley & Sons, Inc.